AF856657

An Introduction to Optical Wireless Mobile Communications

For a listing of recent titles in the *Mobile Communications library*,
turn to the back of this book.

An Introduction to Optical Wireless Mobile Communications

Harald Haas

Mohamed Sufyan Islim

Cheng Chen

Hanaa Abumarshoud

ARTECH HOUSE

BOSTON | LONDON
artechhouse.com

Library of Congress Cataloging-in-Publication Data
A catalog record for this book is available from the U.S. Library of Congress.

British Library Cataloguing in Publication Data
A catalogue record for this book is available from the British Library.

Cover design by Charlene Stevens

ISBN 13: 978-1-63081-655-1

© 2021 ARTECH HOUSE
685 Canton Street
Norwood, MA 02062

All rights reserved. Printed and bound in the United States of America. No part of this book may be reproduced or utilized in any form or by any means, electronic or mechanical, including photocopying, recording, or by any information storage and retrieval system, without permission in writing from the publisher.

All terms mentioned in this book that are known to be trademarks or service marks have been appropriately capitalized. Artech House cannot attest to the accuracy of this information. Use of a term in this book should not be regarded as affecting the validity of any trademark or service mark.

Many product and company names that occur in this book are trademarks or registered trademarks of their respective holders. They remain their property, and a mention does not imply any affiliation with or endorsement by the respective holder.

10 9 8 7 6 5 4 3 2 1

Contents

Preface

We have enjoyed three decades of wireless personal communications. The fifth generation (5G) cellular networks opened the door for a paradigm shift towards wireless machine-type communications. This has brought new challenges, such as enabling ultra-low latency, greater resilience, and enhanced security. The digital revolution eventually means that everything will be connected. Up until now, we primarily used the radio frequency (RF) spectrum in wireless communications. A future paradigm shift will be to also use the optical spectrum for wireless connectivity. There have been multiple attempts to commercialize optical wireless communications in the past. Therefore, it is legitimate to ask: what has changed? What really has changed is the need to expand wireless networks into space and subsea, the ever-increasing scarcity of available RF spectrum required to fulfill the demand, the need for new levels of physical layer security, the trend towards net-zero communication networks, and global availability. All of these new circumstances make it imperative to consider other parts of the electromagnetic spectrum. The optical spectrum is the closest to be harnessed as there are plenty of devices and subsystems due to the tremendous advancements in optical fiber communications during the last decades and new developments such as silicon photonics and III-V semiconductors. In addition, a large body of knowledge exists in optical wireless communications (OWC) as this field has been studied for many decades. In recent years, the industrialization of OWC systems gained new momentum with many new companies entering the market. Alongside this trend, standardization efforts have increased significantly across multiple bodies.

Therefore, we felt that it is exactly the right time for a textbook like this. It not only looks at the key technologies and brings together decades of research, but it also sheds light on other developments that are needed to make this technology pervasive such as standardization and technology ecosystem development as well as commercial aspects. We hope that it is a valuable resource for undergraduate and graduate students who wish to get a complete overview of key technologies and understand market drivers. In addition, there are many new courses that are developed on OWC-related aspects such as visible light communication, LiFi, free space optical (FSO) communications, and optical camera communications. We hope that this book will provide inspiration for course instructors. The book should also be a valuable source for development engineers, product designers, and industry-based researchers. Lastly, it aims to benefit new entrepreneurs who wish to participate in the rapidly growing market that is currently being developed around OWC and to ascertain what the unique selling points of their ideas and concepts could be. Such a resource certainly was not available when the lead author of this book set up his own spin-out company, pureLiFi Ltd.

We would like to thank the editors at Artech House for handling the editing and reviewing process. In addition, we would also like to thank the reviewers for providing insightful feedback on our work, which helped us significantly to improve the manuscript. We would like to thank Miss Hannah Brown (the personal assistant to Professor Harald Haas) for her efforts in proofreading this manuscript, which helped us to improve the writing quality of this book significantly. We would like to thank Dr. Trang Nguyen from the University of Strathclyde, Dr. Nikola Serafimovski, and Dr. Ardimas Andi Purwita from pureLiFi Ltd. for their support in writing the manuscript of Chapter 8. We would like to thank Dr. Enrique Poves for his contributions to Chapter 9.

1

Introduction

1.1 Setting the Scene

Wireless communications have undoubtedly become an integral part of our social, professional, and everyday lives. Billions of mobile phones, tablets, laptops, and smart TVs provide us with an unprecedented level of connectivity and data sharing that we could not have imagined a couple of decades ago. Beside connecting people, the internet of things (IoT) and internet-of-everything (IoE) are transforming our world as we know it by connecting machines, data models, physical objects, and entire business processes, fueling the need for enhancing and expanding the wireless communications landscape. The currently emerging fifth generation (5G) wireless networks are characterized by a diverse set of key performance indicators (KPIs) that can be grouped into three major categories, namely: enhanced mobile broadband (eMBB), ultrareliable and low-latency communication (URLLC), and massive machine type communications (mMTC). The approach adopted in 5G to enable these KPIs builds on the exploitation of virtualization and software-defined radio (SDR) together with massive multiple-input multiple-output (MIMO) and ultra-dense small-cell deployments.

Looking forward, innovative and progressive services and applications are on the rise and are expected to reshape our reality. For example, holographic communications that integrate all human sense information (sight, hearing, touch, smell, and taste) are expected to replace audio/video communications, demanding data rates on the order of terabits per second. Also, high-precision automated manufacturing is likely to reduce the need for human intervention in Industry 4.0, which requires wireless connectivity with exceedingly high reliability and extremely low latency. Moreover, vehicle-to-vehicle (V2V) communication will be a key technology to avoid accidents, ease traffic congestion, and protect the environment. This will require the availability of reliable high-rate, low-delay wireless links both in between different vehicles and also between the vehicles and infrastructure. To fulfill the challenging requirements of these new services, the coming future generation (6G) wireless networks are envisioned to be built on a set of enabling paradigms incorporating a new physical layer and a new networking architecture. Moreover, in order to achieve sustainable evolution, wireless system researchers and engineers need to pay close attention to the associated energy consumption. The targeting communications energy efficiency are expected to be in the order of 1 pJ per bit.

Currently, most of the existing communication systems rely on radio frequency (RF) technologies for wireless data transmission. However, with the staggering growth in wireless communication traffic and the demand for high-speed, reliable, flexible, low-latency secure connectivity, there is a general consensus that the RF spectrum resources will not be sufficient to future-proof wireless communications. The annual data traffic reports show an enormous increase in the connectivity demands, with expectations for the number of connected devices to reach 500 billion by 2030 [1]. Figure 1.1 shows the spectrum deficit estimated through to 2040; as reported in [2], it is predicted that the entire RF spectrum will be utilized by 2035 to fulfill the increase in mobile data traffic. Consequently, it is imperative that future wireless networks consider the other parts of the electromagnetic (EM) spectrum for wireless connectivity. In fact, the optical spectrum is already the main enabler for the global internet. Optical fiber communication networks not only connect all continents, but they also form the backbone of modern fiber broadband, which is increasingly to be deployed worldwide to provide high-speed internet connectivity to our homes and offices. Exploiting other parts of the optical spectrum for free-space communications and wireless mobile access seems like a logical approach to enhance and expand the capabilities of today's communications networks and future-proof them for the requirements of the next generations.

It has been evident in the last few years that optical wireless communications (OWCs) will play an important role in satisfying the demands of future wireless networks, which has been manifested by extensive research, case studies, and implementations from academia and industry alike. OWC provides a two-fold advantage to wireless mobile networks, namely: a wider spectrum and higher cell density. The optical spectrum is in terahertz, with the visible light spectrum alone being almost 1,000 times wider than the entire RF spectrum. Although using the entire available optical spectrum is currently not feasible due to physical limitations in the transceivers, advancements in front-end devices are increasingly opening the door for higher data rates and increased capabilities. OWC solutions are currently entering the wireless communications market,

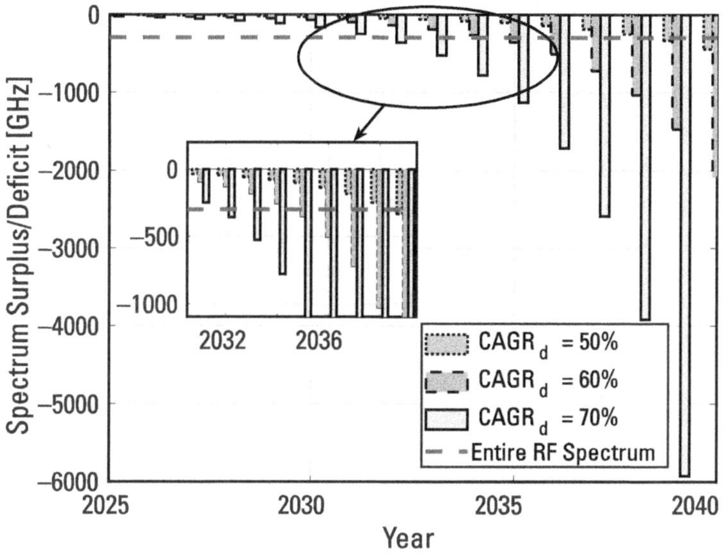

Figure 1.1 Predicted spectrum surplus/deficit through to 2040 for different achievable spectral efficiency as reported in [2].

exploiting the unregulated EM spectrum in the visible light and infrared (IR) bands to complement RF solutions in order to provide ubiquitous connectivity. The main OWC technologies are free-space optical (FSO) communication, ultraviolet (UV) communication, optical camera communication (OCC), visible light communication (VLC), and light fidelity (Li-Fi or LiFi). The functionality of FSO communication systems is the closest to optical fiber communications in the sense that they provide point-to-point links over relatively large distances, primarily using IR sources. The primary application of FSO communication is to establish low-cost wireless backhaul systems. FSO systems typically achieve very long transmission distances in the range of a few kilometers and a very high data rate in the range of a few up to hundreds of gigabit per second. Emerging UV communication enables long-distance non-line-of-sight (NLOS) wireless connectivity through atmospheric scattering, making UV links robust to link blockage and shadowing and, thus relaxing or eliminating the need for perfect transceiver alignment. OCC aims to deliver data from the light sources to the cameras that act as photo-receivers, enabling low-rate, low-range line-of-sight (LOS) links. Similar to OCC, VLC utilizes visible light for connectivity while using photo-detectors at the receiving terminals, offering high-speed, short-range LOS connectivity. Li-Fi, the newest addition to the OWC family, refers to a complete network solution that utilizes optical links for ubiquitous bidirectional connectivity with mobility support and seamless coverage.

Besides offering significant relief for the overcrowded RF spectrum, OWC provides several additional advantages such as enhanced security, resilience to EM interference, a high degree of spatial reuse, and enhanced energy efficiency. However, due to the unique properties of optical wave propagation, unique and innovative solutions are required to be tailored to the actual channel models, the optical front-end systems, and devices and components. Based on this, novel and innovative data transmission techniques, medium access control protocols, interference mitigation, and physical layer security (PLS) paradigms are continuously developed in order to push the boundaries of OWC capabilities.

1.2 Challenges in Optical Wireless Communication Networking

In the following section, we will briefly introduce some of the key challenges that this book attempts to address in different OWC research areas [3].

1.2.1 Devices and Components

While the optical spectrum is three orders of magnitude larger than the entire RF spectrum (see Figure 4.5 in Chapter 4), there are fundamental limitations that currently prevent us from fully harnessing this massive amount of wireless transmission resource, mainly when using light emitting diodes (LEDs) as transmitters. This is because the bandwidth of typical LED devices is limited to tens and hundreds of mega hertz [4]. This limits the data speeds of single LEDs, typically to below 10 Gbps [5]. However, to overcome this limitation, it is possible to use multiple devices of a similar bandwidth but at a different emission spectrum, which is referred to as wavelength division multiplexing (WDM). In [6], it is shown that four standard off-the-shelf LEDs yield an aggregate data rate of 15.7 Gbps despite the partial overlap of their emission spectra. Optimum use of WDM would require subnanometer spectral emission profiles and high device bandwidth. The required device bandwidth can be approximated as follows:

$$B_{\text{device}} \approx c\frac{\Delta\lambda}{\lambda^2} \tag{1.1}$$

where c is the speed of light in vacuum, $\Delta\lambda$ is the emission spectrum of the light source assuming a rectangular spectrum mask, and λ is the center wavelength. Equation (1.1) is plotted in Figure 1.2 as a function of the center wavelength of the emission spectrum. From Figure 1.2, it can be seen that a red light source (600 nm) with an emission spectrum of 0.45 nm would require a device bandwidth of over 350 GHz. First, it is clearly impossible to achieve such a narrow emission spectrum with current LEDs, and second, it is challenging to practically achieve a transmit device bandwidth of 350 GHz with current technology [3,7], let alone

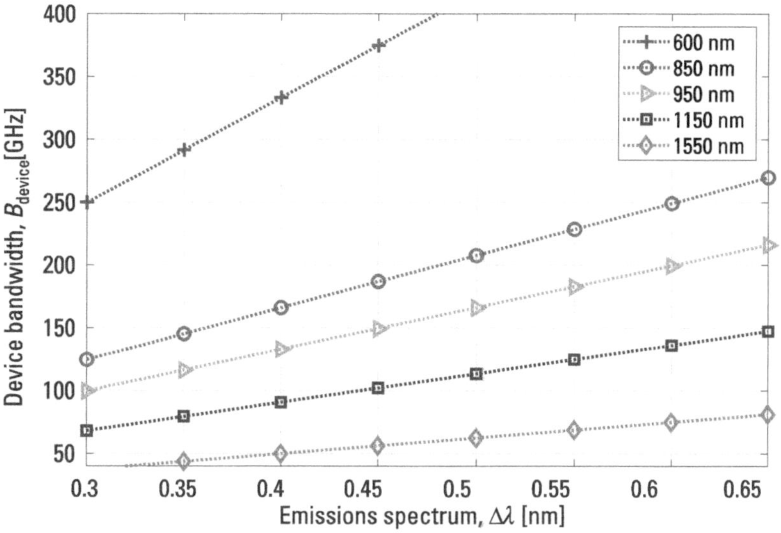

Figure 1.2 Required device bandwidth as a function of emission spectrum [3].

achieving detectors that have a bandwidth of 350 GHz. Overcoming the device limitations is an area of active research. Unlike point-to-point communication, such as in FSO, wireless networking systems have to provide simultaneous wireless connections to a potentially large number of end-users in indoor environments. There is one more challenge (in addition to device bandwidth) that is related to link margins and receiver sensitivity. This is mainly due to the small detector size and the widely spread optical power. In the short to medium term, laser-based lighting may offer the next step towards higher device bandwidths while providing high optical output power to enhance the link data rate performance and maintaining eye-safety requirements. A recent study demonstrated a 20-Gbps Li-Fi link using eye-safe, laser-based, white light sources [8]. In addition, there is a wealth of research papers on using micro LEDs for high-speed OWCs. The major limitation of these devices is the low optical output power, but arrays of micro LEDs can be used to overcome this limitation [3,9]. A recent study reported a data transmission of 11.72 Gbps using a 3 × 3 array of 20-μm-diameter micro LED pixels [10]. Various laser devices, such as vertical-cavity surface-emitting lasers (VCSELs), operating in the infrared spectrum have also been used to build beam-steering, laser-based, optical multiuser systems, which achieved tens of gigabits per second transmission speeds [3,11]. This topic is covered in Chapter 6.

Although inorganic semiconductor devices have primarily been used as optical transmitter and receiver devices, organic semiconductors exhibit properties that make them ideal candidates for optical devices in VLC systems. In particular, their flexibility can be used to bend them around device edges, which can help to increase connectivity when the device is randomly oriented. While it has been shown that it is possible to achieve gigabit per second links using organic LEDs (OLEDs) [12], it has also been demonstrated that organic photovoltaic cells can be used as high-speed data detectors [3,13]. OWC systems primarily use intensity modulation (IM) and direct detection (DD) (IM/DD). The information is encoded in changes in the light intensity. To decode the information, a square-law detector is used—typically, a P-type intrinsic N-type photodiode (PIN-PD) or an avalanche photodiode (APD). APDs have been considered for OWC due to the limited sensitivity of a PIN-PD. These devices operate with a very high electric field so that an electron generated from an incident photon gains very high kinetic energy and generates new electron-hole pairs. This process is called ionization. The newly generated electrons are also subjected to the same high electric field and can generate new electron-hole pairs, creating an avalanche effect. APDs that are biased beyond the breakdown voltage are referred to as single-photon avalanche diodes SPADs. SPADs operate in the Geiger mode and can detect single photons [3,14]. This topic is covered in Chapter 6.

1.2.2 Optical Front-End Systems

The optical front end can be divided into two parts: the transmitter and the receiver. Because the objective is to build bidirectional communication links, both parts must be integrated into a single transceiver unit. Sufficient decoupling of the received and transmitted signals is essential, and various duplexing techniques can be used. Trade-offs to consider in the design of the duplexing techniques are bandwidth, latency, and data rate [15,16]. The optical system design on the transmitter side has to ensure that eye-safety standards are met [17] while achieving quality of service requirements. This can be challenging when the transmitter is a point source. It is, therefore, often desirable to convert the point source into an extended source that relaxes eye-safety limitations.

One of the key challenges in mobile networking is that the mobile terminal randomly changes orientation. Moreover, the link between the mobile terminal and the fixed access point can be obstructed. In a system that heavily relies on LOS link conditions, this becomes a critical challenge. A possible solution to this is to provide link diversity through multiple simultaneously active transmitters whose position and possibly orientation are different. This leads to the concept of angular diversity transmitters [18,19]. At the receiver, the challenge is to capture enough photons to ensure correct detection of the transmitted information. Ideally, the receiver would have a large area to allow the system to collect a large number of photons. However, a large-area detector typically has a low bandwidth because the large area results in a detector that has a high capacitance. Therefore, concentrators are often used. The maximum concentration gain is governed by the conservation of étendue [20], which states that the half-angle of the receiver optics has to be small if a significant gain is desired.

Apart from this condition, the height of the concentrator can also provide some limitations, typically tens of millimeters, which makes it difficult to integrate such a device into a mobile phone. The ideal receiver optical subsystem is flat, uses small detectors, and allows for a half-angle close to 90° to avoid strict alignment. Recently, there has been work on optical concentrators, which is a candidate technology to overcome these challenges [3,21,22]. Moreover, there has been interesting research on ultra-low impedance transimpedance amplifiers, leading to significant bandwidth enhancements [23]. This is because of the reduction of the dominating time constant, which is the product of capacitance and resistance. Therefore, ensuring a low resistance is a promising direction to reduce the time constant and increase the bandwidth. Similar to angular diversity transmitters, there have also been several works on angular diversity receivers to overcome link obstruction and alignment issues [23–25]. This topic is covered in Chapter 6.

1.2.3 Channel Models

An important system element is the optical free-space propagation channel. The optimum design of the transceiver depends on the optical channel characteristics. Since objects act as reflectors, an impulse sent by the transmitter arrives at the receiver via multiple delayed paths. This is referred to as multipath propagation, which may cause intersymbol interference. This means that a transmitted symbol is corrupted by previous symbols. This effect has to be eliminated using digital equalization or avoided by a proper selection of transmission techniques.

Moreover, there may not always be an LOS channel. The receiver has to be able to cope with situations where there are only reflections. It is therefore essential to get a comprehensive understanding of the power included in the reflected paths (in addition to the power received on the LOS path) and the temporal spread of the reflected signals. Furthermore, these properties depend on various factors, such as the indoor space dimension, interior surface material absorption/reflectance, and transmitter-to-receiver separation. A complete understanding of the impacts of these factors in typical deployment scenarios such as homes, offices, and manufacturing plants is paramount to the development of OWC networks [3,26]. This topic is covered in Chapter 5.

1.2.4 Data Transmission Techniques

As stated above, the dominant data transmission mechanism is IM/DD. Since the optical signal must be positive and real, the classic Shannon theory is not applicable and the channel capacity is still unknown. However, upper and lower information-theoretic bounds for such systems were investigated in several studies [27]. Knowledge of the optimum signal distribution is invaluable for the design of optimum modulation techniques.

Although IM/DD is the major transmission technique used in OWC, it is possible to use coherent transmission at the cost of increased transceiver complexity. Recently, there has been an increasing number of works on coherent OWC systems. This is a research direction that this field may also explore in the future. The most basic data transmission technique for IM/DD systems is on-off keying (OOK). However, OOK severely limits the spectrum efficiency to 1 bit/Hz. This results in suboptimal use of the available bandwidth. However, OOK benefits from implementation simplicity, and it is also robust to device nonlinearities. Besides, OOK severely suffers from intersymbol interference when deployed at high data rates. It is desirable to use data transmission techniques that are more spectrum efficient (i.e., transmit multiple bits per channel use while increasing the robustness to multipath propagation). One such technique is orthogonal frequency division multiplexing (OFDM), which sends data symbols simultaneously on orthogonal subchannels. This is achieved by using the inverse fast Fourier transform (IFFT) at the transmitter and the fast Fourier transform (FFT) at the receiver. The parallel transmission means that the OFDM symbol is much longer than the maximum multipath delay. Therefore, a single-tap equalizer can be deployed. Also, every subchannel can carry a complex data symbol out of an alphabet of M total symbols, where every symbol carries $\log_2(M)$ bits. However, since the OFDM time-domain symbol is bipolar, the signal needs to be subjected to a direct current (DC) bias before exposing the signal to the light transmitter, which operates solely in the first quadrant and thus cannot accept negative signals. This measure has an unfavorable impact on energy efficiency as the DC bias power is wasted unless used for other purposes such as illumination. However, while illumination might be desirable if an access point is combined with an illumination device, it is certainly undesirable in a mobile device such as a smartphone. Therefore, new digital modulation techniques have been developed based on multistream or layered data transmission in conjunction with superposition modulation and iterative detection [3,28]. This topic is covered in Chapter 6.

Due to the bandwidth limitation of the devices, it is paramount to consider other dimensions for data encoding, particularly the spatial dimension and the wavelength dimension. MIMO techniques harness the spatial dimension. There are two fundamental techniques in OWC systems: imaging MIMO and non-imaging MIMO. Imaging MIMO is very intolerant to movement and is thus more suitable for FSO-type systems, while nonimaging MIMO can be used in mobile applications. However, it is usually not guaranteed that the MIMO channel matrix is always full rank, which is necessary to fully benefit from maximum spatial multiplexing gains. Besides, MIMO systems suffer from interchannel interference or cross-talk. Iterative interference cancellation techniques can be used to mitigate this effect, but these techniques require significant computational processing power [29]. A MIMO technique that has attracted considerable interest in academia and industry is spatial modulation [30]. Spatial modulation completely avoids inter-channel interference and results in low computational complexity MIMO decoding techniques. This topic is covered in Chapter 6.

1.2.5 Medium Access Control Protocols

In OWC wireless networks, the simultaneous support of many mobile terminals from a single access point (e.g., light in the ceiling) is essential. These mobile terminals may have different service requirements. For example, in a manufacturing plant, the transmission delay may be more significant than the peak data rate. Therefore, it is important to develop optimum multiuser access techniques that avoid multiuser interference and achieve high spectrum efficiency. Ideally, the sum data rate of an access point is greater when supporting many users as compared to a single-user scenario. A candidate technology that can achieve this is non-orthogonal multiple access (NOMA) [31,32]. This topic is covered in Chapter 7.

1.2.6 Interference Mitigation and Mobility Support

A wireless network is characterized by many access points, which are spatially distributed. Every access point covers a specific area. If a mobile device enters the coverage zone of a neighboring access point, the system performs a handover from the currently associated access point to the new access point. This process needs to be fast enough to ensure an uninterrupted wireless service provision. If the coverage of an access point is small, a fast handover sub-10 ms becomes particularly critical. An important advantage of optical wireless networking is that very small cells can be used. A cell is defined by the coverage area of an access point. This is a significant advantage because the same transmission resource can be reused many more times than with large-area cells. The principle of cell shrinkage is the main reason behind the improvements in data rates in RF-based mobile communication during the last decade. However, there is a limit to the continued reduction of cell sizes. This limit is due to increased interference. If access points are spaced too closely together, the transmitted signals overlap (interfere) significantly. This interference enhances the noise and thus reduces the signal-to-noise ratio (SNR), sometimes to levels where it is impossible to maintain an error-free communication link. The benefit of optical wireless networking in comparison with RF wireless networking is that the limit where interference becomes dominant is significantly lower. This means much smaller cells can be generated while interference is kept at low levels. This is because it is much easier to control light waves spatially compared with radio waves. Microwaves, for example, pass through opaque walls. By contrast, light waves are entirely blocked by opaque objects. Also, it is possible to confine the coverage by using simple lenses in an optical subsystem. Nonetheless, due to multipath propagation and random orientation of mobile devices, there is still interference present in optical wireless networks [32]. This topic is covered in Chapter 7.

1.2.7 Networking and Protocols

Lastly, the vision is that optical wireless networks will be seamlessly integrated with existing RF wireless networks [33]. This is seen as an evolutionary process since there are many different RF wireless networking technologies such as 5G cellular networks as well as wireless fidelity (Wi-Fi). Interoperability between these networks is enabled through the concept of heterogeneous networks. Optical wireless networks will be another networking technology that is expected to be seamlessly integrated. However, this requires dynamic network management algorithms. A software-defined network (SDN) is one such dynamic network management technology based on the principles of network virtualization and the establishment of separate control planes and data planes. Optical wireless networks need to be designed to support these paradigms. Hence, it is necessary to develop bespoke SDN agents that act as brokers (hypervisors) between specific optical wireless networking properties and general transport mechanisms and application-specific requirements. These SDN agents will translate these generic requirements via the control plane into network configurations, and they will cooperate to ensure that, at any time, the optimum connectivity to a mobile terminal is provided. For example, an optical wireless SDN agent cooperates with a Wi-Fi agent to ensure that data flows are managed optimally. This, for example, could mean that particular users are served by a Wi-Fi network and by an optical wireless network simultaneously using multipath transmission control protocol (TCP). Here, the network hypervisor directs the specific data flows based on the characteristics of the respective communication links (RF and optical) [34]. This topic is covered in Chapter 7.

1.3 Book Overview

This book aims to provide a comprehensive introduction to free-space light communication, including fundamental technical concepts and recent advancements. We attempt to answer the following questions in this book: How did early OWC technology evolve into Li-Fi? What are the role and function of Li-Fi in relation to other emerging OWC technologies? Why and how can Li-Fi future-proof wireless communications? What essential techniques define a Li-Fi system? What are the recent industrial Li-Fi developments including efforts in standardizing Li-Fi?

The background of Li-Fi will be given, which covers past, present, and potential future Li-Fi development, as illustrated in Figure 1.3(a). First, we cover the history of OWC evolution, including a number of key milestones, such

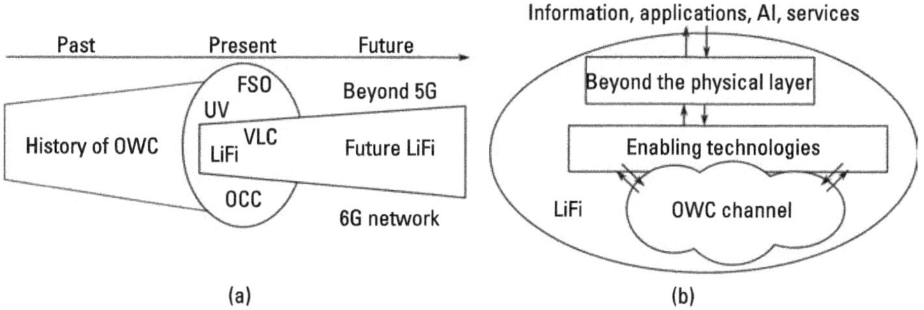

Figure 1.3 (a) Li-Fi background. (b) Li-Fi technical fundamentals.

as the invention of efficient blue LEDs. Next, we consider the current OWC technologies, which can broadly be split into multiple branches. Each of these branches has its own use cases, features, and roles in wireless communications. Li-Fi is one of these branches. We then consider the future vision of Li-Fi in conjunction with the requirements of future cellular communications as, for example, those that are currently formulated under the banner of 6G. We highlight essential properties of Li-Fi and discuss the opportunities and potential applications in future wireless networks.

The most important fundamentals are introduced starting from the physical layer (i.e., devices and components, the channel and modulation techniques) and moving to higher layer techniques (i.e., multiple access, handover and interference management), as illustrated in Figure 1.3(b). The basic difference between RF-based wireless communications and Li-Fi lies in the nature of the communication channel. Therefore, it is not surprising that the most significant differences are noticable in the physical layer. This book is therefore places particular emphasis on channel and physical layer techniques. However, since Li-Fi also enables a new level of small cells that we call optical attocells, there are implications on multiuser access, handover and interference. These aspects are also covered by this book.

Apart from the recent scientific advances in Li-Fi technologies, several standardization and commercialization activities in various organizations and companies have been witnessed. A number of Li-Fi products have been released. A Li-Fi standard will aid the interoperability of products from different companies. With a large-scale adoption of interoperable Li-Fi products, lower-products can be expected. Furthermore, a common industrial vision as articulated by the newly established Light Communication Alliance (LCA) will aid the establishment of the Li-Fi commercial ecosystem from new devices to fully fledged wireless networks. Regarding the third objective, we will focus on reporting the recent standardization and commercialization activities in the past few years, which includes discussions on potential requirements and challenges.

Although there are a number of books on OWC [35,36], most of these focus on adjacent areas such as FSO and VLC. In addition, a lot of the recently published work is not included in these books. The concept of Li-Fi is unique and also shows great potential, with a vast and varied number of possible applications across many industrial verticals. There is still not one dedicated book that introduces this specific subject. Despite the abundant research articles and reports on many aspects of Li-Fi technology, most of these materials are scattered throughout literature in scientific journals and organization archives. In this book, we bring the latest knowledge together as a single piece of work and introduce Li-Fi technologies in a systematic manner. In addition, some new advancements in OWC technologies are also briefly mentioned as a complement to the existing literature in OWC technology.

This book is designed as a learning resource for undergraduate/postgraduate students and researchers in electrical engineering or optical engineering. It is also expected to function as a reference book for self-studying readers who are interested in emerging OWC technologies and recent developments in Li-Fi technology. Regarding any background knowledge necessary, the readers are expected to understand the fundamentals of digital communications, electronics, and signal processing. In addition, readers are also expected to have basic knowledge in calculus, linear algebra, and statistics, in order to understand the mathematical models introduced in Chapters 5, 6, and 7. However, the purpose of this book is to give an overview of OWC technologies, especially the concept of Li-Fi technology.

1.4 Book Structure

We have organized the book into eight main chapters.

In Chapter 2, we discuss the history of OWC and its enabling technologies. The chapter starts with a brief history of light and vision. The corpuscular theory of light, championed by Rene Descartes and Isaac Newton, explained light propagation's basic properties, such as reflection and refraction. However, this theory failed to explain other properties, such as light interference, polarization, and diffraction. The corpuscular theory was abandoned in favor of the waves theory of light, demonstrated by Christiaan Huygens' experiments and later supported by Thomas Young and Jean Fresnel. The photophone, developed by Alexander Graham Bell in 1880, is considered to be the first practical OWC system. Chapter 2 provides a brief history of early OWC systems, such as the optical telegraph, heliograph, and photophone. Furthermore, the recent history of LED-based OWC is discussed in Chapter 2 with a focus on early IR-based systems capabilities and limitations. The invention of the efficient blue LED brought new opportunities to evolve OWC technology to VLC, OCC, and Li-Fi. The inventors of the blue LED, Isamu Akasaki, Hiroshi Amano, and Shuji

Nakamura, were awarded the Nobel Prize in Physics in 2014 for invention of the blue LED. It is interesting to note that the Queen Elizabeth Prize for Engineering was recently awarded to Nick Holonyak, George Craford, Russell Dupuis, Isamu Akasaki, and Shuji Nakamura for their roles in inventing and developing the technology of LEDs. The prize took into consideration the key developments that led to the realization of LED-based white illumination. Holonyak and Craford were awarded this prize for their roles in creating the first red LED in 1962 and yellow LED in 1972, respectively. At the same time, Dupuis was awarded the prize for his contribution to developing the process that industrialized LED technology. Chapter 2 gives a comprehensive overview of the modern history of OWC and its enabling technologies and surveys the recent commercialization and case studies enabled by Li-Fi technology.

In Chapter 3, we provide a comprehensive overview of different emerging technologies in the field of OWC. More specifically, we discuss FSO communications, UV communications, and OCC. The main goal of this chapter is to lay the foundations for the reader to understand the fundamental differences and similarities between these OWC paradigms. The first section focuses on FSO communication technology and presents the most recently reported advancements, including MIMO and multiplexing diversity, as well as their use in unmanned aerial vehicle (UAV)-based networks. We next move on to discuss UV communications and their distinct characteristics, as well as the recent advancements in UV devices and the research efforts in modeling the UV channel. We also discuss some of the open issues related to this technology, such as those related to link efficiency enhancement and the need for more extensive experimental validation. The last part of the chapter focuses on OCC, which relies on the ubiquitous availability of inexpensive image sensors and the existing LED lighting infrastructure, and introduces three different architectures, namely: screen to camera, region of interest and rolling shutter. By the end of this chapter, the reader is expected to have a well-rounded knowledge of the fundamental OWC techniques in terms of their principles of operation, transmission medium, channel models, state-of-the-art front-end devices, and components, as well as their potential applications and practical considerations.

In Chapter 4, we focus on discussing all the features of Li-Fi. On completion of this chapter, readers are expected to have a comprehensive understanding of what Li-Fi technology is; what its advantages are over other technologies; what role it is poised to play in future wireless communications; what the potential applications are; and what the common misconceptions are. The chapter is divided into five sections. In the first section, we discuss the issue of the spectrum crunch and cover the emerging vision of future smart connections based on artificial intelligence (AI). The second section focuses on the main advantages of Li-Fi with illustrations and examples. We attempt to justify that, by taking advantages of these features, potential bottlenecks in future wireless communication can be avoided. In the third section, we cover the potential applications of Li-Fi in future wireless systems, such as cyber-security or vehicular communications. A discussion on various possible use cases is provided.

In Chapter 5, we focus on the subject of indoor OWC channels, as most of the Li-Fi applications are in indoor scenarios. Since the performance of Li-Fi networks depends on the characteristics of indoor OWC channels, it is essential for readers to understand the main features of these channels. This chapter is divided into four parts. In Section 5.2, basic channel effects and evaluation metrics are systematically introduced. The objective of this section is for the readers to understand what phenomena the OWC channel will bring and how to quantify these effects. In Sections 5.3 and 5.4, we introduce the decomposition of indoor OWC channels. Each channel component has its own features and corresponding models. Due to the complexity of the NLOS channel caused by reflections, we present a series of NLOS channel simulation methods in Sections 5.5 to 5.8. The fundamental simulation or, modeling procedures and principles are briefly introduced. At the end, we have also compared the features of different simulation and modeling approaches to provide an intuitive guide for readers to choose a suitable method to obtain the NLOS channel. In Section 5.9, we present the simulation results of a series of indoor OWC channels under various geometric configurations. It is expected that readers will understand the influence of a number of key factors affecting the characteristics of the indoor OWC channels.

Chapter 6 discusses the enabling technologies that allow OWC systems to be realized. The chapter starts with a detailed survey of light emitters based on LED and light amplification by stimulated emission of radiation diode (LD) technologies, then it discusses various photodiode (PD) devices such as PIN-PD, APD, SPAD, and photo-voltaic (PV) detectors. The choice of optical transmitter and receiver devices depends solely on the targeted application. For example, SPAD detectors are favored in applications where the expected level of light incident onto the receiver is modest. Li-Fi is widely expected to be realized using inexpensive off-the-shelf optoelectronic devices such as LEDs. Incoherent light sources limit the signaling in OWC to IM/DD, which, in turn, requires the modulating waveforms to be unipolar and real-valued. An abundance of modulation schemes has been developed for OWC based on frequency, space, and wavelength degrees of freedom (DOF). In addition, unique modulation techniques that rely on the color domain were introduced. A detailed taxonomy of these modulation techniques is provided in Chapter 6. Commercially available luminaries are manufactured in arrays of LEDs in order to achieve an acceptable brightness performance. This allows for the concept of micro-MIMO to be realized in OWC systems. In addition, multiple luminaires can be operated in a macro-MIMO system to enable multiplexing, diversity, and spatial modulation.

Chapter 6 discusses the basic principles and recent developments of MIMO-based OWC systems and the solutions proposed for the decorrelation of the optical MIMO channel.

In Chapter 7, we discuss the networking techniques needed to enable a ubiquitous Li-Fi user experience. We start with an overview of the integration of Li-Fi within heterogeneous networks and discuss the generic two-tier network architecture and how Li-Fi has the potential to provide an additional small-cell layer to support and complement existing RF wireless networks. We also discuss the related resource management and traffic offloading mechanisms in such heterogeneous setups. We then move to discuss multi-access control in Li-Fi systems in Section 7.3. Specifically, we present different orthogonal multiple access (OMA) and NOMA strategies and discuss their principles of operation and the related performance trade-offs. In Section 7.4, we focus on resource allocation in Li-Fi systems and present the different techniques and decision-making strategies reported in the literature. We then move our focus to the issue of interference management in Section 7.5 and discuss various methods of interference cancellation and interference avoidance adopted in Li-Fi networks. In the last section of the chapter, we discuss how Li-Fi supports users' mobility by enabling handover. We discuss different classifications of handover mechanisms (i.e., hard versus soft handover and horizontal versus vertical handover). By the end of this chapter, the reader is expected to have a good understanding of how Li-Fi provides a complete solution for seamless networked connectivity.

In Chapter 8, we discuss the standardization of OWC technology. The OWC market has witnessed an increased product offering from several commercial vendors. However, it is widely believed that OWC products need to be interoperable in order to reach mass market adoption. This allows device manufacturers to integrate Li-Fi optical front-end (OFE) components and associated chipsets and improves the awareness of OWC technology. Chapter 8 discusses four main OWC standards. In particular, Section 8.2 describes the Institute of Electrical and Electronics Engineers (IEEE) 802.15.7 for OCC systems, and the IEEE 802.15.13 standard for industrial multi-Gbps OWC systems. The IEEE 802.15.7 standard has been recently revised to include several OCC scenarios for indoor and vehicular applications. Section 8.3 considers the International Telecommunication Union (ITU) G.9991 for VLC systems, also known as G.vlc. The standard is based on the ITU G.hn home networking standards, based on power line communication (PLC), which makes it easy for G.vlc-compliant devices to use available silicon solutions developed for PLC applications. Finally, Section 8.4 details the IEEE 802.11bb standard for Li-Fi. The IEEE 802.11bb standard is an amendment of the popular Wi-Fi IEEE 802.11 standard suit, and it is aimed at fully networked mobile Li-Fi solutions with handover support. The IEEE 802.11 standard suit allows Li-Fi technology to reuse existing Wi-Fi chipset solutions with minimum modifications and reduced market entry hurdles. As a result, it is expected that the IEEE 802.11bb standard will be a milestone in the development of Li-Fi technology towards mass market adoption. The IEEE 802.11bb standard is still under development at the time of writing this book. Therefore, this section is based on the publicly available material around this standard.

In Chapter 9, we focus on reporting the commercialization activities in the past few years and also provide suggestions on how to further commercialize Li-Fi technology around the broader ecosystem view. In Section 9.2, we provide an overview of the current Li-Fi marketing reports. In this overview, a market segmentation has been considered. In addition, discussions on potential future marketing opportunities in 5G beyond and 6G are followed. In Section 9.3, we argue that Li-Fi commercialization requires a large ecosystem that should include many industries and organizations. In this section, we provide a detailed discussion on the key players and the potential strategy to establish such an ecosystem. In Section 9.4, we provide a number of examples of the commercial use cases of Li-Fi technologies to demonstrate the pathway toward future Li-Fi products. In this part, the considered use cases include demonstrations in a wide variety of use cases. In Section 9.5, we discuss challenges for further commercialization of Li-Fi. In addition, potential solutions to these challenges are briefly discussed.

References

[1] Cisco, "Cisco Edge-to-Enterprise IoT Analytics for Electric Utilities Solution Overview," February 2017. https://www.cisco.com/c/en/us/td/docs/unified computing/ucs/UCS CVDs/Cisco SAS Edge to Enterprise IoT Analytics Platform.html.

[2] Cogalan, T., and H. Haas, "Why Would 5G Need Optical Wireless Communications?" *2017 IEEE 28th Annual International Symposium on Personal, Indoor, and Mobile Radio Communications (PIMRC)*, 2017, pp. 1–6.

[3] Haas, H., J. Elmirghani, and I. White, "Optical Wireless Communication," *Philosophical Transactions of the Royal Society A: Mathematical, Physical and Engineering Sciences*, Vol. 378, No. 2169, March 2, 2020, p. 20200051. https://royalsocietypublishing.org/doi/abs/10.1098/rsta.2020.0051.

[4] O'Brien, D. C., et al., "Visible Light Communications: Challenges and Possibilities," *2008 IEEE 19th International Symposium on Personal, Indoor and Mobile Radio Communications*, 2008, pp. 1–5.

[5] Huang, X., et al., "1.6 Gbit/s Phosphorescent White LED Based VLC Transmission Using a Cascaded Pre-Equalization Circuit and a Differential Outputs PIN Receiver," *Optical Express*, Vol. 23, No. 17, August 2015, pp. 22034–22042. http://www.opticsexpress.org/abstract.cfm?URI=oe-23-17-22034.

[6] Bian, R., I. Tavakkolnia, and H. Haas, "15.73 Gb/s Visible Light Communication with Off-the-Shelf LEDs," *IEEE/OSA Journal of Lightwave Technology*, Vol. 37, No. 10, May 2019, pp. 2418–2424.

[7] M. Burla et al., "500 GHz Plasmonic Mach-Zehnder Modulator Enabling Sub-THz Microwave Photonics," *APL Photonics*, Vol. 4, No. 5, 2019, p. 056106.

[8] Lee, C., et al., "Advanced LiFi Technology: Laser Light," in *Light-Emitting Devices, Materials, and Applications XXIV*, J. K. Kim, M. R. Krames, and M. Strassburg, (eds.), Vol. 11302, International Society for Optics and Photonics/SPIE, 2020, pp. 116–123.

[9] Griffiths, A. D., et al., "Gallium Nitride Micro-Light-Emitting Diode Structured Light Sources for Multi-Modal Optical Wireless Communications Systems," *Philosophical Transactions of the Royal Society A: Mathematical, Physical and Engineering Sciences*, Vol. 378, No. 2169, 2020, p. 20190185. https://royalsocietypublishing.org/doi/abs/10.1098/rsta.2019.0185.

[10] Xie, E., et al., "Over 10 Gbps VLC for Long-Distance Applications Using a GaN-Based Series-Biased Micro-LED Array," *IEEE Photonics Technology Letters*, Vol. 32, No. 9, 2020, pp. 499–502.

[11] Koonen, T., et al., "Ultra-High-Capacity Wireless Communication by Means of Steered Narrow Optical Beams," *Philosophical Transactions of the Royal Society A: Mathematical, Physical and Engineering Sciences*, Vol. 378, No. 2169, 2020, p. 20190192. https://royalsocietypublishing.org/doi/abs/10.1098/rsta.2019.0192.

[12] Yoshida, K., et al., "245 MHz Bandwidth Organic Light-Emitting Diodes Used in a Gigabit Optical Wireless Data Link," *Nature Communications*, Vol. 11, No. 1, 2020, pp. 1–7.

[13] Manousiadis, P. P., et al., "Organic Semiconductors for Visible Light Communications," *Philosophical Transactions of the Royal Society A: Mathematical, Physical and Engineering Sciences*, Vol. 378, No. 2169, 2020, p. 20190186. https://royalsocietypublishing.org/doi/abs/10.1098/rsta.2019.0186.

[14] Kosman, J., et al., "Distortion Losses of High-Speed Single-Photon Avalanche Diode Optical Receivers Approaching Quantum Sensitivity," *Philosophical Transactions of the Royal Society A: Mathematical, Physical and Engineering Sciences*, Vol. 378, No. 2169, 2020, p. 20190194. https://royalsocietypublishing.org/doi/abs/10.1098/rsta.2019.0194.

[15] Oh, C. W., et al., "Low-Crosstalk Full-Duplex All-Optical Indoor Wireless Transmission with Carrier Recovery," *IEEE Photonics Technology Letters*, Vol. 29, No. 6, 2017, pp. 539–542.

[16] Wang, K., et al., "4 12.5 Gb/s WDM Optical Wireless Communication System for Indoor Applications," *IEEE/OSA Journal of Lightwave Technology*, Vol. 29, No. 13, 2011, pp. 1988–1996.

[17] Safety of Laser Products. Equipment Classification and Requirements, British Standards Institution (BSI) Std., Rev. Std. 2014, 2014.

[18] Taylor, M. T., and S. Hranilovic, "Angular Diversity Approach to Indoor Positioning Using Visible Light," *IEEE Globecom Workshops (GC Wkshps)*, 2013, pp. 1093–1098.

[19] Chen, Z., D. A. Basnayaka, and H. Haas, "Space Division Multiple Access for Optical Attocell Network Using Angle Diversity Transmitters," *Journal of Lightwave Technology*, Vol. 35, No. 11, 2017 pp. 2118–2131.

[20] Manousiadis, P. P., et al., "Wide Field-of-View Fluorescent Antenna for Visible Light Communications Beyond the Etendue Limit," *OSA Optica,* Vol. 3, No. 7, 2016, pp. 702–706.

[21] Collins, S., D. C. O'Brien, and A. Watt, "High Gain, Wide Field of View Concentrator for Optical Communications," *Optics Letters,* Vol. 39, No. 7, April 2014, pp. 1756–1759. http://ol.osa.org/abstract.cfm?URI=ol-39-7-1756.

[22] O'Brien, D., S. Rajbhandari, and H. Chun, "Transmitter and Receiver Technologies for Optical Wireless," *Philosophical Transactions of the Royal Society A: Mathematical, Physical and Engineering Sciences,* Vol. 378, No. 2169, 2020, p. 20190182. https://royalsocietypublishing.org/doi/abs/10.1098/rsta.2019.0182.

[23] Alqudah, Y. A., and M. Kavehrad, "Optimum Order of Angle Diversity with Equal-Gain Combining Receivers for Broad-Band Indoor Optical Wireless Communications," *IEEE Transactions on Vehicular Technology,* Vol. 53, No. 1, 2004, pp. 94–105.

[24] Carruthers, J. B., and J. M. Kahn, "Angle Diversity for Nondirected Wireless Infrared Communication," *IEEE Transactions on Communications,* Vol. 48, No. 6, June 2000, pp. 960–969.

[25] Wang, T. Q., C. He, and J. Armstrong, "Performance Analysis of Aperture-Based Receivers for MIMO IM/DD Visible Light Communications," *Journal of Lightwave Technology,* Vol. 35, No. 9, 2017, pp. 1513–1523.

[26] Miramirkhani, F., and M. Uysal, "Channel Modelling for Indoor Visible Light Communications," *Philosophical Transactions of the Royal Society A: Mathematical, Physical and Engineering Sciences,* Vol. 378, No. 2169, 2020, p. 20190187. https://royalsocietypublishing.org/doi/abs/10.1098/rsta.2019.0187.

[27] Chaaban, A., and S. Hranilovic, "Capacity of Optical Wireless Communication Channels," *Philosophical Transactions of the Royal Society A: Mathematical, Physical and Engineering Sciences,* Vol. 37, No. 2169, 2020, p. 20190184. https://royalsocietypublishing.org/doi/abs/10.1098/rsta.2019.0184.

[28] Lowery, A. J., "Spectrally Efficient Optical Orthogonal Frequency Division Multiplexing," *Philosophical Transactions of the Royal Society A: Mathematical, Physical and Engineering Sciences,* Vol. 378, No. 2169, 2020, p. 20190180. https://royalsocietypublishing.org/doi/abs/10.1098/rsta.2019.0180.

[29] Wang, Z., and J. Chen, "Networked Multiple-Input-Multiple-Output for Optical Wireless Communication Systems," *Philosophical Transactions of the Royal Society A: Mathematical, Physical and Engineering Sciences,* Vol. 378, No. 2169, 2020, p. 20190189. https://royalsocietypublishing.org/doi/abs/10.1098/rsta.2019.0189.

[30] Wen, M., et al., "A Survey on Spatial Modulation in Emerging Wireless Systems: Research Progresses and Applications," *IEEE Journal on Selected Areas in Communications,* Vol. 37, No. 9, 2019, pp. 1949–1972.

[31] Ding, Z., et al., "On the Performance of Non-Orthogonal Multiple Access in 5G Systems with Randomly Deployed Users," *IEEE Signal Processing Letters,* Vol. 21, No. 12, 2014, pp. 1501–1505.

[32] Abdalla, I., M. B. Rahaim, and T. D. C. Little, "Interference in Multi-User Optical Wireless Communications Systems," *Philosophical Transactions of the Royal*

Society A: Mathematical, Physical and Engineering Sciences, Vol. 378, No. 2169, 2020, p. 20190190. https://royalsocietypublishing.org/doi/abs/10.1098/rsta.2019.0190

[33] Haas, H., et al., "Introduction to Indoor Networking Concepts and Challenges in LiFi," *IEEE/OSA Journal of Optical Communications and Networking,* Vol. 12, No. 2, 2020, pp. A190–A203.

[34] Alsulami, O. Z., et al., "Optimum Resource Allocation in Optical Wireless Systems with Energy-Efficient Fog and Cloud Architectures," *Philosophical Transactions of the Royal Society A: Mathematical, Physical and Engineering Sciences,* Vol. 378, No. 2169, 2020, p. 20190188. https://royalsocietypublishing.org/doi/abs/10.1098/rsta.2019.0188.

[35] Uysal, M., et al., *Optical Wireless Communications: An Emerging Technology,* New York: Springer, 2016.

[36] Hranilovic, S., *Wireless Optical Communication Systems,* New York: Springer Science & Business Media, 2006.

2

A Brief History of Optical Wireless Communications

2.1 Introduction

The advances in the fields of optics and communications have enabled the concepts of light-based communications to be realized. The idea of using light to convey and transmit information was initially introduced by the pioneering work of Alexander Graham Bell and Charles Sumner Tainter in 1880 [1]. The photophone invention is considered to be the first practical optical wireless communication (OWC) system, which used sunlight to transmit voice messages. Gfeller and Bapst introduced the concept of light emitting diode (LED)-based OWC in 1979 by proposing and investigating an indoor wireless communication system via an infrared LEDs [2]. Modern advances in solid-state lighting (SSL) were enabled by the revolutionary invention of the blue LED by Isamu Akasaki, Hiroshi Amano, and Shuji Nakamura in 1990. This has enabled the concept of communications using the visible light spectrum between wavelength $\lambda = 400$ nm and $\lambda = 700$ nm. The use of LEDs for visible light communication (VLC) appeared early in the research work of Masso Nakagawa [3,4]. This was soon followed by establishing the Visible Light Communication (VLC) Consortium (VLCC) in 2003 by major industrial and research parties in Japan. This chapter briefly reports on the significant contributions that have influenced modern light-based communication systems.

A brief history of optics and vision is explored in Section 2.2. This is followed by a summary of the different theories that shaped our understanding of light from Newton to Einstein in Section 2.3. Early attempts to realize visual-based communications started in ancient China and Greece. The history of early

OWC systems is explored in Section 2.4. Early works on LED-based OWC are presented in Section 2.5. The invention of blue LEDs was a crucial step in enabling white illumination and thus VLC. The development of blue LEDs is presented in Section 2.6. The modern history of the research, development, and commercialization of VLC systems is surveyed in Section 2.7. Finally, the chapter is summarized in Section 2.8.

2.2 A Brief History of Optics and Vision

Humans have been exploiting light since the discovery of fire 500,000 to 1.5 million years ago. Lenses were developed by ancient Egyptians and Mesopotamians centuries before Christ (B.C.). The Nimrud lens is an oval lens made of rock crystal that was made 3,000 years ago [5]. Although its objective remains controversial, it is argued to have been used as a magnifying lens [6].

It is claimed that Archimedes (287–212 B.C.) used polished metal as mirrors to concentrate the heat of the Sun on Roman ships during the siege of Syracuse around 300 B.C. [7]. The reflected Sun rays would disturb the navigation of these ships and would eventually set some on fire. However, this claim is viewed with skepticism by some researchers due to the lack of details [8]. Nowadays, the same principle is being utilized by some commercial companies [9,10]. A large constellation of computer-controlled mirrors is used to collect and concentrate sunlight. This allowed them to capture the heat from the Sun at temperatures as high as 1,000° [9].

Theories of light and vision were developed by ancient Greek philosophers. Two hypotheses were introduced to describe visual perception. The emission theory was introduced by Euclid of Alexandria (325–265 B.C.) and supported by Claudius Ptolemy (AD 100–170) [5]. It claims that vision occurs due to rays being emitted from the eyes of an observer. The contradicting hypothesis is supported by Aristotle (384–322 B.C.) and Galen of Pergamum (A.D. 130–201) [5]. It claims that vision occurs due to rays of light being radiated by the visual object. Ibn al-Haytham (Alhazen), in his *Book of Optics* (*Kitab al-Manazir*, A.D. 1027–1040) introduced the modernly accepted theory of visual perception [11]. He disagreed with the Euclidian theory of vision and proposed that vision occurs when reflections of light from surfaces are observed by human eyes. A geometrical approach to studying optics and vision was introduced by Euclid. However, Alhazen's work greatly influenced the future development of optics. One of his geometrical solutions of light reflections in spherical and parabolic surfaces was only found analytically in 1998 [11,12]. Ptolemy worked on calculating the refraction of optical rays in the water. However, his empirical results were inaccurate and disproved by Alhazen. The law of refraction (Snell law) introduced by Willebrord Snellius in 1621 was also developed independently by Rene Descartes in 1637, and Thomas Harriot in 1602 [11,13]. It is interesting to note that Ibn Sahl (AD 940–1000) derived the mathematical foundations of light refraction six centuries before it was known to Snellius and Descartes [13]. However, the works of Ibn Sahl were not available to Snellius and Descartes since they were not translated into other languages until recently [14].

2.3 History of Light from Newton to Einstein

Isaac Newton (1642–1727) observed that white light can be decomposed into different spectral lights using a prism and that another prism can recombine these colored lights into a white spectrum [11]. He refuted the Cartesian ideas and proved that color is an intrinsic property of the light itself. A mechanically oriented theory of light was initially introduced by Rene Descartes (1596–1650) and further developed and championed by Newton [5]. Light is claimed to be made up of small particles called corpuscles that travel in straight lines with finite velocity and impetus. Johannes Kepler (1571–1630) explained 38 propositions on the properties of light and tried to provide a theoretical analysis of the combined effect of lenses [15]. He is mainly believed to have adopted the mechanical theory of light in his propositions [16].

Robert Hooke (1635–1703) investigated the light interference on a thin film and tried to compare it with the interference of sound waves [15]. Christiaan Huygens (1629–1695) proposed the wave theory of light in 1690 and tried to explain the different speeds of light in dense mediums [11]. Huygens considered the light to be propagating in spherical waves that would add up to form a wavefront. Leonhard Euler (1707–1783) postulated that each color of light represents a sinusoidal wave of a unique frequency and wavelength. However, the corpuscular theory of light survived until the early nineteenth century. Thomas Young (1773–1829) and Augustin-Jean Fresnel's (1788–1827) experiments on light interference confirmed the wave nature of light mainly due to the failure of the corpuscular theory in describing the refraction and polarization of light [11]. Young's efforts mainly aimed at estimating the light wavelength by observing the patterns of interference on a screen behind a closely spaced double slit. He estimated the wavelength of light to vary between 710 nm for the red color and 420 nm for the blue color. Michael Faraday (1791–1867) demonstrated the polarization nature of light by manipulating the light wave with a magnetic field [17,18]. This was followed by the pioneering work of James Clerk Maxwell (1831–1879), which confirmed the electromagnetism nature of optical waves in the 1860s [11]. Heinrich Hertz (1857–1894) discovered the photoelectric effect and the similarities between light and electromagnetic waves in terms of reflection, refraction, and speed of propagation [15].

Max Planck (1858–1947) introduced the concept of *quanta* in which he claimed that light is quantized in nature. Based on this, Albert Einstein (1879–1955) proposed the law of photoelectric effect, which confirmed the quantization nature of light [5]. Gilbert Newton Lewis (1875–1946) coined the term photon for the quantized light [15]. Niels Bohr (1885–1962) demonstrated that the wavelength of the emitted light depends on the *quanta* energy. These developments formed the basis of quantum optics. Nowadays, geometrical wave models of light are only considered as an approximation of wave-particle models when the considered wavelength of the light source is much smaller than the dimensions of the other system components.

2.4 History of Optical Wireless Communication Systems

The early attempts to realize OWCs relied on the visual perception ability of human eyes that can operate as natural optical receivers. However, the human eyes have a limited range and sensitivity. Therefore, the visual signal has to be either large or bright enough to be sufficiently decodable to human eyes. In addition, the signal should be uniquely distinguishable from any other visual signals to avoid confusion. Signaling using fire and smoke is considered to be one of the earliest approaches that has been used for more than 2,200 years.

There were frequent military invasions from the Xiongnu nomadic tribes in ancient China around 200 B.C. In order to defend the northern territories of the Chinese state and empire, the Great Wall of China was built. It now spans over 21,000 km. One of the major functions of the Great Wall is to signal potential invasion activities. To serve this purpose, towers were stationed along the wall, with enough separation in between them for the sentry to see adjacent tower stations. When the soldiers in the tower stations were able to spot any enemy attack, they used fire and smoke to alert the soldiers in the adjacent towers. By using this method of visual signaling from tower to tower, the message can be quickly delivered over hundreds of miles in just a few hours.

A similar long-distance signaling system using fire, known as Phryctoria, was invented by ancient Greek engineers. This system is able to transmit more complicated information by encoding a specific letter in the number of lit torches. Specifically, 24 Greek letters are filled in a 5-by-5 table, and each letter corresponds to a row and a column of the table. Signaling towers were built on top of adjacent mountains, and a message is delivered by using two groups of torches. When transmitting the message of a letter, a certain number of torches in both groups is lit. The numbers of lit torches in the left and right groups match the row and column indices corresponding to that letter, respectively. The message about the fall of Troy was delivered to Mycenae via this system in one night.

Native Americans also used smoke signaling systems to communicate with other friendly and enemy tribes. Common messages include warning of enemies and strangers, requesting help, and reporting news. A unique characteristic of the smoke signaling system is its diverse ability to convey information. For instance, different messages can be encoded by the number of puffs in one column of smoke. A puff or cloud of smoke can be created by temporally covering the fire using a wet blanket and then taking the blanket off to release the smoke. Messages can also be conveyed by means of smoke color variations, which could be created by burning different fuel materials. Furthermore, information can also be conveyed in the elevation of the signaling location. Creating smoke halfway up a hill or at the top of a hill can have different meanings. It is particularly important to use a coding system that is only known by the allies because the signals are visible to enemies as well. This can be viewed as a prototype of encryption.

The optical telegraph developed in 1792 by Claude Chappe (1763–1805) is one of the notable mechanical-based visual communication systems [19]. Messages are encoded to different possible positions of the mechanically controlled semaphore wings. In addition, multiple towers were used to relay messages between two ends using pre-agreed visual codes. A network consisting of 15 semaphores was installed between Paris and Lille in 1794, and the usage of semaphore was instrumental in the battle of Waterloo. The networks of semaphores kept expanding until they began to get replaced by electrical telegraphs during the middle of the nineteenth century [20].

A heliograph is a shutter-aided mirror that was used as an optical transmitter for Morse-coded messages using reflections of sunlight [19]. A night version of the solar telegraph was developed using an oil lamp and a set of shutters, mirrors, and lenses [21]. The communication distance of the heliograph is limited to a few kilometers, depending on the weather conditions.

William Grylls Adams (1836–1915) discovered the photovoltaic property of selenium when exposed to light in 1876 [22]. Alexander Graham Bell (1847–1922) utilized the light-sensitivity of selenium in his pioneering joint invention of the photophone with Charles Sumner Tainter (1854–1940) in 1880 [1]. Alexander Graham Bell considered the photophone as his greatest invention, even greater than the telephone [21]. The photophone was demonstrated at a distance of 213m meters and is considered to be the first practical OWC system [1]. The air pressure induced by the voice variations is applied to a flexible mirror at the transmitter part of the photophone. This would alternately make the mirror concave and convex, which would, in turn, condense and scatter the sunlight. These variations in sunlight intensity would be collected at the receiver end, which consists of a selenium cell at the focal point of a parabolic mirror [1]. The photovoltaic property of the selenium cell would allow it to detect and translate the variations of light intensity to an electrical signal that is heard back at the receiver using a headphone. Photophone-based systems were eventually outperformed by their radio-based counterparts mainly due to their limited ranges and their dependencies on sunlight availability and weather conditions.

2.5 Recent History of LED-Based Optical Wireless Communication

At the end of 1979, two researchers, Fritz R. Gfeller and Urs Bapst, presented their pioneering work on indoor OWC [2]. It was a period where a growing number of computers were being deployed in offices, workstations, and many other indoor environments. In order to interconnect these computers, the conventional method was to use wired networks with cables. The wire installation, however, was complicated and also inconvenient for future maintenance. Therefore, Gfeller and Bapst proposed an IR-based wireless system as an alternative. They considered deploying a satellite interface in the center of a room and equipping each network node with an optical terminal interface. The interfaces are composed of an IR LED and a P-type intrinsic N-type photodiode (PIN-PD). Thus, the satellite interface and the network nodes exchange information through an optical wireless connection. This initial study primarily focuses on the investigation of physical layer factors that limit the link performance. For example, the transmission symbol rate may be limited by the multipath characteristics of the optical wireless channel, including diffused reflections. In addition, ideal scenarios were considered for the optical wireless channel in the simulation model, and a few experimental results on the transmitter and receiver interfaces were presented.

Inspired by Gfeller and Bapst's work, researchers started to consider this indoor wireless network based on optical signals and discuss its advantages over the conventional RF-based wireless networks. A major advantage of adopting an OWC network is its use of unregulated IR spectrum, which is free to use [23]. In addition, IR emitter and detector front-end elements are available at relatively low costs. Furthermore, the OWC network using intensity modulation (IM) and direct detection (DD) (IM/DD) does not suffer from the effects of multipath fading. This is due to the relatively large size of the detector compared to the wavelength of the IR signal. The wavelength range of 780 nm to 950 nm has been found to be suitable for the wireless IR links. This was because of the low cost of LEDs emitting in this spectrum range at that time. In addition, silicon-based photodiodes have peak responsivity in this wavelength range. However, the IR signal is able to go through the cornea. When the incident power is too high, it may cause thermal damage to the retina. Consequently, the radiation power of the LED has to be controlled in a range that is not more than a few hundred of milliwatts. Given the constraint on the emission power of the optical sources, optical concentrators, low-noise receiver circuit design, and optical filters have been considered to improve the link signal-to-noise ratio (SNR).

The early design of wireless IR links has been classified by directionality and the existence of a LOS path, as shown in Figure 2.1(a). Directed links are more power-efficient but require mechanical alignment with tracking capabilities. A nondirected link requires no alignment, but the detectable signal power is lower.

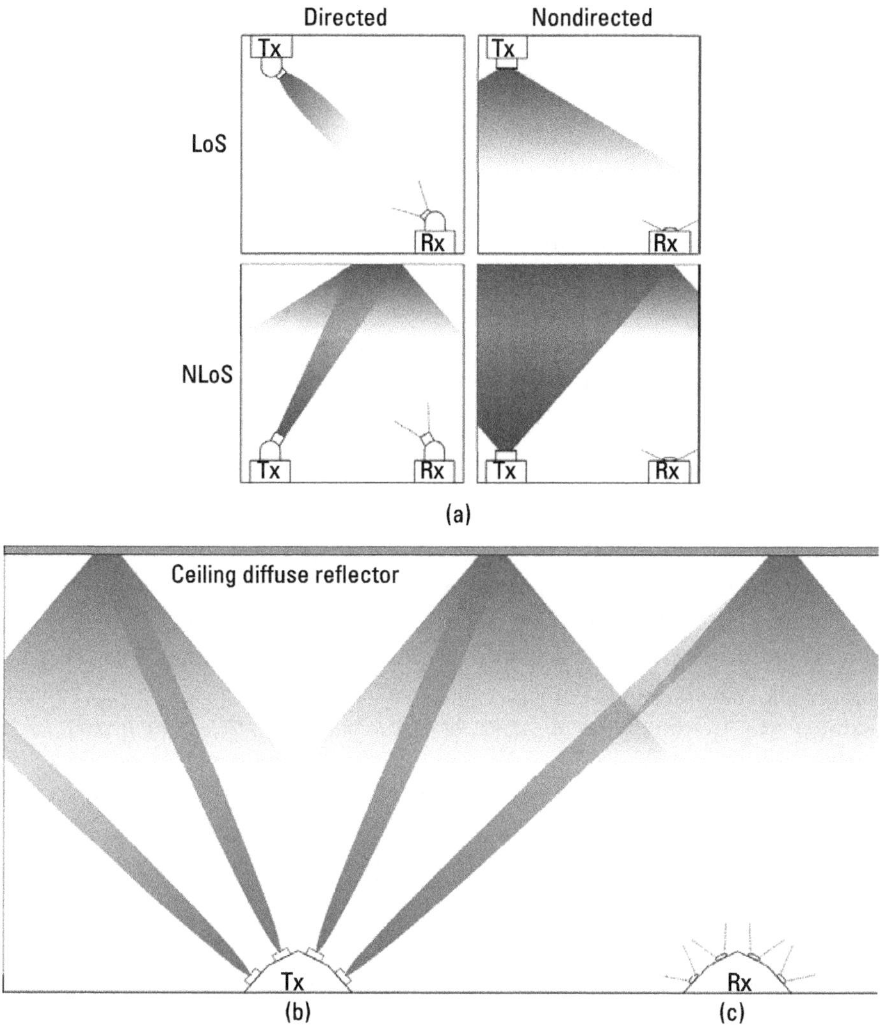

Figure 2.1 (a) Wireless IR link designs. (b) Quasi-diffuse transmitter. (c) Angle-diversity receiver.

A LOS-based link has a high detected signal power, but there is a risk of an outage when the LOS path is obstructed. Lastly, the NLOS-based link is robust against blockage and convenient to use at the cost of very low detectable power. Due to the limitation of transmission power, most of the wireless IR communication systems adopted a directed or hybrid LOS link design because of its low path loss. Such configurations are sufficient for a point-to-point link, but they are proven to be challenging for point-to-multipoint connections such as the transmissions between several devices and an IR access point in a room. Therefore, some unique link designs, such as quasi-diffuse transmitters and angle-diversity receivers, have been proposed. A quasi-diffuse transmitter is designed to be placed on a desktop level and is composed of several narrow beam IR emitters that are directed to a different area of the ceiling [24,25], as shown in Figure 2.1(b). The IR signal is reflected by the ceiling and covers many devices in a large indoor area. This link design effectively increases the coverage area of an access point and, at the same time, maintains the power efficiency of the link. An angle diversity receiver is composed of multiple detectors oriented to different directions with narrow fields of view (FOVs) [26,27], as shown in Figure 2.1(c). It mitigates the power loss caused by large signal incident angles. Moreover, high gain optical concentrators can be used at each detector without sacrificing the overall receiver FOV. It also effectively reduces noise caused by ambient light and interference from other IR transmission. In addition, various combining techniques can be applied to further improve link SNR.

The research and development of wireless IR technology have finally led to the establishment of the industry-driven interest group: Infrared Data Association (IrDA), which defined a complete set of protocols for short-range wireless IR communications [28]. There have been over 130 companies that have joined this standard association. This standard has been primarily used for the last 1-m point-to-point wireless data transfer with low power and a high data rate of up to 4 Mbps (which was considered very fast at that time). The IrDA standard includes an Infrared Physical Layer Specification (IrPHY), which defines the wireless link to be half-duplex using modulation schemes such as pulse position modulation (PPM) and on-off keying (OOK) with a target bit error ratio (BER) in the range of 10^{-8} to 10^{-9}. It also specifies the number of link geometric parameters such as defining the radiation half-power semi-angles to be in the range of 15° to 30° and the link distance to be 1m to 3m. The standard also defines high-layer protocols an Infrared Link Access Protocol (IrLAP) and an Infrared Link Management Protocol (IrLMP). IrLAP defines the detailed operations of processes, including link initialization, address identification, connection, transmission, disconnection, and link termination. This protocol support point-to-point and point-to-multipoint transmissions. It also includes functions that are responsible for negotiation and resolving conflicts. IrLMP establishes a platform for the running of several software applications in each link node. The protocol is composed of service discovery, service multiplexing, and link management functionalities. Based on this standard, point-to-point IR wireless connection capability has been implemented into a lot of commercial portable products in the 1990s, such as mobile phones, pagers, and other types of digital devices. Prototypes of wireless local-area networks (LANs) have also been introduced based on IR transmission, such as the SpectrixLite made by Spectrix Corporation. After 2000, the wireless IR became less popular due to the emergence of wireless fidelity (Wi-Fi) and Bluetooth. However, the invention of the efficient blue LED brought opportunities to evolve the wireless IR technologies.

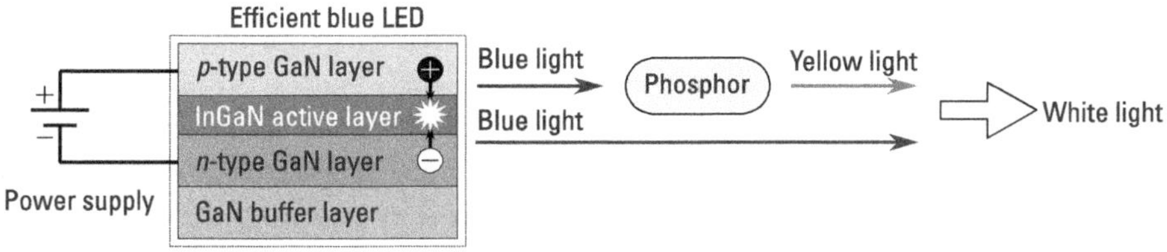

Figure 2.2 Efficient blue LED schematic depiction and white light creation.

2.6 Development of Efficient Blue LED

During the early 1990s, the revolutionary invention of an efficient blue LED established the foundation of SSL. The inventors Isamu Akasaki, Hiroshi Amano, and Shuji Nakamura were awarded the Nobel Prize in Physics in 2014. The major advantage of the efficient blue LED is its significant improvement in electrical-to-optical conversion efficiency. A large proportion of the electrical power is converted to nonradiative heat in incandescent bulbs and fluorescent lamps. This leads to a luminous efficacy (a measure of conversion efficiency) in the range of tens of lm/W. In contrast, an efficient blue LED is able to convert the majority of the power into light, and the resultant luminous efficacy can be improved to about 300 lm/W. The basic structure of an efficient blue LED is shown in Figure 2.2. The LED is composed of a number of layers made from gallium nitride (GaN) or indium gallium nitride (InGaN). Under the biasing from a direct current source, the electrons in the *n*-type GaN layer and the holes in the *p*-type GaN layer are injected into the InGaN active layer. Then the radiative recombination of electrons and holes creates photons. Thanks to the double-heterostructure (DH) LED structure, the recombination process is highly efficient. The creation of white light from a blue LED is illustrated in Figure 2.2. A yellow phosphor material can be used to convert part of the blue light into yellow light, and the combination of the blue light and the yellow light creates the desired white illumination [29].

Before the emergence of SSL in the 1980s, many researchers were interested in the development of an efficient blue LED in the field of applied physics. The main two materials considered to be used in the manufacturing of these LEDs were zinc selenide (ZnSe) and GaN. At that time, most of the researchers favored ZnSe, because it was easy to produce high-quality ZnSe material with very low dislocation density. However, it was very difficult to obtain high-quality GaN material with an acceptable level of dislocation density. Consequently, GaN is less popular in the research field and was expected to become obsolete. Only a few researchers persisted in this direction, including Akasaki, Hiroshi, and Nakamura.

The research was initiated from the development of novel methods to grow high-quality GaN materials. Akasaki and Amano created a novel metal organic chemical vapor deposition (MOCVD) to grow GaN. However, this design has a high carrier-gas velocity requirement and poses difficulties in terms of uniformity, scalability, and reproducibility. One more issue with this approach is to do with the use of high concentrations of aluminum, which causes problems when it comes to the reproduction of GaN materials. In 1991, another novel approach to grow high-quality GaN was proposed by Nakamura. He invented a two-flow MOCVD that uses low carrier-gas flow and is able to grow very high-quality GaN uniformly. With this invention, it becomes possible to create a GaN buffer layer to replace the aluminum nitride (AIN) butter layer.

The next important step is to create *n*-type and *p*-type GaN layers. Akasaki and Amano developed the method of applying low-energy electron beam irradiation (LEEBI) to magnesium-doped GaN, which gave an undesired performance. Three years later, Nakamura developed the theory of hydrogen passivation to explain the phenomenon and proposed thermal anneal to solve the problem, which made the creation of high-quality *n*-type and *p*-type GaN layers possible.

The next challenge for the commercialization of efficient blue LEDs is the low output power, which can be solved by adopting the DH LED structure. In a DH LED, the free carriers are confined within a thin active emitting layer between the *n*-type and *p*-type layers, which significantly increased the carrier concentration, thereby increasing the probability of radiative recombination and efficiency of the LED. This leads to the development of the material for the active layer, and InGaN was found to be a promising candidate. InGaN cannot only narrow the bandgap but also can introduce the ability to change the color of the emission. However, it is very difficult to create high-quality InGaN at room temperature. In addition, precise control over all growth parameters when using a MOCVD is required to make the InGaN effective in a DH LED. This was also solved by the two-flow MOCVD by Nakamura. With all these advancements, Nakamura was able to demonstrate a blue LED with an output power of 1.5 mW and 2.7% external quantum efficiency.

2.7 Modern History of Visible Light Communications

The idea of using LEDs as communication devices was first introduced by Pang et al. in 1998 [30] for intelligent transportation systems (ITSs), followed by the development of a prototype for a traffic light audio broadcasting system by the same research group in [31,32]. In the aforementioned work, intensity modulation was realized by the fast switching of an LED-based traffic light between the on and off states to transmit an audio signal to a receiver located at a distance of 20 m. Utilizing white LEDs for data transmission started to take shape in the early 2000s, when Tanaka et al. from Keio University in Japan developed simulations of LED-based, 10-Mbps wireless home links [33]. The same group studied the effect of optical path differences caused by the existence of multiple transmitting LEDs in a room environment in [3]. The term iLight was then used in [34] to refer to using frequent switching of high brightness LED displays to transmit an audio signal, which had a transmission speed of 128 kbps. Following that, the VLCC [30] was established in 2003 by a number of Japanese companies, with the goal of publicizing and promoting the emerging VLC technology. In this regard, VLCC proposed two VLC standards that were adopted by Japan Electronics and Information Technology Industries Association (JEITA) as JEITA CP-1221 Visible Light Communication System Standard and CP-1222 Visible Light ID System Standard [36]. In 2003, Komine et al. [37] proposed, for the first time, the integration of existing electricity power lines for the purpose of enabling VLC transmissions. In this regard, power line communication (PLC) allows modulating the intensity of the LEDs' transmit power according to the signal waveform transmitted through power lines. A fundamental mathematical analysis of LED-based VLC links was presented in [4], where the effects of multipath reflections, intersymbol interference (ISI), and receiver noise were discussed. It was shown that a VLC network consisting of four LED fixtures in a room environment could provide a 200-Mbps data transmission rate without receiver tracking and up to 10-Gbps rate when receiver tracking was employed. Walewski's team at Siemens launched a research project in 2006 to focus on LED-enabled wireless broadband access [38]. The German company worked on the development of LED-drivers and receivers circuitry using commercially available LEDs made by Osram. In 2008, sponsored by the European Union, a home gigabit access project (OMEGA) was launched with a consortium of 20 European industrial and academic partners to develop physical layer techniques for enabling VLC-based broadcast coverage for home area networks [39]. OMEGA's research delivered a VLC demonstration of a 100-Mbps, high-definition video streaming using 16 LED lamps. The National Science Foundation in the United States worked on the development of hybrid VLC and RF communications networks in partnership with several U.S. universities. The Czech Republic-based Twibright Labs developed the Reasonable Optical Near Joint Access (RONJA) device that is capable of transmitting up to a distance of 1.4 km using red light emissions [40]. A proof-of-concept implementation for an orthogonal frequency division multiplexing (OFDM)-based VLC link was developed by Afgani et al. from the International University Bremen in 2006 [41]; experimental demonstrations and theoretical results proved that OFDM-based IM/DD transmissions are feasible and able to provide significant data rate enhancement for VLC systems. The developed prototype used a single off-the-shelf LED and covered a transmission distance of about 1m with a BER of 10^{-3} without the use of channel or source coding techniques, opening the door for realizing high-speed VLC systems based on OFDM transmissions. In 2008, a 40-Mbps VLC link was realized over a distance of 2m using 16 LEDs with resonant equalization techniques and OOK modulation [42]. An experimental demonstration of 100-Mbps VLC link using blue filtering in combination with simple post-equalization was presented in [43].

Since both filtering and equalization take place at the receiver side, the proposed technique did not require extra hardware circuitry or computations at the LEDs.

In [44], a data rate of 3.22 Gbps was realized over a transmission distance of 0.25m using commercially available red, green, and blue (RGB) LEDs. For this purpose contention access period (CAP) modulation and wavelength division multiplexing (WDM) were employed to maximize the capacity of the system. In 2014, a 3-Gbps link was realized from a single micro-LED [45]. To achieve this transmission speed, direct current (DC) biased optical OFDM (DCO-OFDM) was employed as a modulation scheme and both pre-equalization and post-equalization techniques as well as adaptive bit and energy loading were applied. Following that, several novel modulation schemes were proposed to enhance the spectrum and power efficiency of Li-Fi transmissions. This includes OFDM-based techniques such as layered/enhanced asymmetrically clipped optical OFDM (L/e-ACO-OFDM) [46,47] and enhanced unipolar OFDM (eU-OFDM) [48], augmented spectral efficiency discrete multitone (ASE-DMT) [49] as well as color-based modulation techniques such as enhanced color shift keying (CSK) [50], combined CSK, and PPM [51]. In 2014, a 10-Gbps bidirectional VLC LAN based on a star topology was proposed by Wang et al. in [52]. In this experimental demonstration, a frequency division multiplexing (FDM) fiber link was employed for the backbone network, while time division multiplexing (TDM) was used for the VLC transmission. Moreover, the aforementioned work demonstrated an 8-Gbps full-duplex transmission based on a network configuration of eight VLC access points (APs). A MIMO VLC demonstration operating at a data rate of 1 Gbps was presented in [53], where two-dimensional individually addressable arrays of micro-LEDs were used at the transmitter while the receiver employed CMOS integrated photodiode (PD) arrays. A multiband CAP was experimentally demonstrated for the first time in 2015 in [54]. It was shown that increasing the number of CAP subcarriers results in higher achievable data rates and that for a total of 10 subcarriers a data rate of 31.5 Mbps was achieved with a spectral efficiency of 4.85 b/s/Hz. An RGB LED-based 11.28-Gbps VLC link was reported in [55] based on experimental demonstrations of WDM transmissions with rate-adaptive OFDM modulation. The highest data rate achieved using LED-based transmissions was reported in [56] where a 15.73-Gbps link was achieved over a distance of 1.6 m. The proposed setup used four single-color LEDs based on WDM and DCO-OFDM with adaptive bit loading.

In 2015, the use of light amplification by stimulated emission of radiation diodes (LDs) for the purpose of communications was first proposed by Hussein et al. in [57] to overcome the bandwidth limitation in LED-based VLC systems. A data rate of 5 Gbps was achieved with the use of an imaging receiver while combining the imaging receiver with a delay adaptation approach yielded a data rate of 10 Gbps. Chun et al. demonstrated in [58] that a data rate of 6.25 Gbps could be achieved by white light transmission generated by means of a blue LD combined with a remote phosphor, whereas a rate of 10 Gbps was demonstrated by employing a two-channel imaging approach. Janjua et al. [59] and Tsonev et al. [60] demonstrated the use of RGB LDs for enabling VLC links with transmissions rates up to of 4 Gbps and 14 Gbps, respectively. The work in [61] presented a simulation-based 20-Gbps mobile indoor VLC system based on the use of LD transmitters combined with beam steering and imaging receivers. In this regard, the beam-steering approach was used to improve the received SNR by directing the light beams towards the receiver, based on a location estimation algorithm. The proposed method was shown to provide SNR improvement of 15 dB in the worst-case scenario. Lu et al. proposed the use of vertical-cavity surface-emitting laser (VCSEL) for energy-efficient and high-speed wireless connectivity [62]. An 11.1-Gbps data rate was achieved over a distance of 1.2m using a 682-nm VCSEL with 1-GHz modulation bandwidth. An adaptive inverse fast Fourier transform (IFFT) and fast Fourier transform (FFT) size discrete Fourier transform (DFT) spread OFDM VLC system was experimentally demonstrated in [63] using a 450-nm blue-light LD. It was shown that the adaptive size approach can lead to data rate enhancement from 5.46 Gbps to 5.95 Gbps over a transmission distance of 5m. Moreover, this approach was also shown to improve the BER from 2.05×10^{-2} to 6.8×10^{-3} compared to nonadaptive DFT spread OFDM. A white-light phosphor-coated 460-nm blue LD VLC system was demonstrated in [64]. A transmission rate of 1,062.5 to 1,250 Mbps over a range of 1m to 1.5m was achieved based on OOK-modulated signal without the need for blue color filtering at the receiver.

In 2016, an optical camera communication (OCC) system based on 1,024-ary quadrature amplitude modulation (QAM) signal transmission was experimentally demonstrated in [65]. The presented results showed that the proposed system, which employs a dual LED and a 50-fps commercial camera receiver, can support a data rate of 500 bps over a transmission span of 1.5m. A mobile phone camera-based VLC system with a rotation compensation technique was proposed in [66]. Experimental results demonstrated that the proposed system can achieve a data rate of 1,280 bps using a 120-fps mobile camera. A novel phone camera-based receiver was proposed in [67], where the receiver can suppress the path loss, ISI, and signal distortion by means of a computer vision-based algorithm used to observe the environment by detecting the information in a series of received images. The presented results showed that the proposed mechanism could achieve a success rate of 98% in recovering the transmitted images. An indoor visible light positioning system based on OCC was experimentally demonstrated in [68], where the location coordinates' data was received and demodulated by a camera, and the position was estimated based on the geometrical relations between the different images received on the camera's sensor. The proposed scheme provided location estimation with a mean error of 5.0 cm and 6.6 cm for distances of 120 cm and 180 cm, respectively.

Table 2.1
Li-Fi Products Available in the Market

Product	Company	DL Speed	UL Speed	Multi-Access
TruLiFi	Signify	150 Mbps	150 Mbps	Multiple users
1 Gbit/s LiFi System	Fraunhofer HHI	1 Gbps	1 Gbps	Multiple users
Gigabit Light Antenna	pureLiFi	1 Gbps	1 Gbps	Multiple users
LiFi ASIC	pureLiFi	86.4 Mbps	86.4 Mbps	Up to 16 users
LiFiMAX	Oledcomm	100 Mbps	40 Mbps	Up to 16 users
LumiNet Pad	VLNComm	108 Mbps	53 Mbps	Up to 15 users
LiFi-XC	pureLiFi	43 Mbps	43 Mbps	Up to 8 users
LiFi-X	pureLiFi	42 Mbps	42 Mbps	Multiple users
MyLiFi	Oledcomm	13 Mbps	10 Mbps	Single user
LumiNet Desktop	VLNComm	23 Mbps	23 Mbps	Up to 7 users
GeoLiFi	Oledcomm	1 Mbps	1 Mbps	Single user

The term Li-Fi was first introduced by the University of Edinburgh's Professor Harald Haas in his talk "Wireless Data from Every Light Bulb" at TED Global in 2011 [69]. In this talk, Haas demonstrated for the first time to a general audience that an off-the-shelf LED lightbulb can transmit wireless data for high-definition video streaming. Since then, the term Li-Fi has been widely used to refer to fully networked bidirectional wireless systems based on nanometer-wave communications. In 2013, the Li-Fi Research and Development Center (LRDC) was established at the University of Edinburgh, aiming to take the advancements in Li-Fi research into mainstream applications and accelerating its development as a major global industry through collaborations with industry and key research bodies around the world. Haas and his team established pureLiFi Ltd., a spin-off company based on their research. The Scotland-based company has since pioneered the commercialization of Li-Fi technology worldwide. Since then, many global firms have started to work on the development of Li-Fi products, such as the French company, Oledcomm, and the American company, VLNComm and the Dutch world company Signify (former Philips Lighting). Table 2.1 summarizes some of the Li-Fi products available today in the market.

In the following, we list some of the Li-Fi use cases that have been recently implemented. In 2015, the Carrefour retail chain in France, in cooperation with Philips Lighting and IBA, equipped one of their stores with a VLC system. The LED lights within the store connect customers to the shop's database through an interactive mobile application to make it easier to check the locations of products and availability. In 2016, a similar use case was implemented in an E.Leclerc retail store where the shop's LEDs fixtures were used as VLC transmitters to communicate with shopping trolleys equipped with Li-Fi tracker modules. The collected information was sent to a Li-Fi cloud unit and used to track and analyze customers' shopping behavior. In 2017, Babcock, a leading provider of complex engineering services, established a Li-Fi based smart connected facility where Li-Fi can provide reliable and secure connectivity for the wireless sensor networks. In 2018, the U.K. mobile network operator O2, in partnership with pureLiFi, started a Li-Fi network trial in their Slough headquarters. The system consists of nine LED fixtures and can provide high-speed bidirectional network connectivity. The Scottish Kyle Academy has become the first school to adopt a Li-Fi network connection on its premises in 2018, providing the classrooms with unprecedented access to digital resources. More Li-Fi real-life applications are being constantly introduced as this technology is being shaped to face the challenges of the 5G era and beyond.

2.8 Summary

This chapter provides a brief history of OWC. A long history of innovations and scientific advancements has led to the development of light-based communications. The understanding of light and the exploitation of its uses has advanced the field of OWC. The development of the blue LED has not only enabled energy-efficient white illumination. This has also enabled VLC. The recent history of VLC is rich with multidisciplinary advancements, from the simple switching of white LEDs between on and off states to transmit low-rate information signals to the realization of tens of gigabits per second seamless transmissions by means of advanced multi-carrier modulation techniques. Based on the accelerated advancements in mobile OWC, it is evident that Li-Fi innovations will continue to shape up as a major global industry. Advancements in Li-Fi technology are expected to impact many aspects of the modern world, particularly when the demand for data usage outgrows the available supply from existing technologies.

References

[1] Bell, A. G., "Selenium and the Photophone," *Nature*, Vol. 22, No. 569, 1880, pp. 500–503.

[2] Gfeller, F. R., and U. Bapst, "Wireless In-House Data Communication via Diffuse Infrared Radiation," *Proceedings of the IEEE*, Vol. 67, No. 11, November 1979, pp. 1474–1486.

[3] Tanaka, Y., et al., "Indoor Visible Communication Utilizing Plural White LEDs as Lighting," *Proc. of the 12th IEEE International Symposium on Personal, Indoor and Mobile Radio Communications*, Vol. 2, San Diego, CA, September 2001, pp. 81–85.

[4] Komine, T., and M. Nakagawa, "Fundamental Analysis for Visible-Light Communication System Using LED Lights," *IEEE Transactions on Consumer Electronics*, Vol. 50, No. 1, February 2004, pp. 100–107.

[5] Guarnieri, M., "The Rise of Light – Discovering Its Secrets," *Proc. IEEE*, Vol. 104, No. 2, February 2016, pp. 467–473.

[6] Uluc, K., G. C. Kujoth, and M. K. Baskaya, "Operating Microscopes: Past, Present, and Future," *Neurosurgical Focus*, Vol. 27, No. 3, 2009, p. E4.

[7] Claus, A. C., "On Archimedes' Burning Glass," *Applied Optics*, Vol. 12, No. 10, October 1973, pp. A14–A14. http://ao.osa.org/abstract.cfm?URI=ao-12-10-A14.

[8] Stavroudis, O. N., "Comments on: On Archimedes' Burning Glass," *Applied Optics*, Vol. 12, No. 10, October 1973, pp. A16–A16. http://ao.osa.org/abstract.cfm?URI=ao-12-10-A16.

[9] "Bill Gates-Backed Heliogen Targets 1,500 Degrees Celsius Solar Thermal," *The Engineer*, November 2019. https://www.theengineer.co.uk/bill-gates-heliogen-solar-thermal/.

[10] Coventry, J., and C. Andraka, "Dish Systems for CSP," *Solar Energy, Progress in Solar Energy Special Issue: Concentrating Solar Power (CSP)*, Vol. 152, 2017, pp. 140–170. http://www.sciencedirect.com/science/article/pii/S0038092X17301573.

[11] Guarnieri, M., "Two Millennia of Light: The Long Path to Maxwell's Waves," *IEEE Industrial Electronics Magazine*, Vol. 9, No. 2, June 2015, pp. 54–60.

[12] Neumann, P. M., "Reflections on Reflection in a Spherical Mirror," *The American Mathematical Monthly*, Vol. 105, No. 6, 1998, pp. 523–528. http://www.jstor.org/stable/2589403.

[13] Smith, A. M., *From Sight to Light: The Passage from Ancient to Modern Optics*, Chicago, IL: University of Chicago Press, 2017. https://books.google.co.uk/books?id=hsbmAQAACAAJ.

[14] Rashed, R., "A Pioneer in Anaclastics: Ibn Sahl on Burning Mirrors and Lenses," *Isis*, Vol. 81, No. 3, 1990, pp. 464–491. http://www.jstor.org/stable/.

[15] Vohnsen, B., "A Short History of Optics," *Physica Scripta*, Vol. T109, 2004, p. 75. https://doi.org/10.1238%2Fphysica.topical.109a00075.

[16] Straker, S. M., *Kepler's Optics: A Study in the Development of Seventeenth-Century Natural Philosophy*, Indiana University, 1970.

[17] Faraday, M., "X. The Bakerian Lecture—Experimental Relations of Gold (and Other Metals) to Light," *Philosophical Transactions of the Royal Society of London*, No. 147, 1857, pp. 145–181.

[18] Mansuripur, M., "The Faraday Effect," *Optics & Photonics News*, Vol. 10, No. 11, November 1999, pp. 32–36. http://www.osa-opn.org/abstract.cfm?URI=opn-10-11-32.

[19] Dimitrov, S., and H. Haas, *Principles of LED Light Communications: Towards Networked Li-Fi*. Cambridge, U.K.: Cambridge University Press, 2015.

[20] Selleri, S., "Claude Chappe and the First Telecommunication Network (Without Electricity)," *URSI Radio Science Bulletin*, Vol. 2017, No. 360, March 2017, pp. 96–101.

[21] Bouchet, O., and P. -N. Favennec, *Wireless Optical Communications*. New York: Wiley, 2012.

[22] Adams, W. G., and R. E. Day, "V. The Action of Light on Selenium," *Proceedings of the Royal Society of London*, Vol. 25, No. 171-178, 1877, pp. 113–117. http://rspl.royalsocietypublishing.org/content/25/171-178/113.short.

[23] Kahn, J. M., and J. R. Barry, "Wireless Infrared Communications," *Proceedings of the IEEE*, Vol. 85, No. 2, February 1997, pp. 265–298.

[24] G. Yun and M. Kavehrad, "Spot-diffusing and fly-eye receivers for indoor infrared wireless communications," 1992 IEEE International Conference on Selected Topics in Wireless Communications, 1992, pp. 262–265, doi: 10.1109/ICWC.1992.200761.

[25] A. P. Tang, J. M. Kahn and Keang-Po Ho, "Wireless infrared communication links using multi-beam transmitters and imaging receivers," Proceedings of ICC/SUPERCOMM '96 - International Conference on Communications, 1996, pp. 180–186 vol. 1, doi: 10.1109/ICC.1996.540267.

[26] J. B. Carruther and J. M. Kahn, "Angle diversity for nondirected wireless infrared communication," in IEEE Transactions on Communications, vol. 48, no. 6, pp. 960–969, June 2000, doi: 10.1109/26.848557.

[27] A. M. R. Tavares, R. J. M. T. Valadas, and A. M. de Olveira Duarte, "Performance of an optical sectored receiver for indoor wireless communication systems in presence of artificial and natural noise sources," in Proc. SPIE Conf. on Wireless Data Transmission, vol. 2601, Philadelphia, PA, Oct. 23–25, 1995, pp. 264–273.

[28] S. Williams, "IrDA: past, present and future," in IEEE Personal Communications, vol. 7, no. 1, pp. 11–19, Feb. 2000, doi: 10.1109/98.824566.

[29] Nakamura, S., "Background Story of the Invention of Efficient InGaN Blue-Light-Emitting Diodes (Nobel Lecture)," *Angewandte Chemie International Edition*, Vol. 54, No. 27, 2015, pp. 7770–7788.

[30] Pang, G., et al., "Dual Use of LEDs: Signalling and Communications in ITS," *5th World Congress on Intelligent Transport Systems*, 1998.

[31] Pang, G., et al., "LED Traffic Light as a Communications Device," *Proceedings 1999 IEEE/IEEJ/JSAI International Conference on Intelligent Transportation Systems* (Cat. No.99TH8383), October 1999, pp. 788–793.

[32] Pang, G., et al., "Visible Light Communication for Audio Systems," *IEEE Transactions on Consumer Electronics*, Vol. 45, No. 4, November 1999, pp. 1112–1118.

[33] Tanaka, Y., S. Haruyama, and M. Nakagawa, "Wireless Optical Transmissions with White Colored LED for Wireless Home Links," *Proceedings 11th IEEE International Symposium on Personal Indoor and Mobile Radio Communications (PIMRC 2000)* (Cat. No.00TH8525), Vol. 2, September 2000, pp. 1325–1329.

[34] Pang, G., et al., "LED Wireless," *IEEE Industry Applications Magazine*, Vol. 8, No. 1, January 2002, pp. 21–28.

[35] "Visible Light Communications Consortium (VLCC)," 2007. http://www.vlcc.net/modules/xpage1/?ml lang=en.

[36] Japan Electronics and Information Technology Industries Association (JEITA), https://www.jeita.or.jp/english/.

[37] Komine, T., and M. Nakagawa, "Integrated System of White LED Visible-Light Communication and Power-Line Communication," *IEEE Transactions on Consumer Electronics*, Vol. 49, No. 1, February 2003, pp. 71–79.

[38] Conti, J. P., "What You See Is What You Send," *Engineering & Technology*, Vol. 3, No. 19, pp. 66—67, November 2008.

[39] Javaudin, J., et al., "OMEGA ICT Project: Towards Convergent Gigabit Home Networks," *2008 IEEE 19th International Symposium on Personal, Indoor and Mobile Radio Communications*, September 2008, pp. 1–5.

[40] Garber, L., "Turning on the Lights for Wireless Communications," *Computer*, Vol. 44, No. 11, November 2011, pp. 11–14.

[41] Afgani, M., et al., "Visible Light Communication Using OFDM," *2nd International Conference on Testbeds and Research Infrastructures for the Development of Networks and Communities (TRIDENTCOM)*, Barcelona, Spain, March 1–3, 2006, pp. 6–134.

[42] Le Minh, H., et al., "High-Speed Visible Light Communications Using Multiple-Resonant Equalization," *IEEE Photonics Technology Letters*, Vol. 20, No. 14, July 2008, pp. 1243–1245.

[43] Le Minh, H., et al., "100-Mb/s NRZ Visible Light Communications Using a Postequalized White LED," *IEEE Photonics Technology Letters*, Vol. 21, No. 15, August 2009, pp. 1063–1065.

[44] Wu, F. -M., et al., "3.22-Gb/sWDM Visible Light Communication of a Single RGB LED Employing Carrier-Less Amplitude and Phase Modulation," *Optical Fiber Communication Conference and Exposition and the National Fiber Optic Engineers Conference (OFC/NFOEC)*, March 2013, pp. 1–3.

[45] Tsonev, D., et al., "A 3-Gb/s Single-LED OFDM-Based Wireless VLC Link Using a Gallium Nitride LED," *IEEE Photonics Technology Letters*, Vol. 26, No. 7, April 2014, pp. 637–640.

[46] Islim, M. S., D. Tsonev, and H. Haas, "On the Superposition Modulation for OFDM-Based Optical Wireless Communication," *2015 IEEE Global Conference on Signal and Information Processing (GlobalSIP)*, December 2015, pp. 1022–1026.

[47] Wang, Q., et al., "Layered ACO-OFDM for Intensity-Modulated Direct-Detection Optical Wireless Transmission," *OSA Optics Express*, Vol. 23, No. 9, May 2015, pp. 12382–12393. http://www.opticsexpress.org/abstract.cfm?URI=oe-23-9-12382.

[48] Tsonev, D., S. Videv, and H. Haas, "Unlocking Spectral Efficiency in Intensity Modulation and Direct Detection Systems," *IEEE Journal on Selected Areas in Communications*, Vol. 33, No. 99, 2015, pp. 1758–1770.

[49] Islim, M. S., and H. Haas, "Augmenting the Spectral Efficiency of Enhanced PAM-DMT-Based Optical Wireless Communications," *Optical Express*, Vol. 24, No. 11, May 2016, pp. 11932–11949. http://www.opticsexpress.org/abstract.cfm?URI=oe-24-11-11932.

[50] Singh, R., T. O'Farrell, and J. P. R. David, "An Enhanced Color Shift Keying Modulation Scheme for High-Speed Wireless Visible Light Communications," *IEEE/OSA Journal of Lightwave Technology*, Vol. 32, No. 14, July 2014, pp. 2582–2592.

[51] Luna-Rivera, J. M., et al., "Combined CSK and Pulse Position Modulation Scheme for Indoor Visible Light Communications," *Electronics Letters*, Vol. 50, No. 10, May 2014, pp. 762–764.

[52] Wang, Y., et al., "Network Architecture of a High-Speed Visible Light Communication Local Area Network," *IEEE Photonics Technology Letters*, Vol. 27, No. 2, January 2015, pp. 197–200.

[53] Rajbhandari, S., et al., "High-Speed Integrated Visible Light Communication System: Device Constraints and Design Considerations," *IEEE Journal on Selected Areas in Communications*, Vol. 33, No. 9, September 2015, pp. 1750–1757.

[54] Haigh, P. A., et al., "A Multi-CAP Visible-Light Communications System with 4.85-b/s/Hz Spectral Efficiency," *IEEE Journal on Selected Areas in Communications*, Vol. 33, No. 9, September 2015, pp. 1771–1779.

[55] Chun, H., et al., "LED Based Wavelength Division Multiplexed 10 Gb/s Visible Light Communications," *Journal of Lightwave Technology*, Vol. 34, No. 13, July 2016, pp. 3047–3052.

[56] Bian, R., I. Tavakkolnia, and H. Haas, "15.73 Gb/s Visible Light Communication with Off-the-Shelf LEDs," *IEEE/OSA Journal of Lightwave Technology*, Vol. 37, No. 10, May 2019, pp. 2418–2424.

[57] Hussein, A. T., and J. M. H. Elmirghani, "Mobile Multi-Gigabit Visible Light Communication System in Realistic Indoor Environment," *Journal of Lightwave Technology*, Vol. 33, No. 15, August 2015, pp. 3293–3307.

[58] Chun, H., et al., "Visible Light Communication Using Laser Diode Based Remote Phosphor Technique," *2015 IEEE International Conference on Communication Workshop (ICCW)*, June 2015, pp. 1392–1397.

[59] Janjua, B., et al., "Going Beyond 4 Gbps Data Rate by Employing RGB Laser Diodes for Visible Light Communication," *Optical Express*, Vol. 23, No. 14, July 2015, pp. 18746–18753. http://www.opticsexpress.org/abstract.cfm?URI=oe-23-14-18746.

[60] Tsonev, D., S. Videv, and H. Haas, "Towards a 100 Gb/s Visible Light Wireless Access Network," *OSA Optics Express*, Vol. 23, No. 2, January 2015, pp. 1627–1637. http://www.opticsexpress.org/abstract.cfm?URI=oe-23-2-1627.

[61] Hussein, A. T., M. T. Alresheedi, and J. M. H. Elmirghani, "20 Gb/s Mobile Indoor Visible Light Communication System Employing Beam Steering and Computer Generated Holograms," *IEEE/OSA Journal of Lightwave Technology*, Vol. 33, No. 24, December 2015, pp. 5242–5260.

[62] Lu, I., et al., "Utilization of 1-GHz VCSEL for 11.1-Gbps OFDM VLC Wireless Communication," *IEEE Photonics Journal*, Vol. 8, No. 3, June 2016, pp. 1–6.

[63] Chen, Q., et al., "Experimental Research on Adaptive 128/64QAM DFT-Spread IFFT/FFT Size Efficient OFDM with a High SE in VLLC System," *IEEE Photonics Journal*, Vol. 9, No. 1, February 2017, pp. 1–8.

[64] Yeh, C., C. Chow, and L. Wei, "1250 Mbit/s OOK Wireless White-Light VLC Transmission Based on Phosphor Laser Diode," *IEEE Photonics Journal*, Vol. 11, No. 3, June 2019, pp. 1–5.

[65] Luo, P., et al., "Experimental Demonstration of a 1024-QAM Optical Camera Communication System," *IEEE Photonics Technology Letters*, Vol. 28, No. 2, January 2016, pp. 139–142.

[66] Cahyadi, W. A., et al., "Mobile Phone Camera-Based Indoor Visible Light Communications with Rotation Compensation," *IEEE Photonics Journal*, Vol. 8, No. 2, April 2016, pp. 1–8.

[67] Boubezari, R., et al., "Smartphone Camera Based Visible Light Communication," *Journal of Lightwave Technology*, Vol. 34, No. 17, September 2016, pp. 4121–4127.

[68] Lin, B., et al., "An Indoor Visible Light Positioning System Based on Optical Camera Communications," *IEEE Photonics Technology Letters*, Vol. 29, No. 7, April 2017, pp. 579–582.

[69] Haas, H., "Wireless Data from Every Light Bulb," *TED Talks*, August 2011. https://www.ted.com/talks/harald_haas_wireless_data_from_every_light_bulb.

3

Emerging Technology in Optical Wireless Communications

3.1 Introduction

The recent advances in optical devices and SSL have enabled various types of emerging technologies to be researched and developed within the realm of OWC. These technologies were motivated by the evolving needs and increasing demands for wireless connectivity during the past two decades. A taxonomy of OWC technologies is illustrated in Figure 3.1. Each technology can be categorized based on four basic metrics: transmission speeds, transmission range, duplex mode, and communication mode. The most popular OWC technologies include free-space optical (FSO) communication, ultraviolet (UV) communication, optical camera communication (OCC), and Li-Fi. FSO communication is designed to establish a wireless optical transmission between two fixed locations in an outdoor scenario. Its primary application is for establishing an alternative low-cost wireless backhaul system. Therefore, it typically achieves a very long transmission distance of more than 1 km and a very high data rate in the range of a few or hundreds of gigabits per second. UV communication enables long-distance NLOS wireless connectivity through atmospheric scattering, which makes UV links robust to link blockage and shadowing. As a result, UV communication systems are particularly desirable for relaxing or eliminating the need for pointing and tracking techniques. OCC technology aims to accelerate the adoption of OWC by circumventing the need for custom-built photoreceivers. OCC utilizes ubiquitous cameras as photoreceivers in addition to reusing the lighting infrastructure. OCC-based links are quick and easy to establish when compared to other OWC technologies. OCC offers a low-speed OWC solution with data rates below 10 Mbps [1]. The transmission speeds

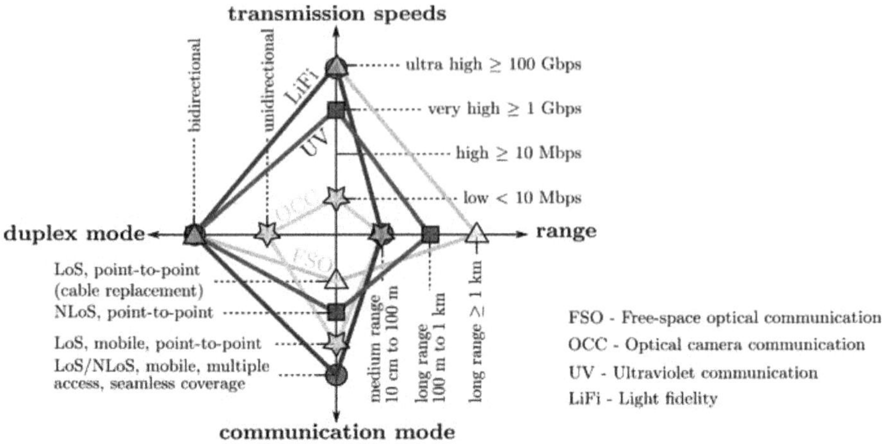

Figure 3.1 OWC taxonomy.

of OCC are mainly limited by the frame rate of the camera. The communication link is unidirectional and the range is below 100 m between a panel of LEDs or a display screen and the camera [2]. Li-Fi technology can be treated as a networked version of a visible light communication (VLC) system with more powerful and diverse functionalities, such as bidirectional transmission, mobility support, and seamless coverage. A Li-Fi system is typically deployed in an indoor environment with a transmission distance in the range of a few to tens of meters based on a LOS or NLOS channel. Due to the extremely high spatial reuse rate and the advances in optical front ends, the achievable data rate can go up to hundreds of gigabits per second.

The FSO communication technology is discussed in Section 3.2. Several fundamental topics, such as applications, system model, and channel characteristics, are introduced in Section 3.2.1. A series of advanced techniques in FSO communications are briefly discussed in Sections 3.2.2 through 3.2.5. The UV technology is discussed in Section 3.4.2.1. State-of-the-art UV devices are discussed in Section 3.3.1 while the channel modeling techniques are presented in Section 3.3.2. Challenges and future trends related to UV communications are discussed in Section 3.3.3. The principles of OCC technology are discussed in Section 3.4. The types of image sensors that can enable OCC are identified in Section 3.4.1. Different types of OCC systems are discussed in Section 3.4.2 such as screen-to-camera communications and OCC systems based on the rolling shutter mechanism. Some potential applications of OCC are discussed in Section 3.4.3. Finally, the challenges and practical considerations are identified in Section 3.4.4. The topics of Li-Fi technology research and development are the main subject of this book and will be covered in later chapters.

3.2 Free-Space Optical Communications

3.2.1 Fundamentals of FSO Communications

FSO communication is an emerging OWC technology for long or ultra-long-range transmissions, which can be defined as a high data rate OWC link between two fixed points over a distance of up to several kilometers [3]. In this section, we present a brief introduction to the FSO technology including some related recent research work. Due to the use of coherent laser sources in the IR spectrum region, up to tens of gigahertz of modulation bandwidth is available for wireless data transmission, which is significantly wider compared to most of the current RF spectrum used for wireless communications [3]. The confinement of the transmitted signal makes FSO systems have both an extremely high spatial spectrum reuse rate and high security against information interception. In addition, the use of the optical spectrum is license-free and causes no electromagnetic interference to existing RF communication systems. As a point-to-point transmission, installing FSO communication systems is more convenient and cost-effective compared to laying optical fibers. However, due to the large divergence of laser beams relative to the size of the optical detector, FSO links pose significant power loss compared to optical fiber links. More importantly, the FSO channel is likely to be affected by various atmospheric effects and weather conditions [4]. Therefore, the major challenge of using FSO communication systems is overcoming the link reliability issues caused by these effects.

3.2.1.1 Development of FSO Communications

In 1960, scientists from Bell Labs attempted to send pulse signals between towers with a 25 miles separation using a ruby laser. With the help of a filter and a sensitive detector, they have successfully spotted the pulse signal [5]. Later in 1962, researchers in MIT Lincoln Labs have implemented an FSO transmission of a TV signal. A link distance of 30 miles has been achieved using an LED made of gallium arsenide (GaAs) [3]. In late 1963, an IBM group demonstrated an FSO link that avoids the atmospheric scintillation effects using a GaAs injection laser with pulse-rate modulation [6]. After the initial attempt of FSO links, researchers found that it is not reliable to transmit an analog signal over an FSO link under the interference of atmospheric turbulence. Due to the nonlinearity issue, optical frequency modulation (FM), subcarrier FM, and pulse code modulation (PCM) started to gain popularity in the late 1960s, providing a dynamic range of up to 55 dB. In the 1970s, the research on FSO was less popular due to the challenges of conquering the atmospheric effects and the emergence of optical fiber. The following use cases of FSO communication are mainly for military and space applications [7]. In recent years, due to the significant increase in the number of smart devices and wireless data traffic, wireless communication using higher frequencies (millimeter wave, OWC) started gaining popularity. The requirement of higher wireless data transmission capacity and supporting various types of services motivated the development of the small-cell/heterogeneous networks. However, how to establish a high-performance and cost-effective backhaul network between the significantly increased number of base stations (BSs) and the core network became a critical challenge for researchers [8]. FSO communication became an attractive solution for last-mile connections between end-users and fast fiber connection to the data center.

3.2.1.2 Applications of FSO Communications

The following are applications of FSO communication:

- *Backhaul connections for cellular systems:* Conventionally, wired optical fiber links are used for cellular network backhaul connections. In the next generation heterogeneous networks, large numbers of small cells will be deployed to achieve data traffic offloading [9]. This trend will inevitably increase the density of deployed BSs and therefore significantly increase the number of backhaul connections between the small cell BSs and the core fiber network linked to the data center. Furthermore, the higher standard for the future cellular quality of service (QoS) requires high capacity and low latency for these last-mile backhaul connections. Deploying optical fibers is a possible solution but has been identified as very cost-inefficient. In this situation, FSO communication would be a potential compromise to achieve high transmission data rates and low cost simultaneously, as shown in Figure 3.2(a).
- *Temporary links for disaster recovery and broadcasting:* In some situations, a temporary high-speed wireless link is required between two fixed locations. During natural disasters, such as earthquakes and flooding, conventional wireless communication systems are likely to be damaged or destroyed and unable to operate properly, as shown in Figure 3.2(b). In this situation, temporary replacements using FSO communication would be a practical solution to recover the wireless network with convenient installations. In the case of broadcasting live events in a remote location, where the broadcasting vehicle is not reachable, a high capacity FSO link can be established between the broadcasting vehicle and the camera in the site of the live event to guarantee the transmission of a high-resolution video signal [3], as shown in Figure 3.2(c).
- *Dedicated links for high-speed transmission and security:* In today's companies, schools, and universities, high-speed secure transmission may be required in their internal networks such as LAN-to-LAN connections between two buildings [4]. Using existing external networks may pose security concerns or insufficient bandwidth while deploying a dedicated

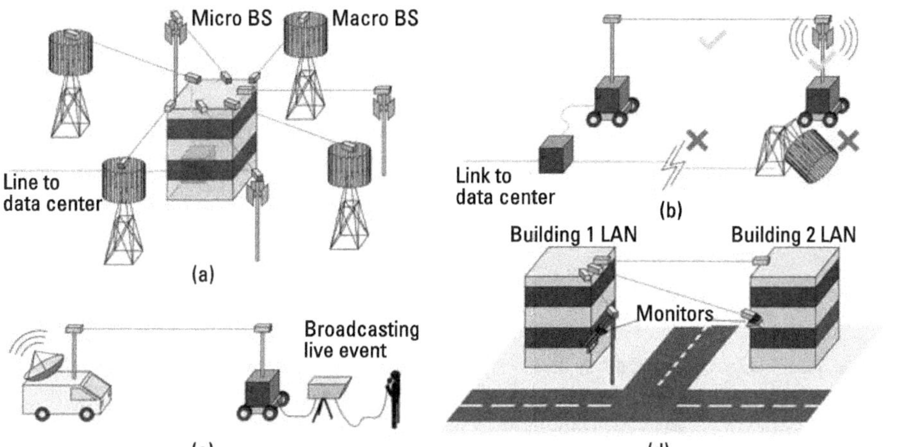

Figure 3.2 (a) FSO communication for cellular backhaul connections. (b) FSO communication for disaster recovery. (c) FSO communication for live broadcasting. (d) FSO communication for local-area network (LAN)-to-LAN link and video monitoring.

wired connection would be expensive. In contrast, deploying an FSO system would be a more viable and powerful alternative, as illustrated in Figure 3.2(d). In addition, FSO technologies can also be used for high-definition video signal transmission from surveillance and monitoring cameras. Other similar applications include the combination of FSO systems with quantum cryptography systems [3].

3.2.1.3 FSO Communication Systems

In a typical FSO communication link, information bits are modulated and converted to an optical signal by an FSO transmitter. The optical signal propagates and is distorted in the FSO channel and captured by the FSO receiver. After a detection and demodulation process, the transmitted information bits can be successfully retrieved.

- *FSO transmitter:* An FSO transmitter is composed of an optical source, a modulator, an optical amplifier (optional), and optics, as illustrated in Figure 3.3(a, b). An optical source is used to convert electrical power to optical power. Light amplification by stimulated emission of radiation diodes (LDs) are the most commonly used optical sources for modern FSO communication. They offer high output power, coherent light, and a very collimated beam. Regarding the operation spectrum of optical sources, the most popular carrier wavelengths are 850 nm and 1,550 nm, which have the lowest atmospheric attenuation in the IR spectrum range [3]. Popular laser sources operating at these two wavelengths include

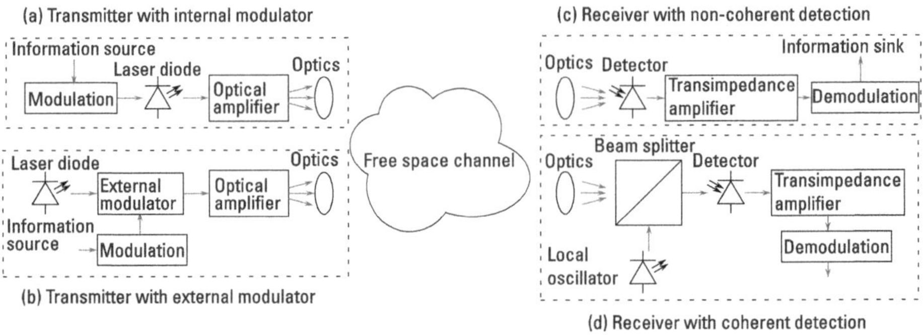

Figure 3.3 FSO communication block diagram. (a) Transmitter with internal modulator. (b) Transmitter with external modulator. (c) Receiver with incoherent detection. (d) Receiver with coherent detection.

the vertical-cavity surface-emitting laser (VCSEL) (850 nm), Fabry-Perot (FP) lasers (1,550 nm), and distributed feedback (DFB) lasers (1550 nm). An optical amplifier can be used when the radiated optical power from the optical source is insufficient. Optics at the transmitter side are used to form the focused beam towards the receiver. As illustrated in Figure 3.3, two types of modulators exist: an internal modulator and an external modulator. An internal modulator converts information bits into an electrical signal and uses this electrical signal to drive the optical source directly via a driving circuit. This method is simple and inexpensive to implement, but the transmission suffers from nonlinear distortion and slow response speed of the optical source [10]. In addition, driving an LD with an alternating current (AC) will shorten the lifetime of the laser device. Utilizing the electro-optic/acousto-optic effects of materials, an external modulator can achieve modulation function optically by altering the refractive index, polarization, and other transmission characteristics of the incoming light from the optical source [10]. This modulator can convey the information on the electromagnetic wave of the optical carrier. Therefore, amplitude modulation (AM) and phase modulation (PM) are possible with an external modulator. In addition, it has better linearity and is able to utilize the full power of the source. However, an external modulator is more expensive and suffers from coupling losses.

- *FSO receiver:* An FSO receiver is composed of a detection module, an amplifier, and a demodulation module. There are two types of detection schemes: incoherent detection and coherent detection (see Figure 3.3(c, d)). In incoherent detection, an optical detector is used to directly sense the incident light intensity, which is also known as direct detection (DD) [10]. Incoherent detection is simple to implement but is unable to capture phase information. Therefore, only intensity modulation (IM) is supported. In coherent detection, an LD local oscillator (LO) is included. The received optical signal is combined with an optical carrier signal from the LO via a beam splitter in the field. Then an optical detector is used to capture the combined signal. By increasing the power of the LO, the thermal noise can be sufficiently suppressed so that the receiver can operate in the shot-noise limited regime. Therefore, coherent detection methods improve the detection sensitivity and also support modulations with complex signals such as quadrature amplitude modulation (QAM) and phase shift keying (PSK). The downside of coherent detection is the requirements of the LO, including the extra power needed to drive the LO. In addition, it is challenging to match the spatial phases between the LO and the received beam when using coherent detection [11]. The received beam pattern is significantly distorted by the atmosphere, which makes the phase matching difficult to be implemented. Photodiodes (PDs) are the most commonly used device for optical detectors. When the link performance is limited by thermal noise, an avalanche photodiode (APD) can be used to further improve the link signal-to-noise ratio (SNR) [10]. After the detection process, the signal is normally forwarded to a transimpedance amplifier to overcome the thermal noise. Eventually, the amplified signal is demodulated and the information bits are recovered.

3.2.1.4 FSO Channel Characteristics

In the process of FSO signal propagation, the optical signal experiences a series of losses and distortions in free space, which is the major limitation to the performance of FSO communication and link distance [3]. The characteristics of the FSO channel have been thoroughly investigated by researchers and the major channel effects include the following.

- *Geometric and misalignment losses:* In FSO systems, the receiver lens aperture has a limited size. After the optical signal travels a distance of hundreds or thousands of meters, even a small divergence angle may lead to a big beam footprint on the receiver side relative to the lens aperture size. Consequently, only a fraction of the optical signal power can be captured by the receiver and the remaining power is lost in the free space, as illustrated in Figure 3.4 ①. Due to the elasticity of the transceiver fitting structures and building structures, the positions and orientations of the FSO transceivers may slightly vary. This will eventually lead to a beam pointing error [12], as illustrated in Figure 3.4 ②. For the FSO systems with tracking capabilities, this misalignment may also happen due to estimation errors in the tracking system. Therefore, minor beam divergence is useful to combat the misalignment at the cost of lower

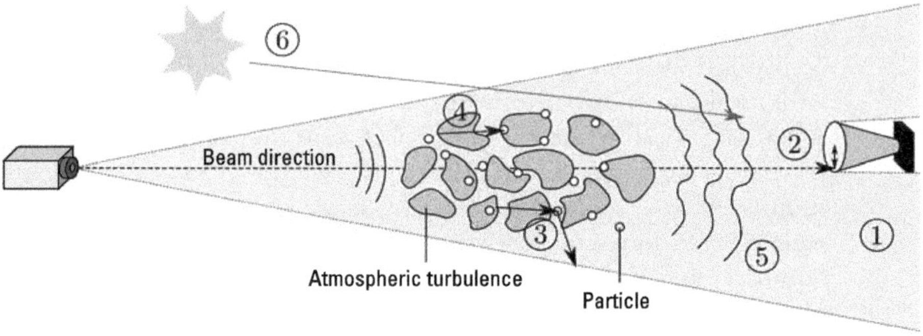

Figure 3.4 FSO channel characteristics. 1: geometric loss. 2: misalignment loss. 3: atmospheric loss (scattering). 4: atmospheric loss (absorption). 5: fading due to atmospheric turbulence. 6: background radiation.

received light intensity. This motivates the studies on investigating the FSO beam width to achieve optimal performance [13].

- *Atmospheric attenuation:* During some weather conditions, such as rain, fog, and snow, a large amount of liquid or solid particles exist in the atmosphere. These particles may absorb or scatter the incident light, thereby causing significant attenuation to the optical power that reaches the receiver [14], as illustrated in Figure 3.4 ③and ④. Especially when the particle size is similar to the wavelength of the light, the scattering effect will be significant. Such a condition happens during the weather with fog. In the cases of rain and snow, the size of water/snow particles is significantly large compared to the optical signal wavelength and the FSO transmission is less affected. Apart from these weather conditions, smoke and dust may also cause similar attenuation issues.
- *Atmospheric turbulence caused scintillation:* Atmospheric turbulence is formed by the variation of air pressure and temperature due to wind and heating effects from the Sun. Consequently, the refractive index of the air along the signal propagation path of the FSO link fluctuates with time. This phenomenon introduces a fading effect to the FSO channel even in clear weather conditions, as demonstrated in Figure 3.4 ⑤. This fading effect may significantly degrade the performance of an FSO link, especially for long-distance transmissions. Fading due to atmospheric turbulence is a dominant factor of FSO channel instability. Therefore, this phenomenon has been extensively studied and modeled in the open literature [15].
- *Background radiation:* An FSO detector aperture collects not only the desired optical signal but also some undesired background light, such as direct sunlight and scattered sunlight [16]. This background radiation causes extra shot noise that lowers the achievable link SNR. When

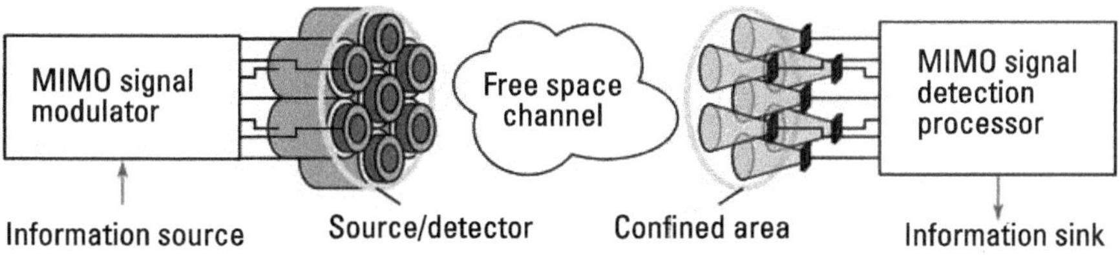

Figure 3.5 Block diagram of an FSO MIMO system.

background radiation is too high, it may saturate the optical detector and cause link outage.

3.2.2 FSO Communications with MIMO Techniques

MIMO technology has been exhaustively studied in the community of RF wireless communications [17]. It is able to enhance the performance of wireless communications in terms of link reliability and transmission data rate. Motivated by this benefit, various MIMO techniques have been considered in FSO communications. The block diagram of an FSO MIMO system is depicted in Figure 3.5. Compared to a single-link FSO communication, a MIMO system uses multiple transmitting optical sources and multiple receiving optical detectors, introducing the spatial degrees of freedom (DOF) into the communication system. Typically, these optical sources/detectors are confined in a finite spatial area. Theoretically, the multiple channels between each pair of sources and detectors are uncorrelated. In practice, the near-field effect and spatial correlation exist, which may somehow lower the effectiveness of MIMO techniques [4]. As mentioned in Section 3.2.1.4, atmospheric turbulence leads to instability in FSO links, which is the major limiting factor of the performance of FSO systems. Therefore, utilizing MIMO techniques (also known as spatial diversity) to mitigate turbulence-induced fading has become a popular research topic.

3.2.2.1 Receive Diversity

There are two types of receive diversity as illustrated in Figure 3.6(a, b). The first type considers using a large receiver aperture to combine incident light with various fading conditions optically [3], which is known as aperture averaging in FSO. This approach has the lowest complexity and has been demonstrated to be effective in weak and moderate turbulence conditions. However, this requires a larger aperture size, which is expensive. The other type of receive diversity uses multiple smaller lenses and detectors placed on various region of the receiver plane. Thus, the receiver will collect multiple copies of the detected signal with different fading conditions and carry out the weighted combining electrically. Equal gain combining (EGC) has the lowest complexity [18]. However, maximum ratio combining (MRC) offers the optimal performance, but requires channel information.

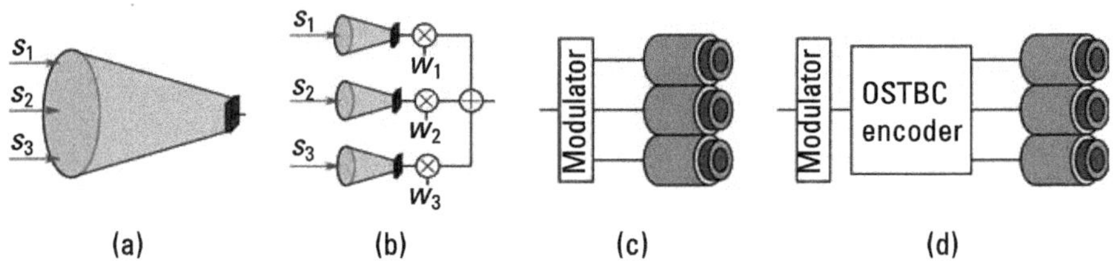

Figure 3.6 Techniques of spatial diversity: (a) aperture averaging, (b) weighted combining, (c) repetition coding, and (d) OSTBC.

3.2.2.2 Transmit Diversity

In RF wireless communications, sending the same signal from multiple antennas leads to no diversity gain. This is because the coherent superposition of multiple copies of the same signal with different channels may lead to a destructive combination. In the case of IM/DD-based FSO systems, the incoherent signals are combined in power. As long as the transmitters are sufficiently separated to make the channels independent from each other, the full diversity gain can be achieved. Therefore, sending same signals from all transmitters in an IM/DD-based FSO system is a viable solution, which is known as repetition coding (RC) [19]. Another approach of transmit diversity is achieved by using an orthogonal space-time block code (OSTBC) to mimic the process of applying space-time coding in RF wireless communications [20]. However, this approach has been demonstrated to be less efficient and more complex than the RC technique [21].

3.2.3 FSO Communications with Orbital Angular Momentum (OAM)

Providing high transmission data rates is another important characteristic of FSO technology. To pursue high spectral efficiency with a limited bandwidth, various multiplexing techniques have been considered and used in FSO systems, such as wavelength division multiplexing (WDM), polarization division multiplexing (PDM), and space division multiplexing (SDM). A special type of SDM is known as mode division multiplexing (MDM), which uses multiple orthogonal spatial modes to separate multiplexing channels [22]. In recent years, OAM, one of the spatial modal basis sets, has gained popularity in FSO communications. OAM can be imagined as the twisting of light beams. By using different beam twisting rates, various orthogonal multiplexing channels can be formed. Among various MDM schemes, FSO multiplexing systems based on OAM are easier to implement [23]. There are various methods to generate and detect OAM beams. A block diagram of an OAM FSO multiplexing system is depicted in Figure 3.7. The modulated coherent optical beams are forwarded to spiral phase plates (SPPs). The output OAM beams are combined via a multiplexer and launched from the same aperture. The simplest multiplexer can be constructed using 50% beam splitters as shown in Figure 3.7. On the receiver side, the combined beam is first

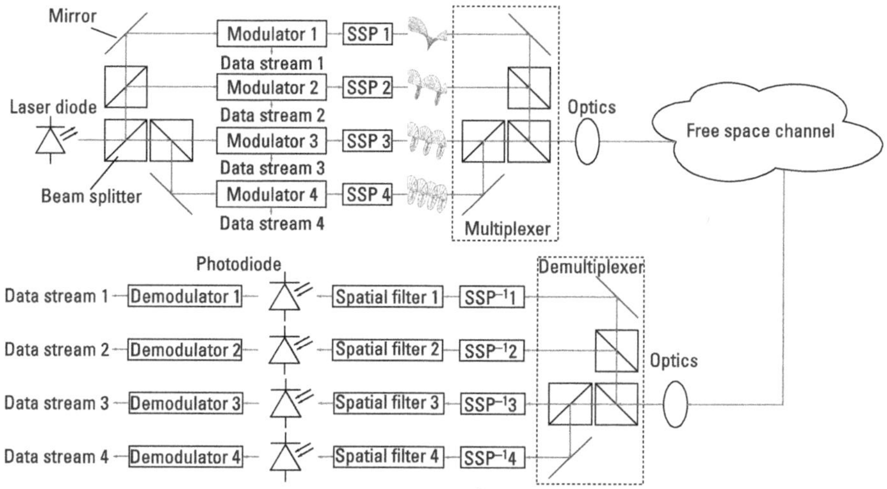

Figure 3.7 OAM multiplexing system block diagram.

split into multiple copies via a demultiplexer, and each copy is forwarded to an inverse phase hologram/conjugate SPP, which removes the helical phase of the beams. Eventually, after a spatial filter, the message of each beam can be detected and demodulated. There are also other ways to form OAM beams, such as using a spiral phase mirror (SPM), spatial light modulator (SLM), metasurfaces, and q-plate liquid-crystal-based device [22]. There exist also a series of approaches to generate OAM from a laser resonant cavity. On the receiver side, alternative approaches also exist such as a mode sorter and photonic integrated circuits [22]. When using OAM in FSO links, the channel turbulence may distort the desired helical phase front, thereby causing interference between different spatial modes. Fortunately, various interference mitigation techniques can be used to deal with this issue, such as using adaptive optical systems, blind MIMO equalization, and zero-forcing beamforming [22,24]. It has been experimentally demonstrated that an FSO system using WDM and OAM is able to achieve a data rate of 500 Gbps over strong turbulence channel conditions [25].

3.2.4 FSO Communications Based on a Kramers-Kronig (KK) Receiver

With the development of FSO technology, there is a trend of shifting from incoherent systems to coherent systems in commercial FSO products [3]. In order to lower the complexity and cost of the coherent FSO system, researchers started to consider a coherent FSO system without using an LO on the receiver side. There are few such designs, and a KK receiver has been found to be the most promising one [26]. A block diagram of a KK receiver-based FSO system can be found in Figure 3.8(a). In addition to a modulated signal with a certain bandwidth, a continuous-wave field is also transmitted concurrently. The spectral impulse

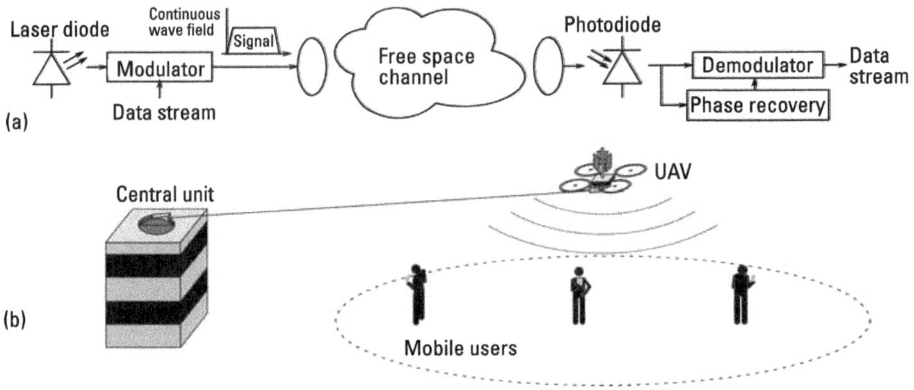

Figure 3.8 (a) Block diagram of coherent FSO system based on KK receiver. (b) FSO link for unmanned aerial vehicle (UAV) BS.

of the continuous-wave field is placed on the left-side edge of the modulated signal spectrum. On the receiver side, the PD is only able to capture the intensity of the signal. Utilizing the characteristics of the transmitted signal and Hilbert transform, the phase information of the signal can be reconstructed. Eventually, the information carried by the complex signal can be decoded. Compared to the methods of a similar type, the KK receiver is more efficient as it uses less optical spectrum and the power ratio of the continuous-wave field to the modulated signal is lower. A recent experimental study has successfully demonstrated a 200-Gbps FSO link using WDM and a KK receiver for a link distance of 50m [27].

3.2.5 FSO Communications in UAV-Based Networks

With the popularity of UAV civil applications, UAV has been proposed as mobile BSs for wireless communication services [28]. Due to the nature of the wireless traffic load, UAV BSs can be quickly deployed into regions with high loads. In the event of a disaster or power outage, the existing BSs may not be able to function normally, UAV BSs can be deployed as a temporary replacement. Since a UAV is in the air, any wired backhaul solution is not possible and traditional RF links cannot provide sufficient data rates for backhaul. A backhaul link for UAV BS using an FSO link is illustrated in Figure 3.8(b). In this condition, the FSO backhaul becomes a promising solution. This idea motivates researchers to consider relevant research problems, such as the FSO channel in the UAV scenario [29].

3.3 UV Communications

UV light is electromagnetic radiation with wavelengths ranging from 10 nm to 400 nm, which corresponds to the frequency range from 30 PHz to 750 THz. UV

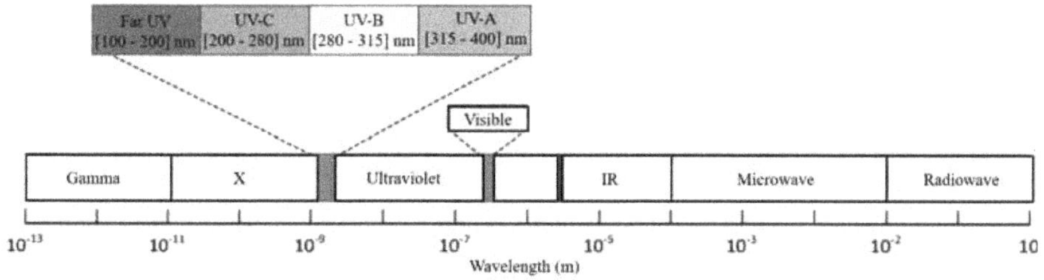

Figure 3.9 The UV sub-bands and the electromagnetic spectrum.

radiation can be classified into four spectral regions, namely: UV A-band (UV-A) (wavelengths between 400 nm and 320 nm), UV B-band (UV-B) (wavelengths between 320 nm and 280 nm), ultraviolet C-band (UV-C) (wavelengths between 280 nm and 200 nm), and far UV wavelengths between 200 nm and 100 nm), as shown in Figure 3.9. UV light was first introduced as a wireless communication medium in the mid-1960s. More specifically, UV-C was utilized for NLOS transmissions in a bandwidth-limited OWC system. It is noted that radiations within the UV band exhibit some unique characteristics compared to visible light; first, the fact that the majority of solar UV radiations get blocked by the ozone layer means that there is negligible UV light at the Earth's surface. This implies that UV communication links exhibit almost noiseless channels [30]. Second, the presence of suspended atmospheric molecules from fog, aerosols, haze, and other particles cause strong UV light scattering because of the short UV wavelengths. This creates multiple communication paths from the source to the destination, which, in turn, facilitates link enhancement by using highly sensitive large FOV receivers to gather the scattering radiation [31]. Consequently, NLOS UV links can be easily established when the LOS link is blocked. Third, UV channels typically exhibit strong attenuation, which results in limiting the transmission range to hundreds of meters, as well as reducing the achievable data rates. As a result, UV communications typically require dense network configurations [32]. However, with the recent advancements in optical front ends, transmission data rates of more than 1 Gbps have been demonstrated [33,34].

3.3.1 UV State-of-the-Art Devices

In this section, we briefly list the latest advancements in UV sources and detectors.

3.3.1.1 UV Sources

The UV communication systems of the 1960s used lasers, low and medium pressure mercury lamps, and xenon flash lamps as sources for UV radiation. These traditional devices were typically bulky in size and characterized with high power consumption as well as limited bandwidth. Recent advancements

Table 3.1
Latest Reported UV Source Specifications

Wavelength (nm)	Driving Current (mA)	Output Power (mW)	Reference
265	350	12	[38]
280	350	250	[38]
285	350	30	[38]
285	350	45	[38]
365	350	90	[39]
265	200	59	[39]
265	850	150	[40]
275	350	30	[41]
265	350	30	[42]
275	350	30	[42]

in semiconductor materials and device fabrication in the past few decades have led to the emergence of low-cost, high-performance UV LEDs. The utilization of such modern UV devices for the purpose of communications offers many advantages over traditional sources, including a smaller size, higher modulation bandwidth, lower receiver noise levels, better reliability, and longer operating lifetime [35].

For communication systems based on UV radiations, the transmit power level of the transmitting devices is an extremely important factor since it determines the achievable communication range. External quantum efficiency (EQE) is typically used to measure the percentage of injected carrier charge that are being converted by the device and emitted as photons. Thus, higher EQE values mean better device power efficiency. As a result, enhancing EQE values of UV sources is a key challenge to enable better communications functionalities. Overexposure to UV radiations is associated with skin damage, skin cancer, and eye damage. Therefore, the adoption of suitable safety measures is particularly critical for optical communications systems based on UV transmissions.

Typical EQE values for current state-of-the-art UV-A LEDs are above 30% [36], which results in up to 12-W UV-A transmit power. Current UV-B and UV-C LEDs, however, are still limited to a low EQE of about 1%–3%, which corresponds to a transmit power of a few milliwatts [37]. Table 3.1 summarizes some of the recently reported UV source specifications.

3.3.1.2 UV Detectors

Apart from their function as optical communication receivers, UV PDs also span a wide range of applications. This includes monitoring, detection, and analysis in industrial, chemical, environmental, and biological fields. UV detectors that are sensitive to UV radiations with wavelengths lower than 280 nm (i.e., UV-C and far UV) are referred to as solar-blind detectors. This is because solar radiations with these wavelengths cannot penetrate through the atmosphere and thus do not reach the Earth's surface. Also, UV detectors that are sensitive to wavelengths less than 400 nm (i.e., UV-A, UV-B, UV-C and far UV) are called visible-blind detectors since these cannot detect visible light wavelengths.

The performance of a UV PD is typically measured by its responsivity to received radiations (i.e., the capability to convert the received optical power into an electric current). Detectors' responsivity depends on different factors such as the receiver active area, the in-band transmission, the out-of-band rejection, and the internally generated false count (also referred to as dark current). For NLOS UV communications systems, the optical antenna design is required to have a large FoV so as to increase the scattering receiving area of the UV signal. Moreover, omni-directional large field-optical antennas are usually required since the relative positions of the transmitting and receiving devices change with time. Therefore, an optimal FoV design is typically essential in order to enhance the received signal power while keeping noise levels as low as possible [43].

High-performance UV detectors are typically required to satisfy the 5Ss requirements: high SNR, high sensitivity, high spectral selectivity, high speed, and high stability [44]. UV-enhanced Si photo-detectors deploy highpass filters and a phosphorus coating in order to stop low energy photons. This, however, leads to low efficiency and high dark current. In the last few years, a new generation of wide bandgap semiconductor UV PDs has emerged based on GaN-based systems, diamond, or SiC-based systems. These PDs show a much better performance, particularly in high temperatures because their operation does not require the use of filters [45].

3.3.2 UV Channel Modeling

When UV rays propagate in the atmosphere, the percentage of photons that successfully reach the receiver depends on some key atmospheric mechanisms, namely: scattering, absorption, and turbulence. As a result, the received optical power could be highly degraded. Therefore, the UV link path loss modeling is highly dependent on the order of the photons' scattering. As a result, an exact generic path loss expression is not available. Current research focuses on the derivation of an estimated path loss model either analytically or by means of numerical simulations. Generally, the UV path loss model can be classified according to the number of scattering photons into single scattering and multiple scattering models. A homogeneous atmosphere, in which the gaseous composition and density is constant with height, is usually characterized by Rayleigh scattering (caused by by molecules much smaller than the wavelength of the UV radiation) and Mie scattering (caused by particles that have a diameter equal to or larger than the wavelength of the UV radiation).

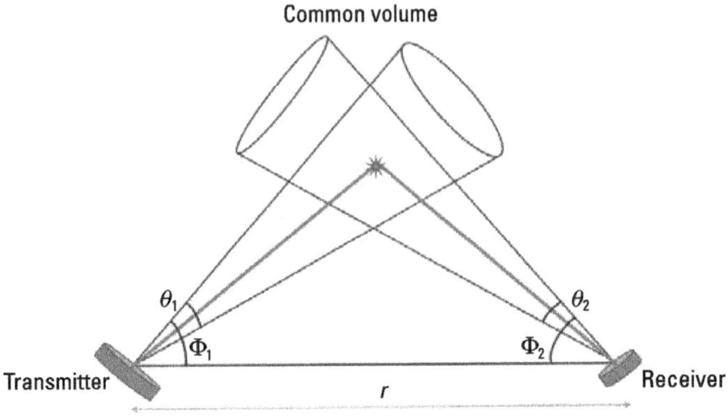

Figure 3.10 NLOS UV single scattering model.

3.3.2.1 Single Scattering Path Loss

Single scattering path loss is typically used to model short-range UV links (i.e., less than 200m), for the purpose of simplicity and tractable analysis. A single scattering assumption was imposed to develop the model in [46]. In this model, a UV source emits a light signal with a divergence angle θ_1 and an elevation angle Φ_1. At the detector's side, the scattered light beam is received with a FOV angle θ_2 and an elevation angle Φ_2 as shown in Figure 3.10.

A generalized NLOS single scattering propagation model was proposed in [47] based on the assumption that both the transmitter and receiver are located in a common plane. The received optical power at the PD was then obtained by an integral form. Many special-purpose propagation models have then been proposed based on that pioneering single scattering propagation model. For example, the work in [48] presented an expression for the received optical power based on assumption of a small common volume V. The work in [49] proposed a simplified approximation for the single scattering path loss channel model by means of a closed-form expression for tractable analysis. Similar expressions that are particularly fit for large FOVs were presented in [50,51], while a closed-form expression for the case of narrow transmitter beams and narrow FOVs was presented in [52].

3.3.2.2 Multiple Scattering Path Loss

The multiple scattering propagation channel model, shown in Figure 3.11 is typically considered when studying long-distance UV communication links. This is due to the fact that the probability of scattering increases with high-range propagation as well as with higher atmospheric particle density. As in the case of the single scattering channel, there is no exact simple closed-form expression for the multiple scattering path loss. Consequently, simulation-based approximations have been developed to capture the multiple scattering effect in

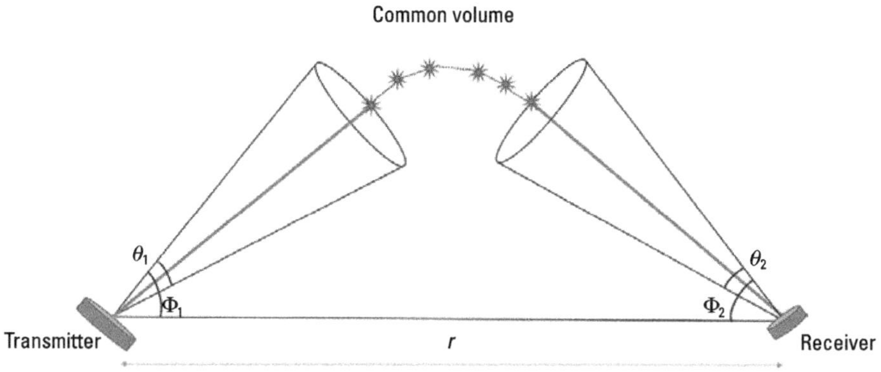

Figure 3.11 NLOS UV multiple scattering model.

UV links. By means of Monte Carlo simulations, the effect of photon emission, propagation, scattering, and reception can be statistically captured. The path loss model and the channel impulse response can then be obtained by repeating the random process for a large number of photons and calculating the arrival time and probability of each photon. The work in [53] proposed an extension of the Monte Carlo-based multiple scattering channel that takes the effect of polarization into account. Recently, a simplified channel model for two-order and three-order scattering was presented in integral form in [54] by considering a narrow transmitting beam and FOV.

3.3.2.3 Optical Turbulence

Turbulence effects result from fluctuations in the refractive index caused by varying temperature-generating irregularities in the atmosphere. When passing through these irregularities, the UV radiation can be distorted. The magnitude of the distortion depends on the severity of the turbulence and the length of the optical path. Generally, turbulence effects can be ignored at relatively short transmission ranges and clear weather conditions. However, as the communication range of UV links increases, they become more susceptible to additional impairments caused by turbulence such as irradiance fluctuation and beam wandering [55].

Various analytical, numerical, and experimental models have been developed for optical turbulence based on field measurements and wave propagation theory. The work in [56] combined Monte Carlo simulations and analytical methods to track the trajectory of the emitted photonsby means of simulations, while the received irradiance distribution was calculated analytically. The effect of turbulence on the received signal energy distribution was investigated in [57], and the work in [58] incorporated the joint effect of multiple scattering and turbulence on channel behavior.

3.3.3 Open Issues and Future Trends

UV communications have advanced rapidly in recent decades and are a promising candidate to play an important role as a complementary solution that adds to the heterogeneity of mobile wireless networks. Nevertheless, a number of challenges exist that need to be further investigated and addressed in order to enable more reliable and efficient use of such technologies. In the following, we discuss some of these open issues.

3.3.3.1 Link Efficiency

NLOS UV communication systems are both power-limited and bandwidth-limited. The power limitation results from the low transmit power of UV sources as well as the significant path loss, while the bandwidth limitation is attributed to multiple order scattering since it can spread the photon arrival time by orders of magnitude. To compensate for this low power and bandwidth efficiency, enhancement mechanisms are required at the transmitter and receiver sides as well as for signal propagation. This includes advanced modulation techniques, pulse shaping, and beam shaping techniques with proper optical design. Moreover, channel analysis and characterization are vital to reach a more accurate power decay and signal fluctuation profile. At the receiver side, low complexity signal recovery and error correction at the receiver can enhance link reliability and extract the desired signal. This requires noise cancellation techniques, ISI cancellation, and channel post-equalization.

3.3.3.2 Multihop Network Issues

Due to the fact that NLOS UV communication links are limited to relatively short distances, the concept of multihop networks is often applied. In this regard, the information signal propagates from source to destination by passing through multiple relays. In this direction, a three-hop UV link with orthogonal node cooperation was presented in [59]. It was demonstrated that the location of relays and the geometry of the system (i.e., symmetry, FOVs, and elevation angles), have a significant effect on the link performance. Furthermore, it was shown in [60] that a proper design of the multihop system geometry can lead to lowering the turbulence effect compared to LOS links. However, it is worth noting that such configurations can suffer from high multi-user interference (MUI), which could form a bottleneck in the system performance in some cases. Dealing with this interference requires the implementation of interference estimation and nodes coordination techniques. Another problem that is typically associated with multihop networks is high node density. Since the communication range is short, a high number of nodes is likely to be deployed in a small area, which adds to the interference and connection coordination issues. In most multihop networks, the multiple nodes are deployed in a random or semi-random manner. This means that it is unlikely for a node to have a priori information about its neighboring nodes. To this end, neighbor discovery is an active research topic in multihop UV networks. A number of effective neighbor discovery protocols were proposed and examined in the open literature, taking into account the UV channel behavior, scattering, and turbulence effects [61].

3.3.3.3 Extensive Experimental Validation

Rich theoretical results on UV channel modeling and transmission/reception have been reported in the literature. However, the experimental demonstrations are still very limited. As a result, the majority of the developed theoretical models and methodologies are not experimentally verified yet. For example, among the various analytic and Monte Carlo-based models presented for the modeling of path loss, turbulence effects, and single or multipoint scattering, it is not known which ones are more accurate in real-life scenarios and under which environmental circumstances. Comprehensive measurement-based investigations are thus needed to evaluate the effect of different geographic, weather, and operational-time factors. Data mining techniques can then be applied to analyze and model the correlation between different events and link performance.

3.4 Optical Camera Communications

OCC relies on the ubiquitous availability of inexpensive image sensors and LED-based lighting infrastructure. Digital cameras have become an indispensable element of modern smartphones. These cameras can work well as spatial and color photoreceivers. The predicted number of mobile phone users is expected to reach 5.7 billion in 2022 (71% of the population) [62]. OCC uses the widely available complementary metal oxide semiconductor (CMOS) cameras as a form of photoreceivers. In comparison, a vast number of photoreceivers such as P-type intrinsic N-type photodiode (PIN-PDs) and APDs can be used in Li-Fi, as will be discussed in Chapter 6. The Institute of Electrical and Electronics Engineers (IEEE) 802.15.7-2011 OWC standard is revised by the IEEE 802.15.7m task group to focus on establishing a physical layer that supports OCC [63,64]. Lately, this standard was released as IEEE 802.15.7-2018 [65].

3.4.1 Image Sensors

The image sensor is composed of multiple PDs that convert the light intensity of an image to a voltage that is proportional to the incident photons. This is then converted into a binary output. Most of the commercially available cameras can be categorized into: global shutter charge coupled device (CCD) image sensors and rolling shutter CMOS image sensors [66]. The pixels of a global shutter image sensor are all exposed to the image simultaneously, whereas a single line of pixels in a rolling shutter image sensor is exposed at each time instant. The rolling shutter sequentially sweeps across the image sensor by exposing one line of the image at a fixed rate. When capturing a modulated LED, some columns would capture the light source when it is on and the rest of the columns would capture the light when it is off. For example, the rolling shutter CMOS camera would ideally result in an image with bands of dark and bright lines when capturing an LED modulated with an on-off keying (OOK) signal. The width of the bands would ideally depend on the frequency of the light.

The image sensor is composed of millions of pixels that convert the collected photons to a photocurrent during the exposure time. A read-out circuit that consists of transistors, amplifiers, and an analog-to-digital converter (ADC) is used to convert the photocurrent to a binary output. A micro-lens is usually used in front of the pixels to improve the light collection. An array of colored filters is used to enable the color sensitivity in cameras. A color filter area (CFA) is a grid of colored filters that are spatially arranged to cover a pixelated array of photodiodes. The mosaic of colors is usually composed of red, green and blue (RGB) filters. Alternatively, vertically stacked photodiodes with different spectral responses can be used to enable the multiwavelength sensitivity [66]. Additionally, imaging lenses can be used in cameras to allow for extended communication distances.

3.4.2 Principle of OCC

The transmitter of an OCC system can be an LED, a matrix of LEDs, or a display screen. The information bits can be modulated in the light position, intensity, wavelength, or a combination of these. The light can be captured from multiple transmitters using an image sensor such as a CMOS camera.

Three distinct types of OCC are identified in Table 3.2 [2]. Screen-to-camera communication is a short-range solution that is best suited for point-to-point communications. It uses display screen pixels to transmit multiple streams of data. The photoreceivers within the camera's image sensor are used to capture the multiple streams. The data rate of screen-to-camera communications increases with the increase of the used number of pixels. Screen-to-camera communication is discussed in Sections 3.4.2.1 and 3.4.3.3. The massive number of pixels in a camera allows the OCC system to establish multiple parallel links. This is defined as the region of interest (ROI) since it is enabled by separating different regions of pixels within the captured image. This type of OCC is discussed in Section 3.4.2.2.

The data rates achieved in OCC are mainly limited by the slow frame rate of cameras. Most of the commercially available cameras operate at around 30 frames per second (fps) [67]. Therefore, the light source should work at half that rate to satisfy the Nyquist rate condition. However, the critical flicker frequency for the human eye is 100 Hz and any modulation at lower speeds causes a flicker in the light that can be perceived by the human eye [68]. In order to deal with

Table 3.2
Taxonomy of OCC Systems

	Screen-to-camera	Region of interest	Rolling shutter
IEEE 802.15.7-2018	PHY VI	PHY IV	PHY V
Data rate	Up to 10 Mbps	Tens of kbps	Up to 10 kbps
Distance	Up to 10 m	Hundreds of meters	Few meters
Transmitter	Display screen	Multiple SSL sources	A single SSL source
Receiver	Typical camera	ROI camera	Rolling shutter camera

Source: [2].

the mismatched sampling rates, two solutions were investigated. Rolling shutter cameras offer a different demodulation mechanism that is enabled by the line-scanning process of the CMOS-based cameras. This type of OCC systems is best suited for short-range indoor applications due to degradation of the data-rate performance with increased link distance [2]. Rolling shutter-based OCC is discussed in Section 3.4.2.3. Alternatively, undersampled OCC techniques were developed to allow the clock rate of the OCC transmitter to be at multiple times of the sampling rate of the camera [69,70]. The undersampling-based techniques are not useful in increasing the data rate. However, they are beneficial in dealing with the low camera frame rate and its implication on light flickering.

3.4.2.1 Screen-Based OCC

Digital screens such as liquid crystal displays (LCDs) can function as massive MIMO transmitters. The transmitted patterns can either be visible or hidden to the human eye. Digital modulation techniques such as the 2-dimensional OFDM (2D-OFDM) can be used to transmit multiple parallel streams of information [71]. The spatial streams can be captured by the massive MIMO pixels of a camera using an imaging lens. The information symbols will be encoded in the spatial dimension instead of the frequency dimension of the modulation signal as in conventional optical OFDM. The signaling would still be based on IM/DD. Therefore, Hermitian symmetry is used to maintain a real-valued signal in a similar way to other OFDM-based IM/DD techniques. A direct current (DC) bias can be used to reduce the likelihood of signal clipping. The DC part of the signal conveys no information. Thus, spatial DC biased optical OFDM (DCO-OFDM) (S-DCO-OFDM) is deemed inefficient in terms of energy consumption. However, it can be claimed that the DC part would be required whenever the screen is displaying some information to the viewers. Alternative methods of generating unipolar signals can be adopted from the literature of OFDM-based IM/DD techniques such as spatial asymmetrically clipped optical OFDM (ACO-OFDM) (S-ACO-OFDM). These techniques are power-efficient and particularly useful when the brightness of the screen is required to be dimmed.

Alternative transformations such as fast Haar wavelet transform (FHWT) can be used to convey information in the spatial dimension of an LED array. This transformation results in real-valued images without requiring any Hermitian symmetry. However, the encoded symbols prior to the FHWT transformation can only be real-valued such as in M-ary pulse-amplitude modulation (PAM) (M-PAM). Therefore, the spectral efficiency of the FHWT-based modulation technique is similar to that in S-DCO-OFDM. An experimental study demonstrated an FHWT-based OCC at 42.7 kbps over 45m for an intelligent transportation system (ITS) [72].

Spatial transformations make it possible to classify the information in terms of priority, which allows for different levels of service based on the available SNR and link distance [72]. For example, safety information with higher priority can be encoded in the lower spatial frequency components. This allows the link distance to be extended, which is particularly important for applications such as ITS.

3.4.2.2 Region of Interest Signaling

Multiple spatially separated transmitters can be captured by a single camera. This transmission mode is defined as ROI signaling and it is mainly enabled by the frame rate of the cameras. Image processing techniques can be used to track the different signals. For example, the streams transmitted by the rear lights of multiple vehicles and roadside units (RSUs) such as traffic lights can all be captured by a high-resolution camera. ROI signaling requires high-speed industrial cameras that can operate at frequencies higher than the critical flicker frequency of the human eye. The number of links that can be established is massive but practically limited by the spatial resolution of the camera.

3.4.2.3 Rolling Shutter OCC

The rolling shutter mechanism of a CMOS camera exposes one line of the pixels to the image at a time. The sequential reading is performed to reduce the power consumption and to simplify the design of the camera read-out circuitry [73]. The rolling shutter results in image distortions when capturing a fast-moving object. However, this phenomenon is viewed as an enabling feature for OCC.

The on and off states of an LED modulated with an OOK waveform can be captured by a rolling shutter camera. However, the period of the modulation signal should be longer than the scanning time of the rolling shutter and shorter than the frame period. At least one line of pixels would be exposed to the LED on state. This results in a bright line in the captured image. Subsequent lines of pixels capture the rest of the transmitted symbols. As a result, the captured image of a modulated LED would consist of dark and bright bands that correspond to the transmitted modulation signal [66]. The width of the bands would be proportional to the transmitted symbol rate and the scanning rate of the rolling shutter. Image processing techniques can be used to decode the received images and to extract the transmitted zeros and ones.

Digital modulation techniques in OCC are required to maintain a flicker-free white illumination. Typically, the light flickering should not be visible to the human eye unless bursts of zeros or ones coincide within the modulation signal. However, this problem can be easily solved by encoding the incoming information bits in a way that would control the probability of any burst bits. Alternatively, the information bits can be encoded in the frequency of the LED in frequency shift keying (FSK), which results in different widths for the dark and bright bands in the captured photos. In addition, the commercially available cameras can capture the color of an image due to the use of the CFA [66]. Different colored LEDs can be used in color shift keying (CSK) with different driving currents to produce the desired colors from the Commission Internationale de l'Eclairage (CIE 1931) color space [70]. RGB-LEDs can be used to encode different symbols in the rapid variations of colors while maintaining a white illumination on time average. The color variations would be fast enough to be perceived by human eye, but slow enough to be captured by the rolling shutter of a CMOS camera.

Image processing techniques can be used to estimate the color at each band of the image using color reference levels. Higher orders of CSK result in higher data rates. However, they can deteriorate the performance if the receiver is unable to distinguish the used colors. The design of the CSK modulation technique should adapt the thresholds of the colors based on the variations in the CFA of different camera manufacturers.

3.4.3 Applications

3.4.3.1 Indoor Navigation

Indoor positioning systems have the potential to assist shopping malls, museums, and exhibition centers in guiding customers and visitors around as illustrated in Figure 3.12. Wi-Fi and other RF-based indoor positioning systems are limited in terms of the location accuracy (measured in meters) [74]. LED-based localization systems are motivated by the fact that the number of LEDs in an indoor environment is much higher than the number of Wi-Fi routers [66]. This allows visible light navigation systems to have a higher positioning accuracy. VLC-based solutions for indoor positioning are mainly based on the received signal strength (RSS). In addition, they require additional hardware in terms of receivers in comparison with OCC-based navigation systems [75]. A number of LED luminaires are used in conjunction with a mobile phone camera to determine the location and orientation of a user in [75]. Each LED luminaire broadcasts a unique signal that reveals the coordination of that luminaire. The centroid of the luminaires captured in an image is determined and a frequency representation of the received image is estimated. A look-up table is used to link the location of

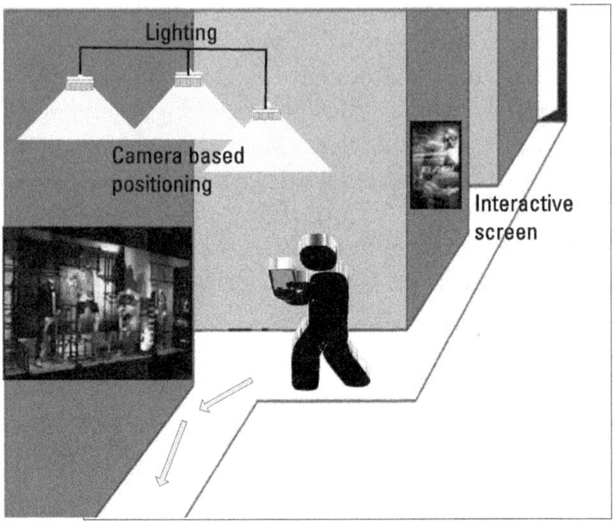

Figure 3.12 An illustration of indoor navigation using OCC.

the luminaires to the received beacon signals. This system was demonstrated at a location error that is less than 100 cm and an orientation error less than 3° [75].

3.4.3.2 Intelligent Transportation Systems

Nowadays, cameras are indispensable elements of modern vehicles. They are being used in conjunction with other systems such as light detection and ranging (LIDAR) for applications that range from park-assist to autonomous driving and platooning. VLC is discussed as an enabling technology for ITS in Chapter 4. OCC is an alternative solution that requires no additional hardware components. The data rates achieved using OCC are modest in comparison with VLC-based vehicular solutions. For example, 10 Mbps was achieved based on OCC [76], whereas it is possible to reach a few gigabits per second using VLC [77]. Nevertheless, the spatial diversity of the massive number of pixels allows the cameras to filter noise and to distinguish between multiple transmitters using image processing techniques. For example, it is possible to establish links to RSU such as traffic lights and street lights (vehicle-to-infrastructure (V2I)) and to other vehicles (vehicle-to-vehicle (V2V)), as illustrated in Figure 3.13 [78]. A multihop network can work seamlessly in crowded roads where RF-based vehicular communication systems performance would deteriorate due to the increased interference.

3.4.3.3 Short-Range Screen Signaling

Screen-to-camera communication offers a simple solution for device-to-device communication that requires minimal initiation process [1]. The information

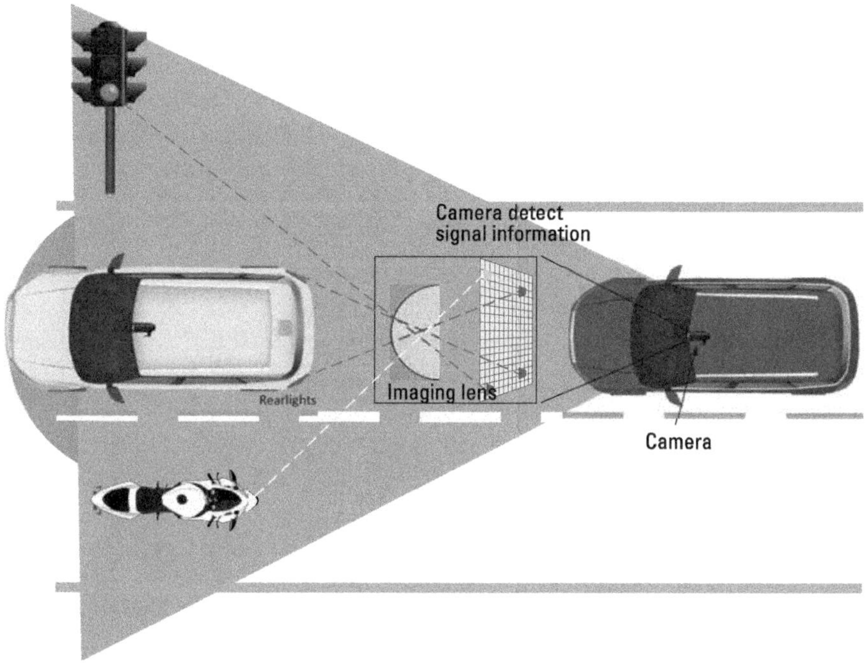

Figure 3.13 An illustration of a V2V and V2I in ITS scenario [78].

bits can be encoded in the brightness or the of the display pixels. However, the transmitted signals are required to maintain an acceptable display quality and an average brightness. The system performance generally depends on the displayed content on the screen. Therefore, adaptive coding is beneficial to optimize the OCC system performance while minimizing the distortion of the displayed content [79]. The data rates of the communication between the screen and the camera can range from a few kilobits per second to a few megabits per second [66].

A screen-to-camera communication system offers a method to transfer information in highly congested areas such as conferences and trade shows. In addition, it enables museums to embed transferable information modulated onto the screen that is used to show some textual information about a displayed item. However, it is argued that screen-to-camera communication should consider prioritizing the quality of the displayed content due to the increasing demand for a higher consumer streaming quality [66].

3.4.3.4 Internet of Things

The performance of RF-based communication technologies is limited in crowded environments due to the increased interference. Moreover, the use of RF is restricted in certain environments such as hospitals and petrochemical plants.

OWC offers a higher spectral efficiency per unit area, which allows for a high number of connected devices. The data rates achieved in OCC are low in comparison with Li-Fi and other OWC-based technologies. However, OCC is an attractive solution for IoT applications. This is because of the spatial diversity of the image sensor pixels and their large FOV. A single camera can be used to receive information from multiple light sources in a crowded environment. In addition, OCC systems require a simplified alignment process and system design in comparison with other OWC-based technologies. OCC reuses the existing lighting infrastructure in a similar way to Li-Fi. On top of that, OCC has the potential to reuse the camera surveillance infrastructure, if available. For example, the status lights of multiple production line machines can be used to transmit information to a single surveillance camera.

The development of smart wearable gadgets presents further opportunities for OCC technology [80]. Smart glasses display interactive information that would be superimposed on the glasses FOV. Smart glasses can be equipped with RF-based communication technologies such as Wi-Fi or Bluetooth. Alternatively, Li-Fi can be used to unlock augmented reality applications that have higher bandwidth requirements. The role of OCC in smart glasses is rather different from Li-Fi and RF-based technologies. Camera-enabled smart glasses allow for instant access to information and analytics using OCC. For example, the status lights and display screens in restricted environments can be used to transmit data. An engineer in a manufacturing environment can use smart glasses to capture these signals by simply looking at the status light or the LCD screen. OCC allows the user to instantaneously access essential information in a control room that could be crowded with thousands of devices.

OCC is also a promising solution for defense and healthcare IoT devices. For example, a medical consultant aided with smart glasses can instantly read out the data of a wearable sensor by simply capturing its status light. The data could be graphically displayed on the display of the smart glasses. This information could be vital in scenarios where surgeons need to take quick evidence-based decisions.

3.4.4 Challenges and Practical Considerations

The spatial and wavelength domains are considered to improve the data rate of OCC, given the limited frame rates of the commercially available cameras. Optical spatial modulation (OSM) [81] is a promising technique for OCC due to the spatial diversity of the camera pixels [68]. An array of LEDs is used at the transmitter of an OSM system. However, one LED is activated at each time instance. The incoming bits at the transmitter are jointly encoded in the light intensity of the activated LED and its position. The camera can detect the position of the activated LED due to the use of an imaging lens in front of the image sensor. Afterward, the intensity of the light is estimated to demodulate the received signal.

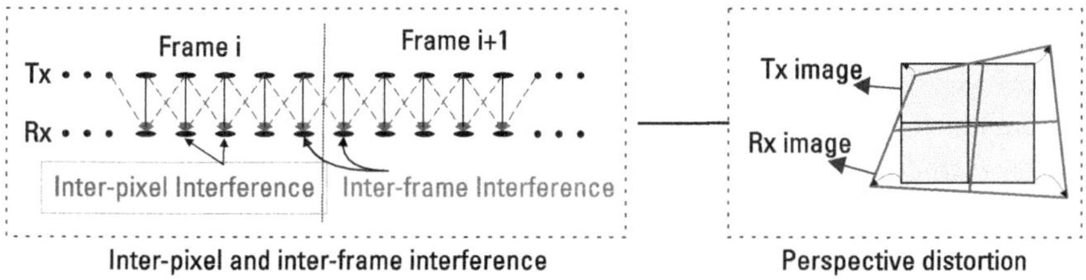

Figure 3.14 An illustration of the interference caused by perspective distortion.

Increasing the number of LEDs at the transmitter results in a logarithmic increase in the data rate, $\log_2 M + \log_2 N$, where M is the number of intensity levels and N is the number of LEDs at the transmitter. However, the increase in the number of LEDs leads to an increase in the complexity at the receiver.

The performance of rolling-shutter OCC degrades with the mobility of one or both of the optical front ends due to the spatial distortion [64]. Moreover, the projection of an image onto the photosensor is dependent on the camera perspective. Any distortion to the shape or size of the image results in an inter-pixel interference (IPI) and interframe interference (IFI) [1,71] as illustrated in Figure 3.14. The photosensors remain exposed to the light for a duration termed as exposure time. This duration is mainly reliant on the rolling shutter mechanism of the camera. Exposure time is determined by the fill factor, the quantum efficiency, and the light-source intensity [66]. Long exposure time results in a lower ability to distinguish the contrast of the transmitting light source, which results in a decreased SNR and an increased bit error ratio (BER). The major limitation in OCC is the instability of the frame rate. This is mainly due to the variation of exposure time and the post-processing of camera firmware [66]. The lack of synchronization between the transmitting light sources and the camera results in decoding errors. Moreover, commercially available cameras are incapable of capturing continuous photos because they require time to process each separate image. This results in an inter-frame data loss for any information transmitted in between two successively captured images. Finally, the absence of proper focusing results in a blur effect that leads to a lower SNR due to the reduced spatial resolution. Classic image processing techniques can be applied to improve the image prior to demodulation [66]. Alternatively, artificial intelligence (AI)-based image processing techniques can be investigated to improve the performance of OCC [82].

3.5 Summary

In this chapter, we have introduced the background, applications, and other fundamentals of FSO communications. In addition, a number of recent advances in FSO communication technology that improve the system capacity and reliability have been introduced, including MIMO diversity, OAM multiplexing, KK-receivers.

We have discussed the distinct characteristics of UV communication links, the recent advancements in UV devices, and the research efforts in modeling the UV channel. We have also discussed some of the open issues related to this technology such as those related to link efficiency enhancement, multihop networks managements and the need for more extensive experimental validation.

The principle of OCC has been introduced based on three different system architectures. Screen-to-camera communications enables a short-distance link that requires a minimal initiation process. Cameras are natural MIMO devices with spatial and color diversity. Multiple transmitters can be simultaneously demodulated at different ROI within the pixel matrix of the image sensor. The unsteady frame rate of commercially available cameras limits the performance of practical OCC. The rolling shutter enables a different means of demodulation for CMOS-based image sensors. Most of the published research on OCC assumes the use of a typical or a commercially available high-speed camera. However, optimizing the designs of typical cameras to enable a higher frame rate can lead to faster transmissions in OCC systems.

Heterogeneous solutions that integrate different OWC technologies present further opportunities that build upon the advantages of all used technologies and circumvent the drawbacks of using a single OWC technology solution. Such hybrid systems require a level of adaptability to allow the most suitable technology to be used for the most suited scenario.

References

[1] Perli, S. D., N. Ahmed, and D. Katabi, "PixNet: Interference-Free Wireless Links Using LCD-Camera Pairs," *Proceedings of the 16th Annual International Conference on Mobile Computing and Networking, Ser. MobiCom '10.* New York: Association for Computing Machinery, 2010, p. 137–148.

[2] Nguyen, T., et al., "Technical Issues on IEEE 802.15.7m Image Sensor Communication Standardization," *IEEE Communications Magazine*, Vol. 56, No. 2, 2018, pp. 213–218.

[3] Khalighi, M. A., and M. Uysal, "Survey on Free Space Optical Communication: A Communication Theory Perspective," *IEEE Communications Surveys Tutorials*, Vol. 16, No. 4, Fourth Quarter 2014, pp. 2231–2258.

[4] Uysal, M., et al., *Optical Wireless Communications: An Emerging Technology*, New York: Springer, 2016.

[5] Hecht, J., "Laser Evolution," *SPIE Professional Magazine*, April 1, 2009. https://spie.org/news/spie-professional-magazine-archive/2009-april/laser-evolution.

[6] Goodwin, F. E., "A Review of Operational Laser Communication Systems," *Proceedings of the IEEE*, Vol. 58, No. 10, October 1970, pp. 1746–1752.

[7] Kaushal, H., and G. Kaddoum, "Applications of Lasers for Tactical Military Operations," *IEEE Access*, Vol. 5, 2017, pp. 20736–20753.

[8] Ge, X., et al., "5G Wireless Backhaul Networks: Challenges and Research Advances," *IEEE Network*, Vol. 28, No. 6, November 2014, pp. 6–11.

[9] Ghosh, A., et al., "Heterogeneous Cellular Networks: From Theory to Practice," *IEEE Communications Magazine,* Vol. 50, No. 6, June 2012, pp. 54–64.

[10] Gagliardi, R. M., and S. Karp, *Optical Communications*, New York: John Wiley & Sons, Wiley Series in Telecommunications and Signal Processing, 1976.

[11] Huang, S., and M. Safari, "Free-Space Optical Communication Impaired by Angular Fluctuations," *IEEE Transactions on Wireless Communications*, Vol. 16, No. 11, 2017, pp. 7475–7487.

[12] Kedar, D., and S. Arnon, "Optical Wireless Communication Through Fog in the Presence of Pointing Errors," *Applied Optics*, Vol. 42, No. 24, 2003, pp. 4946–4954.

[13] Farid, A. A., and S. Hranilovic, "Outage Capacity Optimization for Free-Space Optical Links with Pointing Errors," *Journal of Lightwave Technology*, Vol. 25, No. 7, 2007, pp. 1702–1710.

[14] Kedar, D., and S. Arnon, "Urban Optical Wireless Communication Networks: the Main Challenges and Possible Solutions," *IEEE Communications Magazine*, Vol. 42, No. 5, May 2004, pp. S2–S7.

[15] Andrews, L. C., and R. L. Phillips, "Laser Beam Propagation Through Random Media," *SPIE*, 2005.

[16] Bloom, S., et al., "Understanding the Performance of Free-Space Optics," *Journal of Optical Networking*, Vol. 2, No. 6, 2003, pp. 178–200.

[17] Tse, D., and P. Viswanath, *Fundamentals of Wireless Communication*, Cambridge, U.K.: Cambridge University Press, 2005.

[18] Lee, E. J., and V. W. S. Chan, "Part 1: Optical Communication over the Clear Turbulent Atmospheric Channel Using Diversity," *IEEE Journal on Selected Areas in Communications,* Vol. 22, No. 9, November 2004, pp. 1896–1906.

[19] Navidpour, S. M., M. Uysal, and M. Kavehrad, "BER Performance of Free-Space Optical Transmission with Spatial Diversity," *IEEE Transactions on Wireless Communications*, Vol. 6, No. 8, August 2007, pp. 2813–2819.

[20] Simon, M. K., and V. A. Vilnrotter, "Alamouti-Type Space-Time Coding for Free-Space Optical Communication with Direct Detection," *IEEE Transactions on Wireless Communications*, Vol. 4, No. 1, January 2005, pp. 35–39.

[21] Safari, M., and M. Uysal, "Do We Really Need OSTBCs for Free-Space Optical Communication with Direct Detection?" *IEEE Transactions on Wireless Communications,* Vol. 7, No. 11, November 2008, pp. 4445–4448.

[22] Willner, A. E., et al., "Optical Communications Using Orbital Angular Momentum Beams," *Advances in Optics and Photonics*, Vol. 7, No. 1, 2015, pp. 66–106.

[23] Zhao, N., et al., "Capacity Limits of Spatially Multiplexed Free-Space Communication," *Nature Photonics*, Vol. 9, No. 12, 2015, p. 822.

[24] Huang, S., and M. Safari, "Spatial-Mode Multiplexing with Zero-Forcing Beamforming in Free Space Optical Communications," *2017 IEEE International Conference on Communications Workshops (ICC Workshops)*, 2017, pp. 331–336.

[25] Qu, Z., and I. B. Djordjevic, "500 Gb/s Free-Space Optical Transmission over Strong Atmospheric Turbulence Channels," *Optics Letters*, Vol. 41, No. 14, 2016, pp. 3285–3288.

[26] Mecozzi, A., C. Antonelli, and M. Shtaif, "Kramers–Kronig Coherent Receiver," *Optica*, Vol. 3, No. 11, 2016, pp. 1220–1227.

[27] Lorences-Riesgo, A., et al., "200 G Outdoor Free-Space-Optics Link Using a Single-Photodiode Receiver," *IEEE/OSA Journal of Lightwave Technology*, Vol. 38, No. 2, January 2020, pp. 394–400.

[28] Ramdhan, N., M. Sliti, and N. Boudriga, "Codeword-Based Data Collection Protocol for Optical Unmanned Aerial Vehicle Networks," *International Symposium on High Capacity Optical Networks and Enabling Technologies (HONET)*, October 2016, pp. 35–39.

[29] Najafi, M., et al., "Statistical Modeling of FSO Fronthaul Channel for Drone-Based Networks," *2018 IEEE International Conference on Communications (ICC)*, May 2018, pp. 1–7.

[30] Vavoulas, A., et al., "A Survey on Ultraviolet C-Band (UV-C) Communications," *IEEE Communications Surveys Tutorials*, Vol. 21, No. 3, 2019, pp. 2111–2133.

[31] Drost, R. J., and B. M. Sadler, "Survey of Ultraviolet Non-Line-of-Sight Communications," *Semiconductor Science and Technology*, Vol. 29, No. 8, June 2014, p. 084006.

[32] Xu, Z., and B. M. Sadler, "Ultraviolet Communications: Potential and State-of-the-Art," *IEEE Communications Magazine*, Vol. 46, No. 5, May 2008, pp. 67–73.

[33] He, X., et al., "1 Gbps Free-Space Deep-Ultraviolet Communications Based on III-Nitride Micro-LEDs Emitting at 262 nm," *Photonics Research*, Vol. 7, No. 7, 2019, pp. B41–B47.

[34] Alkhazragi, O., et al., "Gbit/s Ultraviolet-C Diffuse-Line-of-Sight Communication Based on Probabilistically Shaped DMT and Diversity Reception," *Optics Express*, Vol. 28, No. 7, 2020, pp. 9111–9122.

[35] Deng, Z. L., and M. Y. Zhao, "Analyzing and Developing Transmitter of UV Communication Test Bed," *Applied Mechanics, Mechatronics Automation & System Simulation*, Ser. Applied Mechanics and Materials, Vol. 198, Trans Tech Publications Ltd., October 2012, pp. 117–122.

[36] Hirayama, H., et al., "231– 261 nm AlGaN Deep-Ultraviolet Light-Emitting Diodes Fabricated on AlN Multilayer Buffers Grown by Ammonia Pulse-Flow Method on Sapphire," *Applied Physics Letters*, Vol. 91, No. 7, August 2007, p. 084006.

[37] Grandusky, J. R., et al., "High Output Power from 260 nm Pseudomorphic Ultraviolet Light-Emitting Diodes with Improved Thermal Performance," *Applied Physics Express*, Vol. 4, No. 8,

[38] SMD Package DUV-LED: Technical Data Sheet, NIKKISO, Tokyo, Japan, Tech. Rep., October 2016

[39] Inoue, S. -I., et al., "Light Extraction Enhancement of 265 nm Deep-Ultraviolet Light-Emitting Diodes with over 90 mW Output Power Via an AlN Hybrid Nanostructure," *Applied Physics Letters*, Vol. 106, No. 13, 2015, p. 131104.

[40] Inoue, S., N. Tamari, and M. Taniguchi, "150 mW Deep-Ultraviolet Light-Emitting Diodes with Large-Area AIN Nanophotonic Light-Extraction Structure Emitting at 265 nm," *Applied Physics Letters*, Vol. 110, No. 14, 2017, p. 141106.

[41] S-S35D-F2-275-01-3-180: Technical Data Sheet, SETi, Mountain View, CA, Tech. Rep., 2016.

[42] "Klaran Deep UV LEDs: Technical Data Sheet," Crystal, Duluth, MN, Tech. Rep., 2017.

[43] Zhang, X., et al., "Design of an Omnidirectional Optical Antenna for Ultraviolet Communication," *Applied Optics*, Vol. 53, No. 15, 2014, pp. 3225–3232.

[44] Liao, M., et al., "Comprehensive Investigation of Single Crystal Diamond Deep-Ultraviolet Detectors," *Japanese Journal of Applied Physics*, Vol. 51, No. 9R, 2012, p. 090115.

[45] Toda, T., et al., "Operation at 700 C of 6H-SiC UV Sensor Fabricated Using N+ Implantation," *Japanese Journal of Applied Physics*, Vol. 43, No. 1A, 2003, p. L27.

[46] Reilly, D. M., and C. Warde, "Temporal Characteristics of Single-Scatter Radiation," *Journal of the Optical Society of America A*, Vol. 69, No. 3, March 1979, pp. 464–470. http://www.osapublishing.org/abstract.cfm?URI=josa-69-3-464.

[47] Luettgen, M. R., J. H. Shapiro, and D. M. Reilly, "Non-Line-of-Sight Single-Scatter Propagation Model," *Journal of the Optical Society of America A*, Vol. 8, No. 12, 1991, pp. 1964–1972.

[48] Yan-Hong, W., and W. Jing-Zhi, "Detection Efficiency of Non-Line-of-Sight Ultraviolet Communication System," *4th IEEE/IFIP International Conference on Central Asia on Internet*, September 2008, pp. 1–5.

[49] Xu, Z., et al., "Analytical Performance Study of Solar Blind Non-Line-of-Sight Ultraviolet Short-Range Communication Links," *Optics letters*, Vol. 33, No. 16, 2008, pp. 1860–1862.

[50] Yin, H., et al., "Analytical Model of Non-Line-of-Sight Single-Scatter Propagation," *Journal of the Optical Society of America A*, Vol. 27, No. 7, 2010, pp. 1505–1509.

[51] Gupta, A., M. Noshad, and M. Brandt-Pearce, "NLOS UV Channel Modeling Using Numerical Integration and an Approximate Closed-Form Path Loss Model," *Laser Communication and Propagation Through the Atmosphere and Oceans*, Vol. 8517, International Society for Optics and Photonics, 2012, p. 851709.

[52] Zuo, Y., et al., "Closed-Form Path Loss Model of Non-Line-of-Sight Ultraviolet Single-Scatter Propagation," *Optics Letters*, Vol. 38, No. 12, 2013, pp. 2116–2118.

[53] Yin, H., et al., "Vectorized Polarization-Sensitive Model of Non-Line-of-Sight Multiple-Scatter Propagation," *Journal of the Optical Society of America A*, Vol. 28, No. 10, 2011, pp. 2082–2085.

[54] Yuan, R., et al., "An Integral Model of Two-Order and Three-Order Scattering for Non-Line-of-Sight Ultraviolet Communication in a Narrow Beam Case," *IEEE Communications Letters*, Vol. 20, No. 12, 2016, pp. 2366–2369.

[55] Ding, H., et al., "Turbulence Modeling for Non-Line-of-Sight Ultraviolet Scattering Channels," *Atmospheric Propagation VIII*, Vol. 8038, International Society for Optics and Photonics, 2011, p. 80380J.

[56] Liu, W., and Z. Xu, "Characteristics of Optical Scattering and Turbulence Communication Channels," *48th IEEE Asilomar Conference on Signals, Systems and Computers*, 2014, pp. 1876–1879.

[57] Liao, L., et al., "Turbulence Channel Test and Analysis for NLOS UV Communication," *Laser Communication and Propagation Through the Atmosphere and Oceans III*, Vol. 9224, International Society for Optics and Photonics, 2014, p. 92241A.

[58] Liao, L., et al., "UV LED Array Based NLOS UV Turbulence Channel Modeling and Experimental Verification," *Optics Express*, Vol. 23, No. 17, 2015, pp. 21825–21835.

[59] Haghighi Ardakani, M., and M. Uysal, "Relay-Assisted OFDM for Ultraviolet Communications: Performance Analysis and Optimization," *IEEE Transactions on Wireless Communications*, Vol. 16, No. 1, 2017, pp. 607–618.

[60] Ardakani, M. H., A. R. Heidarpour, and M. Uysal, "Performance Analysis of Relay-Assisted NLOS Ultraviolet Communications over Turbulence Channels," *Journal of Optical Communications and Networking*, Vol. 9, 2017, pp. 109–118.

[61] Li, Y., et al., "Neighbor Discovery for Ultraviolet Ad Hoc Networks," *IEEE Journal on Selected Areas in Communications*, Vol. 29, No. 10, 2011, pp. 2002–2011.

[62] Cisco, "Cisco Annual Internet Report, 2018-2023," 2020, https://www.cisco.com/c/en/us/solutions/collateral/executive-perspectives/annual-internet-report/white-paper-c11-741490.pdf.

[63] Serafimovski, N., et al., "Light Communications for Wireless Local Area Networking," *IEEE 5G Tech Focus*, Vol. 2, No. 2, May 2018. https://5g.ieee.org/tech-focus/may-2018/light-communications-for-wireless-local-area-networking.

[64] Nguyen, T., et al., "Current Status and Performance Analysis of Optical Camera Communication Technologies for 5G Networks," *IEEE Access*, Vol. 5, 2017, pp. 4574–4594.

[65] "IEEE Standard for Local and Metropolitan Area Networks–Part 15.7: Short-Range Optical Wireless Communications," IEEE Std. 802.15.7-2018 (Revision of IEEE Std 802.15.7-2011), 2019, pp. 1–407.

[66] Wang, Z., et al., *Optical Camera Communication: Fundamentals*, New York: Wiley-IEEE Press, 2018, pp. 239–290.

[67] Hu, W., H. Gu, and Q. Pu, "LightSync: Unsynchronized Visual Communication over Screen-Camera Links," *Proceedings of the 19th Annual International Conference on Mobile Computing & Networking*, Ser. MobiCom '13, New York: Association for Computing Machinery, 2013, p. 15–26.

[68] Liu, W., and Z. Xu, "Some Practical Constraints and Solutions for Optical Camera Communication," *Philosophical Transactions of the Royal Society A: Mathematical, Physical and Engineering Sciences*, Vol. 378, No. 2169, 2020, p. 20190191. https://royalsocietypublishing.org/doi/abs/10.1098/rsta.2019.0191.

[69] Roberts, R. D., "Undersampled Frequency Shift ON-OFF Keying (UFSOOK) for Camera Communications (CamCom)," *2013 22nd Wireless and Optical Communication Conference*, 2013, pp. 645–648.

[70] Chen, H., et al., "Color-Shift Keying for Optical Camera Communication Using a Rolling Shutter Mode," *IEEE Photonics Journal*, Vol. 11, No. 2, 2019, pp. 1–8.

[71] Nguyen, T., M. D. Thieu, and Y. M. Jang, "2D-OFDM for Optical Camera Communication: Principle and Implementation," *IEEE Access*, Vol. 7, 2019, pp. 29405–29424.

[72] Nagura, T., et al., "Improved Decoding Methods of Visible Light Communication System for Its Using LED Array and High-speed Camera," *IEEE 71st Vehicular Technology Conference*, 2010, pp. 1–5.

[73] Jung, S., S. Hann, and C. Park, "TDOA-Based Optical Wireless Indoor Localization Using LED Ceiling Lamps," *IEEE Transactions on Consumer Electronics*, Vol. 57, No. 4, 2011, pp. 1592–1597.

[74] Sing Liu, H., and G. Pang, "Positioning Beacon System Using Digital Camera and LEDs," *IEEE Transactions on Vehicular Technology*, Vol. 52, No. 2, 2003, pp. 406–419.

[75] Kuo, Y. -S., et al., "Luxapose: Indoor Positioning with Mobile Phones and Visible Light," *Proceedings of the 20th Annual International Conference on Mobile Computing and Networking*, Ser. MobiCom '14, New York: Association for Computing Machinery, 2014, pp. 447–458.

[76] Takai, I., et al., "LED and CMOS Image Sensor Based Optical Wireless Communication System for Automotive Applications," *IEEE Photonics Journal*, Vol. 5, No. 5, 2013, pp. 6801418–6801418.

[77] Lee, C., et al., "Advanced LiFi Technology: Laser Light," *Light-Emitting Devices, Materials, and Applications XXIV*, J. K. Kim, M. R. Krames, and M. Strassburg, (eds.), Vol. 11302, International Society for Optics and Photonics/SPIE, 2020, pp. 116–123.

[78] Chowdhury, M. Z., et al., "A Comparative Survey of Optical Wireless Technologies: Architectures and Applications," *IEEE Access*, Vol. 6, 2018, pp. 9819–9840.

[79] Nguyen, V., et al., "High-Rate Flicker-Free Screen-Camera Communication with Spatially Adaptive Embedding," *IEEE INFOCOM 2016 - The 35th Annual IEEE International Conference on Computer Communications*, 2016, pp. 1–9.

[80] Hossan, M. T., et al., "Human Bond Communication with Head-Mounted Displays: Scope, Challenges, Solutions, and Applications," *IEEE Communications Magazine*, Vol. 57, No. 2, 2019, pp. 26–32.

[81] Mesleh, R. Y., et al., "Spatial Modulation," *IEEE Transactions on Vehicular Technology*, Vol. 57, No. 4, 2008, pp. 2228–2241.

[82] Pham, T. L., et al., "An Artificial Intelligence-Based Error Correction for Optical Camera Communication," *Eleventh International Conference on Ubiquitous and Future Networks (ICUFN)*, 2019, pp. 137–140.

4

Applications and Future-Proofing Wireless Communications

4.1 Introduction

Li-Fi offers an inexpensive solution to the inevitable spectrum crunch due to the availability of low-cost front-end devices. Li-Fi is a unique communication method in scenarios where the RF propagation is restricted or considered to be hazardous.[1]

Li-Fi enables a paradigm shift towards connected lights by integrating SSL, digital communications, and IoT technologies [1]. Li-Fi has the potential to pull the lighting industry to the information and communication technology (ICT) market and shift the concept of pay-per-lux to become pay-per-bit. Li-Fi will consolidate the concept of light-as-a-service (LaaS) not only to enable internet-controlled lighting, but also to allow internet service providers (ISPs) to provide internet and other services to the users.

The motivations behind Li-Fi and its advantages against other communications technologies are provided in this chapter. We aim to clarify some of the misconceptions that surround Li-Fi such as outdoor, uplink, and dimming support. In Section 4.2, we envision the architecture of the future world of artificial intelligence (AI) and review the evolution of wireless networks. Then a discussion about challenges in future wireless communication systems is presented, thereby exposing the necessity of enhancing their capacity and reliability. In Section 4.3, we discuss in detail the numerous advantages that Li-Fi

[1] It should be noted that RF shielding of the front-end electronic circuitry is required as it is susceptible to RF interference.

offers and how it can contribute to the advancements of wireless connectivity. Particularly, we explain how Li-Fi technology holds great potentials for the realization of artificially intelligent systems and services. In Section 4.4, several unique future-proofing applications of Li-Fi are introduced. Finally, we try to clear up several common misconceptions related to this emerging technology in Section 4.5.

4.2 Artificial Intelligence and the Spectrum Crunch

In this section, we challenge the common perception that AI can be fully characterized by advancements in computer science and machine learning algorithms and shed the light on the other requirements needed for the full realization of AI.

4.2.1 The Role of Wireless Communications in Autonomous Systems

Ever-evolving wireless networks hold great prospects to provide users with new services and applications. Diverse types of wireless traffic have emerged over the last few decades and are continuing to evolve. To this end, future IoT promises cohesive real-time connectivity, data collection, and sharing that effectively merge the digital and physical worlds. IoT applications span diverse use cases such as smart healthcare, smart infrastructure management, smart waste handling, autonomously driven cars, remote robots operations, and so on. In order for this to happen, IoT relies on the autonomous interaction between three main components, namely: physical devices, intelligent processors, and wireless networks. Figure 4.1 shows an illustrative block diagram of IoT enabling technologies and illustrates how the architecture of smart IoT systems follows the analogy of the human body. In this analogy, built-in sensors represent the human senses, machine-learning engines take the function of the brain, actuators represent the limbs while wireless links constitute the nervous system that connects all body parts. To enable smart and autonomous IoT operations, large numbers of embedded sensors are needed to collect real-time data from the environment. Examples include, but are not limited to, temperature sensors, pressure sensors, proximity sensors, motion detection sensors, accelerometer sensors, and gyroscope sensors. These sensors transform their measurements into digital data streams for transmission to a central, cloud-based computing entity where data and signal processing engines decide on which useful data needs to be extracted for further processing. It is noted that the collected data comes in all types of structured and unstructured formats and with large volumes. In addition to the increasing volumes and varieties, data streams at high speeds and could be highly inconsistent with periodic peaks. All of these factors make it extremely challenging to link, match, analyze and transform data across

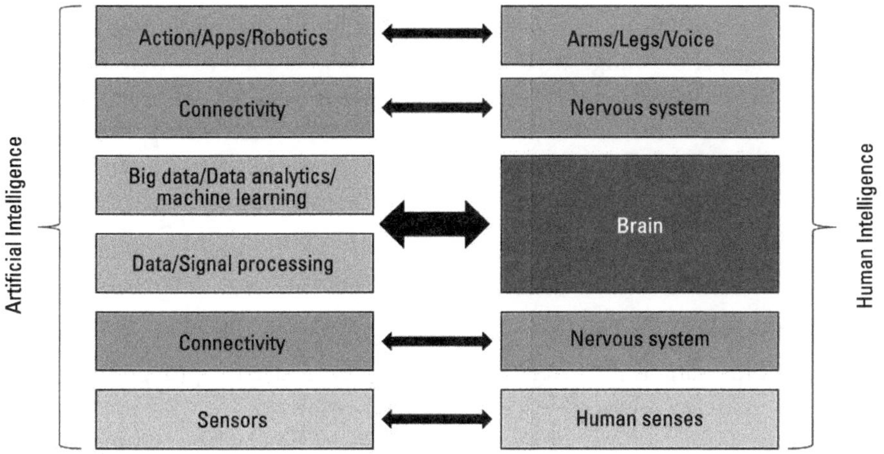

Figure 4.1 AI extends beyond machine learning and data analytics to also include the devices and transmission medium.

systems in real time. This is where big data comes into play by analyzing the connections and patterns hidden in the collected data so as to gain valuable insights that might otherwise stay undiscovered. Moreover, advancements in machine learning techniques allow devices to gain intelligence by learning from repeated patterns in the data collected over a sufficient time period. Based on this, machines become smart autonomous entities that take decisions and perform tasks without being explicitly programmed. These decisions are then transformed into tangible actions by various means such as software applications or physical actuators. Examples include adjusting the temperature of a room, sending a medical report, stopping a self-driven car, or responding to natural and man-made hazards.

To realize such smart autonomous IoT operations, it is evident that the concept of AI extends beyond data analytics and machine learning techniques to also include sensing and responding capabilities, as well as the communication links that facilitate real-time interactions between different entities. In fact, the implementation of smart IoT on a large nonfragmented scale requires parallel growth and development of many emerging technologies [2]. For example, sensing devices need to fit the increasing demands on availability, reliability, size, and energy efficiency. Standardization efforts of IoT architecture need to expand so as to cope with the increase in the numbers of internet-connected objects. Also, innovative protocols are required to govern the communications between highly diverse entities such as living things, appliances, mobile devices, and vehicles. Another critical requirement for enabling smart IoT is the availability of wireless networks that support seamless interaction and communications between all devices and controllers. To achieve this, IoT applications utilize various communications technologies, including wired communications for fixed terminals, short-range wireless communications such as wireless local-area network (LAN), and Bluetooth as well as mobile communications including 4G networks, 5G networks, and beyond. Therefore, it is evident that providing high-speed, secure, and reliable wireless connectivity is essential for the advancements of future smart technologies. Indeed, the biggest threat to such advancements is the looming spectrum crisis.

4.2.2 Network Traffic Trends

Global mobile data traffic trends reveal unprecedented rapid growth in the demands on wireless connectivity over the past few decades [3,4], which indicates that enhancing the capacity of wireless networks will become even more crucial in the immediate future. In the early 1980s, the 1G wireless networks allowed the transmission of analog voice-based traffic, to be followed by digitally transmitted traffic in the 2G networks in the early 1990s. Voice-based traffic was then overtaken by user-driven data traffic in early 2000s with the introduction of the 3G networks. The development of 4G services, in 2012, has made it possible to realize considerably higher connection speeds and lower latency, resulting in an enormous increase in the demand on video-based traffic, which constituted 75% of all IP traffic in 2017 [5]. Moving forward, machine-to-machine (M2M) and IoT traffic is expected to dominate 5G wireless data demand in the near future [6]. Thanks to the availability of affordable smart handheld wireless devices, the number of connected things already exceeds the world's population, with new devices in increasing capabilities and intelligence being continuously introduced to the market. According to the Cisco Visual Networking Index [5], mobile devices and connections grew from 7.9 billion to 8.6 billion in 2017 and are expected to reach 12.3 billion by 2022. Figure 4.2 shows the global mobile devices and connections growth, where smartphones and phablets constitute more than 50% of the total number of devices and connections. Moreover, M2M connections exhibit the most noticeable growth in the market with about a 32% increase in the number of connected machines from 2017 to 2022. Examples of these M2M connections include communications links between medical appliances, environmental sensors, and positioning devices. The growth in the number and diversity of wireless devices is predicted to result in monthly global mobile data traffic of 77 exabytes by 2022 as shown in Figure 4.3, leading to annual traffic reaching almost 1 zettabyte of data. Moreover, it is estimated that, by 2022, 4G broadband cellular networks will support about 71% of mobile data traffic, while evolving 5G networks will support 12% with their extra high bandwidth and ultralow latency. In fact, 5G is perceived to drive new demands on wireless connectivity. For this reason, Cisco projections indicate that mobile data traffic offloading will be even higher in future 5G networks as seen in Figure 4.4,

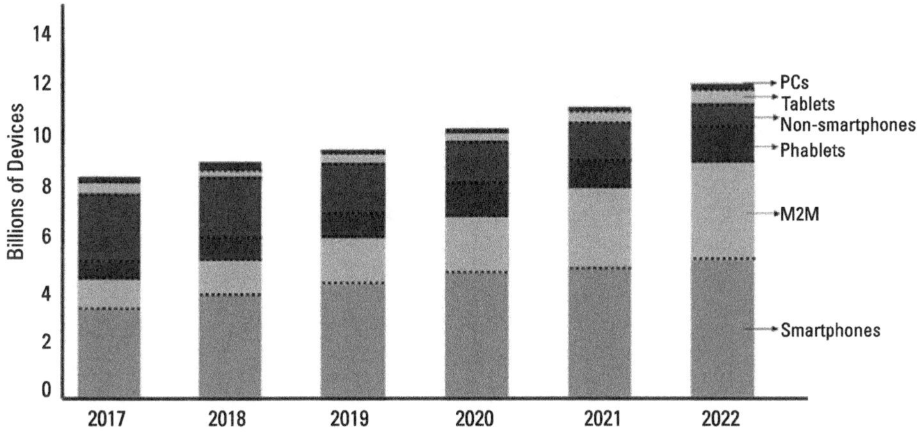

Figure 4.2 Global mobile devices and connections growth. (Source: Cisco Visual Networking Index Mobile, 2019 [5].)

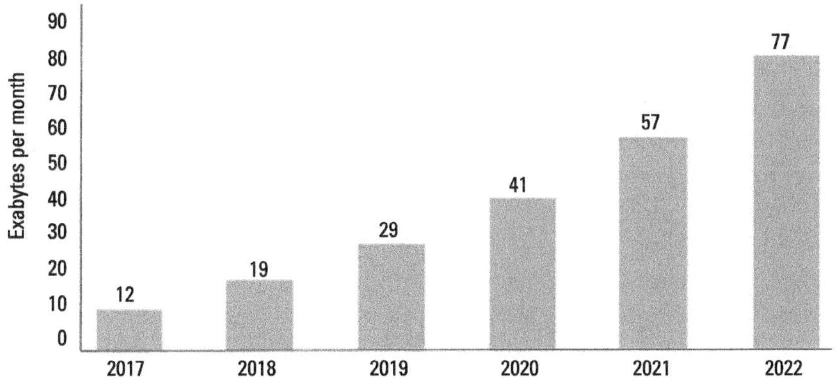

Figure 4.3 Cisco forecasts 77 exabytes per month of mobile data traffic by 2022. (Source: Cisco Visual Networking Index Mobile, 2019 [5].)

where offloaded traffic is expected to form 71% of total traffic in 5G compared to 59% in 4G by 2022.

4.2.3 5G and Beyond

The unprecedented high volume of wireless data traffic puts tremendous strains on the existing wireless networks. In fact, it is predicted that wireless networks will fall short of the required connectivity demands by a factor of approximately 4 every 5 years [7]. This network scalability problem forms a bottleneck that risks the advancements in future smart technologies. This is mainly due to the increasing demand on a decidedly limited resource: the RF spectrum. The RF spectrum is the part of the electromagnetic spectrum with frequencies ranging from 30 Hz to 300

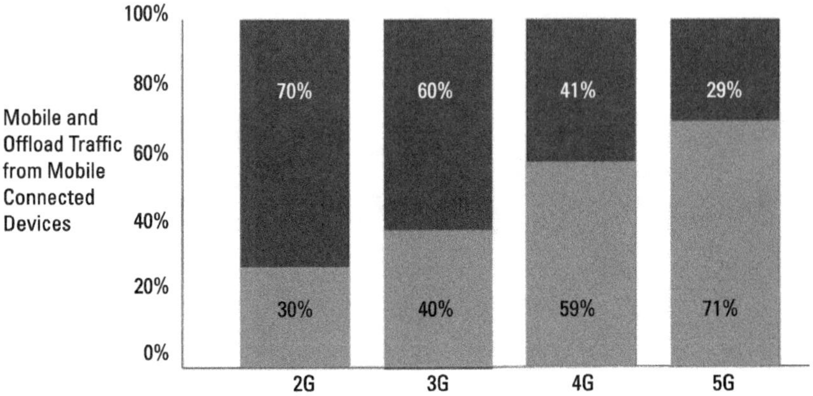

Figure 4.4 Mobile data traffic and offload traffic, 2022: Cisco Visual Networking Index Mobile, 2019 [5].

GHz and is the main transmission medium for the majority of current wireless communications systems. This finite resource is becoming extremely crowded and will soon fall short in providing ubiquitous connectivity for future dense IoT networks. To this end, industry, academia, and standardization bodies are working on solutions for future 5G and beyond wireless networks. 5G supports three service categories according to the International Telecommunication Union (ITU), namely: mobile broadband, massive machine-type communications, and ultrareliable and low-latency communication (URLLC) [8]. To satisfy the diverse requirements of different types of devices and applications, future wireless networks are envisioned to provide the following requirements [9]:

- *Unprecedented high data rates*: High-definition (HD) video streaming, pervasive conferencing, and virtual reality applications, as well as the proliferation of connected mobile devices and tablets are driving the need for never-seen-before high-speed connectivity. The general consensus is that 5G will provide 1,000-fold increase in the aggregate data rate compared to 4G. Moreover, 5G is expected to provide cell-edge rate range of 100 Mbps up to 1 Gbps. It is worth noting that the effects of RF signal propagation loss and intercell interference make supporting such high-rate requirements extremely challenging for cell-edge users [10].
- *Ultralow latency*: Latency-sensitive services such as tactile internet, multi-player online gaming, and autonomous driving require packet transmission time in future wireless networks to be in the order of tens of hundreds of microseconds, compared to tens of milliseconds, in current long-term evolution (LTE) networks [11].
- *High energy efficiency (EE)*: 5G networks will support massive numbers of IoT low-power, low-cost devices. As a result, reducing the energy

consumption in these devices can result in huge savings in the costs of capital and operational expenditures. Moreover, due to their limited power capabilities, the majority of IoT-connected devices cannot afford energy-demanding processes such as highly computational complex decoding, error correction, or synchronization. Also, the existence of huge numbers of dense small cells require all processing at the access point (AP) to be energy-efficient as well, since small energy saving at each AP translates to huge savings on a large scale.

- *High scalability*: 5G architecture is required to accommodate the upsurge in the number of connected devices. Such high scalability necessitates an upgrade across the different network layers. This includes the ability to establish higher numbers of physical communication links, managing the interference between these links, and designing efficient medium access control (MAC) protocols. Also, future wireless networks are required to provide efficient routing, handovers, and mobility support for the large numbers of connected devices. To this end, network densification, virtualization, and software defined networks (SDN) are considered key enablers for 5G. Moreover, high network scalability requires automating routine processes by enabling a degree of intelligence in the physical layer. This allows the communications front ends to interact with the propagation environment so as to achieve the best possible performance with minimal human intervention [12].
- *High link reliability*: Reliability of packet transmission is required to reach $1 - 10^{-5}$ in 5G URLLC service category, and should be as high as $1 - 10^{-7}$ in critical applications such as remote surgery and self-driven cars [13]. These rigid requirements necessitate high accuracy channel estimation, advanced channel coding, or repetitive transmission schemes using time, frequency, or spatial resources.
- *Improved security*: Future wireless networks need to develop unified standards to address security and privacy requirements for dense networks connecting massive numbers of devices. Since many of these connections hold sensitive data related to industries, governments, and day-to-day life, potential breaches could have severe and irreversible consequences. Given that IoT devices are typically resource-constrained in terms of power and memory, there is a need to design cost-efficient security mechanisms, including authentication, authorization, encryption, and physical layer security (PLS) techniques [14].

In order to achieve the supreme requirements on capacity, latency, efficiency, scalability, reliability, and security, a significant paradigm shift is required to further strengthen the wireless communications networks. Such performance gain is expected to be achieved through the combination of various technologies in one of the following three categories [10]:

1. Enhancing the communications bandwidth by various means such as dynamic resource allocation and the use of cognitive radio paradigms. However, it is noted that such solutions, might not be always sufficient to accommodate the required performance gain in future wireless networks. A promising effective approach is to tap into new spectrum resources, such as the millimeter-wave spectrum and the optical spectrum [15,16].
2. Wireless AP/base station (BS) densification to improve the area spectral efficiency. The major benefit of densification is to allow more frequent reuse of the communications bandwidth within the same geographic area. In this regard, the concept of heterogeneous networks and small cells play an important role to accommodate the accelerating demand for wireless connectivity. The main idea is to offload the traffic from the cellular macrocell to smaller cells (i.e., microcells, picocells, and femtocells) [17]. To this end, the reduced distance from the AP to end-users results in enhanced link performance. Moreover, due to their small coverage areas, small cells allow for efficient and interference-free reuse of the available power and spectrum resources.
3. Improved spectral efficiency per AP/BS, which could be achieved by various techniques that support higher bits per second per Hertz (bps/Hz) per node. This includes advanced MIMO techniques, such as massive MIMO, which are recognized as one of the most effective approaches [18]. Another spectrally-efficient solution for multi-user scenarios is nonorthogonal multiple access (NOMA), which allows multiple cell users to reuse the entire time and bandwidth resources [19].

4.3 Li-Fi Motivations

With the explosive increase in the number of mobile devices and various IoT machines, user density in service areas is approaching the region where there is more than one mobile user per square meter. For example, hundreds of passengers could be squeezed in a limited space in the cabin of an aircraft or a train. Other similar high-density indoor environments include train stations, airports, shopping malls, and conference venues. The continuous increase in the level of expected service quality for user applications such as ultra high-resolution video streaming requires providing users with data rates of several gigabits per second, which proves to be extremely challenging in dense environments. Therefore, the next question is how to design a wireless network with sufficient capacity to meet such super high wireless traffic demand. The usage of RF cellular bands for communications in such dense small cells implies that signals could travel far beyond the cell boundaries, resulting in severe intercell cochannel interference

(CCI). Typically, cell edge users would exhibit reduced QoS due to the lower link quality as well as being susceptible to CCI from neighboring cells. To this end, exploiting other spectrum options becomes a viable solution.

In fact, utilizing the optical part of the electromagnetic (EM) spectrum for wireless connectivity is an attractive solution due to its wide range of frequencies and short travel distance compared to RF waves. To this end, Li-Fi arises as a promising solution to add a new dimension to spectrum heterogeneity [20]. For example, each lightbulb in a room or a corridor could act as an AP with a few meters coverage area, causing very little to no interference to neighboring Li-Fi cells and zero interference to the Wi-Fi network. The Li-Fi network thus can be formed by multiple interconnected short-range optical APs with each AP serving a few users. In this regard, the downlink (DL) communications between the user equipment (UE) and its associated AP is realized by means of visible light communication (VLC) while the uplink (UL) can be established via wireless IR. When the user moves in the service area, the Li-Fi communications link can be maintained by seamless handover between adjacent Li-Fi APs. In this section, we show that wireless networks densification and bandwidth enhancement can be extensively realized by Li-Fi technology, which makes it a promising candidate to satisfy the requirements of 5G networks and beyond.

4.3.1 Higher Bandwidth in Optical Spectrum

The conventional RF spectrum used for wireless communication purposes is mainly in the range of several hundred megahertz to a few gigahertz. To keep up with the significant wireless traffic increase, the spectrum in the range of tens of gigahertz up to 300 GHz has been considered for wireless communication in the past 10 years [15]. In the future wireless world, frequency resources in the millimeter-wave region might also become saturated or inefficient for use in certain conditions. In this case, the abundantly available bandwidth in the optical spectrum region can be explored for wireless communication. The bandwidth in the visible light and short-wavelength IR region is about 633.3 THz, which is over 2,000 times wider than the entire spectrum in the RF region as shown in Figure 4.5. Similar to the millimeter-wave spectrum, the use of the optical spectrum is license-free.

However, the use of the optical spectrum has its own challenges. In conventional wireless communication systems, the data symbols are first modulated in the baseband with a certain modulation bandwidth. Then the baseband signal is shifted to the passband centered at a carrier frequency in RF-based wireless systems [21]. Different RF systems use different carrier frequencies to avoid spectrum overlap. Thus, the occupied bandwidth equals the baseband modulation bandwidth. In optical wireless systems, IM with DD is primarily used due to the incoherent nature of optical sources such as LED [22]. Consequently,

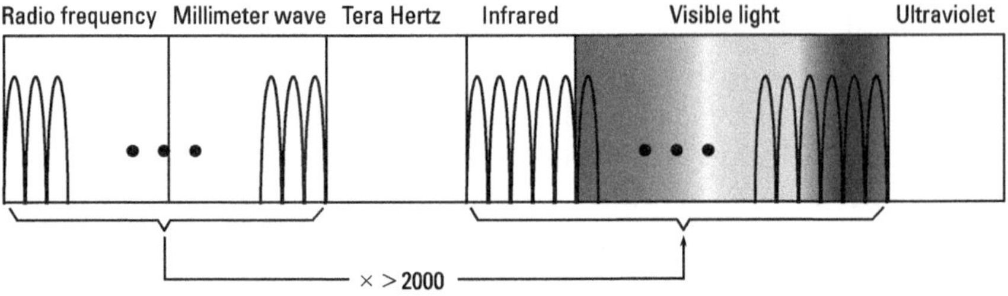

Figure 4.5 Comparison between RF spectrum and optical spectrum.

the occupied bandwidth is equal to the wide spectrum of the optical source. This means that the vast amount of optical spectrum has not been fully utilized yet. Nevertheless, it is possible to overcome this challenge by narrowing each subchannel in the optical spectrum. One of the methods to achieve this is to use a coherent optical source with a narrow spectrum (hundreds of Hertz) in conjunction with a narrow passband optical filter [23]. Another method is to investigate the possibility of coherent modulation and detection. With advances in future optical front-end development, it is expected that such solutions will be practically possible in the immediate future.

4.3.2 Extreme Densification to the Next Level

Cell densification is an effective method to improve network capacity, which has been practiced over several cellular generations. Ideally, reducing the cell size is always beneficial until each AP serves a single user [10]. However, in areas with extremely high user density, reducing the cell size to such a level means that the inter-AP spacing reduces to about 1 to 5m. It is noted that in such near-field scenarios, the power-law path-loss model no longer applies, which causes significant excessive CCI to users [24]. As shown in Figure 4.6(a), the level of CCI can become very high to the extent that the desired user link is seriously interfered with transmissions from multiple APs in the area. In this case, interference management techniques or orthogonal channel access techniques are needed. Such techniques, however, come at the cost of reduced spectral efficiency, which diminishes the gain from the extreme densification. This is the main reason why extreme densification of Wi-Fi APs cannot effectively increase the capacity of data offloading.

It is noted that, due to the high propagation attenuation, millimeter-wave communication is suitable in indoor applications. By deploying 1-millimeter-wave AP per room, the CCI from adjacent APs is considerably lower than RF due to the significant signal attenuation caused by walls. By using beamforming to overcome the high path loss, each AP is able to serve many users in the room [25]. However, the situation changes in extremely user-dense scenarios.

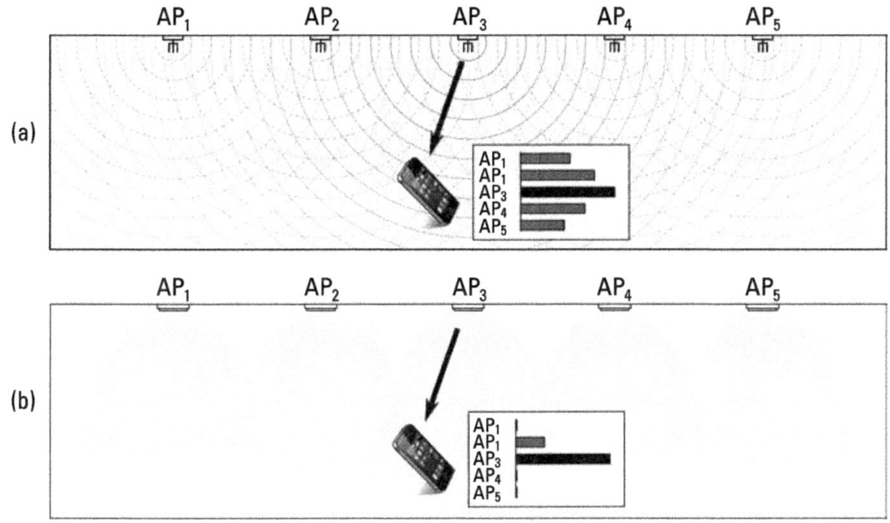

Figure 4.6 CCI characteristics of: (a) RF-based network and (b) Li-Fi network with extreme densification.

If many millimeter-wave APs are deployed in the same room, the beamforming not only needs to amplify the signal to the desired user, but also must reduce the CCI to all remaining users in the room. This will require a larger antenna array with a higher angular resolution. In addition, the computational cost, and transmission overhead increase significantly with a greater number of users in the room. Therefore, extreme densification of millimeter-wave APs is also challenging.

In Li-Fi networks, light source radiation is directional. With ceiling deployment and downward orientation, each light source has a light footprint on the floor. Only the user in the footprint region can detect the signal from the Li-Fi AP. Thus, each AP covers a limited number of users without the need for beamforming. As a result, each Li-Fi AP forms a small cell of 1 to 10 m^2 area. This small transmission range characteristic might be disadvantageous in terms of the size of coverage area per AP, but it is perfect when extreme network densification is desired. Due to the limited range of optical signal propagation, CCI affecting Li-Fi users is limited compared to the case in RF transmissions. As shown in Figure 4.6(b), only the users located in the intersection area of adjacent footprints receive CCI from one or two neighboring APs. In this case, less interference management is required as the level of CCI is much lower. Regarding the perspective of mobility in small cell networks, it is worth noting that IM/DD in Li-Fi has the benefit of not suffering from the negative effect of small-scale fading, which affects RF transmissions. This means that the coherence time of the optical channel is much longer, which is advantageous in terms of link reliability against mobility [26].

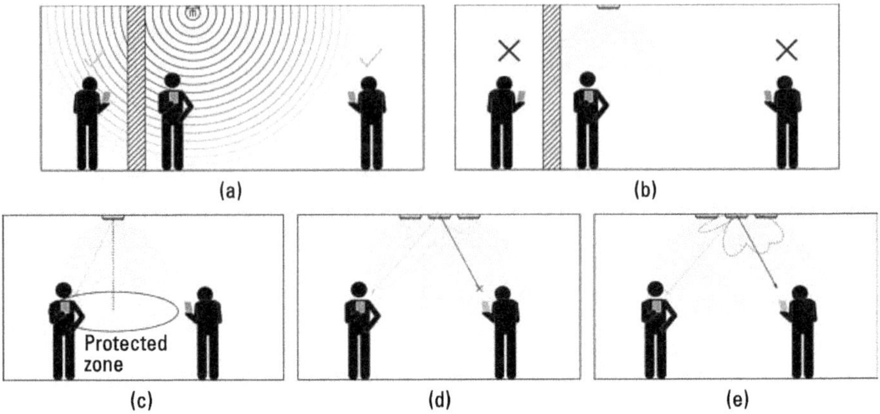

Figure 4.7 (a) Confidential information can be easily intercepted via the RF-based network. (b) Li-Fi transmission provides enhanced security in the physical layer. (c) Protected zone. (d) Null-steering. (e) Artificial noise transmission.

4.3.3 Physical Layer Security (PLS)

Light signals cannot penetrate opaque objects, which not only avoids potential CCI between Li-Fi communications links in adjacent rooms, but also offers enhanced PLS for wireless access. As shown in Figure 4.7(a), an RF signal can be easily intercepted by eavesdroppers from behind a wall or from a distant location. In the case of a Li-Fi link, however, this is no longer possible and the leakage of confidential information can be avoided, as shown in Figure 4.7(b) [27]. In addition to this basic proximity protection, a number of security measures can be implemented to further enhance the protection of confidential messages and ensure they could only be detected by legitimate users. In the following, we list examples of Li-Fi PLS measures.

- One of the strategies, referred to as the protected zone, can be implemented in Li-Fi networks [28]. A circular safety area centered at the serving AP with a specified security radius is defined, as shown in Figure 4.7(c). The radius is sufficiently large enough to cover the legitimate user. The radius of this safety zone can be controlled by adjusting the Li-Fi AP signal transmission power level. Since the Li-Fi AP is aware of all nearby users, if any other user enters the safety area, the transmission of confidential information is paused.
- Another approach to enhance Li-Fi security is known as null-steering, which is based on a system that is able to coordinate the transmissions of multiple APs or a single AP with multiple LEDs to form a multi-user multiple-input single-output (MISO) system [29]. Basically, the method is to use precoding to project the legitimate user's channel onto the null space of the unauthorized user's channel, as shown in Figure 4.7(d).

This, however, requires the channel information of the unauthorized user to be known at the AP. Therefore, this technique can be implemented for scenarios where the unauthorized user is a registered user in the network but does not have the permission to receive confidential information that is meant for the legitimate user.

- In the case of a malicious eavesdropper where the unauthorized user's channel information is unknown to the AP, an approach known as artificial noise transmission can be implemented [29]. The AP creates and projects a jamming signal onto the null space of the legitimate user's channel, as shown in Figure 4.7(e). Since it is unlikely that the authorized user would have the same channel, the reception of confidential messages by the unauthorized user would be distorted.

4.3.4 Other Advantages

4.3.4.1 RF Restricted Areas

In health care facilities such as hospitals or clinics, RF transmission may interfere with sensitive medical equipment or monitoring devices. Thus, policies are made to prohibit the use of wireless communication devices in some restricted areas such as operating rooms or intensive care units [30]. Moreover, in petro-chemical plants where there is a potential risk of explosion, electromagnetic radiation could be a possible hazard of ignition. Therefore, wireless communication devices are restricted in certain areas [31]. In these special application scenes, Li-Fi could be a promising substitution, as shown in Figure 4.8. Automatic operation, monitoring, and control are expected to be widely used in future factories and warehouses. In order to realize this, large numbers of sensors, cameras, and other smart equipment will be necessary. In this regard, the communication between the control room and these objects needs ultrareliability and low latency. It will be challenging to use the RF transmission to fulfill the communication requirement and it will be costly and inflexible to use wired communication. In contrast, Li-Fi will be a promising solution, as shown in Figure 4.8(b). Details about the applications regarding this characteristic will be presented in Section 4.4.4.

4.3.4.2 Integration of Lighting and Wireless Communications Infrastructures

With the trend of extreme cell densification, wireless communication APs will need to be densely deployed in indoor spaces. In this case, there might be a conflict between the layouts of the lighting devices and communication devices. Since Li-Fi technology aims to utilize the existing lighting infrastructure to deliver wireless connectivity on top of the illumination functionalities, Li-Fi forms a compact solution to provide users with both illumination and communications services in an efficient way.

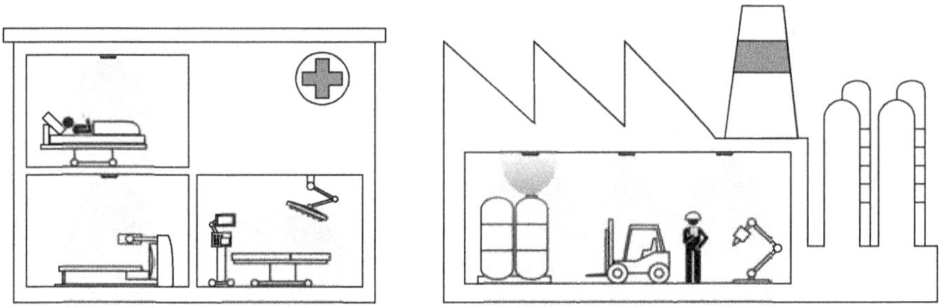

(a) (b)

Figure 4.8 (a) Li-Fi application in health care facilities. (b) Li-Fi application in petro-chemical plants and factories.

4.3.4.3 Visibility of Li-Fi Signal

One more unique characteristic of light signals is their visibility to the human eye. High-power wireless systems may cause health issues under some conditions, but avoiding high RF radiations is not straightforward as the human eye cannot detect the waves. Since light signals are visible to mobile users, the human eye detects any high intensity and instinctively looks away, which makes it safer for users. Another advantage of the visibility of Li-Fi signals is that the user can easily know where to locate itself to get a better signal quality by following the light, whereas, for conventional RF systems, users have to attempt to move around to have better link quality.

4.4 Applications

Mobile smartphones are now an indispensable element of modern human life. The number of applications of mobile communications has evolved since the development of the first mobile telephone by Martin Cooper. The straightforward Li-Fi application is focused on enabling high-speed indoor connectivity. However, it is anticipated that the number of applications of Li-Fi will grow from a few connectivity-based services to a large number of applications. This section aims at discussing a number of the main applications of Li-Fi.

4.4.1 Defense and Cybersecurity

Li-Fi is a good candidate for covert communications since the light confinement property allows for improved physical layer security. Section 4.3 briefly discussed some of the PLS methods proposed for Li-Fi, while in this section, we focus on introducing a number of promising applications of Li-Fi for the purpose of security enhancement, such as optical geo-fences, military communication, and wireless access in times of disasters.

The small cells concept and the extreme densification of Li-Fi APs allow for geolocation-based monitoring of users' access and mobility patterns. For example, the Li-Fi APs in a certain establishment can be categorized based on their fixed locations within the building. The users' connection policies can be devised based on the attocell APs categories. A user with a general communication-access privilege can be assumed to connect through the office optical APs or any other common-space access point inside the establishment. Any abnormal access behavior can be detected when such a user is connected through an optical attocell AP located in a restricted area of the building. This concept can be developed further to enable the creation of optical geo-fences for users. A machine learning algorithm can be devised to adapt the optical geo-fence of users based on their location access and mobility patterns.

There has been an increased interest from the defense industry in Li-Fi [32–35]. Li-Fi was trailed abroad the *USS Carl Vinson* by the Space and Navy Warfare Systems Command in a large, in-sea field experiment at Trident Warrior 2018 [35]. The United States Navy is considering Li-Fi for in-ship, ship-to-ship, and ship-to-submarine communications. Li-Fi is investigated for in-ship broadcasting to the sailors. Tactical LOS optical networks are investigated using Li-Fi to enable ship-to-ship and ship-to-shore communications.

The huge demand for wireless access in times of disaster could disrupt the traditional RF-based wireless networks. Li-Fi is an attractive communications technology for disaster management. The durability of Li-Fi was tested in a subway terrorist attack drill at the Disaster Preparedness and Tactical Training Center in Georgia, United States [36]. Li-Fi was demonstrated to be a resilient technology that can maintain a real-time, bidirectional communication network between the emergency services and the central command [36]. Li-Fi is also being investigated by the United States Department of Defense as a secure and reliable communications solution for tactical shelters to reduce the network installation time. The aim is to replace the military wired and Wi-Fi networks with a Li-Fi network that could offer multi-user, multichannel communications for classified and nonclassified voice, data, and video transmission [37]. A mesh network of unmanned aerial vehicles can serve as Li-Fi APs that are quick to establish and set up in disaster scenarios and tactical fields.

4.4.2 Aerospace and Aeronautical Communications

High-speed, RF-based, in-cabin communications have been proven to be challenging due to the high user density and the strong RF reflections caused by the material of the airplane structure [38]. Li-Fi is an interesting high-speed solution for providing in-cabin communications by means of utilizing the existing in-cabin lighting infrastructure. To this end, Li-Fi is attractive for the aerospace industry due to its high-speed, usage of existing infrastructure, and its improved PLS [39]. Li-Fi is a lightweight technology in comparison with other candidate communications technologies. The usage of license-free spectrum bands simplifies the deployment of Li-Fi in the aerospace industry regardless of the geographical location [38]. In addition, Li-Fi does not interfere with other electronic systems in an aircraft, which is an attractive advantage to the aerospace industry. Li-Fi is a strong candidate for low-latency-high-speed communications within the cockpit due to the light confinement property discussed in Section 4.3. Li-Fi signals within the cockpit will be physically shielded from users inside the cabin by the opaque cockpit door and walls.

The U.S. National Aeronautics and Space Administration (NASA) is considering Li-Fi for in-space shuttles communications [40]. Notable recent deployments of optical wireless aeronautical communications include Project Loon by Google and the Internet Project by Facebook [41,42]. Both projects aim to tackle the digital divide by connecting those in remote or sparsely populated areas [42]. The Internet Project by Facebook aimed to use a constellation of high-altitude unmanned aerial vehicles to connect and feed the internet using IR-based OWC [42]. Project Loon aims to utilize a network of interconnected balloons in the stratosphere. The balloons will connect user terminals using RF-based communications. However, the inter-balloon communications were using a high-speed, long-range, free-space OWC, which is attractive due to the weak weather-dependent attenuation and atmospheric turbulence in the stratosphere. A trial of an OWC link over a distance of 100 km between 20-km altitude balloons was demonstrated at a 130-Mbps speed by Google [41].

4.4.3 Underwater Communications

There is a growing demand for high-speed underwater communication technologies to enable undersea exploration, tracking of pollution, and monitoring of off-shore equipment such as oil and gas rigs. The attenuation of RF radiation in underwater media is considerably high. For example, the attenuation of an underwater electromagnetic wave at 2.4 GHz reaches up to 189 dB/m [43]. Acoustic waves have been used for underwater communications for a long time. However, acoustic waves offer a solution for underwater communications that is low in speed but long in transmission range. Acoustic waves enable communication at a few hundreds of kilobit-per-second speeds that extend to a few tens of kilometers [44]. Nevertheless, acoustic waves suffer from high delays and low bandwidth, which hinder their ability to provide high-speed communications for a networked array of underwater sensors.

Visible light signals experience low attenuation in water. The light absorption of water varies based on the location of the considered water body and the water turbidity. However, blue and green spectra (between 450 nm and 550 nm) have been favored in the majority of the recent studies since they achieve the minimum attenuation compared with other spectra [43]. There has been an interest in the defense applications of underwater optical wireless

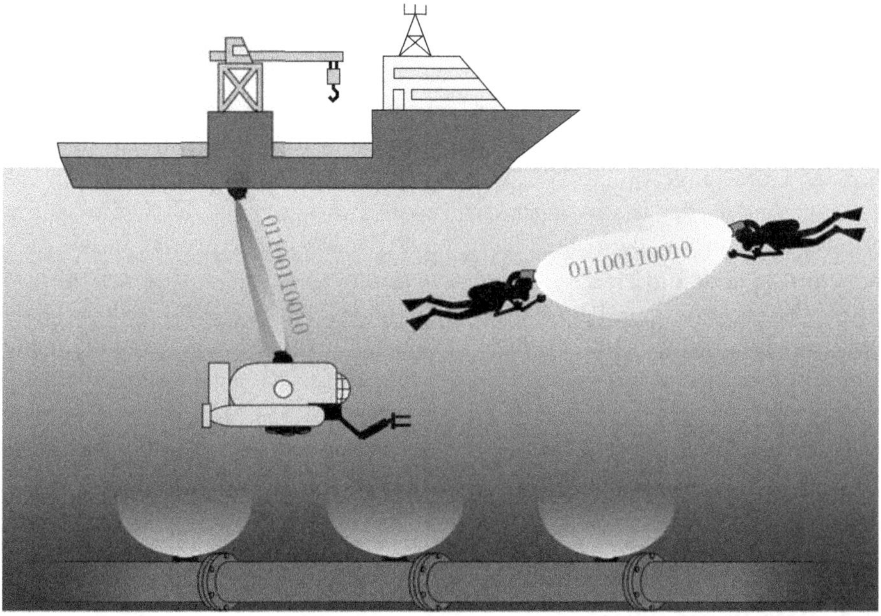

Figure 4.9 A UOWC showing diver-to-diver visible light communications, underwater sensor network unicast to a AUV, and a UOWC link between a ship and a AUV.

communication (UOWC), as discussed in Section 4.4.1. Additionally, there has been a number of studies and experiments to investigate the performance of VLC in underwater scenarios [44]. UOWC offers a high-speed alternative to acoustic communications that is low in cost and small in size. It is an attractive short to medium-range solution for undersea monitoring and inspection, ship-to-submarine, and submarine-to-submarine communications. Depending on the implementation topology and the system components, UOWC can support transmission speeds that range from a few hundreds of megabits per second to a few gigabits per second at transmission distances that extend up to a few tens of meters. A recent study reported a 2.5 Gbps at 60m underwater VLC link using blue light amplification by stimulated emission of radiation diode (LD) [45]. Figure 4.9 shows a number of seabed sensors that are monitoring a pipe. These sensors can store the observed data and sporadically burst the information into surface buoys. Alternatively, a number of autonomous underwater vehicles (AUVs) can swarm the area under survey to collect the data and perform additional inspections. The AUVs can relay the information back to a monitoring station aboard a ship. In another example, diver-to-diver communication can enable vital health and safety information transmission via torchlight as demonstrated in Figure 4.9. The deployment of dense underwater sensor networks imposes strict low-delay and high-speed requirements that cannot be satisfied by acoustic communication technology. A major drawback of UOWC in comparison with acoustic communications is the reduced communications range. However, photodetectors with high sensitivity such as avalanche photodiodes (APDs) and single-photon avalanche diodes (SPADs) can be employed to address this issue. Alternatively, a hybrid acoustic-optical communications system can be devised to extend the range of UOWC systems [44]. Alternatively, a multihop relay of networked nodes is investigated to extend the range of UOWC. Multiple relaying and routing algorithms can enable a successful transmission between a transmitting node and a receiving node over a long transmission distance [43]. The main objective is to extend the transmission range while maintaining the high data rate advantages achieved by UOWC. The trade-off between routing algorithms involves connectivity, adaptability, cost, and complexity [43]. Machine learning algorithms are considered for dynamic UOWC nodes [46]. This extends the network lifetime, improves the energy efficiency of the underwater nodes, and allows for the routing algorithms to be adapted based on the mobility of the nodes.

4.4.4 Data-Driven Cities

A large number of IoT devices such as sensors, actuators, robots, and household equipment would require to be connected to perform remote sensing and control in future smart and data-driven cities. This is the nervous system that is crucial to enable machine learning and AI as discussed in Section 4.2.1. Li-Fi is an attractive technology to enable high-speed internet connectivity in highly dense areas such as airports and stadiums due to the extreme densification of Li-Fi access points as discussed in Section 4.3.2.

The fourth industrial revolution refers to the transformation of industry-based mechanical and electronic technologies into a digital interconnected technology. This transformation is enabled by the massive utilization of ICT in industrial establishments. For example, a factory would employ a dense network of sensor nodes at each stage of the production chain. These sensors would closely monitor a certain aspect of the production and transfer these observations to a control unit. This can help in assessing and improving the quality of the production. M2M communications are required between the different factory modules and the IoT sensor nodes. However, the dense deployment of IoT devices in an RF crowded space will result in increased interference, which may lead to communications failure and a disruption in the device and production function. Moreover, the use of RF can be restricted in radiation-sensitive areas as it might trigger a spark or affect the functionality or reading accuracy of other electronic devices as discussed in Section 4.3.4.1. Therefore, an alternative ubiquitous spectral and energy-efficient communication system that is low in cost would be essential to connect the diverse variety of omnipresent IoT devices.

The adoption SSL has already allowed for significant energy savings. The emergence of LaaS offers further energy savings and an exciting opportunity. Lights can now be connected with IoT devices to be remotely controlled among other different functionalities. For example, SSL are now integrated with occupancy sensors and indoor positioning to allow optimal energy consumption. VLC-enabled IoT devices are easy to set up, simple to reconfigure, and low in cost [47]. In addition, sporadic and opportunistic transmission simplifies the system design and improves the energy efficiency of IoT Li-Fi nodes [48]. Information security threats could pose serious implications not only on the profitability and competitiveness of the business but also on national security and staff safety. Li-Fi can address these security concerns by containing the communication within the secure-vetted area where it should be accessed from, as discussed in Section 4.3.3.

Patients' vital health information should be secure against eavesdropping, tampering, and any other risk that could affect patient safety or privacy. Li-Fi offers an attractive solution for these security concerns. In addition, Li-Fi enables indoor positioning for mobile sensors, actuators, and robots, which helps with surrounding-awareness and hazard avoidance. A VLC indoor positioning and navigation system for a hospital robotic delivery has been proposed and investigated [49]. Similar algorithms can be useful for industrial robots in big warehouses and household equipment such as vacuum robots. Hospitals require an RF radiation-free communications system to connect various types of IoT devices. Additionally, there is a tendency to avoid using RF radiation inside hospitals to reduce the likelihood of interference with medical electronic equipment and to reduce patient exposure to RF. Li-Fi attocell APs can communicate with a wireless on-body sensor network to enable remote health monitoring for health professionals to gain real-time access and monitor the situation of a patient.

A wealth of off-the-shelf optical front-end devices can enable diverse IoT applications. A hybrid Li-Fi, optical camera communication (OCC), and RF IoT system are investigated in [50] to enable illumination, communication, localization, and energy harvesting. The aim is to serve a wide range of IoT applications. The dual functionality of illumination and communications is alleviated to multi-functionalities that also include power harvesting and indoor positioning [50]. Photovoltaic solar cells can be used as photodiode (PDs) in addition to their main usage in energy harvesting [51]. This can enable long-range, free-space optical (FSO) links by using roof-top solar cells as Li-Fi receivers. A recent study reported a deployment of an off-the-shelf, solar cell-based, back-haul solution on a remote Scottish island in the Orkney Islands, Graemsay [52]. A solar-Li-Fi distribution hub is used to connect multiple network nodes with solar cell receivers over a distance of 80m. In comparison with traditional FSO systems, solar cell receivers have a lower bandwidth but a higher detection area. This reduces the likelihood of outage due to blockage and allows for the simple alignment of solar cell receivers-based Li-Fi systems. In addition, solar cell Li-Fi receivers offer improved mitigation against the degradation caused by weather effects such as fog and snow. A simplified scenario of a smart city street lamp with a Li-Fi AP and a backhaul solar cell receiver is shown in Figure 4.10. A Li-Fi-based distribution hub can connect multiple street lamps. In addition, photovoltaic solar cells are attractive for wireless mobile IoT devices. The simultaneous communication and energy harvesting allow for energy-neutral wireless IoT nodes. This simplifies and enables the deployment of dense IoT nodes that would no longer require an external power supply.

Li-Fi-enabled lighting systems can enable indoor positioning and navigation. Indoor positioning can assist blind and disabled people in navigation. Tracking the mobility of users in shopping malls can help the marketing team to understand the customer behaviors as shown in Figure 4.10. Multimedia guides with interactive maps are essential elements in modern museums [53]. Li-Fi can enable localization, navigation, and interaction with displayed artifacts and high-speed internet access in smart museums. There are further applications beyond these use cases. For example, the localization data can be lodged to monitor the visitor density around certain locations within the museum. This will not only help in understanding the occupancy information of each space within the museum, but also help in the further planning of resources to accommodate the varying number of visitors across time and space. These types of use cases are dependent on the integration of sensing, lighting, connectivity, and decision-making.

4.4.5 Vehicular Communications

There has been a huge interest from research, governmental, and commercial bodies in transforming transportation systems into ITSs [54]. Vehicular communications, sensing, control, and AI are the key enabling technologies for ITS. The main objectives of ITS are reducing fuel consumption and CO_2 emission and avoiding collisions and reducing their impact on road users. For example, an urgent traffic-related message can be broadcasted to relevant cars and roadside units whenever an emergency vehicle plans to pass through a relevant route. The traffic lights can be controlled to adaptively change their timing settings to allow for shorter timings for emergency vehicles. Being able to broadcast information regarding the route of the emergency vehicle on a road before the siren can be seen or heard would allow the vehicle to reach its destination faster. Research into autonomous and driverless vehicles is no longer in its infancy. A number of car manufacturers already offer a form of autonomy in their products.

Vehicular communications could take the form of vehicle-to-vehicle (V2V) and vehicle-to-infrastructure (V2I). The infrastructure can be in the form of roadside units such as traffic lights and traffic signs. Vehicles can take knowledge-based decisions by communicating critical sensor and positioning information with other surrounding vehicles. For example, vehicles could relay safety, traffic, and infotainment messages in a multihop relay network. There has been considerable research effort directed towards investigating RF-based vehicular communication systems [55]. However, the wide coverage of the RF-based vehicular communications makes it vulnerable in terms of data security and reliability. The performance of the RF-based vehicular communication systems deteriorates with the increase of traffic density due to the increase of RF interference [54]. This performance degradation cannot be acceptable for safety-related ITS applications.

VLC, has been proposed as a low-cost solution to the vulnerability of the RF-based vehicular communicators. A vehicle's headlights and taillights can form the front end of a VLC system. Photodetectors could be distributed along the perimeter of the vehicle to enable a wide FOV reception. Semiconductor-based LEDs are already being used in headlights and taillights by car manufacturers. Additionally, LDs are recently being considered to be the next generation of semiconductor-based headlights [56]. Enabling LD-based headlights is not only attractive from a lighting point of view, but also it offers additional possibilities from a VLC point of view as it can extend the transmission ranges and significantly improve transmission speeds. There has been a growing interest in evaluating the performance of VLC in vehicular communications [55]. The inherent beamforming of the LED-based headlights allows directing the signal to the intended recipients. This improves the system security against eavesdropping and denial-of-service attacks. In addition, it allows cars to control the transmission distance and coverage and to minimize interference and increase the spatial and spectral reuse factor. In addition, the usage of the visible spectrum guarantees the freedom to operate internationally in a license-free band for all car manufacturers. VLC is an attractive solution for platooning, where a group of vehicles is driven in a convey with a very short separation distance. The cooperation between the vehicles in a platoon cannot be achieved without a reliable low-latency communication system.

The directionality of the headlights and taillights of the vehicle restricts the network topology to multihop and ad hoc links. This allows LOS communications between a vehicle and its surrounding vehicles in a V2V regime and between a vehicle and other roadside units such as traffic lights and road signs in a V2I regime as shown in Figure 4.10. Vehicles can exchange safety data to control and manage sensor readings, vehicle location, and distance to surrounding vehicles, speed, and acceleration. VLC can enable a relative level of accuracy for vehicle localization problems in areas where the global positioning system (GPS) fails to function, such as tunnels. The roadside units used by the vehicle to gain access to a certain service can be identified to localize the vehicle [54]. VLC has a shorter transmission range compared to its RF vehicular communication counterparts.

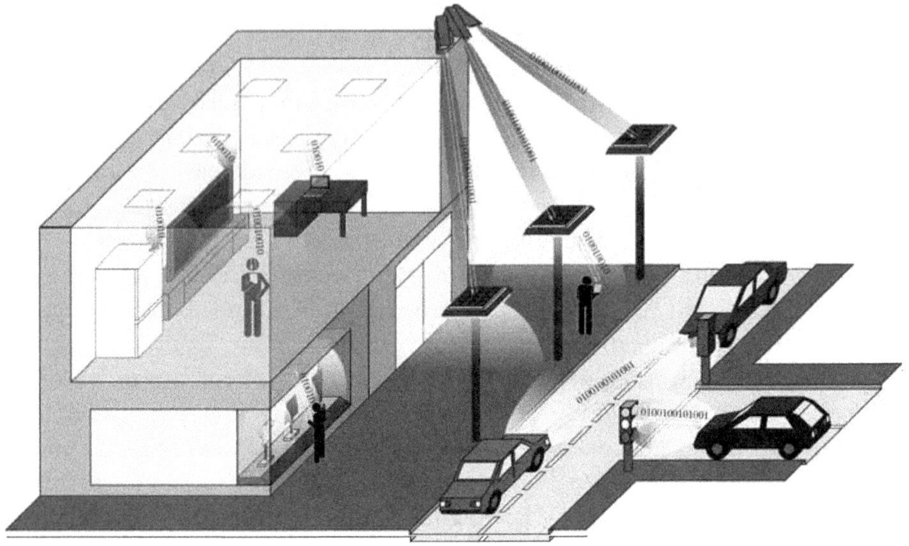

Figure 4.10 An illustration of a Li-Fi-enabled data-driven smart-city scenario demonstrating applications such as vehicular communications, photovoltaic solar-cell Li-Fi receivers, internet connectivity, and household Li-Fi-enabled IoT.

However, this allows VLC to reduce the data congestion, which therefore reduces the latency resulting in vehicles acting faster.

The range of VLC-based vehicular communication is decreased by weather conditions such as fog and snow. Hybrid RF-VLC approaches are considered as a solution to the range limitation of the VLC-based vehicular communication [57]. Background light and sunlight impose some degradation on the performance of the VLC-based communication link. However, optical filtering and automatic gain-controlled detectors can be used to limit this degradation as discussed in Section 4.5.4. There are still some challenges in the design of fully networked Li-Fi based vehicular communication systems due to the dynamic changes of network topology and the difficulty in maintaining alignment with the rapid mobility of network nodes [58]. Therefore, VLC-based vehicular communication requires a novel network layer design.

4.5 Li-Fi Misconceptions

Many common misconceptions about Li-Fi exist. For example, one may think of Li-Fi as a competitive technology to other wireless communication technologies such as Wi-Fi. Others may think Li-Fi will not work during the daytime because of background light. In this section, we consider and try to debunk these major misconceptions in order to help the readers correctly understand this technology.

4.5.1 Will Li-Fi Take over Wi-Fi?

Li-Fi technology has proven to hold great potential to provide unprecedented high-speed wireless connectivity. This, however, does not imply that current wireless networking technologies, such as Wi-Fi, will be entirely replaced by Li-Fi or become obsolete. In fact, Li-Fi could be better perceived as a complementary technology that adds a new layer to network heterogeneity in order to enable wireless networks to support diverse cases and applications. In this regard, both Wi-Fi and Li-Fi could work in conjunction to offload the macrocell traffic. Compared to Wi-Fi, Li-Fi APs have a relatively shorter range. This implies higher signal strength and better link reliability. It also implies more frequent handovers for fast-moving users [59]. To this extent, optimized load balancing can best utilize the advantages of Li-Fi high-rate links and Wi-Fi wide coverage areas so as to provide network users with the highest possible QoS [60]. Another scenario where Li-Fi can be used is in locations where the usage of RF transmissions is not possible or undesired, such as in aircraft and operating theaters. Moreover, due to the local confinement of light signals, Li-Fi provides highly secure communication links that are not susceptible to eavesdropping from outside the rooms, which is different from Wi-Fi signals that can easily penetrate through walls. This makes Li-Fi a viable solution for highly confidential applications such as military communications.

4.5.2 Will Signal Blockage Cause Serious Link Disruption?

The electromagnetic waves with higher frequency exhibit increased atmospheric attenuation and less diffraction during the process of propagation in free space [10]. In addition, they are less capable of penetrating solid objects such as building structures. When a high-frequency millimeter-wave or an optical signal penetrates a wall, it either experiences significant path loss or the signal is completely unable to propagate beyond the wall. These factors make wireless transmissions based on signals in the millimeter-wave, terahertz, and optical range primarily rely on the existence of LOS paths. Consequently, the wireless link quality of this type of systems is likely to be significantly deteriorated or interrupted in the event of the LOS path obstruction caused by human bodies or other objects. This signal blockage issue proved to cause frequent link interruption in the millimeter-wave cellular systems in many scenarios [61]. In order to maintain seamless user connectivity in such systems, an increased rate of handover operations has to be conducted.

The link blockage issue also exists in Li-Fi networks. However, thanks to the dense deployment of Li-Fi APs, this challenge can be overcome in a unique manner. Analytical studies show that the probability of a blockage decreases when the distance between the user and the serving AP becomes smaller [62]. As discussed in Section 4.3, Li-Fi is a wireless system with extremely dense APs

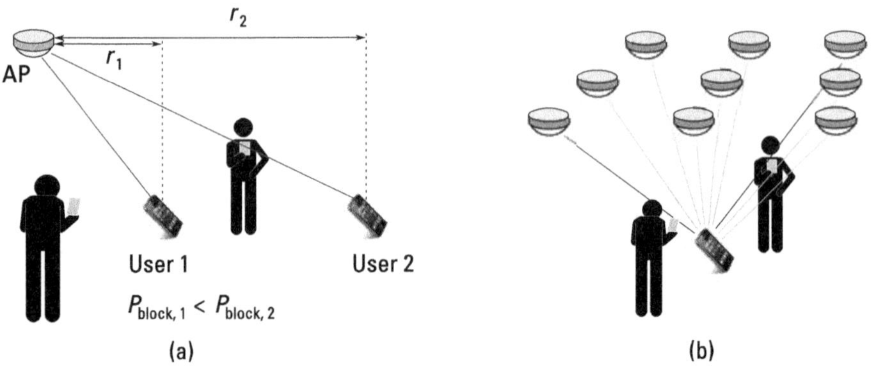

Figure 4.11 (a) The users closer to the AP have a lower probability of experiencing LOS path blockage. (b) Dense AP deployment create multiple LOS paths that make the Li-Fi link robust against blockage.

deployment. As a result, the LOS link between a user and a Li-Fi AP is always shorter than that in a millimeter-wave system. Consequently, the probability of LOS blockage in Li-Fi is lower than that in millimeter-wave networks. Due to the advantage of dense AP deployment, it is also possible to implement a system using coordinated transmission from multiple APs [63]. As shown in Figure 4.11, if one or two LOS paths are obstructed, the wireless transmission can be maintained with the remaining LOS paths. This coexistence of multiple LOS paths is able to effectively reduce the outage probability of the Li-Fi links.

4.5.3 Is There an Approach to Realize a UL Connection?

In Li-Fi networks, it is intuitive to use VLC techniques to establish DL transmissions. However, as a bidirectional wireless access technology, a UL from the mobile user side back to the serving AP is also essential in order to transmit acknowledgment, upload files, and have live streaming. As discussed in Section 4.3.2, Li-Fi networks have the potential to offload heavy wireless traffic in dense mobile user environments in future wireless systems. Compared to the DL transmission data traffic, the traffic load on UL transmission is relatively lower. By taking advantage of this load imbalance, one of the possible solutions to realize the Li-Fi UL is to use existing RF-based transmissions [64]. Studies show that the combination of visible light DL plus RF UL is capable of effective wireless DL data traffic offloading [65].

An alternative promising method to establish Li-Fi UL is by using wireless IR communication, as shown in Figure 4.12. This approach is motivated by two factors. First, lighting function is not required in the UL. Therefore, IR signal, which is invisible to the human eye, causes no disturbance to mobile users. In addition, the IR spectrum is also license-free and does not interfere

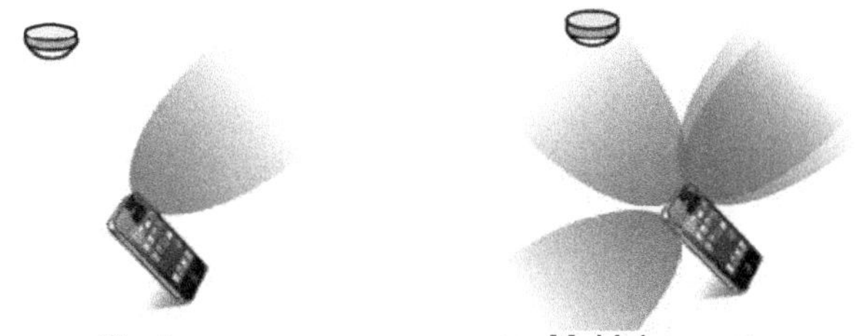

Figure 4.12 (a) Single-source IR UL is likely to experience misalignment. (b) Improved UL channel can be obtained by deploying multiple IR UL sources.

with visible light DL or RF-based wireless systems. Despite these advantages, the source directionality poses a major challenge. Conventional IR optical sources are directional, similar to a flashlight. Uncertainty of the position and direction of mobile devices may lead to misalignment of the UL source and detector at the AP side. Consequently, this issue increases the probability of a low-quality UL channel. A potential solution to this challenge is to use multiple IR sources on different sides of a mobile device to increase the probability of acquiring a well-aligned UL channel, as shown in Figure 4.12(b). It is also possible to remove the channel uncertainty due to the mobile device' random orientation by adopting a unique configuration [66].

4.5.4 Can This Technology Work Outdoors?

An experimental study was devised to assess the impact of solar irradiance on VLC and disprove the misconception that Li-Fi cannot work in an outdoor scenario [67]. It was found that sunlight does not interfere with the Li-Fi signals. Sunlight causes a strong shot noise at the photodetector. Nevertheless, optical filtering can be used to limit the effect of sunlight-induced shot noise, which would improve the overall Li-Fi system performance. The optical filtering is enabled by the fact that most of the white off-the-shelf LEDs are blue LEDs with a yellow phosphor coating. The solar irradiance in Antofagasta, Chile, at 12:00 PM on December 20, 2016, is shown in Figure 4.13 with an example of a typical blue LED irradiance. Antofagasta has been claimed to have the highest solar radiation on Earth [68]. The simple model of the atmospheric radiative transfer of sunshine (SMARTS) is used to obtain the direct solar irradiance [69,70]. The effect of optical filtering on the spectral response of a silicon photodetector is also shown in Figure 4.13 to significantly limit the collected sunlight at the photodetector. Optical filtering was shown to be a simple and cost-effective method to limit the degradation caused by sunlight [67].

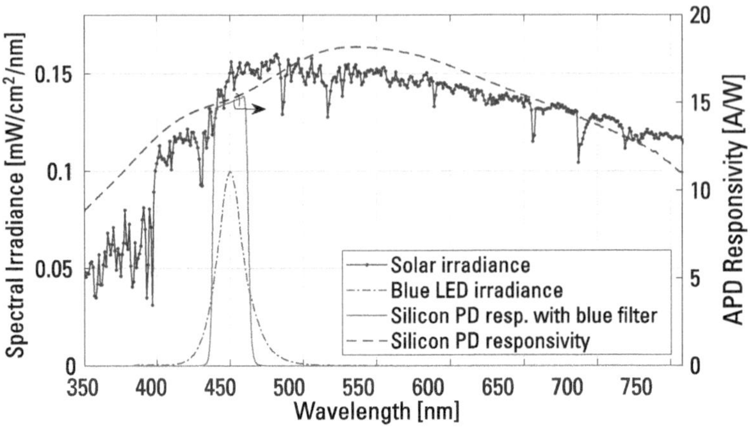

Figure 4.13 The solar irradiance, blue LED irradiance, and the effect of optical filtering on the responsivity of silicon photodetectors.

4.5.5 Can the Lights Be Dimmed While Transmitting?

Data are encoded in Li-Fi by the rapid changes of the light intensity. The modulated signals are bipolar in most of the pulsed and analog Li-Fi modulation techniques. Therefore, a DC bias voltage is required to allow for the full swing of the modulated signal [71]. The selection of the bias voltage is restricted by the modulation depth and the dynamic range of the SSL source. Limited dimming levels can be achieved by varying the modulation depth of the transmitted waveform. Applying dimming with these levels results in performance degradation of the Li-Fi system. There is no information carried on the power-consuming illumination-essential DC bias. There has been significant effort made to enable unipolar modulation techniques that would improve the energy efficiency of the Li-Fi systems [71]. The unipolar modulation techniques are suitable for Li-Fi applications where a low dimming of the light intensity is required. However, there has been an interest in enabling modulation techniques that would require a higher DC bias [71]. This is appealing to the lighting industry as they would favor biasing the light sources at the highest reliable bias levels in order to enable the highest illumination level. There has been a lot of interest in analog pulsed modulation techniques with a higher liberty on dimming levels [72], as will be discussed in detail in Chapter 6.

4.5.6 Does Li-Fi Cause Light Flickers?

The visible flicker of an LED is defined as the frequent changes of light brightness that can be perceived by the human eye. The human biological response to flicker is mainly a function of the wavelength of the light source, flicker frequency, modulation depth, and light source brightness [73]. For example, the increase in the modulation depth can make the visible flicker more pronounced. The IEEE PAR1789 working group has assessed the biological effects of flickering to mitigate the health and safety risk to viewers [73]. It was found that a visible flicker within the range of 3 to 70 Hz can trigger epileptic seizures [73]. However, it was also found that invisible flicker between 100 and 200 Hz can still be resolved by the human retina [73]. Thankfully, the SSL devices are modulated at frequencies that are greatly beyond these levels in Li-Fi. The rapid changes of light intensity in Li-Fi are in the region of a few megahertz to a few gigahertz. This makes it impossible for human eyes to detect any flicker caused by Li-Fi.

4.6 Summary

In this chapter, we reconsidered the definition of AI in the context of a connected world. Despite the highest importance of data analysis and processing functionality, the efficient delivery of data between sensors, processing units, and actuators is extremely critical. This is where wireless communication networks come into play. However, with the evolution of wireless communication from 1G to 5G, spectrum crunch implies that wireless connectivity could be the bottleneck of future AI systems. By reviewing the strategy of 5G technology, it can be concluded that extreme densification and new spectrum resources are two major approaches to acquire the desired performance gain. With this idea in mind, we presented the concept of the Li-Fi network, which applies to both of these approaches. First, the optical spectrum is license-free and has a much wider bandwidth compared to the RF spectrum. However, it is suitable to be used in the extreme densification scenario, since the performance of Li-Fi networks will not be limited by CCI even in the near field. In addition, Li-Fi networks provide extra PLS with proximity-based protection and other techniques. Next, we showed that Li-Fi technology has many application scenes such as defense/cyber-security, aerospace/aeronautical communications, underwater/vehicular communications, smart cities, and short-range backhaul communication. Finally, several remarks related to common misconceptions related to Li-Fi were provided to clarify that Li-Fi will not replace Wi-Fi but works as a data offloading complement besides Wi-Fi. Li-Fi can work during the daytime, because the background light does not introduce significant signal distortion and an energy-efficient modulation scheme can be used to support a dimming mode operation. Also, we discussed that Li-Fi transmissions do not cause flicker as the frequency of signaling is much higher than the range that is perceptible to human eyes. Moreover, we showed that the probability of link interruption caused by link blockage can be mitigated by the coordination of multiple Li-Fi APs or handover between them. We also clarified that a Li-Fi UL system can be realized by various technologies such as wireless IR communication. Wireless IR UL channel equality can be improved by using multiple IR emitters on the UE.

References

[1] Serafimovski, N., et al., "Light Communications for Wireless Local Area Networking," *IEEE 5G Tech Focus*, Vol. 2, No. 2, May 2018. https://5g.ieee.org/tech-focus/may-2018/light-communications-for-wireless-local-area-networking.

[2] Al-Fuqaha, A., et al., "Internet of Things: A Survey on Enabling Technologies, Protocols, and Applications," *IEEE Communications Surveys Tutorials*, Vol. 17, No. 4, 2015, pp. 2347–2376.

[3] *IEEE Industry Connections Ethernet Bandwidth Assessment*, IEEE 802.3 Ethernet Working Group Communication, Tech. Rep., July 2012.

[4] Tkach, R. W., "Scaling Optical Communications for the Next Decade and Beyond," *Bell Labs Technical Journal*, Vol. 14, No. 4, Winter 2010, pp. 3–9.

[5] Cisco, *Cisco Visual Networking Index: Forecast and Trends, 2017-2022*, Tech. Rep., February 2019.

[6] Shafiq, M. Z., et al., "A First Look at Cellular Machine-to-Machine Traffic: Large Scale Measurement and Characterization," *SIGMETRICS Performance Evaluation Review*, Vol. 40, No. 1, June 2012, pp. 65–76.

[7] Winzer, P. J., and D. T. Neilson, "From Scaling Disparities to Integrated Parallelism: A Decathlon for a Decade," *IEEE/OSA Journal of Lightwave Technology*, Vol. 35, No. 5, March 2017, pp. 1099–1115.

[8] Akyildiz, I. F., et al., "5G Roadmap: 10 Key Enabling Technologies," *Computer Networks*, Vol. 106, pp. 17–48, 2016. http://www.sciencedirect.com/science/article/pii/S1389128616301918.

[9] Parkvall, S., et al., "NR: The New 5G Radio Access Technology," *IEEE Communications Standards Magazine*, Vol. 1, No. 4, December 2017, pp. 24–30.

[10] Andrews, J. G., et al., "What Will 5G Be?" *IEEE Journal on Selected Areas in Communications*, Vol. 32, No. 6, June 2014, pp. 1065–1082.

[11] "The Tactile Internet," ITU-T Technology Watch Report, Tech. Rep., August 2014.

[12] Di Taranto, R., et al., "Location-Aware Communications for 5G Networks: How Location Information Can Improve Scalability, Latency, and Robustness of 5G," *IEEE Signal Processing Magazine*, Vol. 31, No. 6, November 2014, pp. 102–112.

[13] Schulz, P., et al., "Latency Critical IoT Applications in 5G: Perspective on the Design of Radio Interface and Network Architecture," *IEEE Communications Magazine*, Vol. 55, No. 2, February 2017, pp. 70–78.

[14] Schinianakis, D., "Alternative Security Options in the 5G and IoT Era," *IEEE Circuits and Systems Magazine*, Vol. 17, No. 4, Fourth Quarter 2017, pp. 6–28.

[15] Pi, Z., and F. Khan, "An Introduction to Millimeter-Wave Mobile Broadband Systems," *IEEE Communications Magazine*, Vol. 49, No. 6, June 2011, pp. 101–107.

[16] Elgala, H., R. Mesleh, and H. Haas, "Indoor Optical Wireless Communication: Potential and State-of-the-Art," *IEEE Communications Magazine*, Vol. 49, No. 9, September 2011, pp. 56–62.

[17] Kamel, M., W. Hamouda, and A. Youssef, "Ultra-Dense Networks: A Survey," *IEEE Communications Surveys Tutorials*, Vol. 18, No. 4, Fourth Quarter 2016, pp. 2522–2545.

[18] Larsson, E. G., et al., "Massive MIMO for Next Generation Wireless Systems," *IEEE Communications Magazine*, Vol. 52, No. 2, February 2014, pp. 186–195.

[19] Ali, K. S., et al., "Non-Orthogonal Multiple Access for Large-Scale 5G Networks: Interference Aware Design," *IEEE Access*, Vol. 5, 2017, pp. 21204–21216.

[20] Haas, H., et al., "What Is LiFi?" *IEEE/OSA Journal of Lightwave Technology*, Vol. 34, No. 6, March 2016, pp. 1533–1544.

[21] Proakis, J. G., and M. Salehi, *Digital Communications*, Vol. 4, New York: McGraw-Hill, 2001.

[22] Dimitrov, S., and H. Haas, *Principles of LED Light Communications: Towards Networked Li-Fi*, Cambridge, U.K.: Cambridge University Press, 2015.

[23] Xiang, C., P. A. Morton, and J. E. Bowers, "Ultra-Narrow Linewidth Laser Based on a Semiconductor Gain Chip and Extended Si3N4 Bragg Grating," *Optics Letters*, Vol. 44, No. 15, August 2019, pp. 3825–3828.

[24] Ramasamy, D., R. Ganti, and U. Madhow, "On the Capacity of Picocellular Networks," *2013 IEEE International Symposium on Information Theory*, July 2013, pp. 241–245.

[25] Ryan, J., G. R. MacCartney, and T. S. Rappaport, "Indoor Office Wideband Penetration Loss Measurements at 73 GHz," *IEEE International Conference on Communications Workshops (ICC Workshops)*, May 2017, pp. 228–233.

[26] Soltani, M. D., et al., "Modeling the Random Orientation of Mobile Devices: Measurement, Analysis and LiFi Use Case," *IEEE Transactions on Communications*, Vol. 67, No. 3, March 2019, pp. 2157–2172.

[27] Mostafa, A., and L. Lampe, "Physical-Layer Security for MISO Visible Light Communication Channels," *IEEE Journal on Selected Areas in Communications*, Vol. 33, No. 9, September 2015, pp. 1806–1818.

[28] Yin, L., and H. Haas, "Physical-Layer Security in Multiuser Visible Light Communication Networks," *IEEE Journal on Selected Areas in Communications*, Vol. 36, No. 1, January 2018, pp. 162–174.

[29] Mostafa, A., and L. Lampe, "Physical-Layer Security for Indoor Visible Light Communications," *2014 IEEE International Conference on Communications (ICC)*, June 2014, pp. 3342–3347.

[30] Tan, K. -S., I. Hinberg, and J. Wadhwani, "Electromagnetic Interference in Medical Devices: Health Canada's Past and Current Perspectives and Activities," *IEEE EMC International Symposium. Symposium Record. International Symposium on Electromagnetic Compatibility* (Cat. No. 01CH37161), Vol. 2, August 2001, pp. 1283–1288.

[31] Kuhl, K., J. Terwoert, and J. Cabecas, "Maintenance and Hazardous Substances: Maintenance in the Chemical Industry," *E-facts*, Vol. 67, 2012, pp. 1–13.

[32] "PureLiFi Works with Astronics on LiFi-Based Airplane Connectivity," *Optics.org*, September 26, 2018. https://optics.org/news/9/9/36.

[33] DISA News, "Li-Fi Technology Offers Benefits in Mobility, Speed, Cost, Security," *CHIPS*, October-December 2017. https://www.doncio.navy.mil/chips/ArticleDetails.aspx?ID=9685.

[34] Benson, J., "Seeing the Light: LiFi Will Revolutionize IT on Mission Command Posts," U.S. Army, November 21, 2018. https://www.army.mil/article/213936/seeing_the_light_lifi_will_revolutionize_it_on_mission_command_posts.

[35] Stinson, T., "Li-Fi Technology in the U.S. Navy," *All Hands*, April 16, 2019. https://allhands.navy.mil/Stories/Display-Story/Article/1815580/li-fi-technology-in-the-us-navy.

[36] Leeming, R., "Li-Fi Tested in US Subway Terrorist Attack Scenario," *Lux Review Magazine*, June 2017. https://luxreview.com/article/2017/06/-li-fi-tested-in-subway-terrorist-attack-scenario.

[37] Department of Defense (DoD), "Li-Fi Network for Tactical Shelters," DoD 2018.1 SBIR Solicitation, November 2017, A18-072. https://www.sbir.gov/sbirsearch/detail/1413927.

[38] Cogalan, T., S. Videv, and H. Haas, "Inflight Connectivity: Deploying Different Communication Networks Inside an Aircraft," *IEEE 87th Vehicular Technology Conference (VTC Spring)*, June 2018, pp. 1–6.

[39] Schmitt, N. P., "Applications and Future Trends in Laser Sensing and Communication at Airbus Group," *Lasers Congress 2016 (ASSL, LSC, LAC)*, Optical Society of America, 2016, p. LM3B.3. http://www.osapublishing.org/abstract.cfm?URI=LSC-2016-LM3B.3

[40] Granath, B., "Light Technology Being Developed for Advanced Communications," NASA, September 2, 2015. https://www.nasa.gov/feature/light-technology-being-developed-for-advanced-communications.

[41] Moision, B., et al., "Demonstration of Free-Space Optical Communication for Long-Range Data Links Between Balloons on Project Loon," in *Free-Space Laser Communication and Atmospheric Propagation XXIX*, H. Hemmati and D. M. Boroson, (eds.), Vol. 10096, International Society for Optics and Photonics/SPIE, 2017, pp. 259–272.

[42] Peyronel, T., et al., "Luminescent Detector for Free-Space Optical Communication," *OSA Optica*, Vol. 3, No. 7, July 2016, pp. 787–792.

[43] Saeed, N., et al., "Underwater Optical Wireless Communications, Networking, and Localization: A Survey," *Ad Hoc Networks*, Vol. 94, 2019, p. 101935. http://www.sciencedirect.com/science/article/pii/S1570870518309776.

[44] Ghassemlooy, Z., et al., "Emerging Optical Wireless Communications-Advances and Challenges," *IEEE Journal on Selected Areas in Communications*, Vol. 33, No. 9, September 2015, pp. 1738–1749.

[45] Lu, C., et al., "60m/2.5Gbps Underwater Optical Wireless Communication with NRZ-OOK Modulation and Digital Nonlinear Equalization," *Conference on Lasers and Electro-Optics (CLEO)*, May 2019, pp. 1–2.

[46] Hu, T., and Y. Fei, "QELAR: A Machine-Learning-Based Adaptive Routing Protocol for Energy-Efficient and Lifetime-Extended Underwater Sensor Networks," *IEEE Transactions on Mobile Computing*, Vol. 9, No. 6, June 2010, pp. 796–809.

[47] Mukku, V. D., S. Lang, and T. Reggelin, "Integration of LiFi Technology in an Industry 4.0 Learning Factory," *Procedia Manufacturing*, Vol. 31, 2019, pp. 232–238. http://www.sciencedirect.com/science/article/pii/S2351978919304019.

[48] Basnayaka, D. A., and H. Haas, "A New Degree of Freedom for Energy Efficiency of Digital Communication Systems," *IEEE Transactions on Communications*, Vol. 65, No. 7, July 2017, pp. 3023–3036.

[49] Murai, R., et al., "A Novel Visible Light Communication System for Enhanced Control of Autonomous Delivery Robots in a Hospital," *IEEE/SICE International Symposium on System Integration (SII)*, December 2012, pp. 510–516.

[50] Demirkol, I., et al., "Powering the Internet of Things Through Light Communication," *IEEE Communications Magazine*, Vol. 57, No. 6, June 2019, pp. 107–113.

[51] Wang, Z., et al., "Towards Self-Powered Solar Panel Receiver for Optical Wireless Communication," *Proc. IEEE International Conference on Communications (ICC)*, Sydney, NSW, June 10–14, 2014, pp. 3348–3353.

[52] Haas, H., et al., "Solar Cell Receiver Free-Space Optical for 5G Backhaul," *Optical Fiber Communications Conference and Exhibition (OFC)*, 2019, pp. 1–3.

[53] Cosmas, J., et al., "5G Internet of Radio Light Services for Musée De La Carte à Jouer," *Global LiFi Congress (GLC)*, February 2018, pp. 1–6.

[54] Uysal, M., et al., "Visible Light Communication for Vehicular Networking: Performance Study of a V2V System Using a Measured Headlamp Beam Pattern Model," *IEEE Vehicular Technology Magazine*, Vol. 10, No. 4, December 2015, pp. 45–53.

[55] Ucar, S., S. C. Ergen, and O. Ozkasap, "IEEE 802.11p and Visible Light Hybrid Communication Based Secure Autonomous Platoon," *IEEE Transactions on Vehicular Technology*, Vol. 67, No. 9, September 2018, pp. 8667–8681.

[56] Nogee, A., "Lasers Light the Way to New Applications," *LEDs Magazine*, January 7, 2014. https://www.ledsmagazine.com/home/article/16699150/lasers-light-the-way-to-new-applications.

[57] Liu, J., et al., "Hybrid Visible Light Communications in Intelligent Transportation Systems with Position Based Services," *IEEE Globecom Workshops*, December 2012, pp. 1254–1259.

[58] Cen, N., et al., "LANET: Visible-Light Ad Hoc Networks," *Ad Hoc Networks*, Vol. 84, 2019, pp. 107–123. http://www.sciencedirect.com/science/article/pii/S1570870518301392

[59] Wu, X., and H. Haas, "Handover Skipping for LiFi," *IEEE Access*, Vol. 7, 2019, pp. 38369–38378.

[60] Ma, W., and L. Zhang, "QoE-Driven Optimized Load Balancing Design for Hybrid LiFi and WiFi Networks," *IEEE Communications Letters*, Vol. 22, No. 11, November 2018, pp. 2354–2357.

[61] Talukdar, A., M. Cudak, and A. Ghosh, "Handoff Rates for Millimeterwave 5G Systems," *IEEE 79th Vehicular Technology Conference (VTC Spring)*, May 2014, pp. 1–5.

[62] Dong, K., X. Liao, and S. Zhu, "Link Blockage Analysis for Indoor 60GHz Radio Systems," *Electronics Letters*, Vol. 48, No. 23, November 2012, pp. 1506–1508.

[63] Li, X., et al., "Users First: User-Centric Cluster Formation for Interference-Mitigation in Visible-Light Networks," *IEEE Transactions on Wireless Communications*, Vol. 15, No. 1, January 2016, pp. 39–53.

[64] Pérez-Nicoli, P., et al., "Uplink Wireless Transmission Overview in Bi-Directional VLC Systems," *2016 IEEE International Conference on Electronics, Circuits and Systems (ICECS)*, December 2016, pp. 588–591.

[65] Rahaim, M. B., A. M. Vegni, and T. D. C. Little, "A Hybrid Radio Frequency and Broadcast Visible Light Communication System," *IEEE GLOBECOM Workshops (GC Wkshps)*, December 2011, pp. 792–796.

[66] Chen, C., R. Bian, and H. Haas, "Omnidirectional Transmitter and Receiver Design for Wireless Infrared Uplink Transmission in LiFi," *2018 IEEE International Conference on Communications Workshops (ICC Workshops)*, May 2018, pp. 1–6.

[67] Islim, M. S., et al., "The Impact of Solar Irradiance on Visible Light Communications," *IEEE/OSA Journal of Lightwave Technology*, Vol. 36, No. 12, June 2018, pp. 2376–2386.

[68] Rondanelli, R., A. Molina, and M. Falvey, "The Atacama Surface Solar Maximum," *Bulletin of the American Meteorological Society*, Vol. 96, No. 3, 2015, pp. 405–418.

[69] Gueymard, C. A., "Parameterized Transmittance Model for Direct Beam and Circumsolar Spectral Irradiance," *Solar Energy*, Vol. 71, No. 5, 2001, pp. 325–346.

[70] Gueymard, C., *SMARTS2: A Simple Model of the Atmospheric Radiative Transfer of Sunshine: Algorithms and Performance Assessment*, Florida Solar Energy Center, Cocoa, FL, 1995.

[71] Islim, M. S., and H. Haas, "Modulation Techniques for Li-Fi," *ZTE Communications*, Vol. 14, No. 2, April 2016, pp. 29–40.

[72] Zafar, F., D. Karunatilaka, and R. Parthiban, "Dimming Schemes for Visible Light Communication: The State of Research," *IEEE Wireless Communications*, Vol. 22, No. 2, April 2015, pp. 29–35.

[73] Wilkins, A., J. Veitch, and B. Lehman, "LED Lighting Flicker and Potential Health Concerns: IEEE Standard PAR1789 Update," *2010 IEEE Energy Conversion Congress and Exposition*, September 2010, pp. 171–178.

5

Optical Wireless Communication Channel

5.1 Introduction

Similar to other wireless communication technologies, the transmission range and capacity of Li-Fi networks are primarily limited by the available communication channels [1]. In order to optimize the link performance, it is important to evaluate and understand the optical wireless communication (OWC) channel characteristics in Li-Fi. In this chapter, we introduce various simulation and modeling approaches that are used to characterize the Li-Fi OWC channel. First, the basic channel effects are discussed and the relevant metrics are introduced. Then channels with front-end elements and channels with optical wireless propagation are separately introduced. In particular, evaluating the NLOS channel requires very high computational complexity, while the characteristics of NLOS channels vary significantly in different scenarios. Therefore, the most popular approaches to simulate or calculate the NLOS channel are introduced, and their advantages and disadvantages are compared and discussed. Finally, we use a number of scenarios in typical indoor environments to evaluate the characteristics of the OWC channels.

5.2 Channel Effects and Metrics

Currently, IM/DD is widely used in various types of OWC systems due to its simplicity and robustness against multipath fading. Despite the improved performance, coherent detection systems require more power and more hardware and are of higher complexity [2]. Coherent detection techniques have been studied and used as high-performance fiber/free-space optical (FSO) systems [3]. In the field of visible light communication (VLC) and Li-Fi systems, only a few

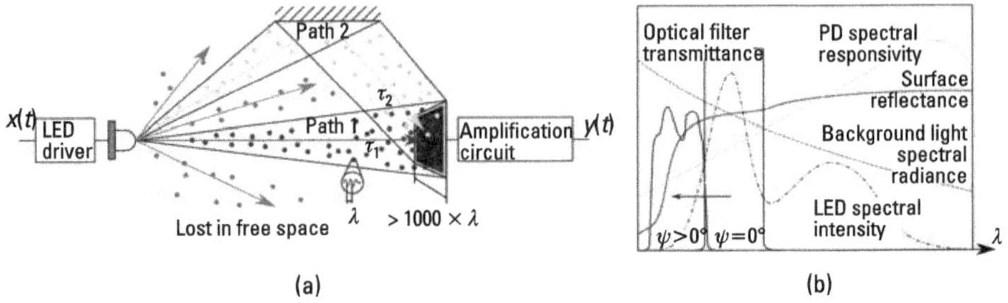

Figure 5.1 (a) An optical link and various channel effects. (b) Demonstration of wavelength-dependent quantities.

studies have considered coherent detection [4]. Therefore, IM/DD channels are considered in this chapter.

5.2.1 Channel Effects

A typical optical wireless link is shown in Figure 5.1(a). An electrical signal is forwarded to a driving circuit, which converts the signal to fit the dynamic range of an optical source, such as an LED. Then the optical source converts the electrical signal to an optical signal and is radiated into the free space. At the receiver end, a fraction of the emitted light signal is captured by an optical detector, such as a PD. Finally, the detected optical signal is converted back to an electrical signal and amplified for further decoding. During this process, the transmission experiences various channel effects which mainly include optical path loss, multipath distortion, and wavelength selectivity. Strictly speaking, there is also a multipath fading effect. However, this effect has been mitigated by spatial averaging to a negligible level [5].

5.2.1.1 Optical Path Loss

Optical path loss is one of the major consequences of the communication channel. Due to the nature of indoor Li-Fi applications, the wireless link is not well aligned due to the random position and orientation of the user equipment (UE). To improve the robustness of this wireless link, a very large divergence angle of the optical source is used. However, this diffused light beam configuration is also a result of lighting and eye-safety requirements [6]. With a large beam divergence angle, only a faction of the emitted light can be captured by the optical detector, and the remaining optical power is lost in the free space, as illustrated in Figure 5.1(a). This effect corresponds to the geometric loss in FSO channels. Compared to FSO channel path loss, the geometric loss in Li-Fi channels is more significant due to the much larger beam divergence angle. In addition, the indoor wireless channels are barely affected by atmospheric effects as the link distance is much shorter than in the case of FSO communications.

5.2.1.2 Multipath Fading

At each time instant, many photons with various phases and frequencies (incoherent sources) are captured by different regions of the active photodetector area. The superposition of these captured photons will lead to a photocurrent [7]. This superposition, which can be either constructive or destructive, leads to the multipath fading effect. A light beam is composed of a huge number of photons with wavelengths in a spectrum range of hundreds of nanometers. In contrast, the diameter of the active area of a typical photodetector for Li-Fi applications is in the range of 1 to 10 mm, which is more than 1,000 to 10,000 times larger than the photon wavelength, as illustrated in Figure 5.1(a) [5]. This means that the photodiode captures a huge number of uncorrelated photons, and the small-scale fading effect (constructive/destructive combing) is spatially averaged. This is why multipath fading does not pose a major issue in an IM/DD optical link.

5.2.1.3 Multipath Distortion

Despite the immunity to multipath fading, the IM/DD channel still experiences the negative effects of multipath distortion. As shown in Figure 5.1(a), the propagation time of photons via different transmission paths varies due to the different lengths of the propagation paths. This leads to a temporal spread of the transmitted optical signal and causes ISI between adjacent time-domain symbols. Consequently, the communication symbol rate or transmission bandwidth is limited. This phenomenon is known as multipath distortion, as explained in [7]. Apart from the light propagation via different paths, the optical elements, including their driving circuits, also cause ISI issues. Due to the limited carrier lifetime and capacitive nature of semiconductors, LEDs and PDs have a relatively long response time [8]. In order to improve the robustness of the received signal against thermal noise, a transimpedance amplifier (TIA) is used to amplify the small signal from an optical detector. However, there is a trade-off between the amplification gain and the operation bandwidth. Consequently, the TIA circuit also causes additional ISI in the transmission. Although these phenomenans are caused by the limited bandwidth of devices, both the multipath propagation and the usage of front-end devices cause ISI. Therefore, we categorize the ISI issues that resulted from both cases as multipath distortion.

5.2.1.4 Wavelength Selectivity

Laser-based optical sources operate in very narrow wavelength ranges (less than a few nanometers), while most of the incoherent sources, such as visible light/infrared LEDs, emit light with a much wider spectrum (tens to hundreds of nanometers) [9]. Therefore, in the wide spectrum case, the received power can be calculated by an integral over the covered spectrum range as:

$$I_{\text{photo}} = \int_{\lambda_{\min}}^{\lambda_{\max}} P_s \mathcal{S}(\lambda) H(\lambda) \mathcal{G}_{\text{of}}(\lambda) \mathcal{R}(\lambda) \mathrm{d}\lambda \tag{5.1}$$

where P_s refers to the total optical power of the source, $\mathcal{S}(\lambda)$ defines the normalized spectrum intensity at wavelength λ, $H(\lambda)$ defines the OWC channel at λ, $\mathcal{G}_{\text{of}}(\lambda)$ is the spectral transmittance of the used optical filter, and $\mathcal{R}(\lambda)$ defines the detector spectral responsivity. The normalized LED spectral intensity has the property: $\int_{\lambda_{\min}}^{\lambda_{\max}} \mathcal{S}(\lambda)d\lambda = 1$. A set of examples of these wavelength-dependent quantities demonstrated in Figure 5.1(b). In addition, the spectral surface reflectance is proportional to the OWC channel gain $H(\lambda)$. Note that the noise caused by the background light is also a wavelength-dependent quantity. Thus, an optical filter is used to minimize this background light noise [7]. This is a very effective approach in an outdoor scenario, such as FSO systems, where background radiation is significant and the optical detector is well-aligned. In a wavelength division multiplexing (WDM)-based wireless optical link, optical filters are essential components to split parallel channels [10]. However, the use of optical filters introduces geometric issues in Li-Fi and many other OWC systems when the link is misaligned. A bandpass optical filter has a spectral transmittance modeled by [10]:

$$\mathcal{G}_{\text{of}}(\lambda) = \begin{cases} G_{\text{T}} & : \lambda \in [\lambda_1(\psi), \lambda_2(\psi)] \\ 0 & : \text{otherwise} \end{cases} \tag{5.2}$$

where G_{T} defines the passband transmittance, ψ refers to the light incident angle, and $\lambda_1(\psi)$ ($\lambda_2(\psi)$) defines the left (right) passband edge of the filter, which is a function of the light incident angle. With an increase of the incident angle ψ, the passband of an optical filter shifts to a range with shorter wavelengths. This passband shift can be approximated by the following function [11]:

$$\lambda_n(\psi) = \lambda_n(0^\circ)\sqrt{1 - \sin^2\psi/\mathfrak{n}_e^2} \tag{5.3}$$

where $\mathfrak{n}_e$ refers to the effective refractive index and $\lambda_n(0^\circ)$ is the wavelength at 0° light incidence. In a Li-Fi link, it is very common that the transmitter and the receiver are not well aligned, which leads to nonzero value of ψ in many cases. Therefore, the optical filter design for Li-Fi links needs to take this issue into consideration. In practice, the transmittance value G_{T} also decreases with ψ [7]. However, special tunable optical filters are available in the market, where transmittance attenuation due to the increased incident angle can be minimized.

5.2.2 Channel Metrics

In order to characterize the channel effects described in Section 5.2.1, various channel metrics have been defined by researchers. In this section, we introduce the most popular channel metrics.

5.2.2.1 Time Domain

An OWC channel can be modeled by a linear time-invariant (LTI) filter system. Similar to other LTI systems, a channel impulse response (CIR) $h(t)$ can be used to characterize the channel in the time-domain, which is a function of the time delay t. It can be interpreted as the time-domain response of the channel system output to a unit impulse input. The magnitude of the CIR could reflect the optical path loss characteristics, while the proportion of the response at different time delays reflects the multipath distortion characteristics. In most cases, the channel has a higher response with shorter delays and lower responses with longer delays. A common way of representing the CIR is to use a vector of the detected optical power with various time delays, where a unity transmitted optical power is assumed. The drawback of this representation is that the CIR magnitude scales with the time resolution. An alternative representation is to normalize the detected optical power by time as that in [12]. Thus, the integral of the CIR becomes dimensionless and the magnitude of the CIR no longer scales with time resolution.

CIR is a two-dimensional (2-D) quantity, which might be inconvenient for visualization or processing. For example, it might be difficult to tell which of the channels has a more serious multipath distortion by comparing their CIRs. Therefore, another simplified one-dimensional (1-D) metric to reflect the multipath distortion is defined, which is known as root mean square (RMS) delay spread τ_{rms}. The RMS delay spread is calculated via the CIR by [7]:

$$\tau_{\mathrm{rms}} = \sqrt{\frac{\int_{-\infty}^{\infty}(t-\mu)^2 h^2(t)\mathrm{d}t}{\int_{-\infty}^{\infty} h^2(t)\mathrm{d}t}} \tag{5.4}$$

where μ is the mean excess delay, which can be calculated by:

$$\mu = \frac{\int_{-\infty}^{\infty} t h^2(t)\mathrm{d}t}{\int_{-\infty}^{\infty} h^2(t)\mathrm{d}t} \tag{5.5}$$

The mean excess delay reflects the weighted average delay of a channel, while the RMS delay spread reflects the significance of the channel delay spread in time.

5.2.2.2 Frequency Domain

In some cases, frequency-domain information is required. By applying the Fourier transform to the CIR, we obtain the channel frequency response (CFR) as:

$$H(f) = \int_{-\infty}^{\infty} h(t)\mathrm{e}^{-2\pi tfj}\mathrm{d}t \tag{5.6}$$

where f represents the frequency and j is the unit imaginary number with $j^2 = -1$. By calculating the absolute value of CFR $|H(f)|$, known as the magnitude response, the response sensitivity of the channel at various frequencies can be evaluated. Alternatively, the magnitude response reflects the level of multipath distortion at various frequencies. In general, the CFRs pose lowpass characteristics. Therefore, CFR helps to determine how much bandwidth can be used for a given OWC channel. Similar to the case of RMS delay spread, a 3-dB bandwidth $f_{3\mathrm{dB}}$ is commonly used as a simplified metric to reflect the multipath distortion in the frequency domain. It is defined based on the following relationship:

$$|H(f_{3\mathrm{dB}})|^2 = \frac{1}{2}|H(0)|^2 \tag{5.7}$$

where the magnitude response at DC $|H(0)|$ has been used as a reference. This is because $|H(0)|$ generally has the highest value due to the non-negativity characteristics of IM/DD OWC systems. The 3-dB bandwidth is commonly used to determine the highest possible symbol rate that a channel is able to support.

5.2.2.3 Channel DC Gain

In accordance with the use of RMS delay spread and a 3-dB bandwidth to reflect the level of multipath distortion in the time and frequency domains, channel DC gain is commonly used to reflect the level of optical path loss. It is defined as the magnitude response at DC [13]:

$$H_{\mathrm{DC}} = \left| \int_{-\infty}^{\infty} h(t)\mathrm{d}t \right| = |H(0)| \tag{5.8}$$

Normally, the achievable SNR is proportional to the square of the channel DC gain. Therefore, it is desired to have a channel DC gain as high as possible. In some OWC studies, researchers attempted to mimic the path loss model used in RF wireless communications and proposed the path loss models in OWC as a function of the horizontal offset [14]. In this case, a variant definition of the channel DC gain on a logarithmic scale is given as:

$$\mathrm{PL} = -10 \log_{10} H_{\mathrm{DC}} \tag{5.9}$$

However, the OWC channel is a more complex function of many other factors, such as optical source/detector orientation, and therefore the fitted path loss models are hard to generalize and are valid only for some specific scenarios.

5.2.3 Channel Decomposition

As shown in Figure 5.1(a), an OWC channel includes a process of wireless propagation in free space and other conversion procedures via LED, PD, and other driving/amplification circuits. The former depends on various geometric factors, while the latter depends on the characteristics of the used device and circuit design. Due to this distinction in nature, we can decompose the OWC channel into an optical wireless channel (free-space propagation) and a front-end channel (LED, PD, circuits). Therefore, the OWC CIR can be rewritten as:

$$h(t) = h_{\rm fe}(t) * h_{\rm ow}(t) \tag{5.10}$$

where $h_{\rm fe}(t)$ is the front-end CIR, $h_{\rm ow}(t)$ is the optical wireless CIR, and $*$ is the convolution operator. In the frequency domain, a similar relationship can be found as:

$$H(f) = H_{\rm fe}(f)H_{\rm ow}(f) \tag{5.11}$$

where $H_{\rm fe}(f)$ and $H_{\rm ow}(f)$ are the Fourier transforms of $h_{\rm fe}(t)$ and $h_{\rm ow}(t)$, respectively. In the following sections, we introduce each of these separately.

5.3 Front-End Channel

As described in Section 5.2.3, the front-end channel is determined by the characteristics of a number of devices connected in series (LED, PD, driving circuit, amplification circuit). The design and models of these devices are diverse and complex, which makes it challenging to find a generic and detailed model for a front-end channel. Consequently, often a flat front-end channel within the modulation bandwidth is assumed by most of the researchers for simplicity [13]. However, many experimental studies have found that the front-end channel is the major limiting factor to the signal modulation bandwidth [15]. Thus, it is inaccurate to omit the effects of the front-end channel. Since the major influence of the front-end channel is on the bandwidth limitation, several simplified analytical models have been proposed to approximate the lowpass characteristics introduced by these devices. Due to the bottleneck effect, the lowpass characteristics are dominated by the device with the lowest bandwidth. In the case of Li-Fi links, LEDs pose the narrowest bandwidth. Therefore, most of these analytical models depend on the parameters of the used LED.

5.3.1 Exponential Function Approximation

The first analytical model considers using an exponential function to approximate the magnitude response of front-end channel in the frequency domain [15]:

$$|H_{\rm fe}(f)| = \exp\left(-\frac{f}{f_{\rm fe}}\right) \tag{5.12}$$

where $f_{\rm fe}$ is a parameter to control the slope of the lowpass sensitivity, which is set to fit the corresponding experimental data. An example response with $f_{\rm fe}$ = 30 MHz is demonstrated in Figure 5.2. The response at DC is 0 dB, as often there is no power loss in the front-end channel. It can be observed that the magnitude response decreases exponentially with an increase in frequency. This approximation has been found to agree with experimental data in several

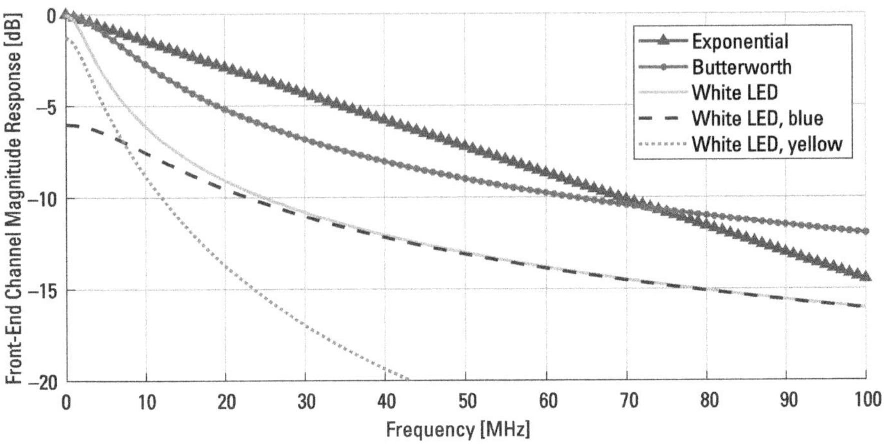

Figure 5.2 Magnitude response of the front-end channel using various analytical model.

studies [16]. The drawback of this model is that it does not have a corresponding TD model.

5.3.2 First-Order Butterworth Filter Model

Since the exponential function approximation has no proper physical explanation, another front-end channel using a circuit analysis approach has been considered. Due to the capacitive nature of the LEDs, the response of the front-end channel is similar to an RC-circuit model [17]. In this model, the impulse response of the LED is defined as an exponential decay function:

$$h_{\text{led}}(t) = \frac{1}{\tau_{\text{s}}} \mathrm{u}(t) \exp\left(-\frac{t}{\tau_{\text{s}}}\right) \tag{5.13}$$

where τ_{s} is the rise time of the optical source and $\mathrm{u}(t)$ is the unit step function. By applying Fourier transform to (5.13), the frequency response of the front-end channel can be found as the following first-order Butterworth function:

$$H_{\text{fe}}(f) \approx H_{\text{led}}(f) = \frac{1}{1 + j2\pi f \tau_{\text{s}}} \tag{5.14}$$

This model has a better physical interpretation as LEDs are capacitive devices. An example magnitude response with $\tau_{\text{s}} = 25$ ns is presented in Figure 5.2, which shows that the magnitude response also decreases with the frequency. However, the decreasing sensitivity drops when the frequency increases.

5.3.3 White LED Model

White LEDs have been widely used for SSL for years. Therefore, they are also a promising candidate for Li-Fi systems. Light radiation from a white LED is composed of a blue light from the LED itself and a yellow light from the phosphor cap excited by part of the blue light [18]. Based on the first-order Butterworth filter model, an extended model is proposed for the front-end channel with a white LED [19]. The front-end CIR can be described as:

$$h_{\mathrm{led,w}}(t) = G_{\mathrm{b}} h_{\mathrm{led,b}}(t) + G_{\mathrm{y}} h_{\mathrm{pho,y}}(t) * h_{\mathrm{led,b}}(t) \quad (5.15)$$

where G_{b} and G_{y} are the power gain factor of the blue LED and yellow phosphor material, which fulfills $G_{\mathrm{b}} + G_{\mathrm{y}} = 1$. The notations $h_{\mathrm{led,b}}(t)$ and $h_{\mathrm{pho,y}}(t)$ are the CIR of the blue LED and yellow phosphor material, respectively, which also follow the exponential decay model in (5.13), but with different rise time parameters $\tau_{\mathrm{s,b}}$ and $\tau_{\mathrm{s,y}}$. Note that the CIR of the yellow light component also depends on the CIR of the blue LED as the yellow light originates from the blue LED. By applying the Fourier transform to (5.15), the corresponding frequency response can be found as:

$$H_{\mathrm{fe}}(f) \approx H_{\mathrm{led,w}}(f) = \frac{1}{1 + j2\pi f \tau_{\mathrm{s,b}}} \left(G_{\mathrm{b}} + \frac{G_{\mathrm{y}}}{1 + j2\pi f \tau_{\mathrm{s,y}}} \right) \quad (5.16)$$

An example response of the white LED model with $G_{\mathrm{b}} = 0.25$, $G_{\mathrm{y}} = 0.75$, $\tau_{\mathrm{s,b}} = 16$ ns, and $\tau_{\mathrm{s,y}} = 62$ ns is shown in Figure 5.2. Note that the magnitude response of the blue and yellow components is also plotted in the figure. Normally, phosphor material has a much slower response time compared to the blue LED. Consequently, it can be observed that the magnitude response of a white LED drops quickly at the low-frequency region and the dropping speed decreases with further increases in frequency.

5.4 Optical Wireless Channel

The optical wireless channel corresponds to the optical signal propagation from the optical source to the optical detector in indoor free space. Similar to RF wireless communications, the optical wireless channel can be formed by multiple propagation paths [1]. It is commonly decomposed into a LOS component and a set of NLOS components, as illustrated in Figure 5.3(a).

5.4.1 LOS Channel

The LOS component corresponds to the channel path from the LED directly to the optical detector without any interaction with other objects in the indoor environment. Consider a single point source with a radiant intensity of I_{s}, which defines the amount of emitted power per solid angle in a specified direction. The Lambertian intensity model is the most widely used LED radiation pattern due to its simplicity and its support of a large source divergence angle. The normalized

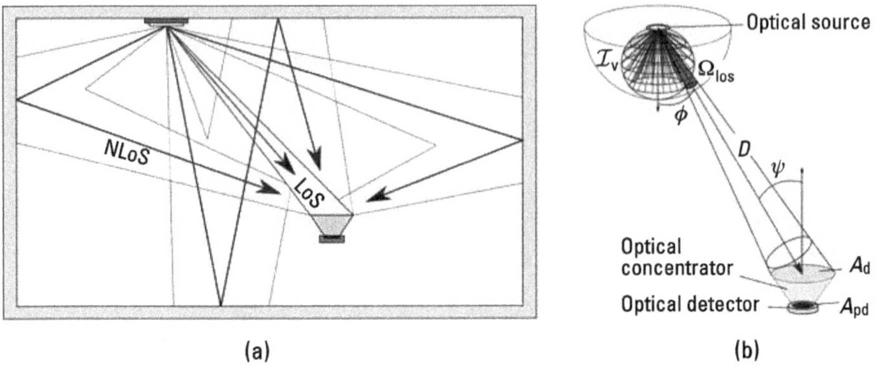

Figure 5.3 (a) Decomposition of optical wireless channel. (b) Geometry of LOS component of optical wireless channel.

Lambertian radiant intensity (1-W optical power) is defined as [12]:

$$\mathcal{J}_s(\phi, m) = \frac{m+1}{2\pi} \cos^m \phi \tag{5.17}$$

where ϕ is the radiant angle. It is defined by the angle between the Lambertian source beam direction and LOS path line segment, as shown in Figure 5.3(b). To characterize the divergence angle of an optical source, a half-power semiangle $\phi_{1/2}$ is defined as the radiant angle at which the intensity is half of that in the beam direction ($\phi = 0°$). The parameter m is known as the Lambertian emission order, which is linked to the half-power semiangle by $m = -1/\log_2(\cos\phi_{1/2})$. On the receiver side, an optical detector, such as a PD, is used to convert the incident optical power to a photocurrent. In order to capture more optical power, the active area of the detector A_{pd} should be as large as possible. However, large-size PDs are expensive and lead to excessive lowpass characteristics on the front-end channel. Consequently, a compromised solution is to use an optical concentrator to guide the light captured by a larger detection area A_d to a PD with a smaller area $A_{pd} < A_d$. The disadvantage of this approach is that a high gain concentrator normally has a narrower FOV and the optical concentrator itself takes some geometric space. In the case of smaller mobile devices, there will be restrictions on usable optical concentrators. The characteristics of the optical concentrator are device-specific. An approximation to the maximum possible concentrator gain G_c has been widely used in the literature [7]:

$$G_c = \begin{cases} \frac{n^2}{\sin^2 \psi_{FOV}} & : \psi < \psi_{FOV} \\ 0 & : \psi \geq \psi_{FOV} \end{cases} \tag{5.18}$$

where n is the refractive index of the concentrator and ψ_{FOV} is the FOV angle. The detection area A_d corresponds to a solid angle with respect to the optical source as shown in Figure 5.3(b). In other words, the light emission within this solid angle will be incident to the detection area. Assume that the Euclidean distance between the optical source and the center of the detection area is D. The small solid angle can be accurately approximated by

$$\Omega = \frac{A_{\rm d} \cos \psi}{D^2} \tag{5.19}$$

as long as D^2 is significantly greater than $A_{\rm d} \cos \psi$, which is typical in cases with indoor OWC links. Finally, the channel DC gain of the optical wireless channel via the LOS path can be calculated by the product of (5.17) and (5.19) as:

$$H_{\rm DC,LOS} = \mathcal{J}_{\rm s}(\phi, m)\Omega = \frac{(m+1)A_{\rm d}}{2\pi D^2} G_{\rm v} \cos^m \phi \cos \psi \tag{5.20}$$

where $G_{\rm v}$ is a visibility factor. It equals 1 when ϕ and ψ are less than 90°. Otherwise, it equals 0. The corresponding CIR can be written as:

$$h_{\rm ow,LOS}(t) = H_{\rm DC,LOS}\delta\left(t - \frac{D}{c}\right) \tag{5.21}$$

where $\delta(u)$ is an unit impulse function and $c = 3.8 \times 10^8$ m/s is the speed of light.

5.4.2 NLOS Channel

In the LOS optical wireless channel, the optical power is transferred from an LED directly to an optical detector. In the case of the NLOS optical wireless channel, the optical power is transferred to the reflective surfaces in the indoor environment. Then part of the power is reflected and captured by the detector. For some of the NLOS components, the optical signal may experience multiple reflections before it reaches the detector. From the communication perspective, the NLOS channel components always have a slightly longer delay, which results in ISI. This introduces extra signal distortion, which increases the chance of signal detection errors. From the frequency-domain point of view, NLOS components may lead to a lowpass channel, which may limit the modulation bandwidth of the Li-Fi link. In conventional RF wireless communication, wireless channels are formed by the superposition of a few or tens of reflected paths via specular reflections. In the case of OWC, due to the very short wavelength, most of the indoor surfaces are uneven relative to light photons. This leads to diffuse reflection in most cases. Consequently, the channel is composed of a much greater number of reflected paths. This makes the computational complexity of the NLOS channel extremely high. Because of this high complexity, it is challenging to capture the characteristics of the NLOS channel with simple models. Therefore, various ray-tracing simulation methods have been proposed to evaluate the NLOS channel.

There are two major categories of NLOS simulation approaches: deterministic simulation and Monte Carlo simulation, as categorized in [20]. We will cover these approaches in Sections 5.5 and 5.6. In addition, some studies aim at capturing the NLOS channel with simpler analytical models such as the model introduced in [21]. Some of these methods will be introduced in Section 5.7. There are also reference models for NLOS channels reported in a technical report [22], which avoids complicated implementations and computations of NLOS channels.

5.4.3 K-Factor

Later in this chapter, it will be shown that the power ratio between the LOS component and the NLOS components plays an important role in the characteristics of optical wireless channels. Therefore, a generic metric is used to quantify the LOS-to-NLOS power ratio, which is known as the Rician K-factor [21]:

$$F_{\mathrm{K}} = \frac{H^2_{\mathrm{DC,LOS}}}{H^2_{\mathrm{DC,NLOS}}} \tag{5.22}$$

where $H_{\mathrm{DC,NLOS}}$ is the channel DC gain of the NLOS components. Note that this ratio refers to the electrical power ratio and the optical power is proportional to the amplitude of the signal. It is intuitive that when the LOS component dominates the optical wireless channel, F_{K} approaches infinity. When NLOS components dominate the optical wireless channel, the F_{K} approaches zero.

5.5 Deterministic NLOS Channel Simulation Approaches

In deterministic simulation approaches, reflecting surfaces in the indoor environment are divided into small square elements with an edge length of ΔL, as shown in Figure 5.4. Each small element is treated as a detector and the power transfer from the LED to the small element is calculated using the LOS calculation method. Then each element is treated as an optical source and the power transfer to the detector and the remaining elements are calculated. This is the basic mechanism of the deterministic approach. With a room size of $L_{\mathrm{w}} \times L_{\mathrm{l}} \times L_{\mathrm{h}}$, the number of reflecting element divisions on each dimension can be calculated as: $N_{\mathrm{w}} = L_{\mathrm{w}}/\Delta L$, $N_{\mathrm{l}} = L_{\mathrm{l}}/\Delta L$, and $N_{\mathrm{h}} = L_{\mathrm{h}}/\Delta L$. Therefore, the total number of elements can be calculated by [12]:

$$N_{\mathrm{e}} = 2(N_{\mathrm{w}}N_{\mathrm{l}} + N_{\mathrm{l}}N_{\mathrm{h}} + N_{\mathrm{h}}N_{\mathrm{w}}) \tag{5.23}$$

The first step of a deterministic approach is to evaluate the interactions between two reflecting elements and the interactions between each element and the LED/PD. The interaction data includes the LOS channel DC gain and the

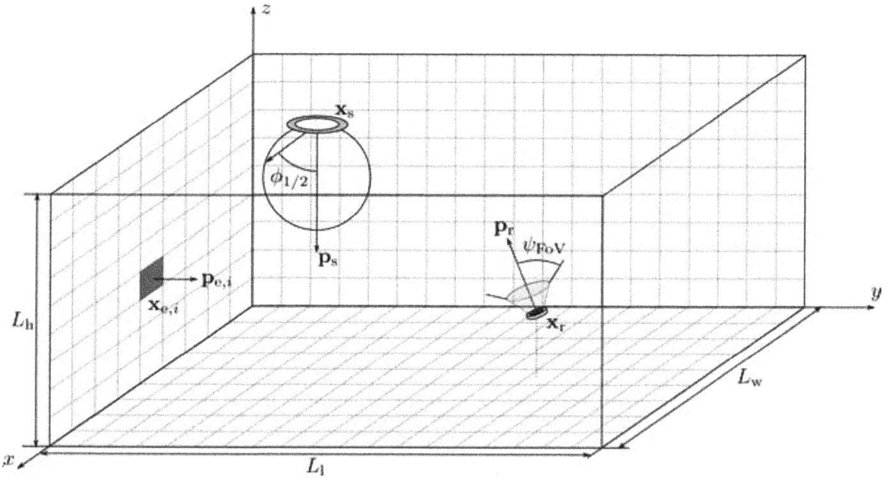

Figure 5.4 Geometry of deterministic NLOS channel simulation approaches.

propagation delay. First, we define the functions calculating these two quantities. We define a transmitting element $\mathfrak{s} = \{\mathbf{x}_\mathfrak{s}, \mathbf{p}_\mathfrak{s}, m_\mathfrak{s}\}$, where $\mathbf{x}_\mathfrak{s}$, $\mathbf{p}_\mathfrak{s}$, $m_\mathfrak{s}$ represent the position, orientation, and Lambertian emission order of the source $\mathfrak{s}$, respectively. In addition, a receiving element $\mathfrak{r} = \{\mathbf{x}_\mathfrak{r}, \mathbf{p}_\mathfrak{r}, A_\mathfrak{r}\}$ is defined, where $\mathbf{x}_\mathfrak{r}$, $\mathbf{p}_\mathfrak{r}$, $A_\mathfrak{r}$ refer to the position, orientation, and detection area of the receiving element $\mathfrak{r}$. Note that $\mathbf{x}_\mathfrak{s}, \mathbf{p}_\mathfrak{s}, \mathbf{x}_\mathfrak{r}, \mathbf{p}_\mathfrak{r} \in \mathbb{R}^{3\times 1}$ are defined in a three-dimensional Cartesian coordinate system. With these parameters, the following quantities can be obtained [12]:

$$D = ||\mathbf{x}_\mathfrak{r} - \mathbf{x}_\mathfrak{s}|| \tag{5.24}$$

$$\cos\phi = \mathbf{p}_\mathfrak{s}^{\mathrm{T}}(\mathbf{x}_\mathfrak{r} - \mathbf{x}_\mathfrak{s})/D \tag{5.25}$$

$$\cos\psi = \mathbf{p}_\mathfrak{r}^{\mathrm{T}}(\mathbf{n}_\mathfrak{s} - \mathbf{n}_\mathfrak{r})/D \tag{5.26}$$

where $||\cdot||$ refers to the Euclidean norm and $\{\cdot\}^{\mathrm{T}}$ is the transpose operation. By inserting (5.24), (5.25), and (5.26) into (5.20) and letting $m = m_\mathfrak{s}$ and $A_\mathrm{d} = A_\mathfrak{r}$, the LOS channel DC gain between $\mathfrak{s}$ and $\mathfrak{r}$ can be calculated by:

$$G(\mathfrak{s}, \mathfrak{r}) = \frac{(m_\mathfrak{s} + 1)A_\mathfrak{r} G_\mathrm{v}}{2\pi ||\mathbf{x}_\mathfrak{r} - \mathbf{x}_\mathfrak{s}||^4} \left(\mathbf{p}_\mathfrak{s}^{\mathrm{T}}(\mathbf{x}_\mathfrak{r} - \mathbf{x}_\mathfrak{s})\right)^{m_\mathfrak{s}} \left(\mathbf{p}_\mathfrak{r}^{\mathrm{T}}(\mathbf{x}_\mathfrak{s} - \mathbf{x}_\mathfrak{r})\right) \tag{5.27}$$

By using the relationship $\tau = D/c$, the propagation delay between $\mathfrak{s}$ and $\mathfrak{r}$ can be calculated by:

$$\tau(\mathfrak{s}, \mathfrak{r}) = ||\mathbf{x}_\mathfrak{r} - \mathbf{x}_\mathfrak{s}||/c \tag{5.28}$$

Next, we use (5.27) and (5.28) to calculate the LOS channel DC gain and propagation delay between each pair of elements and save the calculation results in a number of data matrices. These data matrices will be used repeatedly in the deterministic algorithms to avoid redundant computation. In the following calculations, the case with N_{tx} transmitters and N_{rx} receivers are considered.

- *Transmitter-to-element data matrices*: In this case, the transmitting element is the n_{tx}th transmitter, which is defined as: $\mathcal{S}_{n_{\text{tx}}} = \{\mathbf{x}_{\text{s},n_{\text{tx}}}, \mathbf{p}_{\text{s},n_{\text{tx}}}, m_{n_{\text{tx}}}\}$ with $n_{\text{tx}} = 1, 2, \cdots, N_{\text{tx}}$. The receiving element is the ith reflecting element, which is defined as: $\mathcal{E}_i = \{\mathbf{x}_{\text{e},i}, \mathbf{p}_{\text{e},i}, \Delta L^2\}$ with $i = 1, 2, \cdots, N_{\text{e}}$. Thus, the transmitter-to-element LOS channel DC gain data matrix can be calculated as:

$$\mathbf{G}_{\text{tx-e}} = \begin{bmatrix} G(\mathcal{S}_1, \mathcal{E}_1) & \cdots & G(\mathcal{S}_{N_{\text{tx}}}, \mathcal{E}_1) \\ \vdots & \ddots & \vdots \\ G(\mathcal{S}_1, \mathcal{E}_{N_{\text{e}}}) & \cdots & G(\mathcal{S}_{N_{\text{tx}}}, \mathcal{E}_{N_{\text{e}}}) \end{bmatrix} \in \mathbb{R}^{N_{\text{e}} \times N_{\text{tx}}} \tag{5.29}$$

and the corresponding propagation delay data matrix can be concluded as:

$$\mathbf{D}_{\text{tx-e}} = \begin{bmatrix} \tau(\mathcal{S}_1, \mathcal{E}_1) & \cdots & \tau(\mathcal{S}_{N_{\text{tx}}}, \mathcal{E}_1) \\ \vdots & \ddots & \vdots \\ \tau(\mathcal{S}_1, \mathcal{E}_{N_{\text{e}}}) & \cdots & \tau(\mathcal{S}_{N_{\text{tx}}}, \mathcal{E}_{N_{\text{e}}}) \end{bmatrix} \in \mathbb{R}^{N_{\text{e}} \times N_{\text{tx}}} \tag{5.30}$$

- *Element-to-element data matrices*: In this case, the transmitting element is the i_1th element, which is defined as: $\mathcal{E}_{i_1} = \{\mathbf{x}_{\text{e},i_1}, \mathbf{p}_{\text{e},i_1}, 1\}$. The receiving element is the i_2th reflecting element, which is defined as: $\mathcal{E}_{i_2} = \{\mathbf{x}_{\text{e},i_2}, \mathbf{p}_{\text{e},i_2}, \Delta L^2\}$. Thus, the element-to-element LOS channel DC gain data matrix can be calculated as:

$$\mathbf{G}_{\text{e-e}} = \begin{bmatrix} G(\mathcal{E}_1, \mathcal{E}_1) & \cdots & G(\mathcal{E}_{N_{\text{e}}}, \mathcal{E}_1) \\ \vdots & \ddots & \vdots \\ G(\mathcal{E}_1, \mathcal{E}_{N_{\text{e}}}) & \cdots & G(\mathcal{E}_{N_{\text{e}}}, \mathcal{E}_{N_{\text{e}}}) \end{bmatrix} \in \mathbb{R}^{N_{\text{e}} \times N_{\text{e}}} \tag{5.31}$$

and the corresponding propagation delay data matrix can be calculated as:

$$\mathbf{D}_{\text{e-e}} = \begin{bmatrix} \tau(\mathcal{E}_1, \mathcal{E}_1) & \cdots & \tau(\mathcal{E}_{N_{\text{e}}}, \mathcal{E}_1) \\ \vdots & \ddots & \vdots \\ \tau(\mathcal{E}_1, \mathcal{E}_{N_{\text{e}}}) & \cdots & \tau(\mathcal{E}_{N_{\text{e}}}, \mathcal{E}_{N_{\text{e}}}) \end{bmatrix} \in \mathbb{R}^{N_{\text{e}} \times N_{\text{e}}} \tag{5.32}$$

- *Element-to-receiver data matrices*: In this case, the transmitting element is the ith element, which is defined as: $\mathcal{E}_i = \{\mathbf{x}_{\text{e},i}, \mathbf{p}_{\text{e},i}, 1\}$. The receiving element is the i_2th reflecting element, which is defined as: $\mathcal{R}_{n_{\text{rx}}} = \{\mathbf{x}_{\text{r},n_{\text{rx}}}, \mathbf{p}_{\text{r},n_{\text{rx}}}, A_{\text{d},n_{\text{rx}}}\}$. Thus, the element-to-receiver LOS channel

DC gain data matrix can be calculated as:

$$\mathbf{G}_{\text{e-rx}} = \begin{bmatrix} G(\mathcal{E}_1, \mathcal{R}_1) & \cdots & G(\mathcal{E}_{N_\text{e}}, \mathcal{R}_1) \\ \vdots & \ddots & \vdots \\ G(\mathcal{E}_1, \mathcal{R}_{N_\text{rx}}) & \cdots & G(\mathcal{E}_{N_\text{e}}, \mathcal{R}_{N_\text{rx}}) \end{bmatrix} \in \mathbb{R}^{N_\text{rx} \times N_\text{e}} \quad (5.33)$$

and the corresponding propagation delay data matrix can be calculated as:

$$\mathbf{D}_{\text{e-rx}} = \begin{bmatrix} \tau(\mathcal{E}_1, \mathcal{R}_1) & \cdots & \tau(\mathcal{E}_{N_\text{e}}, \mathcal{R}_1) \\ \vdots & \ddots & \vdots \\ \tau(\mathcal{E}_1, \mathcal{R}_{N_\text{rx}}) & \cdots & \tau(\mathcal{E}_{N_\text{e}}, \mathcal{R}_{N_\text{rx}}) \end{bmatrix} \in \mathbb{R}^{N_\text{rx} \times N_\text{e}} \quad (5.34)$$

By using the above data matrices, we will present a number of deterministic NLOS simulation algorithms in the following sections.

5.5.1 Recursive Algorithm

The recursive algorithm was the first algorithm that was proposed to simulate the indoor NLOS channel in [12], where the CIR is decomposed based on the order of reflections (number of experienced reflections) as:

$$h(t; \mathcal{S}_{n_\text{tx}}, \mathcal{R}_{n_\text{rx}}) = \sum_{k=1}^{\infty} h^k(t; \mathcal{S}_{n_\text{tx}}, \mathcal{R}_{n_\text{rx}}) \quad (5.35)$$

where $h^k(t; \mathfrak{s}, \mathfrak{r})$ refers to the kth-order impulse response between the transmitting element $\mathfrak{s}$ and the receiving element $\mathfrak{r}$. It has been found that the kth order impulse response can be calculated using the $k-1$th order impulse response as:

$$h^k(t; \mathcal{S}_{n_\text{tx}}, \mathcal{R}_{n_\text{rx}}) = \sum_{i=1}^{N_\text{e}} \rho_i h^0(t; \mathcal{S}_{n_\text{tx}}, \mathcal{E}_i) * h^{k-1}(t; \mathcal{E}_i, \mathcal{R}_{n_\text{rx}}) \quad (5.36)$$

where $h^0(t; \mathcal{S}_{n_\text{tx}}, \mathcal{E}_i)$ is LOS CIR, which can be calculated using the data matrices $\mathbf{G}_{\text{tx-e}}$ and $\mathbf{D}_{\text{tx-e}}$ as

$$h^0(t; \mathcal{S}_{n_\text{tx}}, \mathcal{E}_i) = [\mathbf{G}_{\text{tx-e}}]_{i,n_\text{tx}} \, \delta\left(t - [\mathbf{D}_{\text{tx-e}}]_{i,n_\text{tx}}\right) \quad (5.37)$$

where $[\mathbf{A}]_{i_1,i_2}$ refers to the element of matrix $\mathbf{A}$ in the i_1th row and i_2th column. In addition, $h^{k-1}(t; \mathcal{E}_i, \mathcal{R}_{n_\text{rx}})$ can be evaluated by the following recursive equation:

$$h^k(t; \mathcal{E}_{i_1}, \mathcal{R}_{n_\text{rx}}) = \sum_{i_2=1}^{N_\text{e}} \rho_{i_2} h^0(t; \mathcal{E}_{i_1}, \mathcal{E}_{i_2}) * h^{k-1}(t; \mathcal{E}_{i_2}, \mathcal{R}_{n_\text{rx}}) \quad (5.38)$$

and the impulse response $h^{k-1}(t;\mathcal{E}_{i_k},\mathcal{R}_{n_{\text{rx}}})$ can be expanded into the following form:

$$h^{k-1}(t;\mathcal{E}_{i_k},\mathcal{R}_{n_{\text{rx}}}) = \sum_{i_{k-1}=1}^{N_e} \rho_{i_{k-1}} h^0(t;\mathcal{E}_{i_k},\mathcal{E}_{i_{k-1}}) * \left(\sum_{i_{k-2}=1}^{N_e} h^0(t;\mathcal{E}_{i_{k-1}},\mathcal{E}_{i_{k-2}}) \right.$$
$$\left. \times \rho_{i_{k-2}} * \cdots * \left(\sum_{i_1=1}^{N_e} \rho_{i_1} h^0(t;\mathcal{E}_{i_2},\mathcal{E}_{i_1}) * h^0(t;\mathcal{E}_{i_1},\mathcal{R}_{n_{\text{rx}}}) \right) \cdots \right) \tag{5.39}$$

where there are only 0th-order CIRs in the expression. Similar to the calculation in (5.37), the LOS CIRs $h^0(t;\mathcal{E}_{i_1},\mathcal{E}_{i_2})$ and $h^0(t;\mathcal{E}_i,\mathcal{R}_{n_{\text{rx}}})$ can also be calculated via the data matrices as:

$$h^0(t;\mathcal{E}_{i_1},\mathcal{E}_{i_2}) = [\mathbf{G}_{\text{e-e}}]_{i_2,i_1}\, \delta\left(t - [\mathbf{D}_{\text{e-e}}]_{i_2,i_1}\right) \tag{5.40}$$

$$h^0(t;\mathcal{E}_i,\mathcal{R}_{n_{\text{rx}}}) = [\mathbf{G}_{\text{e-rx}}]_{n_{\text{rx}},i}\, \delta\left(t - [\mathbf{D}_{\text{e-rx}}]_{n_{\text{rx}},i}\right) \tag{5.41}$$

Combining (5.35), (5.39), (5.36), (5.38), and (5.39), the NLOS CIR can be calculated using a recursive algorithm, which is concluded in Algorithm 5.1. Due to the discrete nature of the calculation, a discrete vector is defined to store the CIR result as:

$$\mathbf{h} = [h(\Delta t), h(2\Delta t), \cdots, h(N_t \Delta t)] \tag{5.42}$$

where Δt is a discrete time bin. To ensure the smoothness of the resultant CIR, a minimum time bin width of $\Delta t = \Delta / c$ should be guaranteed. The vector $\mathbf{h}$ is initialized with all zeros. During the execution of the algorithm, the power of any CIR is accumulated to the element with the closest propagation delay. For example, when including a new CIR $a\delta(t - \hat{\tau})$ to the vector $\mathbf{h}$, the time index of the corresponding slot is found by $\hat{n}_t = \arg\min_{n_t} |\hat{\tau} - n_t \Delta t|$ and the vector is updated by $h(\hat{n}_t \Delta t) = h(\hat{n}_t \Delta t) + a$. The computation complexity of the kth impulse response is proportional to N_e^k. Therefore, the computation time increases significantly with the order of reflections. It is impossible to reach $k = \infty$. Therefore, the maximum number of reflection order K is defined. The value of K should be small enough to allow the simulation to be completed within an affordable time. Practically, NLOS channels of a room of moderate size with up to the third-order reflection can be simulated.

5.5.2 Iterative Algorithm

The main drawback of the recursive algorithm is its exponential increase in simulation time with the order of reflections. Because of this, it is desirable to consider alternative deterministic approaches. The iterative algorithm is one of the improved methods presented in [23]. Similar to the recursive algorithm, the

Algorithm 5.1 Recursive Algorithm

for $n_{\text{rx}} = 1, 2, \cdots, N_{\text{rx}}$ **do**
  for $k = 1, 2, \cdots, K$ **do**
    Define an index vector $\mathbf{i} = [1, 1, \cdots, 1]^{\mathrm{T}} \in \mathbb{Z}^{(k-1)\times 1}$
    while $\mathbf{i}_{k-1} \leq N_{\mathrm{e}}$ and $k \geq 2$ **do**
      for $l = 1, 2, \cdots, k-1$ **do**
        if $l = 1$ or ($l \geq 2$ and $\mathbf{i}_l > N_{\mathrm{e}}$) **then**
          Calculate $h^l(t; \mathcal{E}_{\mathbf{i}_l}, \mathcal{R}_{n_{\text{rx}}}) =$
          $\sum_{i=1}^{N_{\mathrm{e}}} \rho_i h^0(t; \mathcal{E}_{\mathbf{i}_l}, \mathcal{E}_i) * h^{l-1}(t; \mathcal{E}_i, \mathcal{R}_{n_{\text{rx}}})$
          $\mathbf{i}_l = \mathbf{i}_l + 1$
          if $l \geq 2$ **then**
            $\mathbf{i}_{l-1} = 1$
          end
        end
      end
    end
    for $n_{\text{tx}} = 1, 2, \cdots, N_{\text{tx}}$ **do**
      Calculate $h^k(t; \mathcal{S}_{n_{\text{tx}}}, \mathcal{R}_{n_{\text{rx}}}) =$
      $\sum_{i=1}^{N_{\mathrm{e}}} \rho_i h^0(t; \mathcal{S}_{n_{\text{tx}}}, \mathcal{E}_i) * h^{k-1}(t; \mathcal{E}_i, \mathcal{R}_{n_{\text{rx}}})$
    end
  end
  for $n_{\text{tx}} = 1, 2, \cdots, N_{\text{tx}}$ **do**
    Calculate $h_{n_{\text{tx}}, n_{\text{rx}}}(t) = \sum_{k=1}^{K} h^k(t; \mathcal{S}_{n_{\text{tx}}}, \mathcal{R}_{n_{\text{rx}}})$
  end
end

NLOS CIR is decomposed based on the reflection orders as shown in (5.35), but a different decomposition of the higher order impulse response from the receiver side is used:

$$h^k(t; \mathcal{S}_{n_{\text{tx}}}, \mathcal{R}_{n_{\text{rx}}}) = \sum_{i=1}^{N_{\mathrm{e}}} \rho_i h^{k-1}(t; \mathcal{S}_{n_{\text{tx}}}, \mathcal{E}_i) * h^0(t; \mathcal{E}_i, \mathcal{R}_{n_{\text{rx}}}) \tag{5.43}$$

where $h^0(t; \mathcal{E}_i, \mathcal{R}_{\text{rx}})$ can be calculated using (5.41) and $h^{k-1}(t; \mathcal{S}_{n_{\text{tx}}}, \mathcal{E}_i)$ can be calculated using a similar recursive relationship:

$$h^k(t; \mathcal{S}_{n_{\text{tx}}}, \mathcal{E}_{i_1}) = \sum_{i_2=1}^{N_{\mathrm{e}}} \rho_{i_2} h^{k-1}(t; \mathcal{S}_{n_{\text{tx}}}, \mathcal{E}_{i_2}) * h^0(t; \mathcal{E}_{i_2}, \mathcal{E}_{i_1}) \tag{5.44}$$

Algorithm 5.2 Iterative Algorithm

for $n_{tx} = 1, 2, \cdots, N_{tx}$ **do**
  for $k = 1, 2, \cdots, K$ **do**
    if $k \geq 2$ **then**
      for $i_1 = 1, 2, \cdots, N_e$ **do**
        Calculate $h^{k-1}(t; \mathcal{S}_{n_{tx}}, \mathcal{E}_{i_1}) =$
        $\sum_{i_2=1}^{N_e} \rho_{i_2} h^{k-2}(t; \mathcal{S}_{n_{tx}}, \mathcal{E}_{i_2}) * h^0(t; \mathcal{E}_{i_2}, \mathcal{E}_{i_1})$
      end
      Store the values of $h^{k-1}(t; \mathcal{S}_{n_{tx}}, \mathcal{E}_{i_1})$ for calculation of the case with $k + 1$
    end
    for $n_{rx} = 1, 2, \cdots, N_{rx}$ **do**
      Calculate $h^k(t; \mathcal{S}_{n_{tx}}, \mathcal{R}_{n_{rx}}) =$
      $\sum_{i_1=1}^{N_e} \rho_{i_1} h^{k-1}(t; \mathcal{S}_{n_{tx}}, \mathcal{E}_{i_1}) * h^0(t; \mathcal{E}_{i_1}, \mathcal{R}_{n_{rx}})$
    end
  end
  for $n_{rx} = 1, 2, \cdots, N_{rx}$ **do**
    Calculate $h_{n_{tx}, n_{rx}}(t) = \sum_{k=1}^{K} h^k(t; \mathcal{S}_{n_{tx}}, \mathcal{R}_{n_{rx}})$
  end
end

In the recursive algorithm, the impulse response calculation of different reflection orders are independent from each other. It has been noticed that during the calculation of the kth-order impulse response $h^k(t; \mathcal{S}_{n_{tx}}, \mathcal{R}_{n_{rx}})$, the value of $h^{k-1}(t; \mathcal{S}_{n_{tx}}, \mathcal{E}_i)$ has to be evaluated. When calculating the values of $h^{k-1}(t; \mathcal{S}_{n_{tx}}, \mathcal{E}_i)$, it requires the values of $h^{k-2}(t; \mathcal{S}_{n_{tx}}, \mathcal{E}_i)$, which have already been evaluated during the calculation of the $(k-1)$th order impulse response $h^{k-1}(t; \mathcal{S}_{n_{tx}}, \mathcal{R}_{n_{rx}})$. Therefore, it is possible to reuse the calculation results of the case with $k-1$ to calculate the result of case with k, and the recursive algorithm can be modified into an iterative process as presented in Algorithm 5.2. The iterative algorithm provides the exact same CIR result that can be obtained from the recursive algorithm, but the computation complexity is significantly reduced. The computation time of the impulse response of the kth-order reflection is proportional to N_e^2. In addition, researchers have introduced other types of reflecting surfaces, such as furniture, in addition to the internal surfaces of the room (floor, ceiling, and walls).

5.5.3 DUSTIN Algorithm

The recursive and iterative algorithms decompose the NLOS channel with different numbers of reflection orders. In order to calculate the CIR with a high order of reflections, the computational complexity increases significantly. A novel approach, called the DUSTIN algorithm, has been proposed to avoid such a disadvantage in [24]. In the DUSTIN algorithm, the NLOS channel is no longer decomposed based on the order of reflections. Instead, the following workspace matrix is defined:

$$\mathbf{Q} = \left[\mathbf{q}_1, \mathbf{q}_2, \cdots, \mathbf{q}_{N_\mathrm{e}}\right] \in \mathbb{R}^{N_\mathrm{t} \times N_\mathrm{e}} \tag{5.45}$$

The ith column of $\mathbf{Q}$ corresponds to the ith reflecting element and is denoted as a vector $\mathbf{q}_i \in \mathbb{R}^{N_\mathrm{t} \times 1}$ for $i = 1, 2, \cdots, N_\mathrm{e}$. The n_tth row of $\mathbf{q}$ corresponds to the n_tth time bin with a delay of $\left(n_\mathrm{t} - \frac{1}{2}\right) \Delta t$. The total number of time bin N_t should be large enough to accommodate the majority of the channel delay. Typically, a value to allow a maximum delay of 100 ns is sufficient. The workspace can be imagined as the accumulated power on the reflecting elements at various time delays, which is initialized with all zeros. The first step is to calculate the received power from the n_txth transmitter to the ith element. The time bin index can be calculated by

$$\hat{n} = \arg\min_{n_\mathrm{t} \in \mathbb{Z}} \left| \left(n_\mathrm{t} - \frac{1}{2} \right) \Delta t - [\mathbf{D}_{\text{tx-e}}]_{i, n_\mathrm{tx}} \right| \tag{5.46}$$

Then we assign the power of $[\mathbf{G}_{\text{tx-e}}]_{i, n_\mathrm{tx}}$ to the matrix entry $\left[\mathbf{Q}\right]_{\hat{n}, i}$. The next step is to evaluate the power transfer between the reflecting elements starting from the first row of $\mathbf{Q}$ and scanning from the first to the N_eth element. For each reflecting element, we evaluate the power transfer and delay to all other reflecting elements. For example, when scanning the i_1th element in the n_tth time bin and evaluating its contribution to the i_2th element, the time delay index is calculated by:

$$\tilde{n} = \arg\min_{n_\mathrm{t} \in \mathbb{Z}} \left| \left(n_\mathrm{t} - \frac{1}{2} \right) \Delta t - [\mathbf{D}_{\text{e-e}}]_{i_2, i_1} \right| \tag{5.47}$$

So a transferred power of $\rho_{i_1} \left[\mathbf{Q}\right]_{n_\mathrm{t}, i_1} [\mathbf{G}_{\text{e-e}}]_{i_2, i_1}$ is accumulated to the $(\tilde{n} + n_\mathrm{t})$th row and the i_2th column of $\mathbf{Q}$, as illustrated in Figure 5.5. Repeat this process until the N_tth time bin. The final step is to evaluate the power contribution from all reflecting elements to the n_rxth receiver using $\mathbf{Q}$, $\mathbf{D}_{\text{e-rx}}$, and $\mathbf{G}_{\text{e-rx}}$ as:

$$h_{n_\mathrm{tx}, n_\mathrm{rx}}(t) = \sum_{i=1}^{N_\mathrm{e}} \rho_i \, [\mathbf{G}_{\text{e-rx}}]_{n_\mathrm{rx}, i} \, \mathbf{q}_i * \delta\left(t - [\mathbf{D}_{\text{e-rx}}]_{n_\mathrm{rx}, i}\right) \tag{5.48}$$

Note that each CIR result is stored in a discrete vector $\mathbf{h}$, as that in (5.42). Gathering the above operations, the DUSTIN algorithm is concluded in Algorithm 5.3. For an interelement power transfer, the maximum propagation length is limited by the diagonal length of the room:

$$D_\mathrm{diag} = \sqrt{L_\mathrm{l}^2 + L_\mathrm{w}^2 + L_\mathrm{h}^2} \tag{5.49}$$

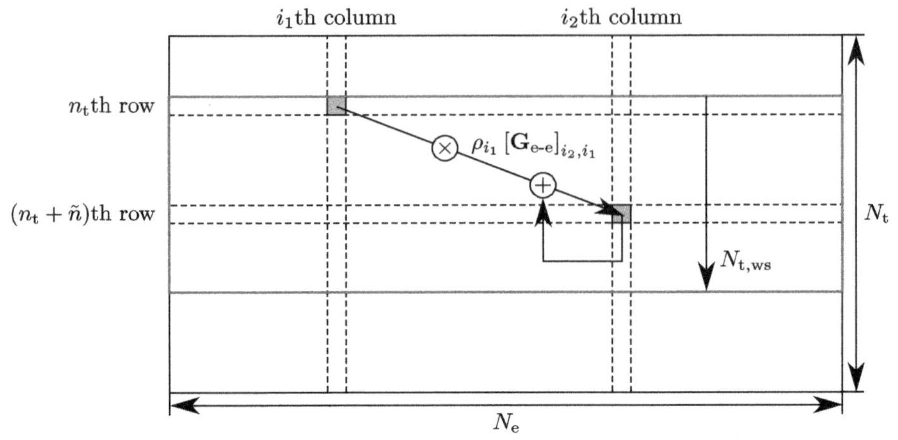

Figure 5.5 DUSTIN algorithm interelement power transfer scanning process.

Algorithm 5.3 DUSTIN Algorithm

---

for $n_{\text{tx}} = 1, 2, \cdots, N_{\text{tx}}$ **do**
  Define a workspace matrix $\mathbf{Q} \in \mathbb{R}^{N_t \times N_e}$
  for $i = 1, 2, \cdots, N_e$ **do**
    $\hat{n} = \underset{n_t \in \mathbb{Z}}{\arg\min} \left| \left(n_t - \frac{1}{2}\right) \Delta t - [\mathbf{D}_{\text{tx-e}}]_{i,n_{\text{tx}}} \right|$
    $[\mathbf{Q}]_{\hat{n},i} = [\mathbf{G}_{\text{tx-e}}]_{i,n_{\text{tx}}}$
  end
  for $n_t = 1, 2, \cdots, N_t$ **do**
    for $i_1 = 1, 2, \cdots, N_e$ **do**
      for $i_2 = 1, 2, \cdots, N_e$ **do**
        $\tilde{n} = \underset{n_t \in \mathbb{Z}}{\arg\min} \left| \left(n_t - \frac{1}{2}\right) \Delta t - [\mathbf{D}_{\text{e-e}}]_{i_2,i_1} \right|$
        $[\mathbf{Q}]_{\tilde{n}+n_t,i_2} = [\mathbf{Q}]_{\tilde{n}+n_t,i_2} + \rho_{i_1} [\mathbf{Q}]_{n_t,i_1} [\mathbf{G}_{\text{e-e}}]_{i_2,i_1}$
      end
    end
  end
  for $n_{\text{rx}} = 1, 2, \cdots, N_{\text{rx}}$ **do**
    $h_{n_{\text{tx}},n_{\text{rx}}}(t) = \sum_{i=1}^{N_e} \rho_i [\mathbf{G}_{\text{e-rx}}]_{n_{\text{rx}},i}\, \mathbf{q}_i * \delta\left(t - [\mathbf{D}_{\text{e-rx}}]_{n_{\text{rx}},i}\right)$
  end
end

Since the power of the current row of $\mathbf{Q}$ contributes to the entries up to the following $N_{t,ws} = \lceil D_{\text{diag}}/c/\Delta t \rceil$ rows, the dimension of the workspace matrix in the original algorithm proposed in [24] is limited to $N_{t,ws} \times N_e$. During the scanning process, the workspace matrix slides in the time dimension from the top to the bottom. Compared to the recursive/iterative algorithm, there are two major advantages. First, it considers reflections of any order in the scanning process. More importantly, the computational complexity no longer depends on the order of reflections but is proportional to $N_t N_e^2$.

5.5.4 Frequency-Domain Algorithm

It is well known that any convolution operation in the time domain can be transformed as a multiplication operation in the frequency domain. By taking advantage of this property, it is possible to convert the recursive/iterative convolution processes in the time-domain algorithms into a matrix multiplication process in the frequency domain, which is the basic motivation of the frequency-domain algorithm proposed in [25]. Intuitively, the frequency-domain algorithm provides simulation results in terms of CFR. One unique aspect of the frequency-domain algorithm is that it is able to simulate NLOS channels among multiple transmitters and receivers simultaneously. Specifically, the simulation results provide the following CFR matrix:

$$\mathbf{H}(f) = \begin{bmatrix} H(f;\mathcal{S}_1,\mathcal{R}_1) & \cdots & H(f;\mathcal{S}_{N_{\mathrm{tx}}},\mathcal{R}_1) \\ \vdots & \ddots & \vdots \\ H(f;\mathcal{S}_1,\mathcal{R}_{N_{\mathrm{rx}}}) & \cdots & H(f;\mathcal{S}_{N_{\mathrm{tx}}},\mathcal{R}_{N_{\mathrm{rx}}}) \end{bmatrix} \in \mathbb{R}^{N_{\mathrm{rx}} \times N_{\mathrm{tx}}} \tag{5.50}$$

where $H(f;\mathcal{S}_{n_{\mathrm{tx}}},\mathcal{R}_{n_{\mathrm{rx}}})$ refers to the NLOS CFR between the n_{tx}th transmitter and the n_{rx}th receiver. Similar to the decomposition of (5.35), $H(f;\mathcal{S}_{n_{\mathrm{tx}}},\mathcal{R}_{n_{\mathrm{rx}}})$ can be decomposed as:

$$H(f;\mathcal{S}_{n_{\mathrm{tx}}},\mathcal{R}_{n_{\mathrm{rx}}}) = \sum_{k=1}^{\infty} H^k(f;\mathcal{S}_{n_{\mathrm{tx}}},\mathcal{R}_{n_{\mathrm{rx}}}) \tag{5.51}$$

where $H^k(f;\mathfrak{s},\mathfrak{r})$ is the CFR due to the kth-order reflection between the transmitting element $\mathfrak{s}$ and the receiving element $\mathfrak{r}$, which is linked to the CIR of the kth-order reflections by:

$$H^k(f;\mathfrak{s},\mathfrak{r}) = \int_0^{+\infty} h^k(t;\mathfrak{s},\mathfrak{r}) \exp(-2\pi f t j)\mathrm{d}t \tag{5.52}$$

Applying the Fourier transform to (5.36) and (5.38), we have the following equations:

$$H^k(f;\mathcal{S}_{n_{\mathrm{tx}}},\mathcal{R}_{n_{\mathrm{rx}}}) = \sum_{i=1}^{N_e} \rho_i H^0(f;\mathcal{S}_{n_{\mathrm{tx}}},\mathcal{E}_i) H^{k-1}(f;\mathcal{E}_i,\mathcal{R}_{n_{\mathrm{rx}}}) \tag{5.53}$$

$$H^k(f;\mathcal{E}_{i_1},\mathcal{R}_{n_{\mathrm{rx}}}) = \sum_{i_2=1}^{N_e} \rho_{i_2} H^0(f;\mathcal{E}_{i_1},\mathcal{E}_{i_2}) H^{k-1}(f;\mathcal{E}_{i_2},\mathcal{R}_{n_{\mathrm{rx}}}) \tag{5.54}$$

To prepare for the following calculation, the following matrices are defined:

$$\mathbf{h}_{\text{tx}}^{n_{\text{tx}}}(f) = [H^0(f;\mathcal{S}_{n_{\text{tx}}},\mathcal{E}_1), H^0(f;\mathcal{S}_{n_{\text{tx}}},\mathcal{E}_2),\cdots,H^0(f;\mathcal{S}_{n_{\text{tx}}},\mathcal{E}_{N_e})]^{\mathrm{T}} \quad (5.55)$$

$$\mathbf{h}_{\text{rx}}^{n_{\text{rx}}}(f) = [H^0(f;\mathcal{E}_1,\mathcal{R}_{n_{\text{rx}}}), H^0(f;\mathcal{E}_2,\mathcal{R}_{n_{\text{rx}}}),\cdots,H^0(f;\mathcal{E}_{N_e},\mathcal{R}_{n_{\text{rx}}})] \quad (5.56)$$

$$\mathbf{H}_{\text{e-e}}(f) = \begin{bmatrix} H^0(f;\mathcal{E}_1,\mathcal{E}_1) & \cdots & H^0(f;\mathcal{E}_{N_e},\mathcal{E}_1) \\ \vdots & \ddots & \vdots \\ H^0(f;\mathcal{E}_1,\mathcal{E}_{N_e}) & \cdots & H^0(f;\mathcal{E}_{N_e},\mathcal{E}_{N_e}) \end{bmatrix} \in \mathbb{R}^{N_e \times N_e} \quad (5.57)$$

$$\mathbf{R} = \operatorname{diag}\left([\rho_1,\rho_2,\cdots,\rho_{N_e}]\right) \in \mathbb{R}^{N_e \times N_e} \quad (5.58)$$

where diag(·) refers to the diagonal matrix of the input vector. In conjunction with (5.55), (5.56), (5.57) and (5.58), expressions (5.53), and (5.54) can be rewritten as:

$$H^k(f;\mathcal{S}_{n_{\text{tx}}},\mathcal{R}_{n_{\text{rx}}}) = \mathbf{h}_{\text{rx}}^{n_{\text{rx}}}(f)\mathbf{R}\left(\mathbf{H}_{\text{e-e}}(f)\mathbf{R}\right)^{k-1}\mathbf{h}_{\text{tx}}^{n_{\text{tx}}}(f) \quad (5.59)$$

for $k = 1, 2, \cdots, +\infty$. By inserting (5.59) into (5.51), we have

$$H_{\text{NLOS}}(f;\mathcal{S}_{n_{\text{tx}}},\mathcal{R}_{n_{\text{rx}}}) = \mathbf{h}_{\text{rx}}^{n_{\text{rx}}}(f)\mathbf{R}\left(\mathbf{I} - \mathbf{H}_{\text{e-e}}(f)\mathbf{R}\right)^{-1}\mathbf{h}_{\text{tx}}^{n_{\text{tx}}}(f) \quad (5.60)$$

where $\mathbf{I}$ is the identity matrix. The expression (5.60) removes the infinite summation and is a more analytically tractable form. However, the complexity of evaluating the matrix inversion is significant. Therefore, the frequency-domain algorithm is still based on the expressions (5.59) and (5.51), and the simulation stops at a specific maximum order of reflection K. Finally, to make the simulation support multiple transmitters and receivers, we substitute the two vectors $\mathbf{h}_{\text{tx}}^{n_{\text{tx}}}(f)$ and $\mathbf{h}_{\text{rx}}^{n_{\text{rx}}}(f)$ with the following two matrices:

$$\mathbf{H}_{\text{tx-e}}(f) = [\mathbf{h}_{\text{tx}}^{1}(f), \mathbf{h}_{\text{tx}}^{2}(f), \cdots, \mathbf{h}_{\text{tx}}^{N_{\text{tx}}}(f)] \quad (5.61)$$

$$\mathbf{H}_{\text{e-rx}}(f) = [\mathbf{h}_{\text{rx}}^{1}(f), \mathbf{h}_{\text{rx}}^{2}(f), \cdots, \mathbf{h}_{\text{rx}}^{N_{\text{rx}}}(f)]^{\mathrm{T}} \quad (5.62)$$

respectively. The matrices $\mathbf{H}_{\text{tx-e}}(f)$, $\mathbf{H}_{\text{e-rx}}(f)$, and $\mathbf{H}_{\text{e-e}}(f)$ are related to data matrices $\mathbf{G}_{\text{tx-e}}$, $\mathbf{G}_{\text{e-rx}}$, $\mathbf{G}_{\text{e-e}}$, $\mathbf{D}_{\text{tx-e}}$, $\mathbf{D}_{\text{e-rx}}$, and $\mathbf{D}_{\text{e-e}}$ by

$$\mathbf{H}_{\text{tx-e}}(f) = \mathbf{G}_{\text{tx-e}} \otimes \exp\left(-2\pi f \mathbf{D}_{\text{tx-e}} j\right) \quad (5.63)$$

$$\mathbf{H}_{\text{e-rx}}(f) = \mathbf{G}_{\text{e-rx}} \otimes \exp\left(-2\pi f \mathbf{D}_{\text{e-rx}} j\right) \quad (5.64)$$

$$\mathbf{H}_{\text{e-e}}(f) = \mathbf{G}_{\text{e-e}} \otimes \exp\left(-2\pi f \mathbf{D}_{\text{e-e}} j\right) \quad (5.65)$$

where $\otimes$ is the Kronecker product operation. The frequency-domain algorithm is concluded in Algorithm 5.4. The frequency-domain algorithm is efficient when the channel transfer function or frequency response is required (evaluating multicarrier modulation systems). The computation complexity for each iteration is proportional to N_e^2. It is also possible to convert the resultant CFR to the

Algorithm 5.4 Frequency-Domain Algorithm

Define matrices:
$\mathbf{H}_{\text{tx-e}}(f) = \mathbf{G}_{\text{tx-e}} \otimes \exp\left(-2\pi f \mathbf{D}_{\text{tx-e}} j\right)$,
$\mathbf{H}_{\text{e-rx}}(f) = \mathbf{G}_{\text{e-rx}} \otimes \exp\left(-2\pi f \mathbf{D}_{\text{e-rx}} j\right)$,
$\mathbf{H}_{\text{e-e}}(f) = \mathbf{G}_{\text{e-e}} \otimes \exp\left(-2\pi f \mathbf{D}_{\text{e-e}} j\right)$ and
$\mathbf{R} = \text{diag}\left([\rho_1, \rho_2, \cdots, \rho_{N_e}]\right)$
Define matrix $\mathbf{S}_0 = \mathbf{H}_{\text{tx-e}}(f)$
for $k = 1, 2, \cdots, K$ **do**
  $\mathbf{S}_k = \mathbf{H}_{\text{tx-e}}(f) + \mathbf{H}_{\text{e-e}}(f)\mathbf{R}\mathbf{S}_{k-1}$
end
Calculate $\mathbf{H}(f) = \mathbf{H}_{\text{e-rx}}(f)\mathbf{R}\mathbf{S}_K$

corresponding CIR by using inverse Fourier transform. However, this requires the CFR evaluation in a large frequency range that is sufficient to reconstruct the original CIR. It was proposed by the author of [25] to apply a smoothing filter for the reconstruction of CIR.

5.5.5 Performance Evaluation

The major advantage of deterministic approaches is their high accuracy with a small value of ΔL. However, the computational complexity increases significantly with the number of reflecting elements N_e. The total number of elements increases when the element size decreases. The number of elements also grows significantly when the room dimensions are increased. In Figure 5.6, a CIR results simulated using an iterative algorithm are presented. It can be observed that the overall NLOS channel is a superposition of the channel components due to different orders of reflections. Generally, the higher the reflection order, the longer the overall time delay and the lower the power contribution to the NLOS channel. The simulation results of other deterministic algorithms are virtually identical. The only difference is the computational complexity. Therefore, we conduct the simulations of the same channel using different deterministic approaches and evaluate their simulation time. Here we use the configuration A in [12] and the corresponding parameters are listed in Table 5.1. The used element edge length is $\Delta L = 0.2$ m.

The simulation time of different algorithms against the highest reflection order is shown in Figure 5.7(a). It can be observed that the simulation time of the recursive algorithm increases significantly when the highest reflection order reaches 3 (more than one day). In contrast, the simulation time of the improved iterative algorithm increases linearly with the highest reflection order. Therefore, even the tenth-order reflection can be simulated using the iterative algorithm. As described in [24], the complexity of the DUSTIN algorithm does not depend

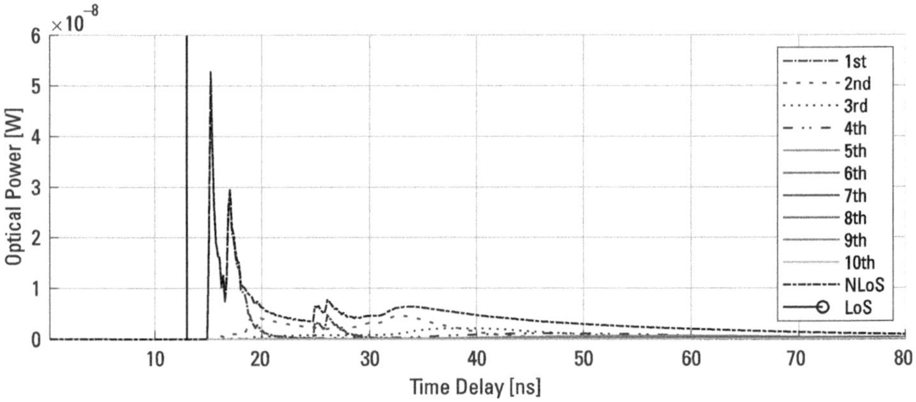

Figure 5.6 CIR results simulated using the iterative algorithm.

Table 5.1
Configuration 1 parameters

Parameter	Symbol	Value
Room dimension	$L_l \times L_w \times L_h$ [m]	$5 \times 5 \times 3$
Reflectance of walls/ceiling/floor	$\rho_w/\rho_c/\rho_f$	0.8/0.8/0.3
Transmitter position	$\mathbf{x}_s$ [m]	[2.5,2.5,3]
Transmitter orientation	$\mathbf{p}_s$ [m]	[0,0,−1]
Receiver position	$\mathbf{x}_r$ [m]	[0.5,1,0]
Receiver orientation	$\mathbf{p}_r$ [m]	[0,0,1]
Half-power semiangle	$\phi_{1/2}$ [degree]	60°
Receiver Fov	ψ_{FOV} [degree]	85°
Detection area	A_d [m^2]	1×10^{-4}

on the order of reflections. However, its complexity is proportional to the total number of time bins, which is also significant. Consequently, the simulation time is much longer than the iterative method but is significantly shorter than the recursive algorithm with the highest reflection order exceeding three. Due to the efficient matrix multiplication operation, the frequency-domain algorithm is very fast compared to the other algorithms. The author of [25] claimed that the matrix multiplication complexity increases linearly with the reflection order. However, we found that the matrix evaluation in (5.63), (5.64), and (5.65) also takes up a significant amount of computation resources. Consequently, the simulation time of the frequency-domain algorithm increases slightly with the highest reflection order. The only issue of the frequency-domain algorithm is that the channel response, with respect to different frequencies, needs to be calculated independently. Therefore, if a large number of samples with different frequencies are required, the calculations have to be repeated by N_f times, where N_f refers to the number of frequency samples. It can be observed that with

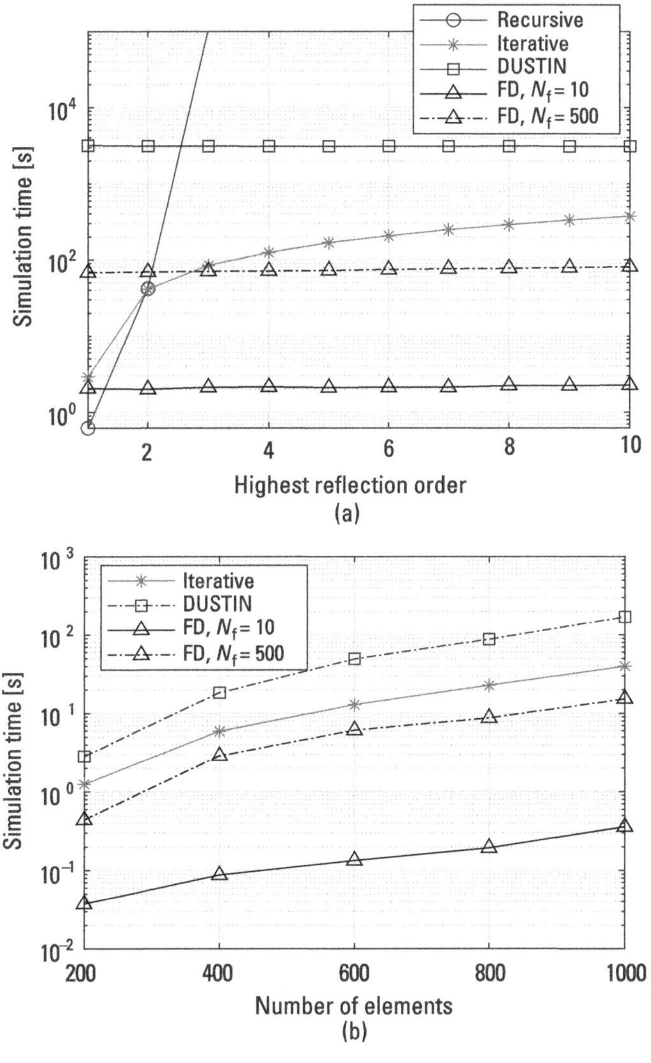

Figure 5.7 (a) Simulation time against the highest order of reflections. (b) Simulation time against the number of reflecting elements. FD: frequency domain.

only 10 frequency samples, the simulation takes a few seconds. In the case of 500 frequency samples, the simulation time increases to more than one minute. Figure 5.7(b) demonstrates that the simulation time is proportional to the square of the number of elements N_e for all deterministic algorithms. In conclusion, the iterative algorithm is the most efficient when calculating time-domain quantities. The frequency-domain simulation algorithm is favored when calculating the frequency-domain quantities. The DUSTIN algorithm is preferred only when the number of time bins is small.

5.6 Monte Carlo NLOS Channel Simulation Approaches

In deterministic simulation approaches, to obtain accurate simulation results, a higher granularity with a greater number of partitioned elements is preferred. However, an increased number of elements significantly increases the computation complexity by the order of magnitude. Furthermore, the number of partitioned elements is also proportional to the surface area of the indoor environment. Consequently, it is very time-consuming to simulate channels with a high reflection order using deterministic approaches for large indoor environments. In order to improve the channel simulation efficiency, a series of Monte Carlo simulation algorithms have been proposed [26]. Unlike the deterministic approaches, which evaluate all possible light propagation paths, Monte Carlo methods simulate randomly generated paths and evaluate their contribution to the NLOS channel. With an increase in the number of random paths, the accumulation will converge to the desired NLOS channel. In this section, several Monte Carlo approaches are introduced in detail.

5.6.1 Photon-Tracing Algorithm

A photon tracing algorithm (PTA) is the simplest and the most straightforward Monte Carlo NLOS channel simulation algorithm as described in [27]. In this algorithm, the propagation process of photons and their interactions with surfaces and detectors are imitated. On the transmitter side, the emitted light is brighter in some directions but darker in others. This phenomenon is equivalent to a statistical photon radiation process in random directions, as illustrated in Figure 5.8(a). In order to make sure the photon radiation process agrees with the light source characteristics, the probability density function (PDF) of an emission direction is set to be proportional to the radiation intensity at that direction. Once a photon leaves the transmitter, it propagates in the indoor free space until it impacts any reflective surface. In a typical indoor environment, we assume that the air in the free space is clear and there are no particles causing absorption or diffraction. Therefore, the photon preserves the energy in the process of free-space propagation. When the photon impacts an opaque surface, there is a chance that the photon is reflected. The probability of reflection is equal to the reflectivity of the surface. If the photon is reflected, the reflection direction depends on the characteristics of the surface. If the surface is diffusely reflective, the reflection direction is randomly selected, as shown in Figure 5.8(b). If the surface is specular, the reflection direction is symmetric to the photon incident direction with respect to the surface normal. Otherwise, when the photon is not reflected, it will be absorbed by the surface and the tracing of this photon ends here. After the reflection of the photon, it may hit another surface and repeat the reflection/absorption process. Alternatively, it may be incident to the target detector and contribute to the simulated channel, as shown in Figure 5.8(c). Note

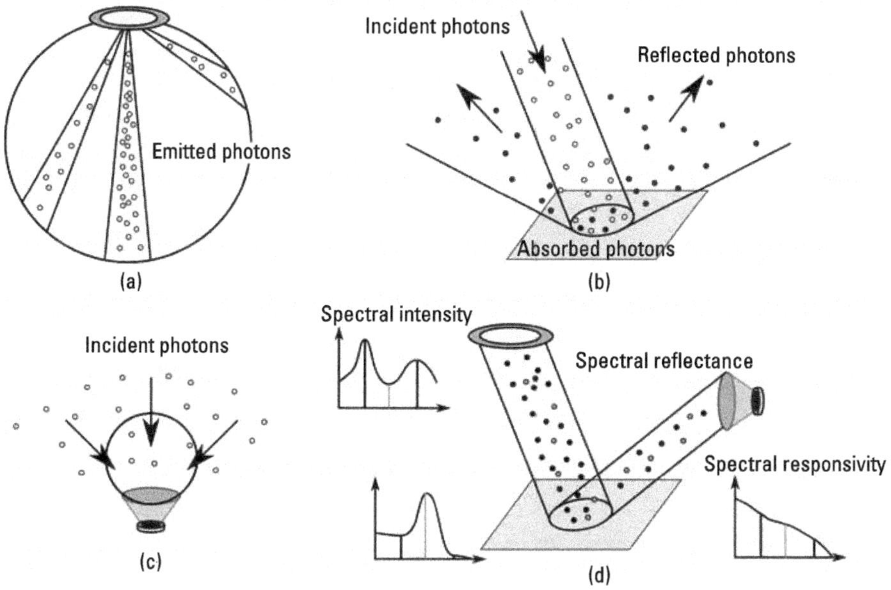

Figure 5.8 (a) Random emission direction generation at the transmitter. (b) Random photon reflection process. (c) Random photon detection process. (c) Wavelength domain random processes.

that the detector has a different sensitivity to the light from different incident directions. Therefore, the contribution of these photons needs to be scaled by the detector sensitivity in their incident directions. By repeating the above photon tracing process and accumulating their contribution to the CIR vector, the result will converge to the desired NLOS CIR eventually. In the case of a system with a wide spectrum, the random photon tracing process can be extended to the wavelength domain. Specifically, the photons with a random wavelength can be generated. The corresponding probability of generating a photon with a certain wavelength is proportional to the spectral intensity distribution of the light source, as illustrated in Figure 5.8(d). Similarly, in the reflection process, the reflection probability of a photon with a specific wavelength follows the reflectivity at that particular wavelength. On the detector side, the photon is also scaled by the corresponding sensitivity at the corresponding wavelength.

Despite the simplicity of the PTA, it has introduced too many random quantities, such as a random emission direction, random reflection direction, random reflection probability and random wavelength generation. In particular, the chance that a photon hits the detector is extremely low, which makes the PTA extremely inefficient. All these factors lead to a very low convergence speed. In order to increase the convergence rate of Monte Carlo approaches, more efficient algorithms have been proposed, and these will be covered in the following sections.

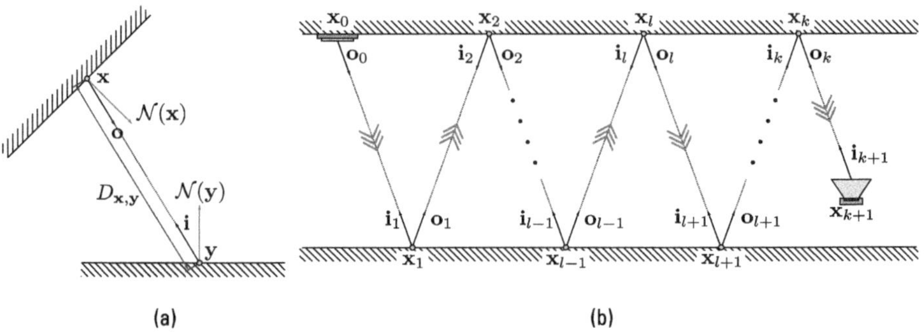

Figure 5.9 (a) Light transportation quantities demonstration. (b) Light transportation in an indoor environment from a transmitter to a receiver via *k* reflections.

5.6.2 Monte Carlo Ray-Shooting and Ray-Gathering Algorithms

A modified Monte Carlo ray-tracing algorithm was first proposed in [26], which is more efficient than the PTA. A similar Monte Carlo ray-shooting (MCRS) algorithm was introduced in [28]. In particular, the authors of [28] presented the MCRS algorithm using a light transportation theory framework, which provides a better mathematical interpretation of the same algorithm. In addition, another backward tracing method, called the Monte Carlo ray-gathering (MCRG) algorithm was also proposed, which is more efficient in specific conditions. In this section, we will focus on the MCRS and MCRG algorithms and provide more details about them both.

First, let's consider two points located at $\mathbf{x}$ and $\mathbf{y}$, as shown in Figure 5.9(a). Light leaves point $\mathbf{x}$ in the direction $\mathbf{o}$ and incidents to the point $\mathbf{y}$ from the direction $\mathbf{i}$, which can be described by the following equations:

$$\mathbf{x} + D_{\mathbf{x},\mathbf{y}}\mathbf{o} = \mathbf{y}, \ \text{or}\ \mathbf{y} + D_{\mathbf{x},\mathbf{y}}\mathbf{i} = \mathbf{x} \tag{5.66}$$

where $D_{\mathbf{x},\mathbf{y}}$ refers to the Euclidean distance between $\mathbf{x}$ and $\mathbf{y}$. It is apparent that $\mathbf{i} = -\mathbf{o}$. The emitted optical power per unit area per solid angle (exit radiance) at position $\mathbf{x}$ in the direction $\mathbf{o}$ is defined as [29]:

$$\mathcal{L}^{\uparrow}(\mathbf{x}, \mathbf{o}) = \frac{\mathrm{d}^2\Phi(\mathbf{x}, \mathbf{o})}{|\mathbf{o} \cdot \mathcal{N}(\mathbf{x})| \mathrm{d}A(\mathbf{x}) \mathrm{d}\Omega(\mathbf{o})} \tag{5.67}$$

where Φ refers to the radiant flux or optical power, $\mathcal{N}(\mathbf{x})$ refers to the normal vector of the surface at $\mathbf{x}$, A refers to emission area, Ω refers to the emission solid angle and the operation $|\mathbf{u}\cdot\mathbf{v}|$ is equivalent to $\max(0, \mathbf{u}\cdot\mathbf{v})$. The operator $\cdot$ denotes the dot product. The max function ensures that the result is non-negative, which agrees with the nature of light transportation. The incident radiance to point $\mathbf{y}$ from direction $\mathbf{i}$ is denoted as $\mathcal{L}^{\downarrow}(\mathbf{y}, \mathbf{i})$, which has a similar definition as the exitant radiance. In addition, the exitant radiance from one point is equal to the incident radiance to the destination point:

$$\mathcal{L}^{\uparrow}(\mathbf{x}, \mathbf{o}) = \mathcal{L}^{\downarrow}(\mathbf{y}, \mathbf{i}) \tag{5.68}$$

as shown in Figure 5.9(a).

5.6.2.1 Channel Impulse Response Expressions and Monte Carlo Integration

Next we consider the case where the number of reflections is fixed by k with a light source at $\mathbf{x}_0$ and an optical detector at $\mathbf{x}_{k+1}$, as shown in Figure 5.9(b). The NLOS channel calculation can start from the transmitter or from the receiver. We first consider a receiver starting point as it is more intuitive. According to the definition of radiance, the optical power captured by the detector can be calculated by [28]:

$$\Phi_k = \int_{\mathcal{A}_\mathrm{d}} \int_{\mathcal{S}(\mathbf{x})} \mathcal{L}^{\downarrow}(\mathbf{x}, \mathbf{i}) \left|\mathbf{i} \cdot \mathcal{N}(\mathbf{x})\right| \mathrm{d}\Omega(\mathbf{i})\mathrm{d}A(\mathbf{x}) \tag{5.69}$$

where $\mathcal{A}_\mathrm{d}$ is the region of detection and $\mathcal{S}(\mathbf{x})$ refers the region of directions that is within the FOV of point $\mathbf{x}$. Typically, the detection region is a very small area around point $\mathbf{x}_{k+1}$. The dimension of the area is significantly smaller than the light propagation distance. Therefore, we can conduct the following approximation to (5.69) as:

$$\Phi_k = A_\mathrm{d} \int_{\mathcal{S}(\mathbf{x}_{k+1})} \mathcal{L}^{\downarrow}(\mathbf{x}_{k+1}, \mathbf{i}_{k+1}) \left|\mathbf{i}_{k+1} \cdot \mathcal{N}(\mathbf{x}_{k+1})\right| \mathrm{d}\Omega(\mathbf{i}_{k+1}) \tag{5.70}$$

where A_d is the area of the detection. In order to evaluate (5.70) further, we introduce the following rendering equation [29]:

$$\mathcal{L}^{\uparrow}(\mathbf{x}_l, \mathbf{o}_l) = \mathcal{L}_\mathrm{e}^{\uparrow}(\mathbf{x}_l, \mathbf{o}_l) + \int_{\mathcal{S}(\mathbf{x}_l)} \mathcal{L}^{\downarrow}(\mathbf{x}_l, \mathbf{i}_l) f_\mathrm{r}(\mathbf{x}_l, \mathbf{i}_l, \mathbf{o}_l) \left|\mathbf{i}_l \cdot \mathcal{N}(\mathbf{x}_l)\right| \mathrm{d}\Omega(\mathbf{i}_l) \tag{5.71}$$

where $\mathcal{L}_\mathrm{e}^{\uparrow}(\mathbf{x}, \mathbf{o})$ is the self-emitted exitant radiance. It is equal to zero if point $\mathbf{x}$ is not an optical source; and $f_\mathrm{r}(\mathbf{x}, \mathbf{i}, \mathbf{o})$ is the bidirectional reflection distribution function (BRDF) at point $\mathbf{x}$ with an incident direction of $\mathbf{i}$ and an exitant direction of $\mathbf{o}$, which can be defined as:

$$f_\mathrm{r}(\mathbf{x}, \mathbf{i}, \mathbf{o}) = \frac{\mathrm{d}\mathcal{L}^{\uparrow}(\mathbf{x}, \mathbf{o})}{\mathcal{L}^{\downarrow}(\mathbf{x}, \mathbf{i}) \left|\mathbf{i} \cdot \mathcal{N}(\mathbf{x})\right| \mathrm{d}\Omega(\mathbf{i})} \tag{5.72}$$

In a case where the surfaces are ideally diffuse reflective, we have

$$f_\mathrm{r}(\mathbf{x}, \mathbf{i}, \mathbf{o}) = \frac{\rho(\mathbf{x})}{\pi} \tag{5.73}$$

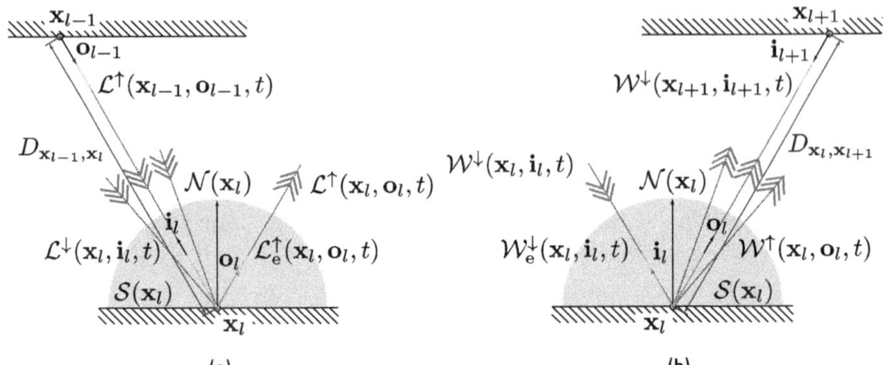

Figure 5.10 (a) Rendering equation geometry. (b) Potential equation geometry.

where $\rho(\mathbf{x})$ is the reflectance at $\mathbf{x}$. In cases of partially specular and partially diffuse surfaces, the Blinn-Phong BRDF can be used [30]:

$$f_r(\mathbf{x}, \mathbf{i}, \mathbf{o}) = \frac{\rho_d(\mathbf{x})}{\pi} + \rho_s(\mathbf{x})\frac{m_s + 2}{2\pi}\left|\frac{\mathbf{o} - \mathbf{i}}{||\mathbf{o} - \mathbf{i}||} \cdot \mathcal{N}(\mathbf{x})\right|^{m_s} \tag{5.74}$$

where ρ_d is the diffuse reflectance, ρ_s is the specular reflectance, and m_s defines the beamwidth of the specular reflection component. In addition to the above two BRDFs, other models are also available which show the flexibility of the Monte Carlo approaches [31]. By observing the geometry shown in Figure 5.10(b), we can find that

$$\mathcal{L}^{\downarrow}(\mathbf{x}_l, \mathbf{i}_l, t) = \delta\left(t - \tau_{\mathbf{x}_{l-1},\mathbf{x}_l}\right)\mathcal{L}^{\uparrow}(\mathbf{x}_{l-1}, \mathbf{o}_{l-1}, t) \tag{5.75}$$

where $\tau_{\mathbf{x}_{l-1},\mathbf{x}_l} = D_{\mathbf{x}_{l-1},\mathbf{x}_l}/c$. Here, we introduced the characteristics of the time delay due to the requirement of the CIR calculation. At the transmitter, the exitant radiance is equal to the self-emitted exitant radiance of the LED: $\mathcal{L}^{\uparrow}(\mathbf{x}_0, \mathbf{o}_0, t) = \mathcal{L}_e^{\uparrow}(\mathbf{x}_0, \mathbf{o}_0)$. In the case of a Lambertian point source, it can be calculated as:

$$\mathcal{L}^{\uparrow}(\mathbf{x}_0, \mathbf{o}_0, t) = \mathcal{L}_e^{\uparrow}(\mathbf{x}_0, \mathbf{o}_0) = \frac{\mathcal{J}_s(\mathbf{x}_0, \mathbf{o}_0)}{|\mathbf{o}_0 \cdot \mathcal{N}(\mathbf{x}_0)|\, dA_{tx}} \tag{5.76}$$

where A_{tx} is the emission area of the transmitter. Since $\mathbf{i}_1$ is directed at the position of the transmitter, the corresponding differential solid angle can be calculated by

$$d\Omega(\mathbf{i}_1) = \frac{|\mathbf{o}_0 \cdot \mathcal{N}(\mathbf{x}_0)|\, dA_{tx}}{D^2_{\mathbf{x}_0,\mathbf{x}_1}} \tag{5.77}$$

By using (5.71) and (5.75) recursively in (5.70) and in conjunction with (5.76) and (5.77), the integral (5.70) can be expanded as:

$$
\begin{aligned}
\Phi_k(t) = A_\mathrm{d} & \int_{\mathcal{S}(\mathbf{x}_{k+1})} \left\{ \int_{\mathcal{S}(\mathbf{x}_k)} \cdots \left\{ \int_{\mathcal{S}(\mathbf{x}_2)} \right.\right. \\
& \times \frac{\mathcal{J}_\mathrm{s}(\mathbf{x}_0, -\mathbf{i}_1)}{D^2_{\mathbf{x}_0,\mathbf{x}_1}} \delta\left(t - \tau_{\mathbf{x}_0,\mathbf{x}_1}\right) f_\mathrm{r}(\mathbf{x}_1, \mathbf{i}_1, -\mathbf{i}_2) \left|\mathbf{i}_1 \cdot \mathcal{N}(\mathbf{x}_1)\right| \\
& \times \delta\left(t - \tau_{\mathbf{x}_1,\mathbf{x}_2}\right) f_\mathrm{r}(\mathbf{x}_2, \mathbf{i}_2, -\mathbf{i}_3) \left|\mathbf{i}_2 \cdot \mathcal{N}(\mathbf{x}_2)\right| \mathrm{d}\Omega\left(\mathbf{i}_2\right) \} \\
& \times \cdots \\
& \times \delta\left(t - \tau_{\mathbf{x}_{k-1},\mathbf{x}_k}\right) f_\mathrm{r}(\mathbf{x}_k, \mathbf{i}_k, -\mathbf{i}_{k+1}) \left|\mathbf{i}_k \cdot \mathcal{N}(\mathbf{x}_k)\right| \mathrm{d}\Omega\left(\mathbf{i}_k\right) \} \\
& \times \delta\left(t - \tau_{\mathbf{x}_k,\mathbf{x}_{k+1}}\right) \left|\mathbf{i}_{k+1} \cdot \mathcal{N}(\mathbf{x}_{k+1})\right| \mathrm{d}\Omega\left(\mathbf{i}_{k+1}\right)
\end{aligned}
\tag{5.78}
$$

where there are k integrals. Note that all exitant directions $\mathbf{o}_l$ are replaced by $\mathbf{i}_{l+1}$. The integrand of (5.78) can be concluded as:

$$
\begin{aligned}
g_k^{\downarrow}(\mathcal{I}, t) = & \frac{A_\mathrm{d}}{D^2_{\mathbf{x}_0,\mathbf{x}_1}} \mathcal{J}_\mathrm{s}(\mathbf{x}_0, -\mathbf{i}_1) \left|\mathbf{i}_{k+1} \cdot \mathcal{N}(\mathbf{x}_{k+1})\right| \\
& \times \delta\left(t - \sum_{l=1}^{k+1} \tau_{\mathbf{x}_{l-1},\mathbf{x}_l}\right) \prod_{l=1}^{k} f_\mathrm{r}(\mathbf{x}_l, \mathbf{i}_l, -\mathbf{i}_{l+1}) \left|\mathbf{i}_l \cdot \mathcal{N}(\mathbf{x}_l)\right|
\end{aligned}
\tag{5.79}
$$

where $\mathcal{I} = \left\{\mathbf{i}_{k+1}, \mathbf{i}_k, \cdots, \mathbf{i}_2\right\}$ is the set of incident direction samples. Each set of direction samples corresponds to a specific propagation path. Then the integration can be written as:

$$
\Phi_k(t) = \int_{\mathcal{S}(\mathbf{x}_{k+1})} \int_{\mathcal{S}(\mathbf{x}_k)} \cdots \int_{\mathcal{S}(\mathbf{x}_2)} g_k^{\downarrow}(\mathcal{I}, t) \mathrm{d}\Omega\left(\mathbf{i}_2\right) \cdots \mathrm{d}\Omega\left(\mathbf{i}_k\right) \mathrm{d}\Omega\left(\mathbf{i}_{k+1}\right)
\tag{5.80}
$$

Next, we consider the transmitter starting point. The optical power that can reach the detector after k reflections can be calculated by:

$$
\Phi_k = \int_{\mathcal{S}(\mathbf{x}_0)} \mathcal{W}^{\uparrow}(\mathbf{x}_0, \mathbf{o}_0) \mathcal{J}_\mathrm{s}(\mathbf{x}_0, \mathbf{o}_0) \mathrm{d}\Omega\left(\mathbf{o}_0\right)
\tag{5.81}
$$

where the $\mathcal{W}^{\uparrow}(\mathbf{x}, \mathbf{o})$ is defined as the exitant potential of point $\mathbf{x}$ to the direction $\mathbf{o}$. Similar to the rendering equation, the potential/importance equation is defined to characterize the light transportation in the reverse direction as [29]:

$$
\mathcal{W}^{\downarrow}(\mathbf{x}_l, \mathbf{i}_l) = \mathcal{W}_\mathrm{e}^{\downarrow}(\mathbf{x}_l, \mathbf{i}_l) + \int_{\mathcal{S}(\mathbf{x}_l)} \mathcal{W}^{\uparrow}(\mathbf{x}_l, \mathbf{o}_l) f_\mathrm{r}(\mathbf{x}_l, \mathbf{i}_l, \mathbf{o}_l) \left|\mathbf{o}_l \cdot \mathcal{N}(\mathbf{x}_l)\right| \mathrm{d}\Omega\left(\mathbf{o}_l\right)
\tag{5.82}
$$

where $\mathcal{W}^{\downarrow}(\mathbf{x}, \mathbf{i})$ is the incident potential at point $\mathbf{x}$ from the direction $\mathbf{i}$ and $\mathcal{W}_{\mathrm{e}}^{\downarrow}(\mathbf{x}, \mathbf{i})$ is defined as the emitted importance which equals 1 when $\mathbf{i}$ is directed to the detector and equals 0 otherwise. By observing the geometry shown in Figure 5.9(b), we can find that

$$\mathcal{W}^{\uparrow}(\mathbf{x}_l, \mathbf{o}_l, t) = \delta\left(t - \tau_{\mathbf{x}_l, \mathbf{x}_{l+1}}\right) \mathcal{W}^{\downarrow}(\mathbf{x}_{l+1}, \mathbf{i}_{l+1}, t) \tag{5.83}$$

Since $\mathbf{o}_k$ is direct to the position of the receiver, the corresponding differential solid angle can be calculated by

$$\Omega(\mathbf{o}_k) = \frac{\left|\mathbf{i}_{k+1} \cdot \mathcal{N}(\mathbf{x}_{k+1})\right| A_{\mathrm{d}}}{D^2_{\mathbf{x}_k, \mathbf{x}_{k+1}}} \tag{5.84}$$

By using (5.82) and (5.83) recursively in (5.81) and in conjunction with (5.84), the integral (5.81) can be expanded as:

$$\begin{aligned}
\Phi_k(t) = & \int_{\mathcal{S}(\mathbf{x}_0)} \left\{ \int_{\mathcal{S}(\mathbf{x}_1)} \cdots \left\{ \int_{\mathcal{S}(\mathbf{x}_{k-1})} \right.\right. \\
& \times \frac{\left|-\mathbf{o}_k \cdot \mathcal{N}(\mathbf{x}_{k+1})\right| A_{\mathrm{d}}}{D^2_{\mathbf{x}_k, \mathbf{x}_{k+1}}} \delta\left(t - \tau_{\mathbf{x}_k, \mathbf{x}_{k+1}}\right) f_{\mathrm{r}}(\mathbf{x}_k, -\mathbf{o}_{k-1}, \mathbf{o}_k) \left|\mathbf{o}_k \cdot \mathcal{N}(\mathbf{x}_k)\right| \\
& \times \delta\left(t - \tau_{\mathbf{x}_{k-1}, \mathbf{x}_k}\right) f_{\mathrm{r}}(\mathbf{x}_{k-1}, -\mathbf{o}_{k-2}, \mathbf{o}_{k-1}) \left|\mathbf{o}_{k-1} \cdot \mathcal{N}(\mathbf{x}_{k-1})\right| \mathrm{d}\Omega(\mathbf{o}_{k-1}) \big\} \\
& \times \cdots \\
& \times \delta\left(t - \tau_{\mathbf{x}_1, \mathbf{x}_2}\right) f_{\mathrm{r}}(\mathbf{x}_1, -\mathbf{o}_0, \mathbf{o}_1) \left|\mathbf{o}_1 \cdot \mathcal{N}(\mathbf{x}_1)\right| \mathrm{d}\Omega(\mathbf{o}_1) \big\} \\
& \times \delta\left(t - \tau_{\mathbf{x}_0, \mathbf{x}_1}\right) \mathcal{J}_{\mathrm{s}}(\mathbf{x}_0, \mathbf{o}_0) \mathrm{d}\Omega(\mathbf{o}_0)
\end{aligned} \tag{5.85}$$

The integrand of (5.85) can be found as:

$$\begin{aligned}
g_k^{\uparrow}(\mathcal{O}, t) = & \frac{A_{\mathrm{d}}}{D^2_{\mathbf{x}_k, \mathbf{x}_{k+1}}} \mathcal{J}_{\mathrm{s}}(\mathbf{x}_0, \mathbf{o}_0) \left|-\mathbf{o}_k \cdot \mathcal{N}(\mathbf{x}_{k+1})\right| \\
& \times \delta\left(t - \sum_{l=1}^{k+1} \tau_{\mathbf{x}_{l-1}, \mathbf{x}_l}\right) \prod_{l=1}^{k} f_{\mathrm{r}}(\mathbf{x}_l, -\mathbf{o}_{l-1}, \mathbf{o}_l) \left|\mathbf{o}_l \cdot \mathcal{N}(\mathbf{x}_l)\right|
\end{aligned} \tag{5.86}$$

where $\mathcal{O} = \left\{\mathbf{o}_0, \mathbf{o}_1, \cdots, \mathbf{o}_{k-1}\right\}$ is the set of exitant direction samples, which also forms a certain propagation path. Then the integration can be written as:

$$\Phi_k(t) = \int_{\mathcal{S}(\mathbf{x}_0)} \int_{\mathcal{S}(\mathbf{x}_1)} \cdots \int_{\mathcal{S}(\mathbf{x}_{k-1})} g_k^{\uparrow}(\mathcal{O}, t) \mathrm{d}\Omega(\mathbf{o}_{k-1}) \cdots \mathrm{d}\Omega(\mathbf{o}_1) \mathrm{d}\Omega(\mathbf{o}_0) \tag{5.87}$$

For the next step, we use the MC integration technique to solve the integrals (5.80) and (5.87). Considering an integral:

$$I = \int_{\mathcal{U}} g(u)\mathrm{d}u \tag{5.88}$$

with an integrand $g(u)$, the integration value can be approximated by:

$$I \approx \hat{I} = \frac{1}{N_s}\sum_{n=1}^{N_s}\frac{g(u_n)}{\mathfrak{f}_u(u_n)} \tag{5.89}$$

where $\mathfrak{f}_u(u)$ is the PDF of the variable u, u_n is the nth random sample of variable u, and N_s is the number of random samples. By applying (5.89) to the integrals (5.80) and (5.87), we are able to conclude the expressions for the MCRG and MCRS algorithms as:

$$\hat{\Phi}_k(t) = \frac{1}{N_s}\sum_{n=1}^{N_s}\frac{g_k^{\downarrow}(\mathcal{I}_n, t)}{\mathfrak{f}_{\mathcal{I}}(\mathcal{I}_n)} \tag{5.90}$$

$$\hat{\Phi}_k(t) = \frac{1}{N_s}\sum_{n=1}^{N_s}\frac{g_k^{\uparrow}(\mathcal{O}_n, t)}{\mathfrak{f}_{\mathcal{O}}(\mathcal{O}_n)} \tag{5.91}$$

where $\mathfrak{f}_{\mathcal{I}}(\mathcal{I})$ and $\mathfrak{f}_{\mathcal{O}}(\mathcal{O})$ are defined as the PDF of the propagation path; $\mathcal{I}_n$ and $\mathcal{O}_n$ refer to the nth random path sample.

5.6.2.2 Sampling Random Propagation Paths

Next, we consider the sampling of random paths in MCRG and MCRS algorithms. The simplest option is to uniformly select the direction in the range of $\mathcal{S}(\mathbf{x}_l)$. Since the solid angle of a hemisphere is 2π, the corresponding PDF of one exitant/incident direction is $\frac{1}{2\pi}$. There is one exception for the incident direction to the detector in the MCRG algorithm, where the solid angle is reduced to $2\pi(1 - \cos\psi_{\mathrm{FOV}})$ with a FOV smaller than 90° and the corresponding PDF is $\frac{1}{2\pi(1-\cos\psi_{\mathrm{FOV}})}$. The generation of incident/exitant directions are independent from each other. Therefore, the joint PDF can be written as the product of the PDF of each direction:

$$\mathfrak{f}_{\mathcal{I}}(\mathcal{I}) = \frac{1}{1-\cos\psi_{\mathrm{FOV}}}\left(\frac{1}{2\pi}\right)^k, \text{ for } \psi_{\mathrm{FOV}} \in (0, \pi/2] \tag{5.92}$$

$$\mathfrak{f}_{\mathcal{O}}(\mathcal{O}) = \left(\frac{1}{2\pi}\right)^k \tag{5.93}$$

Then we specify how to generate the uniform incident/exitant direction vectors in the global Cartesian coordinate system. First, we consider a local

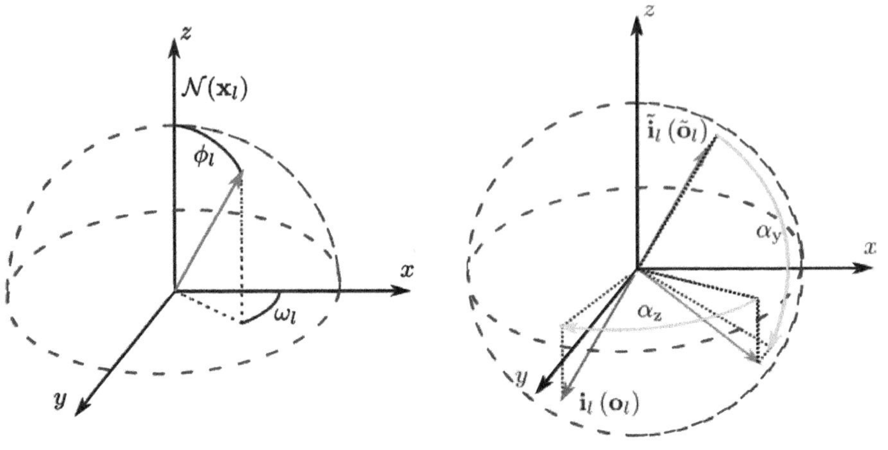

(a) (b)

Figure 5.11 (a) Incident/exitant direction sample geometry in a local Cartesian coordinate system. (b) Direction sample vector rotation to obtain the vector in the room coordinate system.

Cartesian/spherical coordinate system with an upward surface normal or transmitter direction ([0, 0, 1]), as shown in Figure 5.11(a). For each direction, we generate two random numbers u_1 and u_2 that are uniformly distributed in [0, 1). Then a polar angle and an azimuth angle can be calculated by:

$$\omega_l = 2\pi u_1 \tag{5.94}$$

$$\phi_l = \begin{cases} \arccos\left(\cos\psi_{\text{FOV}} + (1 - \cos\psi_{\text{FOV}})u_2\right) & : \text{calculate } \mathbf{i}_{k+1} \\ \arccos(u_2) & : \text{otherwise} \end{cases} \tag{5.95}$$

The direction vector in the local Cartesian coordinate system can be calculated as:

$$\tilde{\mathbf{i}}_l(\tilde{\mathbf{o}}_l) = [\cos\omega_l \sin\phi_l, \sin\omega_l \sin\phi_l, \cos\phi_l] \tag{5.96}$$

The vector in the local coordinate system needs to be rotated about the y-axis and the z-axis subsequently to obtain the vector in the global Cartesian coordinate system, as shown in Fig 5.11(b). With a given normal/orientation vector of $\mathcal{N}(\mathbf{x}_l) = [x_{\text{N}}, y_{\text{N}}, z_{\text{N}}]$ in the global Cartesian coordinate system, the rotation angle about the y-axis is $\alpha_{\text{y}} = \arccos(z_{\text{N}})$ and the rotation angle about the z-axis is $\alpha_{\text{z}} = \text{atan2}(y_{\text{N}}, x_{\text{N}})$. The incident direction vectors can be calculated as:

$$\begin{aligned} \mathbf{i}_l &= \mathbf{R}_{\text{z}}(\alpha_{\text{z}})\mathbf{R}_{\text{y}}(\alpha_{\text{y}})\tilde{\mathbf{i}}_l^{\text{T}} \\ &= \begin{bmatrix} \cos\alpha_{\text{z}} & -\sin\alpha_{\text{z}} & 0 \\ \sin\alpha_{\text{z}} & \cos\alpha_{\text{z}} & 0 \\ 0 & 0 & 1 \end{bmatrix} \begin{bmatrix} \cos\alpha_{\text{y}} & 0 & \sin\alpha_{\text{y}} \\ 0 & 1 & 0 \\ -\sin\alpha_{\text{y}} & 0 & \cos\alpha_{\text{y}} \end{bmatrix} \begin{bmatrix} \cos\omega_l \sin\phi_l \\ \sin\omega_l \sin\phi_l \\ \cos\phi_l \end{bmatrix} \end{aligned} \tag{5.97}$$

where $\mathbf{R}_y(\alpha)$ is the 3-D rotation matrix about the y-axis by a rotation angle of α and $\mathbf{R}_z(\alpha)$ is the 3-D rotation matrix about the z-axis by a rotation angle of α. The exitant direction vectors $\mathbf{o}_l$ are calculated in the same manner. This direction sampling is based on the inversion method. Other sampling techniques can also be used, such as the rejection method [29].

Sampling the incident/exitant directions uniformly is not the only option. According to the concept of importance sampling, if we generate more path samples that matter to the NLOS channel, the precision of the Monte Carlo estimation will increase. Thus, sampling according to the Lambertian pattern could be a potential choice, which leads to the following PDFs:

$$\mathfrak{f}_{\mathcal{I}}(\mathcal{I}) = \frac{\prod_{l=1}^{k} \cos\phi_{l+1}}{(1-\cos\psi_{\mathrm{FOV}})\pi^k}, \text{ for } \psi_{\mathrm{FOV}} \in (0, \pi/2] \tag{5.98}$$

$$\mathfrak{f}_{\mathcal{O}}(\mathcal{O}) = \frac{(m+1)}{2\pi^k} \cos^m \phi_0 \prod_{l=1}^{k-1} \cos\phi_l \tag{5.99}$$

To sample the corresponding incident/exitant directions, two random numbers u_1 and u_2 are generated (uniform distribution in [0, 1)). Then the polar angle and azimuth angle of the incident/exitant direction can be calculated by

$$\omega_l = 2\pi u_1 \tag{5.100}$$

$$\phi_l = \begin{cases} \arccos\left((\cos\psi_{\mathrm{FOV}} + (1-\cos\psi_{\mathrm{FOV}})u_2)^{\frac{1}{2}}\right) & : \text{calculate } \mathbf{i}_{k+1} \\ \arccos\left(u_2^{\frac{1}{m+1}}\right) & : \text{calculate } \mathbf{o}_0 \\ \arccos\left(u_2^{\frac{1}{2}}\right) & : \text{otherwise} \end{cases} \tag{5.101}$$

Finally, (5.96) and (5.97) are used again to obtain the direction vectors in the room coordinate system. Since the sampling based on the Lambertian pattern offers CIR results with lower variance, it is the default sampling strategy in both the MCRG and MCRS algorithms. The MCRS and MCRG algorithms are concluded in Algorithm 5.6 and Algorithm 5.5, respectively. Note that the CIR result is stored in a discrete vector $\mathbf{h}$ as the one in (5.42) and $\tau(\mathcal{O}_n)/\tau(\mathcal{O}_n)$ refers to the propagation delay of the nth path sample.

Compared to the PTA, MCRG and MCRS algorithms removed the MC process at the reflection/absorption points. In addition, the power collected by the detector is calculated in a deterministic fashion instead of the accumulation of random photon arrival events. These differences make the MCRG and MCRS algorithms more efficient and accurate compared to PTA. Since the same tracing path in the MCRG algorithm can be used to evaluate the channels with respect

Algorithm 5.5 MCRG Algorithm

Initialise $\mathbf{h} = [0, 0, \cdots, 0]$
for $n = 1, 2, \cdots, N_s$ **do**
  Generate path sample $\mathcal{I}_n$
  for $k = 1, 2, \cdots, K$ **do**
    Define $\mathcal{I}_n^k$ as the first k elements of $\mathcal{I}_n$
    Calculate $g_k^{\downarrow}(\mathcal{I}_n^k, t)$ and $\mathfrak{f}_{\mathcal{I}}(\mathcal{I}_n^k)$
    Find the delay index $\hat{n}_t = \arg\min_{n_t} \left|\tau(\mathcal{I}_n^k) - n_t \Delta t\right|$
    Update the CIR result $\mathbf{h}\left(\hat{n}_t\right) = \mathbf{h}\left(\hat{n}_t\right) + \frac{g_k^{\downarrow}(\mathcal{I}_n^k, \tau(\mathcal{I}_n^k))}{N_s \mathfrak{f}_{\mathcal{I}}(\mathcal{I}_n^k)}$
  end
end

Algorithm 5.6 MCRS Algorithm

Initialise $\mathbf{h} = [0, 0, \cdots, 0]$
for $n = 1, 2, \cdots, N_s$ **do**
  Generate path sample $\mathcal{O}_n$
  for $k = 1, 2, \cdots, K$ **do**
    Define $\mathcal{O}_n^k$ as the first k elements of $\mathcal{O}_n$
    Calculate $g_k^{\uparrow}(\mathcal{O}_n^k, t)$ and $\mathfrak{f}_{\mathcal{O}}(\mathcal{O}_n^k)$
    Find the delay index $\hat{n}_t = \arg\min_{n_t} \left|\tau(\mathcal{O}_n^k) - n_t \Delta t\right|$
    Update the CIR result $\mathbf{h}\left(\hat{n}_t\right) = \mathbf{h}\left(\hat{n}_t\right) + \frac{g_k^{\uparrow}(\mathcal{O}_n^k, \tau(\mathcal{O}_n^k))}{N_s \mathfrak{f}_{\mathcal{O}}(\mathcal{O}_n^k)}$
  end
end

to multiple transmitters, it is very efficient when calculating MISO channels. Similarly, the same tracing path in the MCRS algorithm can be used to evaluate the channels for multiple receivers. Therefore, it is efficient when calculating SIMO channels.

5.6.3 Markov Chain Monte Carlo Algorithms

By increasing the number of strong path samples, the variance of the CIR result can be effectively reduced. The strength of a path is determined by the incident/exitent angle at each point and the total propagation distance of the path. However, the importance of the sampling technique is only applied to the sampling of incident (exitent) directions in the MCRG (MCRS) algorithm. Therefore, it does not guarantee that the path has a strong contribution to the

NLOS channel. To further improve the efficiency of the Monte Carlo method, Markov chain Monte Carlo (MCMC) algorithms based on the Metropolis Hastings sampling were proposed in [32]. In the MCMC algorithms, the CIR expressions based on light transportation framework and the corresponding Monte Carlo estimators (5.90) and (5.91) are the same as those in MCRG and MCRS algorithms. The improvement lies in the propagation path sampling method. Based on the new path sampling method, Markov chain Monte Carlo ray-gathering (MCMCRG) and Markov chain Monte Carlo ray-shooting (MCMCRS) algorithms have been developed [32].

5.6.3.1 Metropolis Hastings Sampling

Firstly we introduce the Metropolis Hastings sampling technique. Metropolis Hastings sampling uses a Markov chain process to generate the correlated samples following an unknown distribution where direct sampling is challenging [33]. This technique is very useful for distributions of large dimensions. Now, we go back to the integral and the corresponding Monte Carlo estimator (5.88) and (5.89) with a non-negative integrand $g(u)$. This time, the sampling PDF is defined to be proportional to the integrand function $g(u)$

$$\mathfrak{f}_u(u) = \frac{g(u)}{b} \tag{5.102}$$

where b is a scaling factor to ensure that the integration of the PDF in its sampling domain equals 1. In the Markov chain process, when the current sample state is u_1, there is a certain transition probability that the next sample value will transfer to u_2, which is denoted as $\mathbb{P}_{\mathrm{t}}[u_1 \rightarrow u_2]$. Since the target distribution is $\mathfrak{f}_u(u)$, after a generation of sufficient samples to fulfill the ergodicity condition, the following equilibrium should be fulfilled:

$$\frac{\mathbb{P}_{\mathrm{t}}[u_2 \rightarrow u_1]}{\mathbb{P}_{\mathrm{t}}[u_1 \rightarrow u_2]} = \frac{\mathfrak{f}_u(u_1)}{\mathfrak{f}_u(u_2)} \tag{5.103}$$

which means that when being in state u_1, the probability of choosing u_2 as the next sample is the same as that being state u_2 and choosing u_1 as the next sample. Metropolis Hastings sampling decomposes the transition operation into two steps. Assuming the current state sample is u_1, the algorithm proposes a new sample u_2 based on another conditional distribution $\mathfrak{g}_u(u_2|u_1)$ in the first step. In the second step, the algorithm randomly decides to accept the proposed sample with a probability of $\mathbb{P}_{\mathrm{a}}[u_1 \rightarrow u_2]$. Otherwise, the proposed sample is rejected. Thus, the transition probability can be converted as:

$$\mathbb{P}_{\mathrm{t}}[u_1 \rightarrow u_2] = \mathfrak{g}_u(u_2|u_1)\mathbb{P}_{\mathrm{a}}[u_1 \rightarrow u_2] \tag{5.104}$$

By inserting (5.104) into (5.103), we have

$$\frac{\mathfrak{g}_u(u_1|u_2)\mathbb{P}_a[u_2 \to u_1]}{\mathfrak{g}_u(u_2|u_1)\mathbb{P}_a[u_1 \to u_2]} = \frac{\mathfrak{f}_u(u_1)}{\mathfrak{f}_u(u_2)} \tag{5.105}$$

In order to ensure the equilibrium hold, the acceptance probability can be set as:

$$\mathbb{P}_a[u_1 \to u_2] = \min\left(1, \frac{\mathfrak{g}_u(u_1|u_2)\mathfrak{f}_u(u_2)}{\mathfrak{g}_u(u_2|u_1)\mathfrak{f}_u(u_1)}\right) \tag{5.106}$$

Usually, the choice of $\mathfrak{g}_u(u|\tilde{u})$ is symmetric, which means that $\mathfrak{g}_u(u_1|u_2) = \mathfrak{g}_u(u_2|u_1)$. In this case, in conjunction with (5.102), the acceptance probability can be simplified as:

$$\mathbb{P}_a[u_1 \to u_2] = \min\left(1, \frac{g(u_2)}{g(u_1)}\right) \tag{5.107}$$

5.6.3.2 Propagation Path Sampling in MCMC Algorithms

In the MCMC algorithms, a similar sampling approach has been used to decide the transition of propagation path samples. A tracing path transition in MCMC algorithms are named as a path mutation [32]. To complete the Markov chain mutation process, a conditional distribution to propose a new path is required. For the design of the path proposing conditional distribution in MCMC algorithms, two mutation strategies have been used:

- The first strategy is called a large mutation, where the algorithm finds a new path that is independent of the previous one. The incident/exitant directions are selected uniformly from the hemisphere corresponding to each point, as shown in Figure 5.12(a). This is the same as the uniform direction sampling described in Section 5.6.2.2. Therefore, the conditional PDF $\mathfrak{g}_{\mathcal{I}}(\mathcal{I}_n|\mathcal{I}_{n-1})$ follows the expression (5.92) and $\mathfrak{g}_{\mathcal{O}}(\mathcal{O}_n|\mathcal{O}_{n-1})$ follows the expression (5.93). The sampling of the direction azimuth and elevation angles follow the expressions (5.94) and (5.95).
- The second mutation strategy is called small mutation. In this case, the proposed new sample path will be close to the previous path sample. Therefore, the generation of a new incident/exitant direction is uniformly generated from the cone with an apex angle of $2\psi_c$ centered at the previous sample direction. The corresponding path proposing conditional PDF can be found as:

$$\mathfrak{g}_{\mathcal{O}}(\mathcal{O}_n|\mathcal{O}_{n-1}) = \mathfrak{g}_{\mathcal{I}}(\mathcal{I}_n|\mathcal{I}_{n-1}) = (2\pi(1 - \cos\psi_c))^{-k} \tag{5.108}$$

The generation of the azimuth and elevation angles are described as the following: first, the algorithm generates two random numbers u_1 and

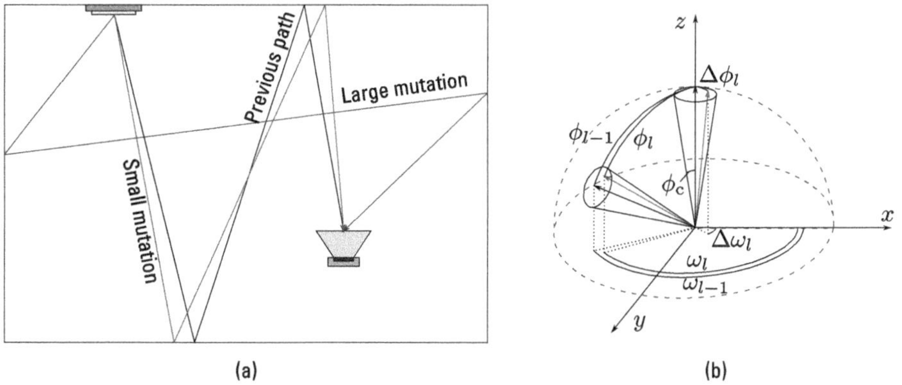

Figure 5.12 (a) Large mutation and small mutation strategies. (b) Small mutation proposed direction implementation geometry.

u_2 that are randomly distributed between 0 and 1. Then the rotation angles about the y-axis and z-axis relative to the previous direction can be calculated by:

$$\alpha_y = \arccos(\cos\psi_c + (1 - \cos\psi_c)u_1) \tag{5.109}$$

$$\alpha_z = 2\pi u_2 \tag{5.110}$$

Then the new direction vector in a local Cartesian coordinate system after four rotations can be calculated as:

$$[x_l, y_l, z_l]^T = \mathbf{R}_z(\omega_{l-1})\mathbf{R}_y(\phi_{l-1})\mathbf{R}_z(\alpha_z)\mathbf{R}_y(\alpha_y)[0, 0, 1]^T \tag{5.111}$$

Then the azimuth angle and the elevation angle of the proposed direction vector can be calculated by:

$$\omega_l = \text{atan2}(y_l, x_l) \tag{5.112}$$

$$\phi_l = \arccos(z_l) \tag{5.113}$$

Finally, the direction vector in the global Cartesian coordinate system can be obtained by using (5.96) and (5.97). An illustration of the small mutation direction implementation is shown in Figure 5.12(b).

When the algorithm finds a strong path, the small mutation strategy helps to explore other potential strong paths that are next to the current one. However, we know that the integrand functions (5.79) and (5.86) have a high dimension of k. If only the small mutation strategy is used, it will be hard for the algorithm to fulfill the ergodicity condition. The use of large mutations is to let the algorithm also explore other distinct paths, which speeds up the convergence of the algorithm. During the iteration, the algorithm randomly decides to use a large mutation or a small mutation. The probability of the algorithm choosing the large mutation

Algorithm 5.7 MCMCRG Algorithm

$\mathbf{h} = [0, 0, \cdots, 0]$
for $k = 1, 2, \cdots, K$ **do**
  Initialize $\mathbf{h}^k = [0, 0, \cdots, 0]$
  $N_L = 0$
  for $n = 1, 2, \cdots, N_s$ **do**
    Generate random numbers $u_1, u_2 \sim \mathbb{U}(0, 1)$
    if $u_1 < \mathcal{P}_L$ **then**
      Propose $\mathcal{I}_n$ using large mutation
      $N_L = N_L + 1$
      else
        Propose $\mathcal{I}_n$ using small mutation
      end
    end
    Calculate $\mathbb{P}_a[\mathcal{I}_{n-1} \to \mathcal{I}_n]$
    if $u_2 < \mathbb{P}_a[\mathcal{I}_{n-1} \to \mathcal{I}_n]$ **then**
      Accept the proposed path $\mathcal{I}_n$
    end
    Calculate delay index $\hat{n}_t = \arg\min_{n_t} |\tau(\mathcal{I}_n) - n_t \Delta t|$
    Update CIR result $\mathbf{h}^k(\hat{n}_t) = \mathbf{h}^k(\hat{n}_t) + 1$
  end
  Calculate value of b_k
  Calculate $\mathbf{h}^k = \frac{b_k}{N_s}\mathbf{h}^k$
  Update the overall CIR vector $\mathbf{h} = \mathbf{h} + \mathbf{h}^k$
end

is denoted as $\mathcal{P}_L$. Then the acceptance probability is calculated using (5.107). The authors of [32] have empirically found that $\mathcal{P}_L = 0.1$ and $\psi_c = 10°$ offer an accurate NLOS channel estimation. In addition, the scaling factor b_k can be calculated using all proposed paths from large mutations as:

$$b_k = \frac{1}{N_L}\sum_{n=1}^{N_L} \frac{\int_0^\infty g_k^{\downarrow}(\mathcal{I}_n^L, t)\mathrm{d}t}{(1 - \cos\psi_{FOV})^{-1}\left(\frac{1}{2\pi}\right)^k} = \frac{1}{N_L}\sum_{n=1}^{N_L} \frac{\int_0^\infty g_k^{\uparrow}(\mathcal{O}_n^L, t)\mathrm{d}t}{\left(\frac{1}{2\pi}\right)^k} \tag{5.114}$$

where N_L is the total number of large mutation path samples; $\mathcal{I}_n^L$ and $\mathcal{O}_n^L$ correspond to the nth path samples proposed in the large mutation process. Gathering the above sampling process, the MCMCRG algorithm is concluded in Algorithm 5.7, while the MCMCRS algorithm is implemented in a similar way. The MCMC algorithms offer an efficient sampling method, which can effectively

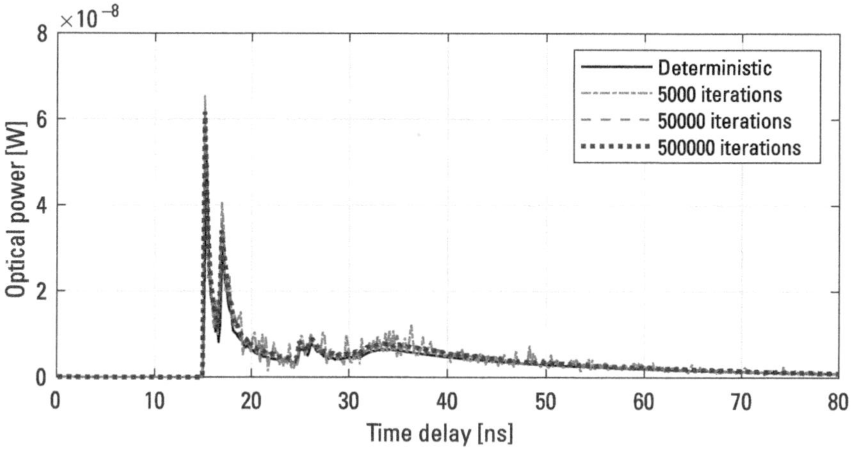

Figure 5.13 The CIR results simulated using MCRG algorithm with different number of samples.

reduce the variance of the simulated NLOS CIR results. A disadvantage of this algorithm is that the MC CIR estimation of various reflection orders has to be conducted separately.

5.6.4 Performance Evaluation

In this section, the performance of the four Monte Carlo approaches is evaluated and compared. Monte Carlo methods require a large number of random samples to converge to the desired result. The CIR results simulated with different numbers of samples are shown in Figure 5.13. The simulation uses MCRG algorithm and configuration A in [12] is used. The parameters are listed in Table 5.1. It is clear that the results simulated using the Monte Carlo method are very close to those which are simulated using the deterministic approach (iterative algorithm). When the number of samples is small (5,000), the result poses a significant noise component. When the sample number increases, the result becomes smoother. In order to quantify the convergence of a Monte Carlo method, two generalized performance metrics were proposed in [34]. The first metric is called relative error, which is defined as:

$$\eta(n_\mathrm{t}) = \sqrt{\frac{\sum_{n=1}^{N(n_\mathrm{t})} C_{n_\mathrm{t},n}^2}{\left(\sum_{n=1}^{N(n_\mathrm{t})} C_{n_\mathrm{t},n}\right)^2} - \frac{1}{N(n_\mathrm{t})}} \tag{5.115}$$

where n_t is the index of the n_tth time slot with time delay of $n_\mathrm{t}\Delta t$, $N(n_\mathrm{t})$ represents the number of path falls in the n_tth time slot and $C_{n_\mathrm{t},n}$ refers to the power contribution from the nth path to the n_tth time slot. Note that in the original definition, the summation term counts only the nonzero contributions. However, even when we include the zero contribution paths, the calculation result is the same. To conclude the relative error to be a single value quantity, the following mean relative error is defined:

$$\bar{\eta} = \frac{\sum_{n_t=1}^{N_t} \mathbf{h}(n_t)\eta(n_t)}{\sum_{n_t=1}^{N_t} \mathbf{h}(n_t)} \tag{5.116}$$

where the relative error is weighted by the corresponding CIR and averaged over the time delay. The performance of MCRG, MCRS, MCMCRG, and MCMCRS algorithms in terms of a mean relative error and the relative error with 5×10^5 samples are shown in Fig 5.14. It can be observed that with an increase in the number of samples, the mean relative error decreases exponentially after the number of samples exceeds around 1,000. The mean relative error decreasing speeds of various algorithms are generally the same. With the same number of samples, the mean relative error achieved by the MCRG algorithm is about one-third of that achieved by the MCRS algorithm. Furthermore, the mean relative error achieved by MCMCRG and MCMCRS algorithms are about one-third of that achieved by MCRG algorithm. The performance of MCMCRG and MCMCRS algorithms are similar. In Figure 5.14(b), the relative error can drop below 0.05 for the MCRS algorithm while the MCMCRG and MCMCRS algorithms achieve a much lower relative error of less than 0.01. From these results, the superiority of the MCMC algorithms can be observed.

5.7 Analytical NLOS Channel Modeling

Deterministic and Monte Carlo approaches provide accurate NLOS channel simulation results. However, they are relatively complex to implement and the simulations can take much longer to execute. Furthermore, the results are purely numerical and do not give any insight into the characteristics of the NLOS channel. Therefore, a number of analytical models have been proposed to approximate and characterize the NLOS channels, such as the method introduced in [21]. Due to the simplified assumptions, these models generally provide less accurate results, but they are convenient to use and provide higher tractability.

5.7.1 Sphere-Integrating/Exponential-Decaying Model

One of the analytical approaches is to approximate the indoor environment as a perfect sphere [21]. When considering the light propagation inside the sphere, the CIR can be simplified as an exponential-decaying function as:

$$h(t) = \frac{g_{\mathrm{NLOS}}}{\tau_0} \exp\left(-\frac{t}{\tau_0}\right) \mathrm{u}(t) \tag{5.117}$$

where g_{NLOS} corresponds to the NLOS path loss and τ_0 is the exponential decay time constant. The corresponding transfer function becomes a first-order

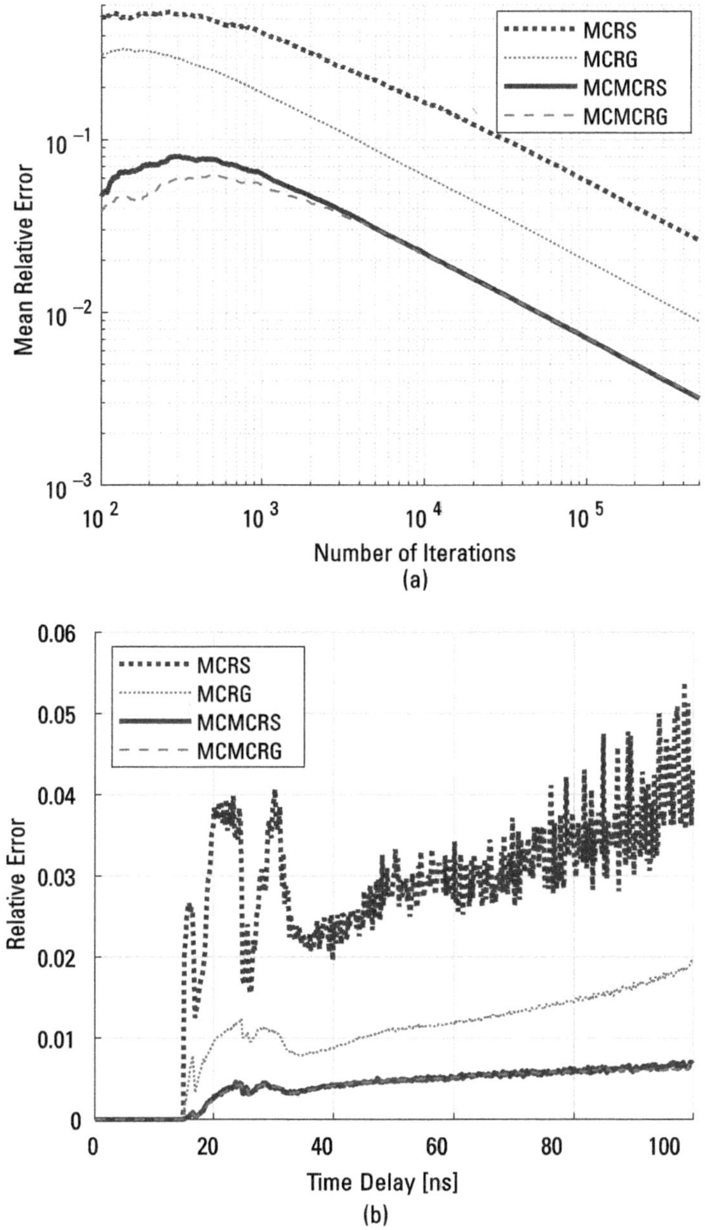

Figure 5.14 (a) Mean relative error against number of samples. (b) Relative error against time delay.

Butterworth lowpass filter:

$$H(f) = \frac{g_{\text{NLOS}}}{1 + 2\pi j \tau_0 f} \tag{5.118}$$

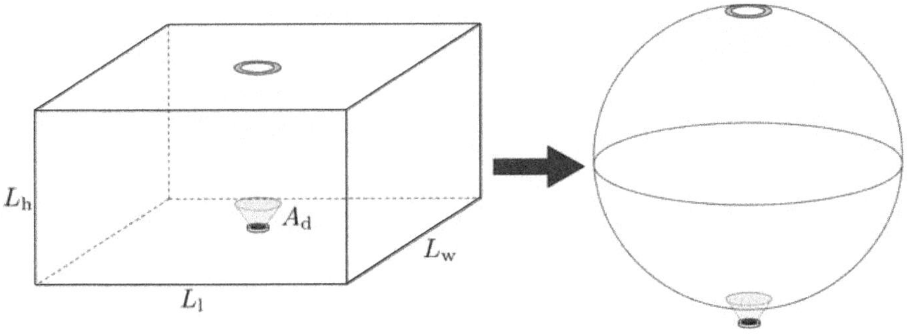

Figure 5.15 Geometry of sphere-integrating/exponential-decaying model.

The sphere-integrating model defines g_{NLOS} and τ_0 as functions of the indoor environment parameters. Let's consider a NLOS channel in a cubic room of dimension $L_l \times L_w \times L_h$, as shown in Figure 5.15. There are six surfaces in the room. Then the internal area and the volume of the room can be calculated as:

$$A_{\text{room}} = 2(L_l L_w + L_l L_h + L_w L_h) \tag{5.119}$$

$$V_{\text{room}} = L_l L_w L_h \tag{5.120}$$

The reflectance and the area of the ith surface are defined as ρ_i and A_i. Then the average room reflectance can be calculated as:

$$\bar{\rho} = \frac{\sum_{i=1}^{6} \rho_i A_i}{A_{\text{room}}} \tag{5.121}$$

With a receiver detector area of A_d, the path loss can be estimated by:

$$g_{\text{NLOS}} = \frac{A_d \bar{\rho}}{A_{\text{room}}(1 - \bar{\rho})} \tag{5.122}$$

The exponential decay time constant can be estimated by:

$$\tau_0 = -\frac{4V_{\text{room}}}{cA_{\text{room}} \ln \bar{\rho}} \tag{5.123}$$

This simple model has been found to offer very close approximations in a number of indoor scenarios [21,25], and more importantly, it is easy to use with just a few geometric parameters: room dimension, reflectance, and detector size.

5.7.2 Ceiling Bounce Model

Another simplified model assumes a unique link scenario, where the transmitter and receiver are both lower than an infinitely large ceiling and both have an upward orientation towards the ceiling, as shown in Figure 5.16(a). The separation

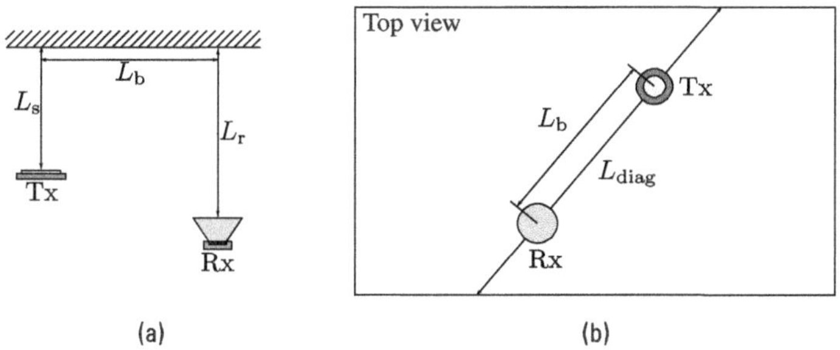

Figure 5.16 (a) Ceiling-bounce model geometry. (b) Geometry of the parameter s: the ratio of transmitter-to-receiver horizontal offset to the length of transmitter-to-receiver diagonal.

between the transmitter (receiver) and the ceiling is denoted as L_s (L_r). The horizontal offset between the transmitter and receiver is denoted as L_b. An expression of this ceiling-bounce model CIR is found as [5]:

$$\hat{h}(t) = \frac{d}{dt}\int_{\mathcal{W}_0(t)} \frac{\rho A_d L_s^2 L_r^2}{\pi\left(L_s^2 + \left(x - \frac{L_b}{2}\right)^2 + y^2\right)^2\left(L_r^2 + \left(x + \frac{L_b}{2}\right)^2 + y^2\right)^2} dx dy \tag{5.124}$$

where

$$\mathcal{W}_0(t) : \frac{\left((L_s^2 - L_r^2) + 2L_b x\right)^2}{c^2t^2\left(L_b^2 + (L_s - L_r)^2\right)} + \frac{4y^2}{c^2t^2 - (L_b^2 + (L_s - L_r)^2)} + \frac{(L_b(L_s + L_r) - 2x(L_s - L_r))^2}{c^2t^2\left(L_b^2 + (L_s - L_r)^2\right) - (L_b^2 + (L_s - L_r)^2)^2} < 1 \tag{5.125}$$

In order to further lower the complexity of (5.124), the authors of [5] derived a closed-form expression for a special case with $L_{tx} = L_{rx} = \tilde{L}$ and $L_b = 0$ as:

$$h_{cb}(t) = \frac{2^7 \rho A_d \tilde{L}^4 u\left(t - \frac{2\tilde{L}}{c}\right)}{\pi c^6 t^7} \tag{5.126}$$

To make the expression more generalized for various link environments, further modification is applied. First, the function is left-shifted by $\frac{2\tilde{L}}{c}$. By letting $\frac{2\tilde{L}}{c} = a$ and replacing $\frac{\rho A_d}{3\pi\tilde{L}^2}$ with g_{NLOS}, the following simplified expression is concluded as:

$$h_{cb}(a, t) = \frac{6a^6 g_{NLOS} u(t)}{(t + a)^7} \tag{5.127}$$

where a and g_{NLOS} are two custom parameters that adapt with the propagation environment. The value of parameter a in the unshadowed and shadowed scenarios are predicted by:

$$a_{\text{unshad}} = 12\sqrt{11/13}(2.1 - 5s + 20.8s^2)\tau_{\text{rms}}\left(\hat{h}(t)\right) \tag{5.128}$$

$$a_{\text{shad}} = 12\sqrt{11/13}(2 + 9.4s)\tau_{\text{rms}}\left(\hat{h}(t)\right) \tag{5.129}$$

where the parameter s is the ratio of the transmitter-to-receiver horizontal offset and the length of the transmitter-to-receiver diagonal, as shown in Figure 5.16(b), $\tau_{\text{rms}}(h(t))$ is the RMS delay spread of the CIR, and $h(t)$ and $\hat{h}(t)$ need to be obtained from the numerical integration (5.124). The path loss parameter g_{NLOS} in the shadowed and unshadowed scenarios with $m = 1$ are estimated by:

$$g_{\text{NLOS,unshadowed},m=1} = 10^{0.18+s}\int \hat{h}(t)\mathrm{d}t \tag{5.130}$$

$$g_{\text{NLOS,shadowed},m=1} = 10^{-0.14+1.02s}\int \hat{h}(t)\mathrm{d}t \tag{5.131}$$

respectively. To include cases that have a different transmitter beam profile, the path loss parameter g_{NLOS} in the unshadowed scenario with $m = 1.85$ is estimated by:

$$g_{\text{NLOS,unshadowed},m=1.85} = 10^{0.37-1.2s+2.55s^2}\int \hat{h}(t)\mathrm{d}t \tag{5.132}$$

The estimation function of a and g_{NLOS} are obtained via a curve fitting process. As a result, the resultant CIRs are matched with the experimental data. The ceiling bounce model can be used to estimate the transmission performance in wireless IR diffused link scenarios [7].

5.7.3 Efficient Analytical Calculation Method

Despite the simplicity of the ceiling bounce model, it cannot be used directly in the VLC scenario where a light source is mounted on the ceiling. Therefore, a more generalized case for the ceiling bounce model with arbitrary transmitter/receiver characteristics (orientation, FOV, beamwidth) has been considered in [35]. The modified ceiling bounce model is called a single-reflection model and it considers a similar geometry as that in the ceiling bounce model, which is illustrated in Figure 5.17(a). We assume an infinitely large reflective surface. An optical source and a detector are positioned L_s and L_r away from the surface, respectively. They also have a horizontal separation of L_b. Different from the ceiling bounce model, the transmitter and the receiver have orientations of $\mathbf{p}_s = [\tilde{x}_s, \tilde{y}_s, \tilde{z}_s]$ and $\mathbf{p}_r = [\tilde{x}_r, \tilde{y}_r, \tilde{z}_r]$, respectively. The orientation vectors are defined in a local

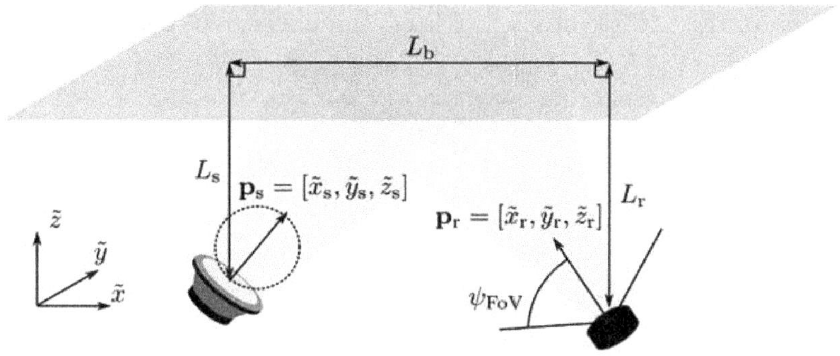

(a)

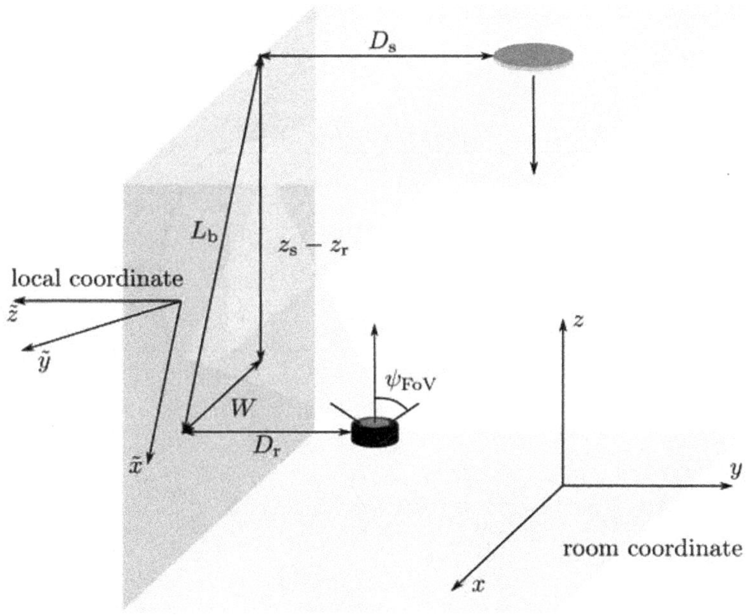

(b)

Figure 5.17 (a) Single reflection model geometry. (b) Transmitter-to-wall-to-receiver channel geometry.

Cartesian coordinate system $\tilde{x}$-$\tilde{y}$-$\tilde{z}$, which has the $\tilde{z}$-axis towards the ceiling. In addition, the $\tilde{x}$-axis is parallel to the reflective surface. From the top view, the x-axis passes through the positions of the transmitter and receiver, as shown in Figure 5.17(a). With a transmitter Lambertian order of m, a receiver FOV of ψ_{FOV}, a receiver detector size of A_{d}, and a surface reflectivity of ρ, an exact expression for the CIR of the single reflection model can be concluded as [36]:

$$\hbar^{\mathrm{p_s},\mathrm{p_r}}_{\{m,A_\mathrm{r},\rho,L_\mathrm{s},L_\mathrm{r},L_\mathrm{b},\psi_\mathrm{FOV}\}}(t) = \frac{\rho(m+1)L_\mathrm{s}L_\mathrm{r}A_\mathrm{d}}{2\pi^2}\mathrm{u}\left(t-\frac{L_0}{c}\right)\int_0^{2\pi}\mathfrak{r}_\theta(t)G(\mathfrak{r}_\theta(t),\theta)\frac{\mathrm{d}\mathfrak{r}_\theta(t)}{\mathrm{d}t}\mathrm{d}\theta \tag{5.133}$$

where

$$G(r,\theta) = \frac{\left(r\left(\tilde{x}_\mathrm{s}\cos\theta+\tilde{y}_\mathrm{s}\sin\theta\right)+\frac{L_\mathrm{s}L_\mathrm{b}\tilde{x}_\mathrm{s}}{L_\mathrm{s}+L_\mathrm{r}}+L_\mathrm{s}\tilde{z}_\mathrm{s}\right)^m}{\left(r^2+\frac{2L_\mathrm{s}L_\mathrm{b}r\cos\theta}{L_\mathrm{s}+L_\mathrm{r}}+L_\mathrm{s}^2+\frac{L_\mathrm{s}^2L_\mathrm{b}^2}{(L_\mathrm{s}+L_\mathrm{r})^2}\right)^{\frac{m+3}{2}}} \times \frac{\left(r\left(\tilde{x}_\mathrm{r}\cos\theta+\tilde{y}_\mathrm{r}\sin\theta\right)-\frac{L_\mathrm{r}L_\mathrm{b}\tilde{x}_\mathrm{r}}{L_\mathrm{s}+L_\mathrm{r}}+L_\mathrm{r}\tilde{z}_\mathrm{r}\right)\mathbf{1}_\mathrm{v}}{\left(r^2-\frac{2L_\mathrm{r}L_\mathrm{b}r\cos\theta}{L_\mathrm{s}+L_\mathrm{r}}+L_\mathrm{r}^2+\frac{L_\mathrm{r}^2L_\mathrm{b}^2}{(L_\mathrm{s}+L_\mathrm{r})^2}\right)^2} \tag{5.134}$$

$$\mathfrak{r}_\theta(t) = \frac{ct\sqrt{\mathfrak{c}_\theta(t)}-\left(c^2t^2-L_0^2\right)\left(L_\mathrm{s}-L_\mathrm{r}\right)L_\mathrm{b}\cos\theta}{2\left(c^2t^2-L_\mathrm{b}^2\cos^2\theta\right)\left(L_\mathrm{s}+L_\mathrm{r}\right)} \tag{5.135}$$

$$\frac{\mathrm{d}\mathfrak{r}_\theta(t)}{\mathrm{d}t} = \frac{2c^2tL_\mathrm{b}\cos\theta\left(L_\mathrm{s}-L_\mathrm{r}\right)\left(L_\mathrm{b}^2\cos^2\theta-L_0^2\right)}{2(L_\mathrm{s}+L_\mathrm{r})\left(c^2t^2-L_\mathrm{b}^2\cos^2\theta\right)^2} - \frac{c\left(c^2t^2+L_\mathrm{b}^2\cos^2\theta\right)\sqrt{\mathfrak{c}_\theta(t)}}{2(L_\mathrm{s}+L_\mathrm{r})\left(c^2t^2-L_\mathrm{b}^2\cos^2\theta\right)^2} + \frac{c^3t^2\left(L_\mathrm{s}-L_\mathrm{r}\right)^2\left(c^2t^2-L_0^2\right)}{\left(L_\mathrm{s}+L_\mathrm{r}\right)\left(c^2t^2-L_\mathrm{b}^2\cos^2\theta\right)\sqrt{\mathfrak{c}_\theta(t)}} + \frac{2c^3t^2L_\mathrm{s}L_\mathrm{r}\left(2c^2t^2-L_0^2-L_\mathrm{b}^2\cos^2\theta\right)}{\left(L_\mathrm{s}+L_\mathrm{r}\right)\left(c^2t^2-L_\mathrm{b}^2\cos^2\theta\right)\sqrt{\mathfrak{c}_\theta(t)}} \tag{5.136}$$

$$\mathfrak{c}_\theta(t) = 4L_\mathrm{s}L_\mathrm{r}\left(c^2t^2-L_\mathrm{b}^2\cos^2\theta\right)\left(c^2t^2-L_0^2\right) + \left(L_\mathrm{s}-L_\mathrm{r}\right)^2\left(c^2t^2-L_0^2\right)^2 \tag{5.137}$$

$$L_0 = \sqrt{L_\mathrm{b}^2+\left(L_\mathrm{s}+L_\mathrm{r}\right)^2} \tag{5.138}$$

In the light path function (5.134), $\mathbf{1}_\mathrm{v}$ is the visibility factor, which equals unity ($\mathbf{1}_\mathrm{v} = 1$) when the light radiant angle is less than 90° and the incident angle is smaller the the detector FOV. This corresponds to the following two conditions:

$$\frac{r(\tilde{x}_\mathrm{r}\cos\theta+\tilde{y}_\mathrm{r}\sin\theta)-\frac{\tilde{x}_\mathrm{r}L_\mathrm{r}L_\mathrm{b}}{L_\mathrm{s}+L_\mathrm{r}}+\tilde{z}_\mathrm{r}L_\mathrm{r}}{\sqrt{r^2-\frac{2L_\mathrm{r}L_\mathrm{b}r\cos\theta}{L_\mathrm{s}+L_\mathrm{r}}+L_\mathrm{r}^2+\frac{L_\mathrm{r}^2L_\mathrm{b}^2}{(L_\mathrm{s}+L_\mathrm{r})^2}}} \geq \cos\psi_\mathrm{FOV} \tag{5.139}$$

$$r(\tilde{x}_{\mathrm{r}} \cos\theta + \tilde{y}_{\mathrm{r}} \sin\theta) + \frac{\tilde{x}_{\mathrm{s}} L_{\mathrm{s}} L_{\mathrm{b}}}{L_{\mathrm{s}} + L_{\mathrm{r}}} + \tilde{z}_{\mathrm{s}} L_{\mathrm{s}} \geq 0 \tag{5.140}$$

If either of the two conditions is not fulfilled, the visibility function equals zero ($\mathbf{1}_{\mathrm{v}} = 0$). Despite the more complicated form of (5.133) compared to the ceiling bounce model, this expression is more general and accurate. In terms of computational complexity, it requires only one numerical integration with respect to an angle with finite limits in the range of $[0, 2\pi)$. Therefore, it is more efficient and more tractable than deterministic or Monte Carlo ray-tracing approaches. With the single reflection model as a basis, it is possible to develop the CIR expressions for part of the VLC/LiFi NLOS channel with several specific scenarios. Considering a typical VLC/Li-Fi channel, a visible light transmitter is deployed on the ceiling with a downward orientation, which is z_{s} above the floor. In addition, a receiver is deployed at a height of z_{r} above the floor with an upward detector orientation.

5.7.3.1 Transmitter-to-Wall-to-Receiver Channel

The first scenario considers the NLOS channel due to the first-order reflection by a single wall, which is called the transmitter-to-wall-to-receiver channel. The geometry of the scenario is illustrated in Figure 5.17(b), which shows similarities with the single reflection model. Therefore, we can directly use the single reflection model to evaluate the transmitter-to-wall-to-receiver channel. In the room coordinate system, the transmitter-to-receiver separation in the z direction equals $z_{\mathrm{s}} - z_{\mathrm{r}}$ and the separation in the x direction is assumed to be W. Then the input to the parameter L_{b} can be found as $\sqrt{W^2 + (z_{\mathrm{s}} - z_{\mathrm{r}})^2}$, as shown in Figure 5.17(b). In addition, the transmitter and receiver orientation vectors in the room coordinate system equal $[0, 0, -1]$ and $[0, 0, 1]$, respectively. With some vector rotation, it can be found that the orientation vectors in the local coordinate system becomes:

$$\mathbf{p}_{\mathrm{s,twr}} = \left[\frac{z_{\mathrm{s}} - z_{\mathrm{r}}}{\sqrt{W^2 + (z_{\mathrm{s}} - z_{\mathrm{r}})^2}}, \frac{W}{\sqrt{W^2 + (z_{\mathrm{s}} - z_{\mathrm{r}})^2}}, 0 \right] \tag{5.141}$$

$$\mathbf{p}_{\mathrm{r,twr}} = \left[-\frac{z_{\mathrm{s}} - z_{\mathrm{r}}}{\sqrt{W^2 + (z_{\mathrm{s}} - z_{\mathrm{r}})^2}}, -\frac{W}{\sqrt{W^2 + (z_{\mathrm{s}} - z_{\mathrm{r}})^2}}, 0 \right] \tag{5.142}$$

Thus, the CIR for the transmitter-to-wall-to-receiver channel can be concluded as:

$$h_{\mathrm{twr}}^{\{D_{\mathrm{s}}, D_{\mathrm{r}}, W, \rho_{\mathrm{w}}\}}(t) = \hbar^{\mathbf{p}_{\mathrm{s,twr}}, \mathbf{p}_{\mathrm{r,twr}}}_{\left\{m, A_{\mathrm{pd}}, \rho_{\mathrm{w}}, D_{\mathrm{s}}, D_{\mathrm{r}}, \sqrt{W^2 + (z_{\mathrm{s}} - z_{\mathrm{r}})^2}, \psi_{\mathrm{FOV}}\right\}}(t) \tag{5.143}$$

where A_{pd} is the detector physical area and ρ_{w} is the reflectance of the wall.

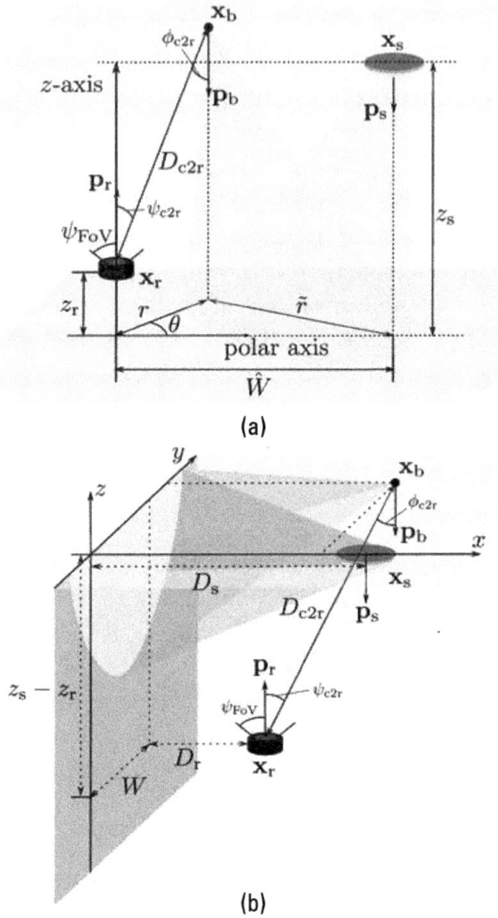

Figure 5.18 (a) Transmitter-to-floor-to-ceiling-to-receiver channel geometry. (b) Transmitter-to-wall-to-ceiling-to-receiver channel geometry.

5.7.3.2 Transmitter-to-Floor-to-Ceiling-to-Receiver Channel

The second considered scenario is part of a second-order reflection channel called the transmitter-to-floor-to-ceiling-to-receiver channel. The corresponding geometry is demonstrated in Figure 5.18(a). We first consider the channel from the transmitter to point $\mathbf{x}_b$ on the ceiling via a reflection by the floor, which is called the transmitter-to-floor-to-ceiling channel. It can be directly evaluated by applying the single-reflection model:

$$\mathrm{d}h^{\mathbf{x}_b}_{\mathrm{tfc}}(t) = \hbar^{[0,0,1],[0,0,1]}_{\{m,\mathrm{d}A(\mathbf{x}_b),\rho_f,z_s,z_s,\tilde{r},90^\circ\}}(t) \qquad (5.144)$$

where ρ_f is the reflectance of the floor and the point position vector $\mathbf{x}_b$ can be redefined by a cylinder coordinate system $[r, \theta, z_s]$. Given a horizontal offset between the transmitter and receiver of $\hat{W}$, the horizontal offset between the transmitter and the point $\mathbf{x}_\mathrm{b}$ can be calculated using the law of cosine as:

$$\tilde{r} = \sqrt{r^2 + \hat{W}^2 - 2r\hat{W}\cos\theta} \tag{5.145}$$

In addition, the term $A(\mathbf{x}_\mathrm{b})$ can be replaced by $r\mathrm{d}r\mathrm{d}\theta$. In order to lower the computational complexity, the following approximation to (5.144) is used:

$$\begin{aligned} \mathrm{d}h_{\mathrm{tfc}}^{\mathbf{x}_\mathrm{b}}(t) &\approx C_{\mathrm{tfc}}\mathrm{u}\left(t - \frac{D_{\mathrm{tfc}}}{c}\right)\hbar_{\{m,r\mathrm{d}r\mathrm{d}\theta,\rho_\mathrm{f},z_\mathrm{s},z_\mathrm{s},0,90^\circ\}}^{[0,0,1],[0,0,1]}(t) \\ &= \frac{(m+1)r\mathrm{d}r\mathrm{d}\theta\rho_\mathrm{f}z_\mathrm{s}^{m+3}C_{\mathrm{tfc}}}{\pi(c/2)^{m+5}t^{m+6}}\mathrm{u}\left(t - \frac{D_{\mathrm{tfc}}}{c}\right) \end{aligned} \tag{5.146}$$

where D_{tfc} denotes the shortest propagation distance in the transmitter-to-floor-to-ceiling channel and C_{tfc} is a scaling coefficient. They are calculated by:

$$D_{\mathrm{tfc}} = \sqrt{4z_\mathrm{s}^2 + \tilde{r}^2} \tag{5.147}$$

$$C_{\mathrm{fcr}} = \beta_{\mathrm{tfc}}\frac{\tilde{r}^2}{z_\mathrm{s}^2} + 1 \tag{5.148}$$

where the coefficient β_{tfc} can be estimated by $\beta_{\mathrm{tfc}} = 0.006242(m + 0.2578)^2 + 0.1328$. By using the approximation (5.146), the CIR for the transmitter-to-floor-to-ceiling-to-receiver channel can be calculated by:

$$\begin{aligned} h_{\mathrm{tfcr}}^{\{\hat{W}\}}(t) &= \int_{-\pi}^{\pi}\int_{0}^{\infty}\frac{\rho_\mathrm{c}(z_\mathrm{s}-z_\mathrm{r})^2A_{\mathrm{pd}}\hat{\mathbf{1}}_\mathrm{v}}{\pi(r^2+(z_\mathrm{s}-z_\mathrm{r})^2)^2}\mathrm{d}h_{\mathrm{tfc}}^{\mathbf{x}_\mathrm{b}}\left(t - \frac{\sqrt{r^2+(z_\mathrm{s}-z_\mathrm{r})^2}}{c}\right) \\ &= \frac{2^{m+6}cA_{\mathrm{pd}}\rho_\mathrm{f}\rho_\mathrm{c}(m+1)}{\pi^2(z_\mathrm{s}-z_\mathrm{r})^{-2}z_\mathrm{s}^{-m-1}}\mathrm{u}\left(t - \sqrt{(3z_\mathrm{s}-z_\mathrm{r})^2+\hat{W}^2/c}\right) \\ &\times \int_{\mathfrak{r}_1}^{\mathfrak{r}_2}\frac{\vartheta\left(\beta_{\mathrm{tfc}}\left(r^2+\hat{W}^2-\frac{2r\hat{W}\sin\vartheta}{\vartheta}\right)+z_\mathrm{s}^2\right)r\hat{\mathbf{1}}_\mathrm{v}\mathrm{d}r}{(r^2+(z_\mathrm{s}-z_\mathrm{r})^2)^2\left(ct-\sqrt{r^2+(z_\mathrm{s}-z_\mathrm{r})^2}\right)^{m+6}} \end{aligned} \tag{5.149}$$

where

$$\mathfrak{r}_1 = \max(r_1, 0), \text{ with } r_1 = \frac{\hat{W}\left(c^2t^2 - \mathfrak{c}_{\mathrm{tfcr},1}\right) - ct\sqrt{\mathfrak{c}_{\mathrm{tfcr},2}}}{2\left(c^2t^2 - \hat{W}^2\right)} \tag{5.150}$$

$$\mathfrak{r}_2 = \frac{\hat{W}\left(c^2t^2 - \mathfrak{c}_{\mathrm{tfcr},1}\right) + ct\sqrt{\mathfrak{c}_{\mathrm{tfcr},2}}}{2\left(c^2t^2 - \hat{W}^2\right)} \tag{5.151}$$

$$\vartheta = \begin{cases} \arccos\left(\frac{2ct\sqrt{r^2+(z_s-z_r)^2}-c^2t^2+\mathfrak{c}_{\text{tfcr},1}}{2r\hat{W}}\right) & : r \geq -r_1 \\ \pi & : r < -r_1 \end{cases} \tag{5.152}$$

$$\mathfrak{c}_{\text{tfcr},1} = \hat{W}^2 + (z_s + z_r)(3z_s - z_r) \tag{5.153}$$

$$\mathfrak{c}_{\text{tfcr},2} = \left(c^2t^2 - \mathfrak{c}_{\text{tfcr},1}\right)^2 - 4(z_s - z_r)^2\left(c^2t^2 - \hat{W}^2\right) \tag{5.154}$$

$$\hat{\mathbb{1}}_v = \begin{cases} 1 & : (z_s - z_r) \geq \sqrt{r^2 + (z_s - z_r)^2}\cos\psi_{\text{FOV}} \\ 0 & : \text{otherwise} \end{cases} \tag{5.155}$$

5.7.3.3 Transmitter-to-Wall-to-Ceiling-to-Receiver Channel

There is one more scenario considered in [36], which is called the transmitter-to-wall-to-ceiling-to-receiver channel, as shown in Figure 5.18(b). Similarly, we define a point on the ceiling $\mathbf{x}_b = [x, y, z_s]$, and calculate the CIR from the transmitter to the point $\mathbf{x}_b$ via the reflection by a wall. Again, the solution can be obtained via the single reflection model as:

$$\mathrm{d}h_{\text{twc}}^{\mathbf{x}_b}(t) = \hbar_{\{m,\mathrm{d}A(\mathbf{x}_b),\rho_w,D_s,x,|y|,90^\circ\}}^{[0,1,0],[0,1,0]}(t) \tag{5.156}$$

In order to lower the computational complexity, another approximation is used to (5.156). First, we modify four of the input parameters D_s, x, $\mathrm{d}A(\mathbf{x}_b)$, and $|y|$, where $\mathrm{d}A(\mathbf{x}_b)$ is replaced by $\mathrm{d}x\mathrm{d}y$, $|y|$ is set to zero, D_s is replaced by

$$\hat{D}_s = \beta_{\text{twc},3}\sqrt{(D_s + x)^2 + y^2} \tag{5.157}$$

and x is replaced by

$$\hat{x} = (1 - \beta_{\text{twc},3})\sqrt{(D_s + x)^2 + y^2} \tag{5.158}$$

respectively. Note that $\beta_{\text{twc},3} = D_s\left((D_s + x)^2 + 4y^2(m^{0.045} - 1)\right)^{-\frac{1}{2}}$. Thus, (5.156) can be approximated by the following expression with a scaling:

$$\mathrm{d}h_{\text{twc}}^{\mathbf{x}_b}(t) \approx C_{\text{twc}}\hbar_{\left\{m,\mathrm{d}x\mathrm{d}y,\rho_w,\hat{D}_s,\hat{x},0,90^\circ\right\}}^{[0,1,0],[0,1,0]} = \frac{2^m\rho_w\hat{D}_s\hat{x}C_{\text{twc}}\mathcal{B}\left(\frac{m+2}{2},\frac{m+2}{2}\right)}{\pi^2 t}$$
$$\times \frac{(m+1)\left(\frac{c^2t^2}{4} + \frac{\left(\hat{D}_s^2-\hat{x}^2\right)^2}{4c^2t^2} - \frac{\hat{D}_s^2+\hat{x}^2}{2}\right)\mathrm{u}\left(t - \frac{\sqrt{(D_s+x)^2+y^2}}{c}\right)\mathrm{d}x\mathrm{d}y}{\left(1 + \hat{D}_s^2\left(\frac{c^2t^2}{4} + \frac{\left(\hat{D}_s^2-\hat{x}^2\right)^2}{4c^2t^2} - \frac{\hat{D}_s^2+\hat{x}^2}{2}\right)^{-1}\right)^{\frac{m-1}{2}}\left(\frac{c^2t^2}{4} - \frac{\left(\hat{D}_s^2-\hat{x}^2\right)^2}{4c^2t^2}\right)^3} \tag{5.159}$$

where $\mathcal{B}(u_1, u_2)$ is the β-function and the scaling factor is defined as:

$$\mathcal{C}_{\mathrm{twc}} = \left(1 + \frac{y^2 + \beta_{\mathrm{twc},1}\frac{y^3|x-D_\mathrm{s}|^3}{(x+D_\mathrm{s})^4}}{(x + \beta_{\mathrm{twc},2}D_\mathrm{s})^2}\right)^{-\frac{1}{2}} \tag{5.160}$$

with $\beta_{\mathrm{twc},1} = 0.8125$ and $\beta_{\mathrm{twc},2} = 0.8088\mathrm{e}^{-0.6878m} + 0.5304\mathrm{e}^{-0.007006m}$.

Then the CIR for the transmitter-to-wall-to-ceiling-to-receiver channel can be calculated by:

$$\begin{aligned} h_{\mathrm{twcr}}^{\{D_\mathrm{s},D_\mathrm{r},W,\rho_\mathrm{w}\}}(t) &= \int_{-\infty}^{\infty}\int_{0}^{\infty} \frac{\rho_\mathrm{c}(z_\mathrm{s}-z_\mathrm{r})^2 A_{\mathrm{pd}}\hat{\mathbf{1}}_\mathrm{v}}{\pi\left((z_\mathrm{s}-z_\mathrm{r})^2 + (x-D_\mathrm{r})^2 + (y-W)^2\right)^2} \\ &\times \mathrm{d}h_{\mathrm{twc}}^{\mathrm{x_b}}\left(t - \sqrt{(z_\mathrm{s}-z_\mathrm{r})^2 + (x-D_\mathrm{r})^2 + (y-W)^2}/c\right) \\ &= \frac{2^m\rho_\mathrm{w}\rho_\mathrm{c}A_{\mathrm{pd}}\mathcal{B}\left(\frac{m+2}{2},\frac{m+2}{2}\right)}{\pi^3(m+1)^{-1}(z_\mathrm{s}-z_\mathrm{r})^{-2}} \int_{\mathfrak{y}_1}^{\mathfrak{y}_2}\int_{0}^{\mathfrak{x}} \frac{\left(\frac{c^2\left(t-\frac{D_{\mathrm{c2r}}}{c}\right)^2}{4} + \frac{\left(\hat{D}_\mathrm{s}^2-\hat{x}^2\right)^2}{4c^2\left(t-\frac{D_{\mathrm{c2r}}}{c}\right)^2} - \frac{\hat{D}_\mathrm{s}^2+\hat{x}^2}{2}\right)}{\left(t-\frac{D_{\mathrm{c2r}}}{c}\right)\left(\frac{(ct-D_{\mathrm{c2r}})^2}{4} - \frac{\left(\hat{D}_\mathrm{s}^2-\hat{x}^2\right)^2}{4(ct-D_{\mathrm{c2r}})^2}\right)^3} \\ &\times \frac{\hat{\mathbf{1}}_\mathrm{v}\hat{D}_\mathrm{s}\hat{x}\mathcal{C}_{\mathrm{twc}}\mathrm{u}\left(t - \frac{1}{c}\sqrt{W^2 + \left(D_\mathrm{s} + \sqrt{(z_\mathrm{s}-z_\mathrm{r})^2 + D_\mathrm{r}^2}\right)^2}\right)\mathrm{d}x\mathrm{d}y}{\left(1 + \hat{D}_\mathrm{s}^2\left(\frac{c^2\left(t-\frac{D_{\mathrm{c2r}}}{c}\right)^2}{4} + \frac{\left(\hat{D}_\mathrm{s}^2-\hat{x}^2\right)^2}{4c^2\left(t-\frac{D_{\mathrm{c2r}}}{c}\right)^2} - \frac{\hat{D}_\mathrm{s}^2+\hat{x}^2}{2}\right)^{-1}\right)^{\frac{m-1}{2}} D_{\mathrm{c2r}}^4} \end{aligned} \tag{5.161}$$

where

$$D_{\mathrm{c2r}} = \sqrt{(z_\mathrm{s}-z_\mathrm{r})^2 + (x-D_\mathrm{r})^2 + (y-W)^2} \tag{5.162}$$

$$\mathfrak{y}_2 = \frac{W\left(c^2t^2 - \mathfrak{c}_{\mathrm{twcr},1} + 2(D_\mathrm{s}+D_\mathrm{r})x\right) + ct\sqrt{\mathfrak{c}_{\mathrm{twcr},2}}}{2\left(c^2t^2 - W^2\right)} \tag{5.163}$$

$$\mathfrak{y}_1 = \frac{W\left(c^2t^2 - \mathfrak{c}_{\mathrm{twcr},1} + 2(D_\mathrm{s}+D_\mathrm{r})x\right) - ct\sqrt{\mathfrak{c}_{\mathrm{twcr},2}}}{2\left(c^2t^2 - W^2\right)} \tag{5.164}$$

$$\mathfrak{x} = \frac{\left(\sqrt{c^2t^2 - W^2} - D_\mathrm{s}\right)^2 - (z_\mathrm{s}-z_\mathrm{r})^2 - D_\mathrm{r}^2}{2\left(\sqrt{c^2t^2 - W^2} - D_\mathrm{s} - D_\mathrm{r}\right)} \tag{5.165}$$

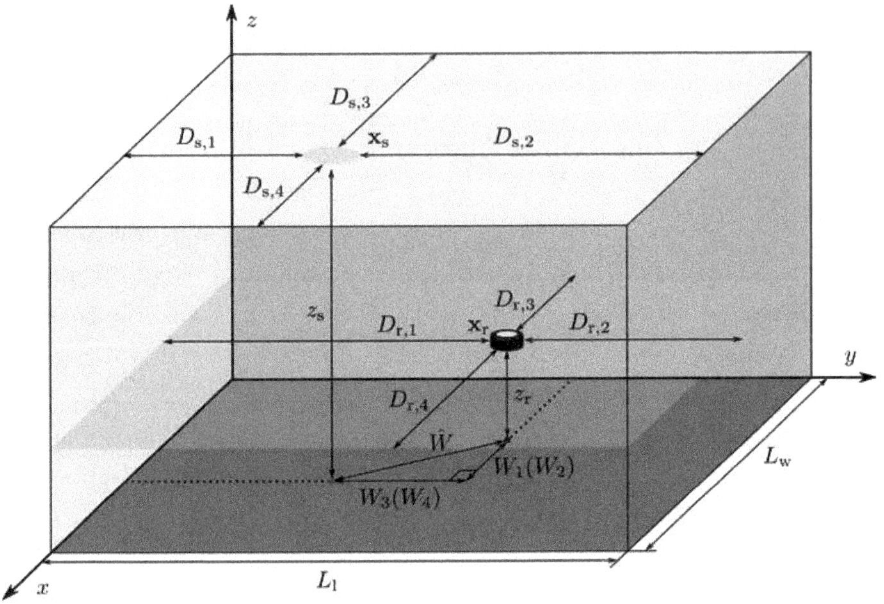

Figure 5.19 Room geometric parameters for an efficient analytical NLOS channel calculation.

$$\mathfrak{c}_{\text{twcr},1} = (z_s - z_r)^2 + W^2 + D_r^2 - D_s^2 \tag{5.166}$$

$$\begin{aligned}\mathfrak{c}_{\text{twcr},2} = &\left(c^2t^2 - \mathfrak{c}_{\text{twcr},1} + 2\left(D_s + D_r\right)x\right)^2 \\ &- 4\left(c^2t^2 - W^2\right)\left(D_s + x\right)^2\end{aligned} \tag{5.167}$$

$$\hat{\mathbb{1}}_v = \begin{cases} 1 & : (z_s - z_r) \geq D_{c2r}\cos\psi_{\text{FOV}} \\ 0 & : \text{otherwise} \end{cases} \tag{5.168}$$

It has been found that the NLOS channel components of the three introduced scenarios dominate the overall VLC/Li-Fi NLOS channel in many cases. Therefore, it is straightforward to use the superposition of CIRs in these three scenarios to estimate the overall NLOS CIR. A VLC/Li-Fi link geometry is shown in Figure 5.19, which shows all required geometric parameters for the CIR calculation. Given the dimension of the room, $L_l \times L_w \times z_s$, and the position of the transmitter and receiver, $\mathbf{x}_s = [x_s, y_s, z_s]$, $\mathbf{x}_r = [x_r, y_r, z_r]$, all parameters can be calculated using the expressions listed in Table 5.2. Thus, the following NLOS CIR expression is concluded:

$$h_{\text{NLOS}}(t) \approx h_{\text{tfcr}}^{\{\hat{W}\}}(t) + \sum_{n=1}^{4} h_{\text{twr}}^{\{D_{s,n}, D_{r,n}, W_n, \rho_{w,n}\}}(t) + h_{\text{twcr}}^{\{D_{s,n}, D_{r,n}, W_n, \rho_{w,n}\}}(t) \tag{5.169}$$

Table 5.2
Calculation of Geometric Parameters

n	1	2	3	4
$D_{s,n}$	y_s	$L_l - y_s$	x_s	$L_w - x_s$
$D_{r,n}$	y_r	$L_l - y_r$	x_r	$L_w - x_r$
W_n	$\lvert x_s - x_r \rvert$		$\lvert y_s - y_r \rvert$	
$\hat{W}$	$\sqrt{(x_s - x_r)^2 + (y_s - y_r)^2}$			

where $h_{\text{twr}}^{\{D_{s,n},D_{r,n},W_n,\rho_{w,n}\}}(t)$, $h_{\text{tfcr}}^{\{\hat{W}\}}(t)$ and $h_{\text{twcr}}^{\{D_{s,n},D_{r,n},W_n,\rho_{w,n}\}}(t)$ are calculated using (5.143), (5.149), and (5.161), respectively. The form of these expressions are very complicated, but the calculation of these expressions requires only one or two numerical integrations, which takes less than a second to obtain the solution. Therefore, this method is much faster than any deterministic or Monte Carlo approach.

5.7.4 Performance Evaluation

Next, we evaluate the performance of various analytical approaches. Generally, the analytical methods are easy to implement and fast to calculate. Therefore, it is trivial to compare their calculation speed. However, they are less accurate compared to deterministic and Monte Carlo simulation approaches due to the use of simplifications. Therefore, we present the CIR results of two configurations to demonstrate the accuracy of different analytical methods, as shown in Figure 5.20. The first configuration is again the configuration A in [7], where the parameters are listed in Table 5.1. For the first configuration, we calculate the sphere-integrating and the efficient analytical methods. The ceiling-bounce model is not considered as it does not apply to this geometry. In addition, we present the simulated result from an MCRS algorithm for comparison. In order to accommodate the ceiling-bounce model, we consider a second applicable configuration. The corresponding parameters are listed in Table 5.3.

It can be observed that the sphere-integrating model offers a very good estimation of the CIR with longer time delays. This is because the approximation used in the sphere-integrating model is more accurate for high-order reflections with a longer time delay. However, it is not able to capture the details at the beginning of each CIR. The efficient analytical approach is able to capture the details of the CIR shape, but it generally underestimates the channel with long-time delays due to the missing high order reflection components. The performance of the ceiling-bounce model is similar to that of the efficient analytical method, which offers better accuracy at shorter time delay regions and underestimates the channel in the long time delay region for the same reason.

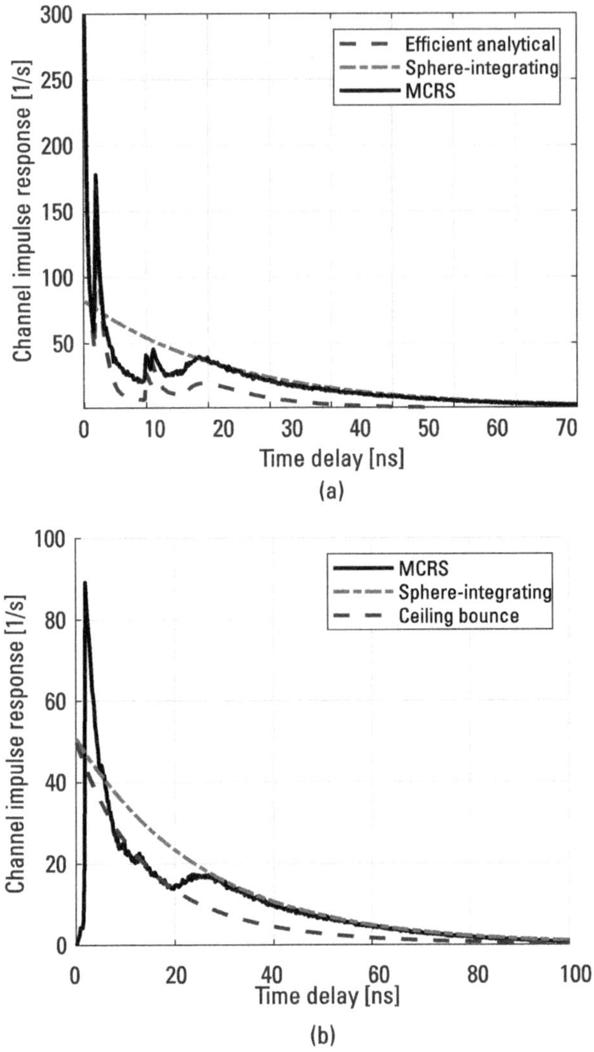

Figure 5.20 (a) CIR results of configuration 1. (b) CIR results of configuration 2.

5.8 Simulation Approach/Modeling Comparison

In this section, we compare the three presented NLOS channel simulations or modeling approaches and discuss their advantages and disadvantages. Table 5.4 concludes the performance of each approach in four different aspects. Intuitively, the simulation accuracy reflects how close the simulated result is to the actual NLOS channel, and the simulation speed reflects how fast the simulation results can be obtained. Implementation complexity indicates how complex or time-consuming the simulator implementation is, while the environmental

Table 5.3
Configuration 2 Parameters

Parameter	Symbol	Value
Room dimension	$L_l \times L_w \times L_h$ [m]	$8 \times 5 \times 3$
Reflectance of walls/ceiling/floor	$\rho_w / \rho_c / \rho_f$	0.8/0.8/0.3
Transmitter position	$\mathbf{x}_s$ [m]	[2,3,1]
Transmitter orientation	$\mathbf{p}_s$ [m]	[0,0,1]
Receiver position	$\mathbf{x}_r$ [m]	[6,1,0.75]
Receiver orientation	$\mathbf{p}_r$ [m]	[0,0,1]
Half-power semiangle	$\phi_{1/2}$ [degree]	60°
Receiver FOV	ψ_{FOV} [degree]	90°
Detection area	A_d [m^2]	1×10^{-4}

Table 5.4
Comparison of Simulation Approaches

Approach	Deterministic	Monte Carlo	Analytical
Simulation accuracy	High	High	Low-medium
Simulation speed	Slow	Medium	Fast
Implementation complexity	High	Medium-high	Low
Environment compatibility	Medium	High	Low-medium

compatibility reflects the flexibility of the approach. Alternatively, it evaluates whether the method can be used in all indoor environments.

- In terms of the simulation accuracy, both deterministic and Monte Carlo approaches are based on ray-tracing techniques to mimic the actual light propagation in reality. Therefore, these two approaches achieve the highest accuracy. In contrast, analytical approaches need to compromise between tractability and accuracy. Consequently, analytical approaches generally use simplification or approximations, which leads to reduced accuracy.
- Regarding the simulation speed, most of the deterministic approaches require multiple layers of iterations or recursion. Furthermore, the simulation time of these approaches is proportional to the square of the number of reflective elements. This means that the simulation time of deterministic approaches also depends on the size of the simulated environment. Therefore, deterministic approaches are generally the slowest. The only exception to this is the frequency-domain algorithm, which is very efficient when frequency quantities are required with a few frequency samples. Monte Carlo approaches require only a single loop

to generate random path samples. Therefore, they are more efficient than the deterministic approaches. The calculation of the analytical approaches only needs to conduct an analytical or numerical evaluation of certain mathematical expressions, which is the fastest.

- Regarding implementation complexity, deterministic approaches have the highest, which need to define positions/orientations of all reflection elements. In addition, they require the implementation of multiple loops of iteration/recursions, which further increases the implementation complexity. In Monte Carlo approaches, only an integrand function and random path generator are required to be implemented. However, the MCMC algorithms require a few more steps for the Markov chain process. Therefore, Monte Carlo approaches have a medium to high implementation complexity. Since analytical approaches just need to deal with expressions, they generally have the lowest implementation complexity.
- Considering the environment compatibility, Monte Carlo approaches are the best among the three categories, as they are able to simulate NLOS channels of any indoor environment, any transmitter/receiver configurations and any type of reflecting surfaces. Note that the state-of-the-art rendering algorithms in computer graphics are based on Monte Carlo approaches. Deterministic approaches are also highly flexible. However, it can only deal with diffuse reflections. In addition, the limited memory space and computational resources restrict the deterministic approaches to simulate large indoor environments. Therefore, we determine it has medium environment compatibility. For analytical approaches, except for the sphere-integrating model, the other two approaches can only be used in specific configurations. For example, the ceiling-bounce model can only be used in the ceiling-bounce diffused link configuration [7].

According to the above discussion, we can conclude that Monte Carlo approaches are preferred when accurate NLOS channel information is required, while analytical approaches are more suitable when a coarse estimation of a large number of NLOS channels are required. However, if frequency-domain metrics are required, such as CFR, the frequency-domain approach is the best option.

5.9 OWC Channel Characteristics

In order to design/optimize a VLC/Li-Fi system, it is important to understand the characteristics of OWC channels in indoor environments. In this section, we use an MCRG simulator to evaluate various performance metrics of various optical wireless channels. In particular, the effects of varying a number of parameters are

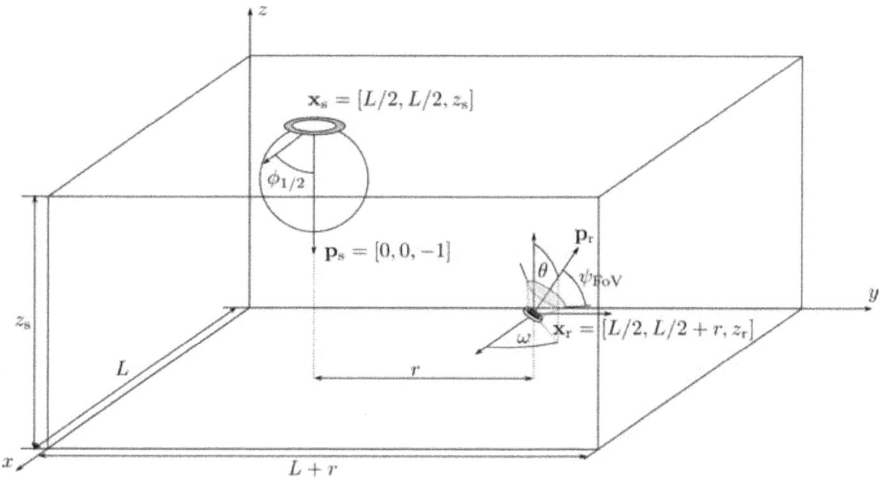

Figure 5.21 Simulation geometry for the study of optical wireless channel characteristics.

investigated. In the end, we include the front-end channel on top of the optical wireless channel to provide more insights.

5.9.1 Optical Wireless Channel Characteristics

In this section, the characteristics of an optical wireless channel are investigated using simulations. The simulation geometry is shown in Figure 5.21, where L is a room dimension length parameter, and a transmitter-receiver horizontal offset r is defined. With a room height/transmitter height of z_s, the room dimension is $L \times (L + r) \times z_s$. Then the transmitter is placed at $\mathbf{x}_s = [L/2, L/2, z_s]$ with a half-power semiangle of $\phi_{1/2}$ and a downward orientation. On the other side, the receiver is placed at $\mathbf{x}_r = [L/2, L/2 + r, z_r]$ with an orientation elevation angle of θ, an orientation azimuth angle of ω, and an FOV of ψ_{FOV}.

5.9.1.1 Effects of Varying Horizontal Offset

In this section, we investigate the effects of varying the horizontal offset between the transmitter and the receiver. In the simulation, we consider four different sets of parameters as listed in Table 5.5. In the first parameter set, we have $L = 5$m, $\phi_{1/2} = 60°$, $\theta = 0°$, $\omega = 0°$, $z_s = 3$m, $z_r = 0.75$m, $A_d = 1 \times 10^{-4}$ m^2, $\psi_{FOV} = 90°$, $\rho_w = 0.8$, $\rho_c = 0.8$, and $\rho_f = 0.3$. In the second, third, and fourth parameter sets, we have changed $L = 3$m, $\phi_{1/2} = 40°$, and $\theta = 45°$, respectively, while other parameters remain the same as those in the first parameter set. With the four parameters sets, we evaluate the channel metrics for different horizontal offsets r in the range of 0m to 5m. We use an MCRG simulator with 5×10^5 iterations to generate the CIR data. The evaluated metrics are the same as those introduced in Sections 5.2.2 and 5.4.3.

Table 5.5
Varying Horizontal Offset Simulation Parameters

Parameter	Symbol	Set 1	Set 2	Set 3	Set 4
Room edge length	L [m]	5	3	5	5
Half-power semiangle	$\phi_{1/2}$	60°	60°	40°	60°
Orientation elevation angle	θ	0°	0°	0°	45°
Orientation azimuth angle	ω	0°			
Room/transmitter height	z_s [m]	3			
Receiver height	z_r [m]	0.75			
Detector area	A_d [m^2]	1×10^{-4}			
Detector FoV	ψ_{FOV}	90°			
Walls/ceiling/floor	$\rho_w/\rho_c/\rho_f$	0.8/0.8/0.3			

The CIR and CFR results are shown in Figure 5.22. The channel DC gain, K-factor, RMS delay spread, and 3-dB bandwidth results are shown in Figure 5.23. Note that in order to avoid overcrowding the figure, only the CIR and CFR results of $r = 0.5$m with parameter sets 1, 2, and 4 are shown in Figure 5.22 for demonstration. Figure 5.22(a) shows that the detector captures a spiky LOS component first, which is then followed by a stream of NLOS channel components after delay of a few nanoseconds delay. The power of the NLOS channel is higher at shorter delays and decays with the increase of time delay. With various configurations, the magnitude and time span of the NLOS vary. For example, the significance of the channel with parameter set 2 is higher than the channels with the other two sets of parameters, as it is in a smaller indoor environment. Figure 5.22(b) shows the channel gain results with and without the NLOS component. The LOS channel gain does not vary with frequency. In the case with NLOS components, the channel gain at low frequencies (including DC) is always the highest. This is due to the non-negativity feature of the optical wireless channel. When the frequency reaches the megahertz region, the channel gain starts to decrease until it reaches a minimum value which is normally slightly below the LOS channel gain level. Then the channel gain starts to fluctuate around the LOS channel gain level with a further increase of frequency. This notch normally corresponds to the minimum value of the channel gain. If the gap between the highest (at DC) and the lowest (the notch) channel gains is greater than 3 dB, it is possible to find a finite 3-dB bandwidth, such as the case in parameter sets 2 and 4. Otherwise, an infinite 3-dB bandwidth is obtained,[1] like the case in parameter set 1.

Figure 5.23 (a) shows that the channel DC generally decreases with an increase in the horizontal offset due to the longer transmission distance and

[1] Note that the effects of the front-end channel are excluded here.

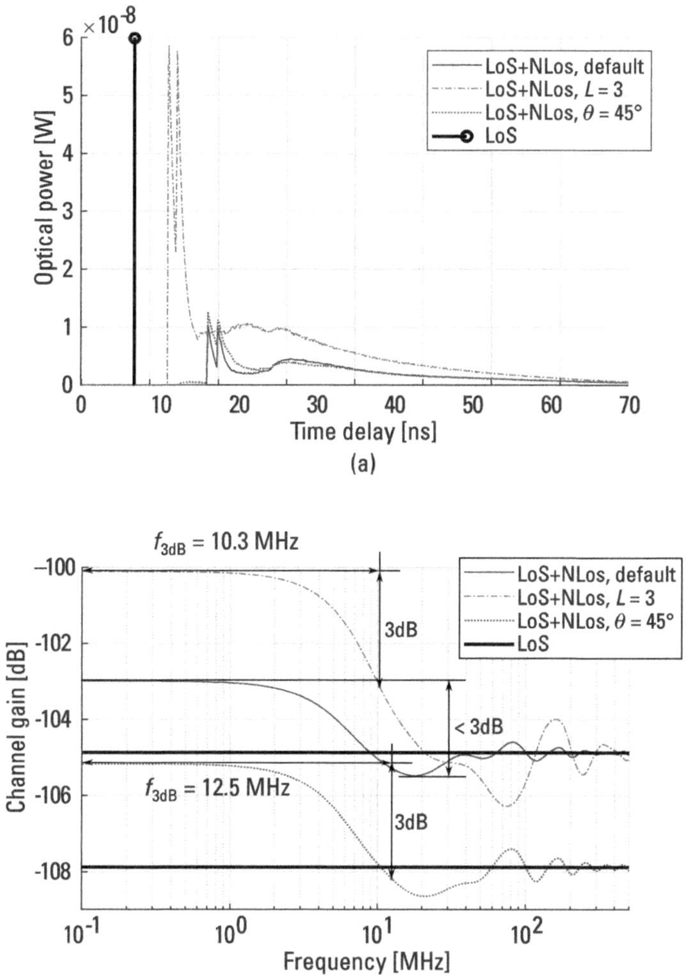

Figure 5.22 Simulation results with $r = 0.5$ and parameter sets 1, 2, and 4 in terms of (a) CIR and (b) channel gain.

misalignment. In addition, the channel DC gain due to the LOS component drops faster than the case with the NLOS components. This is because the LOS component is more sensitive to the link alignment, which becomes worse with an increase in horizontal offset. This also leads to a decrease in the K-factor with an increase in horizontal offset, as shown in Figure 5.23(b). The longer transmission distance, due to the increase in horizontal offset, also leads to an increase of RMS delay spread and a significant decrease of 3-dB bandwidth, as shown in Figure 5.23(c,d). According to the results of different parameter sets, it is worth noting that with a smaller optical source half-power semiangle, the drop rate of the channel DC gain and the K-factor increase. This is because a more focused beam

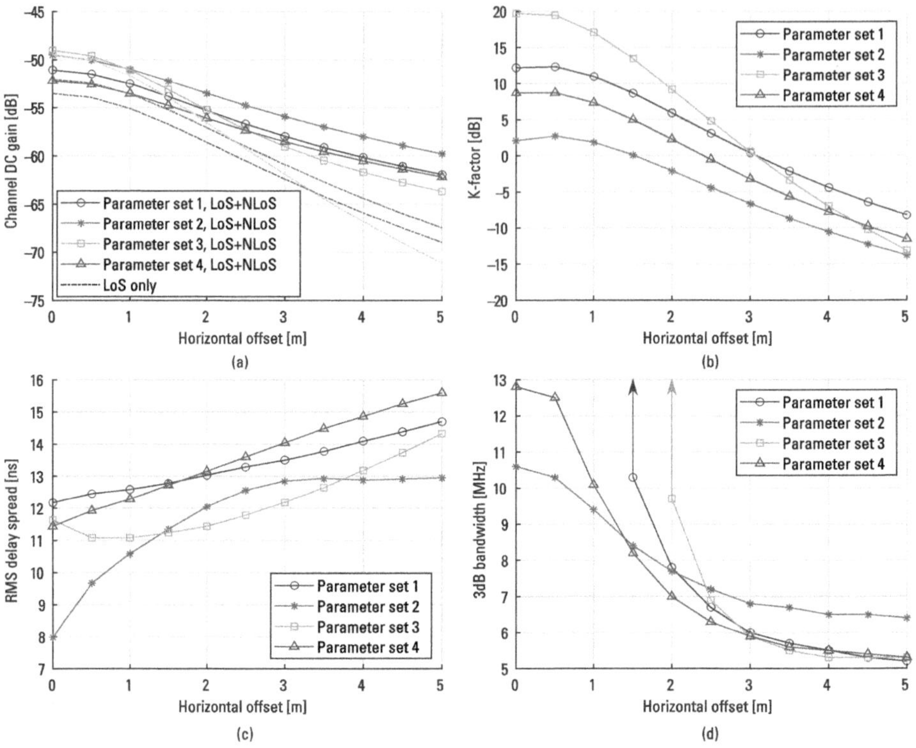

Figure 5.23 Performance metrics vary with horizontal offset. (a) Channel DC gain. (b) K-factor. (c) RMS delay spread. (d) The 3-dB bandwidth.

is more sensitive to misalignment. In addition, a smaller indoor environment enhances the NLOS channel, which leads to a higher channel DC gain. At the same time, the decrease rate of 3-dB bandwidth is also lower. Also note that when $r < 1.5$m with parameter sets 1 and 3, the optical wireless channel has an infinite 3-dB bandwidth. This is because of the LOS domination. The channel gain never drops more than 3 dB with an increase of frequency from the DC level as explained above. Note that the K-factor in this region is higher than 10 dB.

5.9.1.2 Effects of Varying Receiver Orientation

In many cases, the orientation of handheld devices is determined by random user behavior. Consequently, the receiver detector orientation may deviate from the upward direction. Here we investigate the effect of varying the receiver orientation. The simulation geometry is the same as that shown in Figure 5.21. Similar to the previous section, we consider six different sets of parameters listed in Table 5.6. In the first parameter set, we have $L = 5$m, $\phi_{1/2} = 60°$, $\theta = 0°$, $r = 1$m, $z_s = 3$m, $z_r = 0.75$m, $A_d = 1 \times 10^{-4}$ m^2, $\psi_{FOV} = 90°$, $\rho_w = 0.8$, $\rho_c = 0.8$, and $\rho_f = 0.3$, which corresponds to a link with a receiver directed

Table 5.6
Varying Receiver Orientation Simulation Parameters

Parameter	Symbol	Set 1	Set 2	Set 3	Set 4	Set 5	Set 6
Room edge length	L [m]	5	5	5	5	5	3
Half-power semiangle	$\phi_{1/2}$	60°	60°	60°	40°	60°	60°
Orientation elevation angle	θ	0°	30°	60°	30°	30°	30°
Horizontal offset	r [m]	1	1	1	1	2	1
Room/transmitter height	z_S [m]	3					
Receiver height	z_r [m]	0.75					
Detector area	A_d [m^2]	1×10^{-4}					
Detector FoV	ψ_{FOV}	90°					
Walls/ceiling/floor	$\rho_W/\rho_C/\rho_f$	0.8/0.8/0.3					

upwards. In the second and third parameter sets, we attempt to increase the elevation angle from 0° to 30° and 60°, respectively. In the following three parameter sets, we have changed $\phi_{1/2} = 40°$, $r = 2$m, and $L = 3$m, respectively, while the other parameters remain the same as those in parameter set 2. Then we simulate the performance metrics with ω in the range of $-90°$ to 90°.

The results of the channel metrics with a different receiver orientation azimuth angle ω are shown in Figure 5.24. The receiver detector is more oriented towards the transmitter with ω closer to $-90°$, while the receiver detector is more oriented away from the transmitter with ω closer to 90°. Consequently, the channel DC gain and K-factor decrease with an increase of ω except for the case with parameter set 1 ($\theta = 0°$). However, the RMS delay spread also varies slightly with the change of ω. When ω is in the range of 30° to 60°, the NLOS components have a wider spread in time. In addition, the 3-dB bandwidth decreases with an increase of ω. By comparing the results with parameter sets 1, 2, and 3 (different θ), we can see that the channel DC gain and K-factor generally decrease. The values of RMS delay spread and 3-dB bandwidth decrease with an increase of θ. Furthermore, the deviation ranges of metrics also increase with an increase of θ. By inspecting the performance of channels with parameter sets 4, 5, and 6, it can be found that the variance of the channel DC gain, K-factor and RMS delay spread with ω becomes more significant with larger horizontal offset r. The variance of the channel DC gain, K-factor, and 3-dB bandwidth with ω decreases when the indoor environment gets smaller.

5.9.1.3 Effects of Varying Optical Source Half-Power Semiangle and Detector FOV

Another factor of optical wireless channel characteristics is the directionality of the optical source and detector. In this section, channel metrics with a different half-power semiangle and receiver FOV are considered. Six different sets of parameters listed in Table 5.7 are used. In the first parameter set, we have $L = 5$m, $\psi_{FOV} =$

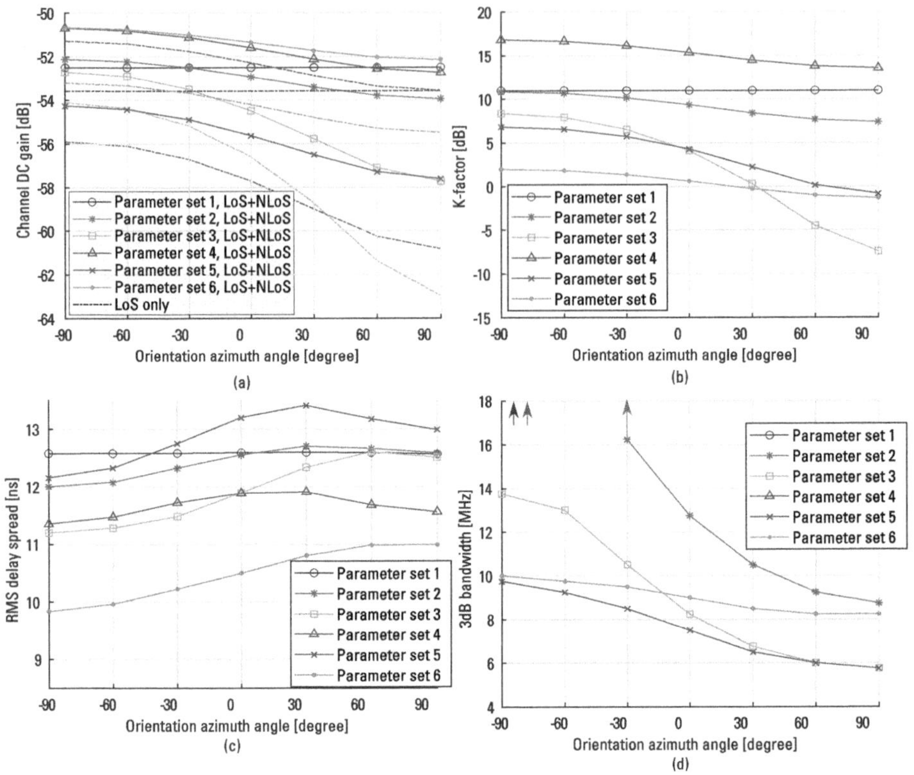

Figure 5.24 Performance metrics vary with receiver orientation azimuth angle. (a) Channel DC gain. (b) K-factor. (c) RMS delay spread. (d) The 3-dB bandwidth.

Table 5.7
Varying Half-Power Semiangle Simulation Parameters

Parameter	Symbol	Set 1	Set 2	Set 3	Set 4	Set 5	Set 6
Room edge length	L [m]	5	5	5	5	3	5
Detector FOV	ψ_{FOV}	90°	70°	50°	90°	90°	90°
Orientation elevation angle	θ	0°	0°	0°	0°	0°	45°
Horizontal offset	r [m]	1	1	1	2	1	1
Orientation azimuth angle	ω			0°			
Room/transmitter height	z_s [m]			3			
Receiver height	z_r [m]			0.75			
Detector area	A_d [m^2]			1×10^{-4}			
Walls/ceiling/floor	$\rho_w/\rho_c/\rho_f$			0.8/0.8/0.3			

90°, $\theta = 0°$, $\omega = 0°$, $r = 1$m, $z_s = 3$m, $z_r = 0.75$m, $A_d = 1 \times 10^{-4}$ m^2, $\rho_w = 0.8$, $\rho_c = 0.8$, and $\rho_f = 0.3$, which corresponds to the case with full FOV. In the second and third parameter sets, we reduce the receiver FOV from 90° to 70° and 50°, respectively. Note that a FOV smaller than 50° is unlikely for

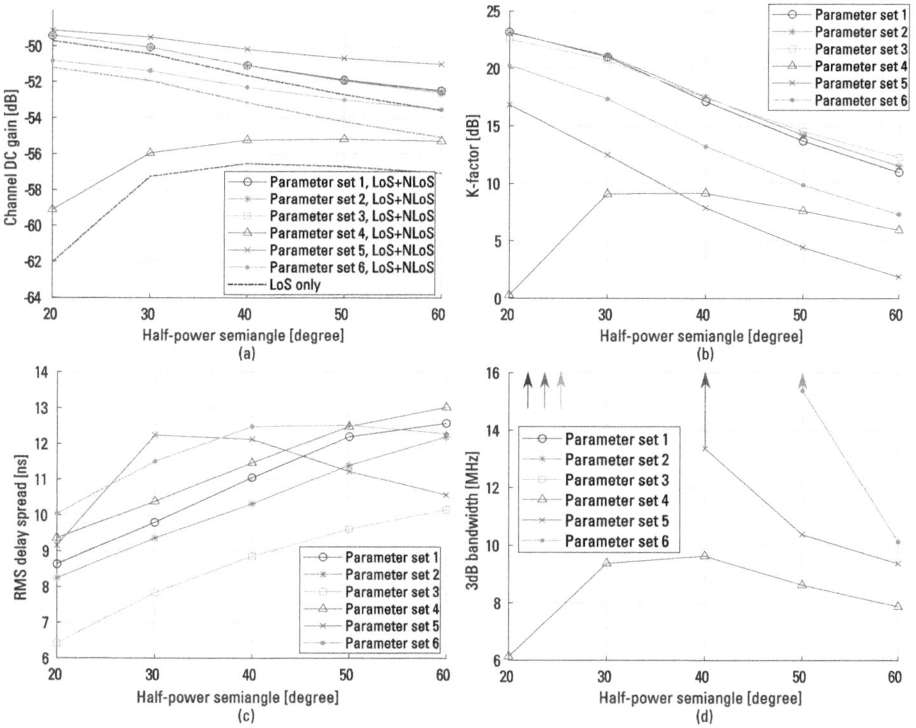

Figure 5.25 Performance metrics vary with optical source half-power semiangle. (a) Channel DC gain. (b) K-factor. (c) RMS delay spread. (d) The 3-dB bandwidth.

Li-Fi/VLC applications. In the following three parameter sets, we have changed $r = 2\text{m}$, $L = 3\text{m}$, and $\theta = 45°$, respectively, while the other parameters remain the same as those in parameter set 1. Then we simulate the performance metrics with $\phi_{1/2}$ in the range of 20° to 60°.

The results of channel metrics with different $\phi_{1/2}$ are shown in Figure 5.25. It is obvious that the channel DC gain, K-factor, and 3-dB bandwidth decrease with an increase of $\phi_{1/2}$. This is because the channel is dominated by the LOS component. The strength of the LOS component decreases with an increase of $\phi_{1/2}$ when there is good alignment. However, there is an exception with parameter set 4 where the horizontal offset r is greater. This is the case with a slightly worse alignment, and increasing $\phi_{1/2}$ weakens the directionality of the optical source, thereby alleviating the misalignment issue. When the misalignment is no longer the dominant factor of the weak LOS component, a further increase in $\phi_{1/2}$ starts to decrease the strength of the LOS component again. With an increase of $\phi_{1/2}$, the light source is more scattered, which leads to a wider spread of light in the indoor environment and increased RMS delay spread. The only exception is the case with parameter set 5, which has a smaller room size. By comparing the results with parameter sets 1, 2, and 3, we note that by decreasing the FOV of

Table 5.8
Varying Indoor Environment Size Simulation Parameters

Parameter	Symbol	Set 1	Set 2	Set 3	Set 4
Half-power semiangle	$\phi_{1/2}$	60°	60°	40°	60°
Orientation elevation angle	θ	0°	0°	0°	45°
Horizontal offset	r[m]	1	2	1	1
Orientation azimuth angle	ω	0°			
Room/transmitter height	z_s [m]	3			
Receiver height	z_r [m]	0.75			
Detector area	A_d [m^2]	1×10^{-4}			
Detector FOV	ψ_{FOV}	90°			
Walls/ceiling/floor	$\rho_w/\rho_c/\rho_f$	0.8/0.8/0.3			

the receiver, the channel DC gain barely changes, the K-factor increases slightly, and the RMS delay spread has an obvious drop.

5.9.1.4 Effects of Varying Indoor Environment Size

An important factor of the NLOS channel strength is the size of the considered indoor environment. In this section, we consider the effects of varying the size of the room on the channel metrics. Four different set of parameters listed in Table 5.8 are used. In the first parameter set, we have $\phi_{1/2} = 60°$, $\theta = 0°$, $\omega = 0°$, $r = 1\text{m}$, $z_s = 3\text{m}$, $z_r = 0.75\text{m}$, $A_d = 1 \times 10^{-4}$ m^2, $\psi_{FOV} = 90°$, $\rho_w = 0.8$, $\rho_c = 0.8$, and $\rho_f = 0.3$. In the other parameter sets, we have changed $r = 2\text{m}$, $\phi_{1/2} = 40°$, and $\theta = 45°$, respectively, while the other parameters remain the same as those in parameter set 1. The performance metrics with the room dimension length L in the range of 3m to 10m are evaluated.

The simulation results of different metrics are shown in Figure 5.26. It can be found that a smaller indoor environment leads to a higher channel DC gain due to the strengthened NLOS components. It is worth noting that the channel DC gain gap between the case with and without an NLOS component is about 2 to 3 dB with $L = 3\text{m}$. This gap is less than 1 dB when $L = 10\text{m}$. Since the LOS component does not vary with changes in the indoor environment, the K-factor becomes smaller. In addition, the RMS delay spread increases when the L is below 4m. Then the RMS delay spread starts to decrease with a further increase of L. In particular, with a narrower half-power semiangle $\phi_{1/2} = 40°$, the RMS delay spread drops faster than the channels with other parameter sets with an increase in L. In addition, when L is small, the channel is dominated by the NLOS channel, and the 3-dB bandwidth is limited to a few or tens of megahertz. When L is greater than 6m, all channels have a high K-factor and infinite 3-dB bandwidth.

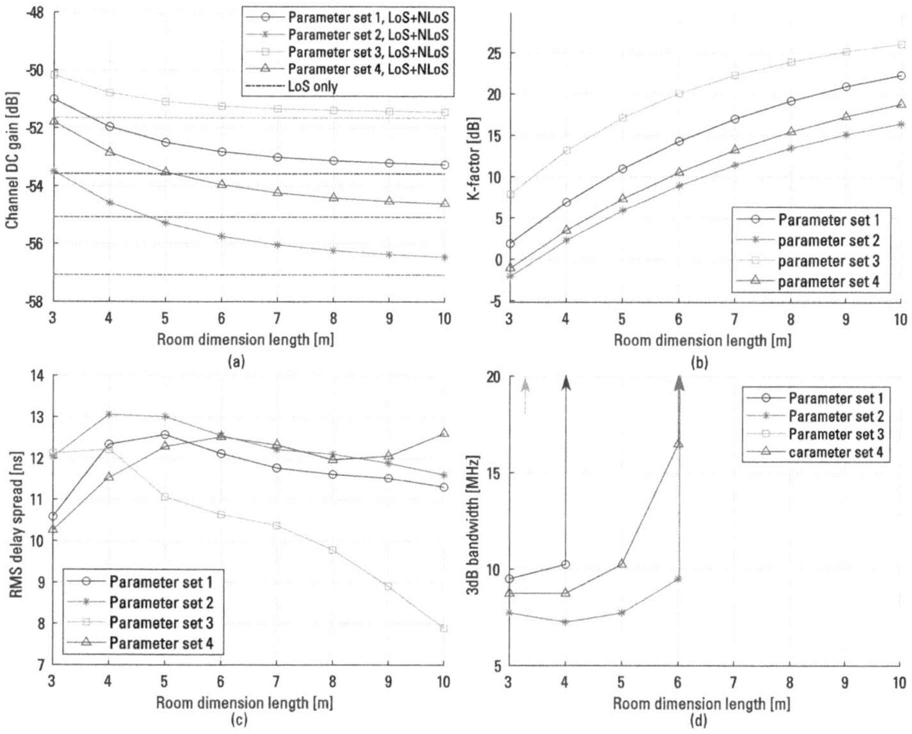

Figure 5.26 Performance metrics vary with indoor environment size. (a) Channel DC gain. (b) K-factor. (c) RMS delay spread. (d) The 3-dB bandwidth.

5.9.1.5 Effects of Varying Surface Reflectance

Another important factor in the significance of an NLOS channel is the reflectance of the indoor surfaces. In this part, we consider the effects of varying the reflectance of the indoor surfaces on the channel metrics. Five different set of parameters listed in Table 5.9 are used. In the first parameter set, we have $L = 5\text{m}$, $\phi_{1/2} = 60^\circ$, $\theta = 0^\circ$, $\omega = 0^\circ$, $r = 1\text{m}$, $z_s = 3\text{m}$, $z_r = 0.75\text{m}$, $A_d = 1 \times 10^{-4}\ \text{m}^2$, and $\psi_{FOV} = 90^\circ$. In the remaining four parameter sets, we have changed $r = 2\text{m}$, $L = 3\text{m}$, $\phi_{1/2} = 40^\circ$, and $\theta = 45^\circ$, respectively. The other parameters are the same as those in parameter set 1. In the simulation, we calculate the metrics with different surface reflectance in the range of 0.1 to 0.9. Note that we use the same value for all surfaces ($\rho_w = \rho_c = \rho_f$) in this simulation.

The simulation results are concluded in Figure 5.27. It can be observed that the channel DC gain increases significantly with the reflectance. The gap between the channel DC gain with and without NLOS components increases from about 0 dB to 3 to 5 dB with an increase of reflectance from 0.1 to 0.9. Intuitively, the K-factor decreases with the increase of reflectance. Consequently, the 3-dB bandwidth is infinite when the reflectance is below 0.5 but quickly decreases after the reflectance becomes higher. With the increase in reflectance, the average

Table 5.9
Varying Surface Reflectance Simulation Parameters

Parameter	Symbol	Set 1	Set 2	Set 3	Set 4	Set 5
Room edge length	L [m]	5	5	3	5	5
Half-power semiangle	$\phi_{1/2}$	60°	60°	60°	40°	60°
Orientation elevation angle	θ	0°	0°	0°	0°	45°
Horizontal offset	r [m]	1	2	1	1	1
Orientation azimuth angle	ω			0°		
Room/transmitter height	z_s [m]			3		
Receiver height	z_r [m]			0.75		
Detector area	A_d [m^2]			1×10^{-4}		
Detector FOV	ψ_{FOV}			90°		

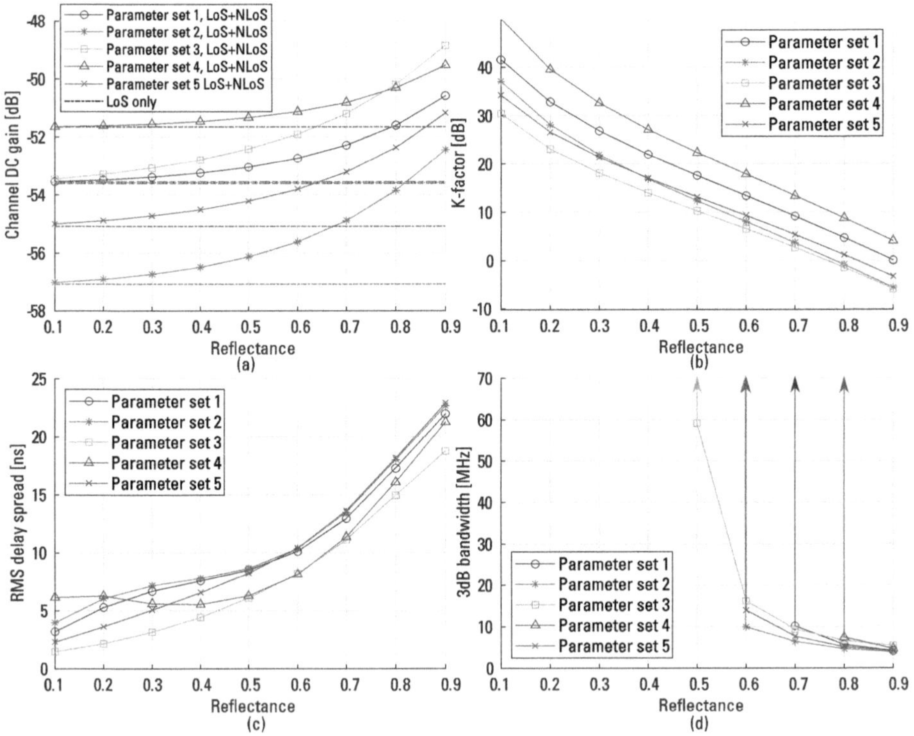

Figure 5.27 Performance metrics vary with surface reflectance. (a) Channel DC gain. (b) K-factor. (c) RMS delay spread. (d) The 3-dB bandwidth.

propagation time of light in the environment increases. Therefore, the RMS delay spread increases with an increase of reflectance in general. The rate of the increase also grows as the reflectance gets higher. The results of parameter set 3 show that the cases with a smaller area are more sensitive to the change in reflectance.

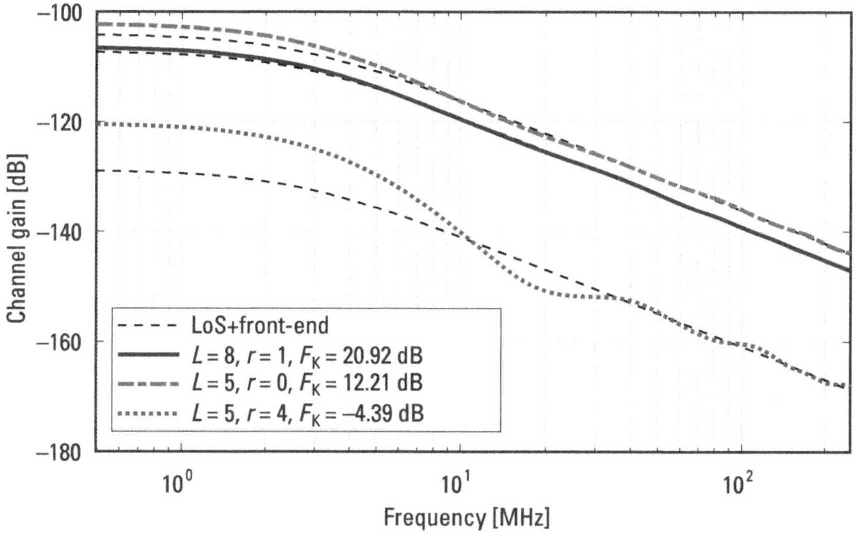

Figure 5.28 Channel gain result with front-end channel included.

5.9.2 Including the Front-End Channel

In the previous section, the characteristics of the optical wireless channel are investigated. Next, we use simulations to study channel characteristics after including the front-end channel introduced in Section 5.3. The simulation geometry shown in Figure 5.21 is also used here. The simulations include three different geometric configurations. In the first configuration, $L = 8$m and $r = 1$m. In the second configuration, $L = 5$m and $r = 0$m. In the last configuration, $L = 5$m and $r = 4$m. The remaining geometric parameters are the same as those specified in Table 5.5. The K-factor of the corresponding optical wireless channels is 20.92 dB, 12.21 dB, and −4.39 dB, respectively. In addition, we use the white LED model as the front-end channel with configuration of $G_b = 0.25$, $G_y = 0.75$, $\tau_{s,b} = 16$ ns, and $\tau_{s,y} = 62$ ns, as that in [19].

The channel gain results of the three OWC channels are shown in Figure 5.28. In order to highlight the effects of the NLOS channel, the results with and without NLOS channels are presented. It can be observed that due to the lowpass characteristics, the channel gain decreases as the frequency increases. Regarding the same channel, the result with the NLOS channel fluctuates around the result without the NLOS channel. At frequencies close to the DC, the result with the NLOS channel is always higher than the one without the NLOS channel. This fluctuation diminishes as the frequency increases. It is worth noting that the significance of the fluctuation depends on the K-factor value. When the K-factor is high (20.92 dB), the maximum difference is less than 1 dB. In the case of a low K-factor (−4.39 dB), the maximum difference increases to about 10 dB. In conclusion, the main effects of the NLOS channel are at low frequency

(<50 to 100 MHz) with a low K-factor, and the effects at a high frequency are negligible. Therefore, if the Li-Fi system uses a narrow bandwidth (10 MHz) with a weak LOS optical wireless channel, the NLOS components will play an important role. If a Li-Fi system with a wide bandwidth is considered (1 GHz), the effects of the NLOS channel can be omitted.

5.10 Summary

In this chapter, we presented OWC channels in indoor environments. The channel effects and popular performance metrics have been concluded. In addition, the OWC channel is composed of a front-end channel and an optical wireless channel, which have been introduced in detail. In particular, three important simulation/modeling approaches for NLOS optical wireless channels have been presented, namely: deterministic, Monte Carlo, and analytical approaches. The performance of each approach was evaluated. The advantages and disadvantages of various simulation/modeling approaches were discussed. We investigated the characteristics of optical wireless channels with various configurations, and the characteristics after including the front-end channel were considered.

References

[1] Proakis, J. G., and M. Salehi, *Digital Communications, Vol. 4,* New York: McGraw-Hill, 2001.

[2] Uysal, M., et al., *Optical Wireless Communications: An Emerging Technology,* New York: Springer, 2016.

[3] Lorences-Riesgo, A., et al., "200 G Outdoor Free-Space-Optics Link Using a Single-Photodiode Receiver," *IEEE/OSA Journal of Lightwave Technology,* Vol. 38, No. 2, January 2020, pp. 394–400.

[4] Tan, Y., X. Wu, and H. Haas, "Performance Comparison Between Coherent and DCO-OFDM LiFi Systems," *ICC 2019 - 2019 IEEE International Conference on Communications (ICC),* 2019, pp. 1–6.

[5] Carruthers, J. B., and J. M. Kahn, "Modeling of Nondirected Wireless Infrared Channels," *IEEE Transactions on Communications,* Vol. 45, No. 10, 1997, pp. 1260–1268.

[6] BS EN 62471:2008, Photobiological Safety of Lamps and Lamp Systems, BSI British Standards Std., September 2008.

[7] Kahn, J. M., and J. R. Barry, "Wireless Infrared Communications," *Proceedings of the IEEE,* Vol. 85, No. 2, February 1997, pp. 265–298.

[8] Xie, E., et al., "High-Speed Visible Light Communication Based on a III-Nitride Series-Biased Micro-LED Array," *IEEE/OSA Journal of Lightwave Technology,* Vol. 37, No. 4, 2019, pp. 1180–1186.

[9] Lee, K., H. Park, and J. R. Barry, "Indoor Channel Characteristics for Visible Light Communications," *IEEE Communications Letters*, Vol. 15, No. 2, 2011, pp. 217–219.

[10] Cui, L., et al., "Analysis of the Multichannel WDM-VLC Communication System," *IEEE/OSA Journal of Lightwave Technology*, Vol. 34, No. 24, 2016, pp. 5627–5634.

[11] Ge, P., et al., "Optical Filter Designs for Multi-Color Visible Light Communication," *IEEE Transactions on Communications*, Vol. 67, No. 3, 2019, pp. 2173–2187.

[12] Barry, J. R., et al., "Simulation of Multipath Impulse Response for Indoor Wireless Optical Channels," *IEEE Journal on Selected Areas in Communications*, Vol. 11, No. 3, 1993, pp. 367–379.

[13] Komine, T., and M. Nakagawa, "Fundamental Analysis for Visible-Light Communication System Using LED Lights," *IEEE Transactions on Consumer Electronics*, Vol. 50, No. 1, February 2004, pp. 100–107.

[14] Dimitrov, S., et al., "On the SIR of a Cellular Infrared Optical Wireless System for an Aircraft," *IEEE Journal on Selected Areas in Communications*, Vol. 27, No. 9, 2009, pp. 1623–1638.

[15] Le Minh, H., et al., "100-Mb/s NRZ Visible Light Communications Using a Postequalized White LED," *IEEE Photonics Technology Letters*, Vol. 21, No. 15, August 2009, pp. 1063–1065.

[16] Chen, C., D. A. Basnayaka, and H. Haas, "Downlink Performance of Optical Attocell Networks," *IEEE/OSA Journal of Lightwave Technology*, Vol. 34, No. 1, January 2016, pp. 137–156.

[17] Schubert, E., *Light-Emitting Diodes*, Cambridge, U.K.: Cambridge University Press, 2006.

[18] Nakamura, S., "Background Story of the Invention of Efficient InGaN Blue-Light-Emitting Diodes (Nobel Lecture)," *Angewandte Chemie International Edition*, Vol. 54, No. 27, 2015, pp. 7770–7788.

[19] Stepniak, G., M. Schüppert, and C. Bunge, "Advanced Modulation Formats in Phosphorous LED VLC Links and the Impact of Blue Filtering," *IEEE/OSA Journal of Lightwave Technology*, Vol. 33, No. 21, 2015, pp. 4413–4423.

[20] Al-Kinani, A., et al., "Optical Wireless Communication Channel Measurements and Models," *IEEE Communications Surveys Tutorials*, Vol. 20, No. 3, 2018, pp. 1939–1962.

[21] Jungnickel, V., et al., "A Physical Model of the Wireless Infrared Communication Channel," *IEEE Journal on Selected Areas in Communications*, Vol. 20, No. 3, April 2002, pp. 631–640.

[22] Uysal, M., F. Miramirkhani, and T. Baykas, IEEE 802.11bb Reference Channel Models for Indoor Environments, IEEE Std., Rev. IEEE 802.11-18/1582r4, November 2018. https://mentor.ieee.org/802.11/dcn/18/11-18-1582-04-00bb-ieee-802-11bb-reference-channel-models-for-indoor-environments.pdf.

[23] Carruthers, J. B., and P. Kannan, "Iterative Site-Based Modeling for Wireless Infrared Channels," *IEEE Transactions on Antennas and Propagation*, Vol. 50, No. 5, May 2002, pp. 759–765.

[24] Lopez-Hermandez, F. J., and M. J. Betancor, "DUSTIN: Algorithm for Calculation of Impulse Response on IR Wireless Indoor Channels," *Electronics Letters*, Vol. 33, No. 21, October 1997, pp. 1804–1806.

[25] Schulze, H., "Frequency-Domain Simulation of the Indoor Wireless Optical Communication Channel," *IEEE Transactions on Communications*, Vol. 64, No. 6, 2016, pp. 2551–2562.

[26] Lopez-Hernandez, F. J., R. Perez-Jimenez, and A. Santamaria, "Ray-Tracing Algorithms for Fast Calculation of the Channel Impulse Response on Diffuse IR Wireless Indoor Channels," *Optical Engineering*, Vol. 39, No. 10, 2000, pp. 2775–2780.

[27] Lee, H. -S., "A Photon Modeling Method for the Characterization of Indoor Optical Wireless Communication," *Progress in Electromagnetics Research*, Vol. 92, 2009, pp. 121–136.

[28] Behlouli, A., et al., "Efficient Simulation of Optical Wireless Channel Application to WBANs with MISO Link," *Procedia Computer Science*, Vol. 40, September 2014, pp. 190–197.

[29] Veach, E., "Robust Monte Carlo Methods for Light Transport Simulation," Ph.D. thesis, Stanford University, 1997.

[30] Blinn, J. F., "Models of Light Reflection for Computer Synthesized Pictures," *SIGGRAPH Computer Graphics*, Vol. 11, No. 2, July 1977, pp. 192–198.

[31] Duvenhage, B., K. Bouatouch, and D. G. Kourie, "Numerical Verification of Bidirectional Reflectance Distribution Functions for Physical Plausibility," *Proceedings of the South African Institute for Computer Scientists and Information Technologists Conference*, 2013, pp. 200–208.

[32] Behlouli, A., P. Combeau, and L. Aveneau, "MCMC Methods for Realistic Indoor Wireless Optical Channels Simulation," *Journal of Lightwave Technology*, Vol. 35, No. 9, 2017, pp. 1575–1587.

[33] Chib, S., and E. Greenberg, "Understanding the Metropolis-Hastings Algorithm," *The American Statistician*, Vol. 49, No. 4, 1995, pp. 327–335.

[34] Gonzalez, O., et al., "Error Analysis of the Simulated Impulse Response on Indoor Wireless Optical Channels Using a Monte Carlo-Based Ray-Tracing Algorithm," *IEEE Transactions on Communications*, Vol. 53, No. 1, 2005, pp. 124–130.

[35] Chen, C., D. Basnayaka, and H. Haas, "Non-Line-of-Sight Channel Impulse Response Characterisation in Visible Light Communications," *2016 IEEE International Conference on Communications (ICC)*, 2016, pp. 1–6.

[36] Chen, C., et al., "Efficient Analytical Calculation of Non-Line-of-Sight Channel Impulse Response in Visible Light Communications," *IEEE/OSA Journal of Lightwave Technology*, Vol. 36, No. 9, 2018, pp. 1666–1682.

6

Enabling Technologies

6.1 Introduction

The advances in the fields of optoelectronics and communications have enabled the concept of Li-Fi to be realized. The emergence of VLC is also motivated by the low cost and widespread use of enabling technologies such as SSL sources. These technologies have enabled the concept of communication using the visible light spectrum to be first introduced between wavelength $\lambda = 400$ nm and $\lambda = 700$ nm. However, visible and invisible light sources are now being considered to be used in Li-Fi for downlink and uplink communication.

Nowadays, a wealth of optical sources and detectors can be used in VLC transmitters and receivers. The choice of optical front ends depends mainly on the considered application. For example, avalanche photodiodes (APDs) and single-photon avalanche diodes (SPADs) are favorable detectors in VLC systems where the background illumination is limited and the expected received optical power is modest. Commercially available LEDs are the main candidates for Li-Fi transmitters due to their vast availability and their low cost. Phosphor-coated LEDs (PC-LED) luminaries are generally manufactured in arrays due to the limited maximum optical power of individual LEDs that fall below the illumination requirements. Arrays of LEDs are also very beneficial from a communication system perspective, as each individual LED can transmit a different stream of data. Wavelength division multiplexing (WDM) improves the capacity of the communications system by enabling multistreams of data to be transmitted on different wavelengths. For example, red, green, and blue LEDs (RGB-LEDs) and photodiodes (PDs) aided with optical filters can be used to enable multistream data communications and at the same time realize white illumination.

Incoherent light sources limit the signaling in VLC to IM/DD. This imposes restrictions on the modulation techniques that can be applied. Pulse-based modulation techniques such as on-off keying (OOK) and pulse-amplitude modulation (PAM) can be used straightforwardly in VLC. However, the performance of such techniques degrades at high speeds due to the increased ISI of the VLC channel [1]. Multicarrier modulation (MCM) techniques used in RF, such as orthogonal frequency division multiplexing (OFDM), are prominent candidates for VLC due to the computationally cost-effective, single-tap equalizers and adaptive bit and energy-loading algorithms. However, RF-based OFDM modulation waveforms are complex and bipolar. Hermitian symmetry can be used in OWC to obtain real-valued OFDM waveforms. In addition, a DC bias can be added to the OFDM bipolar waveform to shift most of its negative samples to become positive. However, the addition of DC bias results in an energy dissipation that increases with the increased spectral efficiency. Many energy-efficient modulation techniques were developed to solve the high-power dissipation in the widely used DC biased optical OFDM (DCO-OFDM). Most of these techniques attempt to realize unipolar OFDM waveforms based on the time and frequency properties of fast Fourier transform (FFT).

MIMO techniques are promising candidates to increase the spectral efficiency of communication systems. These techniques are ideally suited to Li-Fi networks due to the nature of the lighting infrastructure, which features multiluminaries that can satisfy the illumination requirements of the considered space (room, conference hall, airplane cabin). Moreover, each of these luminaries could consist of an array of LEDs, which allows for the realization of various MIMO functions. However, techniques such as spatial modulation make use of the physical arrangement of the lighting infrastructure to offer a simplified approach to realize Li-Fi systems.

In this chapter, the main types of light sources and detectors are surveyed in Section 6.2. The channel noise is briefly reviewed in Section 6.3. A detailed discussion of the main modulation techniques considered in VLC literature is provided in Section 6.4. This includes energy-efficient modulation techniques that are capable of supporting high-speed VLC systems. The principle of MIMO technologies and their recent developments are discussed in Section 6.5. Finally, Section 6.6 summarizes this chapter.

6.2 Front-End Devices and Systems

The equivalent of the antenna in RF communications is the LED at the transmitter and the PD at the receiver in Li-Fi systems. In comparison with RF wireless communications, in which a single antenna radiates signals in an omnidirectional pattern, LEDs and PDs are natural beam-formers that exhibit directionality characteristics by their FOVs. This characteristic is less important in VLC, which describes a wireless connection between two static entities, while it is important in Li-Fi which supports full mobility.

VLC transmitters incorporate a single or multiple light source emitting at single or multiple wavelength bands with the aid of optical components such as reflectors, lenses, and diffusers. On the other end, receivers consist of single or multiple photodetectors with an optical system that may include lenses and filters. Electrical-to-optical (E-O) conversion is performed at the transmitter of a VLC system. SSL sources such as LEDs and light amplification by stimulated emission of radiation diodes (LDs) are the main possible choices for VLC. Silicon-based photodetectors used in optical wired and wireless communications such as P-type intrinsic N-type photodiode (PIN-PDs) and APDs are suitable candidates for VLC due to their relatively high responsivity between wavelength $\lambda = 320$ nm and $\lambda = 1,100$ nm. This section presents descriptions of the main categories of front-end devices in VLC.

6.2.1 Transmitters

The transmitter end of a VLC system is widely expected to be additionally used as a lighting device. Therefore, the VLC transmitter is expected to satisfy the illumination requirements[1] of the considered application. The types of VLC transmitters vary in their radiation pattern, optical power, emission wavelength, and frequency response. Off-the-shelf white LEDs have been widely considered as the primary candidates for VLC transmitters because of their wide availability and low cost. SSL devices have a higher luminance efficacy (lumens per watt) and thus offer an energy-efficient solution to the high-power consumption of conventional lighting sources such as incandescent and fluorescent lamps [2].

6.2.1.1 Basic Optical Source Characteristics

A generalized Lambertian model is usually used to approximate the radiation pattern of an LED. The far-field radiation pattern is described by the normalized intensity versus the longitudinal and azimuth angles, and by the semiangle at half power,[2] as discussed in Chapter 5. A Li-Fi transmitter with a wide beam angle will have a larger coverage area and a lower light intensity (W/m^2) than another Li-Fi transmitter with a narrow beam angle.[3] Small cell sizes allow for a higher capacity per unit area due to the reduced interference. In addition, extremely small beam angles are required for long-distance, point-to-point transmission in

[1] The illumination requirements are defined by the considered application. For example, the lighting system of an airplane must enable dimming when airplanes take off and land.

[2] Angle at which the optical power drops by 3 dB, $\Phi_{1/2}$.

[3] Assuming they both have the same optical power.

VLC. The small beam angle preserves the optical power per unit solid angle, which maximizes the data rate and reduces transmission errors. An optical system is usually used at the Li-Fi transmitter to achieve a desired radiation pattern. Optical lenses, color converters, diffusers, parabolic reflectors, and mirrors can be used to adjust the emission wavelength and radiation pattern of the transmitter. The emission profile of most LDs is usually modeled using Gaussian beams. Beam divergence measures the beam diameter of a LD source at the far field.

The modulation signal is used to drive an LED or an LD of Li-Fi transmitters. Therefore, the electrical-optical transfer function of the Li-Fi transmitter is detrimental to the communication system performance. The current-voltage (I-V) and luminance-current (L-I) characteristic functions are required to have a large and linear dynamic range that allows the modulation signal to swing with little or no degradation at all. However, the turn-on voltage of the used device and the voltage at optical power saturation impose a limit on the lower and upper permissible voltage values for the modulation signal. Modulation waveform values that are below and above these points will be clipped.

The luminance-voltage (L-V) characteristic function of a blue LED (Vishay, VLMB1500) is shown in Figure 6.1. The probability density function (PDF) of a Gaussian distributed input signal is shown in Figure 6.1 as a function of the forward voltage of the LED. The PDF of the output signal is shown as a function of the optical power of the LED. The output PDF is shown to be shaped by the nonlinearity of the light source. The statistics of the output signal can be investigated using the Bussgang theorem [3,4], as discussed in Section 6.4.2.1 The turn-on voltage of the LED results in a clipping distortion for the lower part of the output signal PDF. Similarly, the saturation of the optical power of the LED results in a clipping distortion for the upper part of the output signal PDF.

The nonlinear distortion created by this clipping results in a degradation in the SNR and the bit error ratio (BER) performance. In addition, LED and LD devices exhibit nonlinearities that depend on the I-V and L-I characteristic functions. This has prompted several studies that attempt to understand and mitigate the effects of nonlinearity on the performance of the VLC and Li-Fi systems [5–9]. Signal predistortion was the first method to be considered for mitigating the effects of nonlinearity. However, the nonlinear effects are not static, and they could be different depending on the input signal power level, frequency components, and the temperature of the device [10]. This highlights the importance of thermal management in maintaining a reliable performance of Li-Fi systems.

The spectral width of a Li-Fi transmitter is mainly considered in WDM systems. Devices with a narrow spectral width can be used to minimize interference and to increase the number of parallel transmissions within a fixed spectral band. The data rate of a Li-Fi system is linearly dependent on the

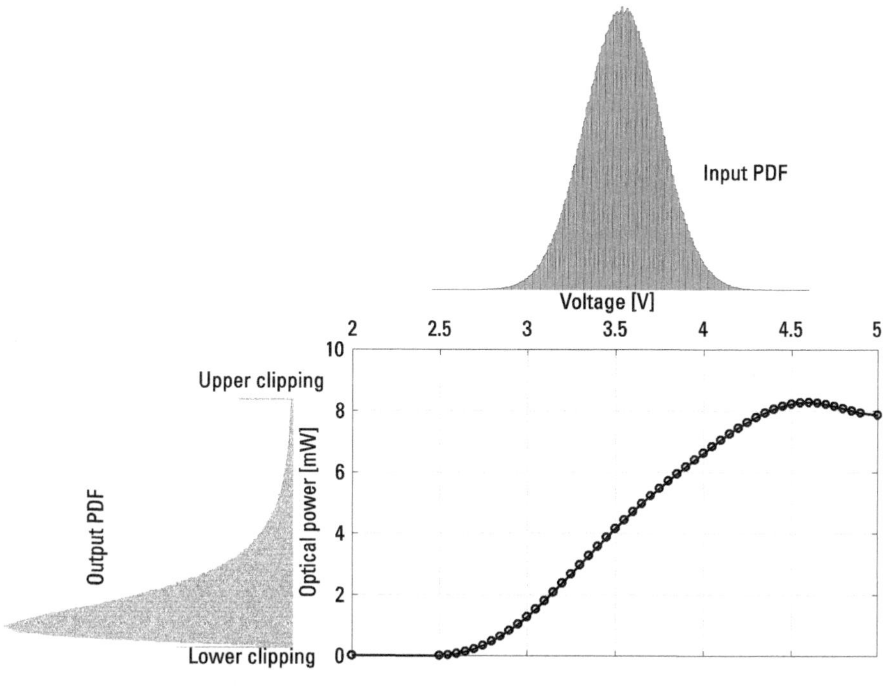

Figure 6.1 The voltage-optical power transfer function of a typical LED. A voltage signal with a Gaussian PDF is used as a driving signal (top). The output optical power has a truncated Gaussian PDF (left).

bandwidth of the used optical front-end devices. The bandwidth of an LED is inversely proportional to the carrier lifetime and the parasitic capacitance of the device. Devices with a relatively flat frequency response are desired to simplify the equalization process and to reduce the computational complexity of the communication system. The 3-dB bandwidth refers to the frequency at which the electrical power drops by 3 dB. This is usually used to quantify the bandwidth of Li-Fi front-end devices. However, the frequency profile of a Li-Fi transmitter may exhibit a slow roll-off that would enable reliable communications beyond its 3-dB bandwidth. This has led researchers to additionally report the 6-dB bandwidth of their devices, which refers to the 3-dB optical bandwidth.[4] The 3-dB bandwidth of commercially available LEDs ranges between a few kilohertz and a few megahertz (typically 2 to 3 MHz) [1], whereas the modulation bandwidth of LDs ranges between 200 MHz and 6.8 GHz [12,13]. A comparison between

[4]The optical bandwidth refers to the frequency at which the optical power drops 3 dB. The optical power is proportional to the injected current, whereas the electrical power is proportional to the squared injected current. Therefore, a drop in the optical power by $1/2$ or -3 dB is equivalent to a drop in the electrical power by $1/4$ or -6 dB.

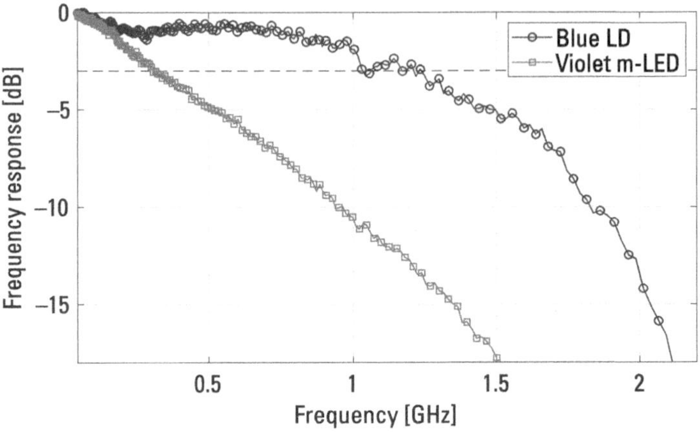

Figure 6.2 A comparison between the frequency response of violet m-LED [10] and a blue LD from Osram, PL450B.

the frequency response of a commercially available blue LD and a violet micro-sized gallium nitride LED (m-LED) [10] is shown in Figure 6.2. The frequency response of the LD is shown to be mostly flat below the 3-dB bandwidth, which is shown at 1.04 GHz. In contrast, the frequency response of the m-LED is shown to drop faster with the 3-dB bandwidth shown at 307 MHz.

Devices with high optical power are desired to relax the alignment constraints and enable longer transmission distances.[5] Generally, this comes at the price of increasing the junction capacitance of the device, which results in reducing the 3-dB bandwidth. However, devices with higher bandwidth are desired as the modulation bandwidth is directly proportional to the achievable data rate. This shows the potential of transmitters based on LDs in VLC systems. Figure 6.3 shows the peak data rate reported for numerous Li-Fi transmitters [14–21] . The trend shows the potential of utilizing all the possible degrees of freedom (DoF) in realizing Li-Fi systems beyond 1 Tbps.

6.2.1.2 LED-Based Devices

LEDs are light sources with narrowband emissions that span across ultraviolet, visible, and IR wavelength bands. LED-based white illumination is now possible because of the innovative work of Isamu Akasaki, Hiroshi Amano, and Shuji Nakamura. Their invention of the efficient blue LED has a global impact on the lighting energy consumption, which warranted them the Nobel Prize in physics in 2014 [22], as detailed in Chapter 2.

---

[5]The increase in the DC optical power does not necessarily translate to an increase in the alternating current (AC) optical power, depending on the impedance mismatch between the LED/LD and the driving circuitry.

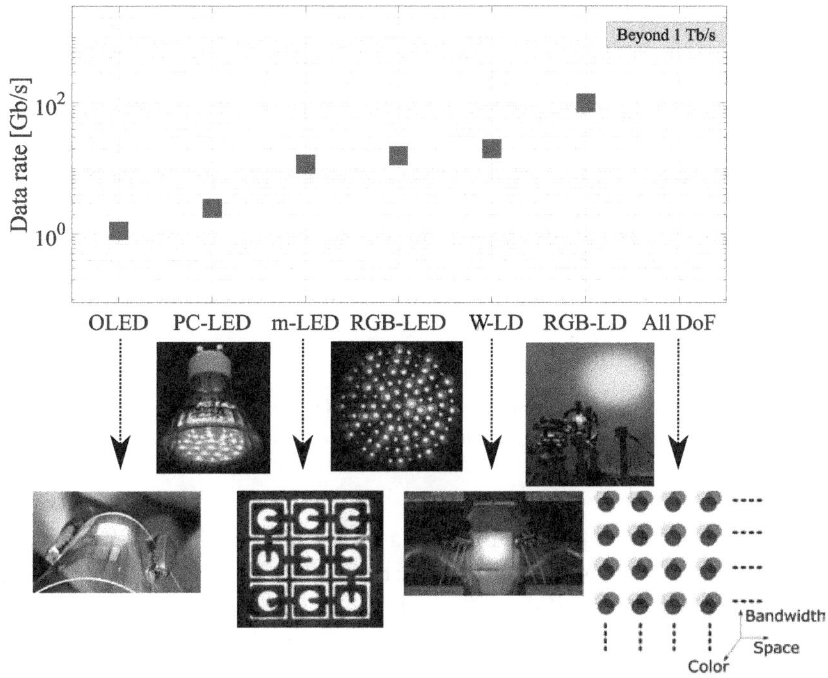

Figure 6.3 The peak data rate reported for each of organic LED (OLED) [14], PC-LED [15], m-LED [16], RGB-LED [17], commercially available white LD based on 2 blue LDs + remote phosphor from SLDLaser [18], red, green, and blue light amplification by stimulated emission of radiation diodes (LDs) (RGB-LD) [19,20], and the recent attempts to utilize all the possible DoF such as frequency, space, wavelength, and polarization [21]. The photo of the m-LED is provided by the courtesy of Dr. Enyuan Xie from Professor Martin Dawson, group at the Institute of Photonics, University of Strathclyde [16].

There are two main methods to produce white illumination using a semiconductor-based LED. The first method is based on coating a violet or a blue LED with a color-converting material. Yellow phosphor is the most popular choice due to its low cost and simple manufacturing process. Blue LEDs emit light with a peak between 425 nm and 475 nm. Part of this emission is absorbed and then reemitted between the green and red wavelength bands after it excites the yellow phosphor. The combined blue and converted light color is perceived as white illumination by the human eyes. The color temperature of the produced light is tuned by varying the thickness of the phosphor. The white illumination of a PC-LED has two spectral components, as illustrated in Figure 6.4: a narrowband blue component and a broadband component between 500 nm and 630 nm. The broadband component has a long carrier lifetime that significantly reduces the overall bandwidth of PC-LEDs. The frequency response of LEDs can be approximated by a lowpass filter, as described in Chapter 5. The 3-dB bandwidth

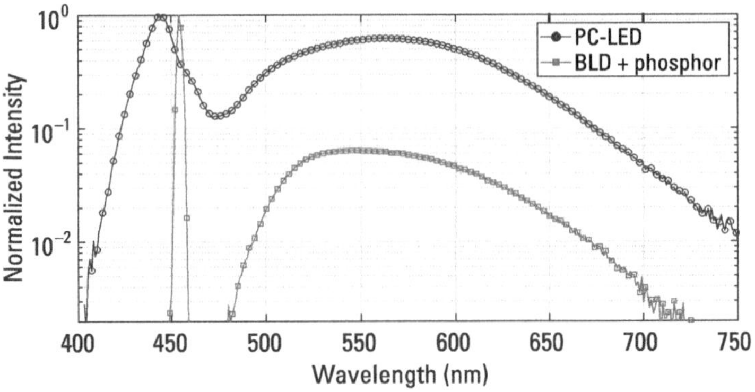

Figure 6.4 The spectral distribution of the normalized intensity of a PC-LED and a commercially available blue LD with a remote phosphor.

of commercially available PC-LEDs is around 3 MHz [11]. However, this value can be increased to 20 MHz when a blue filter is used in front of the photo-receiver [11]. The slow response of the yellow phosphor has motivated the research on fast color converting materials that could approximate the photometric performance of yellow phosphor and, at the same time, enable high-speed VLC communication systems [23,24]. The second method of producing white illumination is based on combining the light of different multicolored LEDs. For example, RGB-LEDs can be used to produce white light. Additional colored LEDs, such as yellow LEDs, can be added to the color mix to improve the quality of the final light output.

A data rate at 2.5 Gbps was recorded for a VLC system using a PC-LED as a transmitter [15]. This result was achieved using hardware pre-equalization techniques to extend the bandwidth of the used PC-LED. Multicolored LEDs allows the Li-Fi system to achieve significantly higher data rates due to the possibility of transmitting data on parallel streams of wavelength channels, which can be separated with optical filters at the receiver. A data rate of 15.7 Gbps has been experimentally demonstrated with off-the-shelf multicolored LEDs [17].

Micro-Sized Gallium Nitride LEDs

Significant research efforts have been devoted to the development of faster LED devices. LEDs with micro-sized pixels have attracted increasing interest in the past few years due to their high 3-dB bandwidth, which exceeds 800 MHz [10,16,25]. A data rate of 7.9 Gbps was reported using a violet m-LED with an active area of 435 μm^2 and a 3-dB bandwidth of 370 MHz [10]. A micrograph showing the design of this m-LEDs array is shown in Figure 6.5 [10]. The reduction in the junction size of m-LEDs results in a reduced resistance-capacitance (RC) time constant. Therefore, the bandwidth of m-LEDs becomes mainly limited

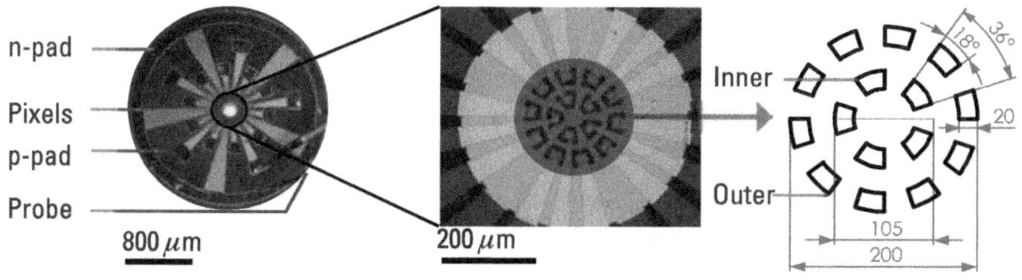

Figure 6.5 Micrograph showing a segmented m-LED array that consists of inner and outer groups of pixels. The inner and outer pixels have active areas of 435 μm^2 and 465 μm^2. This device is designed and fabricated by Professor Martin D. Dawsons research group at the Institute of Photonics, University of Strathclyde [10].

by the carrier lifetime. In addition, the reduction of the m-LED size results in an increased current density, which allows for a short carrier lifetime and thus enables the high 3-dB bandwidth of m-LEDs [26].

The reduction in the sizes of m-LEDs results in reducing their maximum optical power. A possible solution is to use an array of m-LEDs to increase the total optical power. However, this increases the junction temperature, which could degrade the SNR. Recently, a data rate was reported at 11.7 Gbps using a blue m-LED array that has 9 in-series pixels, where the diameter of each of the pixels is given at 20 μm [16].

Organic LEDs

Organic LEDs (OLEDs) are commonly used in display systems due to their simple and low manufacturing cost, flexibility, high quality of color rendering, and wide radiation patterns. The distinct form factor and flexibility of OLED makes them ideal for smart wearables. However, OLEDs are generally considered to have a low brightness level and a short lifespan.

OLEDs are slower than other inorganic LEDs due to their low charge mobility and high capacitance, which limits their modulation bandwidth to a few tens of megahertz [27]. However, it was shown that higher modulation speeds can be achieved using improved designs and fabrication methods [14,27]. A record modulation bandwidth of 245 MHz was recently reported to achieve a data rate of 1.13 Gbps over a 2-m distance [14].

Resonant Cavity LEDs

A resonant cavity LED (RC-LED) is an LED with an active region that is placed in a resonant optical cavity [28]. RC-LEDs have a higher brightness and a lower spectral width than conventional LEDs. In addition, they are often used in plastic optical fiber (POF) due to their narrow emission pattern [29]. Recently, an RC-LED emitting at 650 nm was used to demonstrate a VLC system at a data rate of 8.76 Gbps using an improved frequency-domain equalizer [30].

Super-Luminescent Diodes

Edge-emitting LEDs are designed to improve the coupling efficiency to POF [28]. Superluminescent diode (SLDs) behave like LDs in the sense that they operate at higher current densities and thus can generate a stimulated emission. In stimulated emission, the generation of a photon can be stimulated by the recombination of an electron-hole pair [28]. However, SLDs fail to achieve a high coherence that could classify them as coherent light sources due to the lack of the optical feedback that is usually provided by reflectors in LDs.

SLDs combine the advantages of LEDs and LDs as they can achieve a droop-free high luminance flux and a wide optical spectrum with low speckle [31]. A large 3-dB bandwidth can be achieved using SLDs because of their reduced carrier lifetime [32]. A 3-dB bandwidth in the order of 2.5 GHz was recently reported using a blue SLD at a current density of 30 kA/cm^2 [33]. The high bandwidth performance was mainly attributed to the reduction in carrier lifetime and the ability to operate at high current densities without reaching lasing thresholds [33].

6.2.1.3 LD-Based Devices

The quantum efficiency of LEDs drops at high driving current levels, which puts a limit on the luminance flux that can be achieved using a single LED. A large array of LEDs is usually required to provide a high luminance flux for applications such as outdoor lighting. LD-based lighting is emerging as a solution to the efficiency droop of LEDs [34]. Unlike LEDs, LDs have high efficiency at high current densities and thus can achieve a high luminance flux. White illumination using LDs can be achieved in similar ways to those used in LED-based lighting. Blue LDs can be used to excite a color-converting material to generate a white illumination [34]. Alternatively, multicolored RGB-LDs can be used to achieve white illumination. Figure 6.4 shows the spectral distribution of a PC-LED and a commercially available white LD that is based on a blue LD and a remote phosphor. This shows that the ratio of the light intensity at the blue wavelength to the light intensity of the converted light is shown to be higher for the white LD.

IR LDs have been used for optical fiber communications for decades. The use of visible LDs for VLC was foremost motivated by the abundant modulation bandwidth capabilities of LDs in comparison with LEDs. In addition, LDs have a narrow beamwidth and thus can be preferred in long-range, point-to-point VLC systems. The 3-dB bandwidth of a 410-nm semipolar gallium nitride (GaN) LD was reported at 6.8 GHz [13]. Moreover, it was estimated that the bandwidth of LDs could reach as high as 27 GHz [35]. Visible LDs have a narrow spectral emission in comparison with LEDs [36], and thus can be used to aggregate a larger number of channels than LEDs within the same wavelength band. It was estimated that a 100-Gbps VLC system can be realized using LDs that cover the whole visible spectrum [19].

There has been an increasing interest in using LDs as Li-Fi transmitters in the last few years [37]. Some of the studies considered the development and fabrication of high-bandwidth LDs, which can be used to generate white illumination. For example, a GaN blue LD emitting at 422 nm was used to demonstrate a 2.5-Gbps VLC system over a 1.4-GHz bandwidth in 2013 [38]. Additionally, a single 450-nm LD was used to demonstrate 15 Gbps over a 2.5-GHz bandwidth in 2016 [39]. Soon after, studies on demonstrating simultaneous communication and white illumination using LDs started to appear [37]. In most of these studies, an LD device emitting in the region between 400 nm and 450 nm is used in conjunction with a color-converting material such as the commonly used yellow phosphor. A 2-Gbps VLC system was demonstrated using a GaN blue LD that is exciting a YAG:Ce phosphor [40]. The system operated at 1.1 GHz and achieved a white illumination with good color stability. Other studies considered the use of additional LDs to improve the quality of white illumination and increase the total throughput of the Li-Fi system. A WDM system of RGB-LDs was used to generate a white illumination at 3.43 Gbps in 2015 [19]. A recent study demonstrated 40.67 Gbps by utilizing a polarization division multiplexing of RGB-LDs [20]. However, the use of polarization restricts the ability to use an optical diffuser [19], which is usually required in RGB-LD-based systems to comply with the eye-safety regulations governed by the British standard EN 60825-1:2007 [41]. LD-based transmitters should guarantee a speckle-free white illumination that conforms to eye-safety regulations before they can be publicly accepted. A yellow LED was used with a violet, green, and red LDs in a 34.8-Gbps WDM VLC system [42]. The yellow LED was only used for color tuning of the final light output, whereas the LDs were used for data communications. A microelectromechanical system (MEMS)-based beam-steering mirror was used to enable a wide coverage area for a VLC system with violet, blue, green, and red LDs [43]. A data rate of 35 Gbps was reported at a distance of 4m with a large coverage area of 39 m^2. The coverage area here refers to the supported area where the system can operate. However, tracking systems are required to maintain a link with the narrow FOV beam-steered transmission.

Most of the previously discussed demonstrations were conducted in research labs using low-power LDs to generate a white illumination that can be suitable for point-to-point VLC systems. A commercially available LD-based surface mount device (SMD) emitting a white illumination was recently used to achieve 20 Gbps [18]. The 7 × 7 mm SMD uses two 3-W LDs emitting at 445 nm and 455 nm to excite a remote yellow phosphor inside the same package simultaneously. Each LD can generate 400 to 500 lumens, and the SMD luminance level is reported to be over 1,000 cd/mm^2. The SMD is efficient in delivering higher luminance at targeted directions due to the efficient collimation of light, and thus it can support long-distance VLC applications.

Vertical-cavity surface-emitting laser (VCSELs) are interesting devices for VLC since they have a small form factor and a circular emission pattern. In addition, VCSEL devices can achieve a high modulation bandwidth at low driving currents. Most of the commercially available VCSEL devices operate in the near IR wavelength band due to the maturity of their fabrication using InGaAs and AlGaAs distributed Bragg reflectors [44]. Fabricating VCSEL devices at wavelengths shorter than 680 nm is challenging technically. However, there have been few studies on techniques that can be used to fabricate VCSEL devices in the violet and blue wavelength bands [45–47]. One of these devices was reported to achieve a 3-dB bandwidth of 1 GHz [48].

LD-based transmitters are popular choices for high-speed underwater VLC systems. Data rates over 25 Gbps were achieved over a 5m of highly turbid harbor water using a two-stage injection-locked red VCSEL device. Injection locking is used here to extend the modulation bandwidth of the used LD. In external injection locking, a fraction of the output of a master LD is injected into the gain region of a slave LD. Injection locking can be used to achieve a narrow line width and to stabilize the optical frequency of the slave laser [49]. A 500-Gbps underwater VLC link was recently demonstrated using RGB-LDs employing two-stage injection locking and polarization multiplexing over 100-m free space and 5m of turbid water [50]. Although a significant number of devices are used in this demonstration, this result highlights the capability of VLC systems in supporting ultrahigh-speed underwater applications.

Launching coherent-modulated streams from an optical fiber into free-space has also been considered to enable high-speed OWC systems. An unprecedented data rate was demonstrated at 100 Tbps using optical angular momentum (OAM), wavelength, and polarization division multiplexing of 1,008 channels [21]. The system uses 12 OAM beams, 2 polarizations, and 42 wavelength channels, with 100 Gbps per channel, to achieve a spectral efficiency of 22.3 bits/s/Hz over a 1-m point-to-point, free-space link. However, implementing such systems for mobile wireless access requires pointing, acquisition, and tracking (PAT) systems to direct the narrow IR beams to the user equipment. Beam steering can be usually achieved using spatial light modulators, MEMS mirrors, or diffractive gratings [51]. A recent study demonstrated a passive beam steering of 80 IR channels between 1,529.10 nm and 1,569.80 nm, achieving 112 Gbps per channel over a distance of 2.5m [52,53]. The use of wavelengths beyond 1,400 nm is suggested to be eye-safe when the used optical power is less than 10 mW [51]. In addition, the wavelength band around 1,550 nm benefits from a wealth of optoelectronic devices that have undergone decades of development in optical fiber communications. An arrayed waveguide grating router (AWGR) is used to enable 2-dimensions beam steering. The light output of the AWGR propagates to relatively different locations within the coverage area based on the light wavelength. These locations are mapped to different user equipment using a predefined look-up table. The beam steering can then be controlled by tuning the wavelength of the LD [52,53]. The system achieves a coverage area of $1.68 \times 1.68\ m^2$ with an overall throughput of 8.9 Tbps [53]. Independently, a fiber-wireless-fiber system was proposed for indoor mobile optical wireless access in 2016 [54]. The system operates at 416 Gbps with a 40° FOV and achieves a practical room coverage using beam steering and angle magnification at 3-m distance.

Recently, a system architecture that demonstrates a 2.33-Tbps bidirectional optical wireless mobile communication system was proposed by Wei et al. [55]. The architecture consists of a 10-km fiber-optic backhaul that connects the central units to the cloud and a 1.5-m free-space mid-haul solution that connects the central unit to the access points. At mid-haul, 42 IR channels were utilized in half between 1,575.78 nm and 1,608.33 nm for the uplink, and between 1,530.33 nm and 1,560.61 nm for the downlink. The 54.3-Gbps and a 46.2-Gbps data rates were demonstrated for each of the uplink and downlink channels, respectively [55]. The reported transmission speed is calculated for the aggregate data rate of both the uplink and the downlink. However, the study only demonstrates 2.7 Gbps for the downlink between the access points and the user equipment. This is enabled by a blue LD and a yellow phosphor [55].

6.2.2 Receivers

Photodetectors can be characterized by their detection area, frequency and spectral responses, FOV, and sensitivity. A PD is a diode that generates a photocurrent when it absorbs light in its depletion region. The most commonly used PDs in VLC systems are based on PIN-PDs and APDs due to their high linearity and fast response. Additional types of detectors started to attract interest due to their distinct characteristics. For example, SPADs have recently attracted an increasing interest due to their high sensitivity [56,57]. In addition, detectors capable of optical energy harvesting such as photovoltaic (PV) cells were recently considered for VLC systems [58].

In addition to single-element PDs, imaging sensors, also called camera sensors, can be used as Li-Fi receivers. An imaging sensor typically consists of a matrix of PDs assembled on an integrated circuit. Since most imaging sensors are designed to capture images and record videos with high resolutions, it requires integrating a significantly large number of pixels. This inherently limits the sensor's frame rate and limits the achievable data rates to a few hundred kilobits per second, as detailed in Chapter 3. Photo-multiplier tube (PMTs) are highly sensitive detectors that operate at a very high internal gain, and thus they are optimal for applications where extremely low optical power levels are expected. However, a PMT has low quantum efficiency and requires high bias voltages between 800V and 1,000V [59]. The rest of this section is focused on PDs due to their capabilities in achieving multigigabits per second VLC.

6.2.2.1 PIN-PDs and APDs

Current Li-Fi systems primarily use PDs, such as PIN-PDs and APDs, to convert the received optical intensity into a photocurrent. a PIN-PD consists of a transparent P-type doped contact layer on top of an intrinsic undoped absorbing layer and an N-type doped contact layer underneath. An incident photon can generate an electron-hole pair when the photon energy is larger than the bandgap of the photodiode material. Most of the generated electron-hole pairs contribute to the photocurrent, which is proportional to the incident optical power. The main principle of light detection of an APD is similar to that in a PIN-PD. However, a single detected photon in an APD can create an avalanche photocurrent when the APD is reverse-biased. A high reverse voltage between 100V and 500V is usually required in APDs to achieve a multiplicative gain that would be proportional to the reverse voltage [60]. However, the SNR that can be achieved by an APD-based VLC system is limited by the gain-dependent high excess noise. The high sensitivity of APDs makes them suitable to applications where low optical power is expected at the detector, such as long-range VLC and underwater OWC systems.

In addition to photodetectors, optical receivers require the use of an optical system to collect and filter the incident light and an electronic system of high-gain transimpedance amplifiers (TIAs), electrical filters, and adaptive equalizers [61]. Automatic gain control (AGC) can also be used to adapt the gain of the TIA based on the incident optical power.

The responsivity in PDs is defined as the efficiency of light power conversion to an electrical current. Therefore, it is given as the ratio of the generated current to the incident optical power. This can be given as follows [62]:

$$R = \frac{\eta_q q \lambda}{hc} \approx \frac{\eta_q \lambda[\text{nm}]}{1,240} \tag{6.1}$$

where q is the electron charge, h is the Planck constant, c is the speed of light, and η_q is the quantum efficiency, which is given as the ratio of carrier generation rate to the photon incident rate. The responsivity is illustrated in Figure 6.6 for quantum efficiencies ranging between $10\% < \eta_q < 90\%$.

The responsivity of a PD varies based on the used semiconductor material due to the differences in bandgap energy and light absorption characteristics. APDs have a much higher responsivity than PIN-PDs because of their high multiplicative gain. The responsivity of PDs fabricated using different semiconductor materials is presented in Figure 6.6 for three commercially available PDs from Thorlabs [63]. InGaAs-based PDs are suitable for IR-based optical communications beyond 1,000 nm. Ge-based PDs have a similar wavelength range support to InGaAs-based PDs, but at a lower responsivity. Most of the commercially available PDs in the visible spectrum are based on silicon.

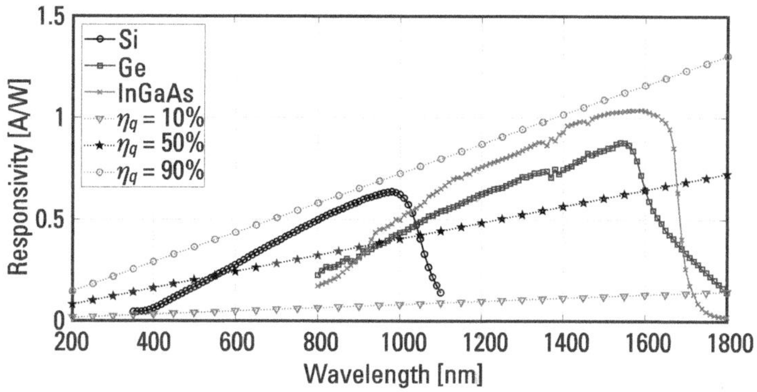

Figure 6.6 The responsivity of PDs at different quantum efficiencies compared with the responsivity of three calibrated PDs from Thorlabs [63]: Si-based FDS100-CAL, Ge-based FDG03-CAL, and InGaAs-based FGA21-CAL. Data extracted from [63].

The responsivity of Si-based PDs ranges from 400 nm to 1,100 nm and peaks between 900 nm and 1,000 nm.

It was shown that reverse-biased LEDs can function as wavelength-selective PDs, but at very low sensitivity levels [64]. An interesting study has recently demonstrated the use of an array of m-LEDs as a multifunctional device for display, data transmission, photo-detection, and energy harvesting [65]. The peak responsivity at 405 nm of a 60-μm diameter m-LEDs array was reported at 0.31 A/W in reverse-biased condition [65]. A high responsivity at short wavelength bands between 400 nm and 460 nm is essential to VLC due to the reliance of SSL devices on these wavelength bands. Figure 6.6 shows the disparity between the responsivity at the blue wavelength band and the responsivity at the peak wavelength. The responsivity at 450 nm is 17.9% of the peak responsivity at 980 nm, which is given at 0.637 A/W. This highlights the need for a new type of PD with a peak responsivity around 450 nm, which could significantly improve the performance of Li-Fi systems.

Micro-photodetectors (m-PDs) based on InGaN/GaN quantum wells were proposed as a solution to the low responsivity of the Si-based PDs around 400 nm [66,67]. Recent results reported a data rate of 7.4 Gbps using multiple InGaN/GaN quantum-well m-PDs with a 3-dB bandwidth of 300 MHz and a responsivity between 360 nm and 420 nm [67]. Although these devices are developed in research labs, their reported responsivity is comparable to the achievable responsivity of commercially available Si-based PD around 400 nm. The increased efforts on such devices have the potential to enable the development of PDs that are optimized for VLC applications.

PD has an inherent spatial diversity feature because the size of the detection area in OWC is relatively larger than the considered wavelength of light. This allows the receivers to avoid multipath fading that is usually present in RF

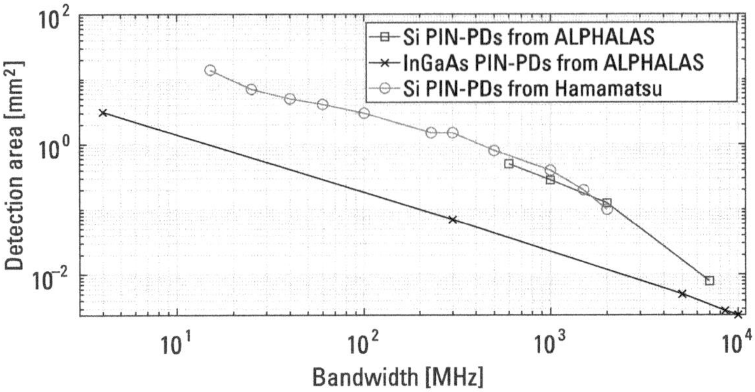

Figure 6.7 The relationship between bandwidth and size of commercially available silicon photodiodes selected from ALPHALAS GmbH [68] and Hamamatsu Photonics [69].

communications. However, there is an inverse relationship between the detection area and the bandwidth of the PD. Increasing the detection area improves the light collection capability of the PD, which improves the reliability of the Li-Fi system. However, this also results in reducing the 3-dB bandwidth due to the increased capacitance. For example, the detection area of a commercially available Silicon PIN-PD that has a 3-dB bandwidth of 0.5 GHz is 0.5 mm^2 (ALPHALAS UPD-500-UP). This is more than 63 times larger than the detection area of a 7-GHz PIN-PD (ALPHALAS UPD-50-UP), which is given at 0.0079 mm^2. This trend is illustrated in Figure 6.7, which shows the relationship between bandwidth and detection area for a few commercially available PDs selected from ALPHALAS GmbH [68] and Hamamatsu Photonics [69]. The small detection area of high-speed PDs reduces the system reliability and practicality since it limits VLC systems to narrow FOV applications, which require complex pointing and tracking systems.

An optical system is usually required to aid the PD in the light collection. The optical system could consist of lenses, optical filters, mirrors, or a combination thereof. The FOV of a VLC receiver can be controlled by the design of its optical system. A large diameter lens can improve the light collection at the price of decreasing the FOV. This is usually suitable for stationary long-range VLC systems. The aggregation of multiple streams on different wavelengths of light is desirable to increase the throughput of a Li-Fi system. However, this comes at the price of increasing the complexity of the optical front ends. The receiver of a WDM system requires the use of optical filters or dichroic mirrors to direct the different data streams to different PDs. Optical filters are also required to mitigate the optical interference from background illumination and other neighboring luminaries [70]. A blue filter is essential at the receiver of a Li-Fi system that uses a PC-LED as a transmitter choice [71]. Filters as narrow as 1 nm are commercially available with a transmittance above 80% [36]. The optical design of the receiver is usually tailored based on the targeted application. For example, neutral density (ND) filters are usually required to reduce the exposure of the detector to high optical power levels. The use of ND filters reduces the likelihood of PD saturation and failure. Alternatively, an optical beam shutter could be used to reduce the PD exposure to light, but at the price of reducing the FOV.

A number of PDs with different orientations can be mounted on a non-planar surface or on different facets of a mobile device [72–74]. This improves the FOV of the Li-Fi receiver and enables different MIMO techniques, as discussed in Section 6.5.3.2. Signal combining schemes such as select best combining (SBC), equal gain combining (EGC), maximum ratio combining (MRC), and optimum combining (OPC) can be used to combine the signals detected using the different PDs [75]. It has been shown that it is possible to realize fully integrated optical APD detectors with standard digital complementary metal oxide semiconductor (CMOS) technology [76,77]. An optical receiver with 49 APDs was demonstrated based on a 180-μm CMOS technology with a 90-MHz bandwidth and a responsivity of 2.61 A/W at 450 nm [77]. The APDs were designed with multiple spatial and spectral properties to improve the diversity of the receiver.

It is desirable that the receiver has a large FOV and a high optical gain. As the optical channel gain scales linearly with the detection area of the PD [36], using PDs with a larger detection area can increase the optical signal strength. However, PDs with a larger detection area typically have a smaller modulation bandwidth. An alternative approach is to decouple the light collection function from the optical detection function of a PD sensor. This can be achieved by using an extra optical element, such as a compound parabolic concentrator (CPC) to concentrate light onto the detection area of the PD, as shown in Fig 6.8(b). The optical gain of a CPC is given by the ratio of its input surface area to the area of its output. Optical elements based on the reflection-refraction, similar to CPCs, are limited by the étendue property [78]. This means that increasing the size of the light collection surface of a CPC results in reducing its FOV. In addition, it means that the gain of such optical elements is inversely proportional to their FOV. An upper bound for the optical gain of a CPC is given as [78]:

$$G_{\mathrm{CPC}} = \frac{n^2}{\sin^2\theta} \tag{6.2}$$

where n is the refractive index of the CPC material and θ is the semi-angle FOV. For example, glass CPCs with a FOV of 25° and 45° can achieve optical gains of 12.6 and 4.5, respectively.

Recent designs have shown that the performance of fluorescent concentrators is not limited by the étendue constraint and that a high optical

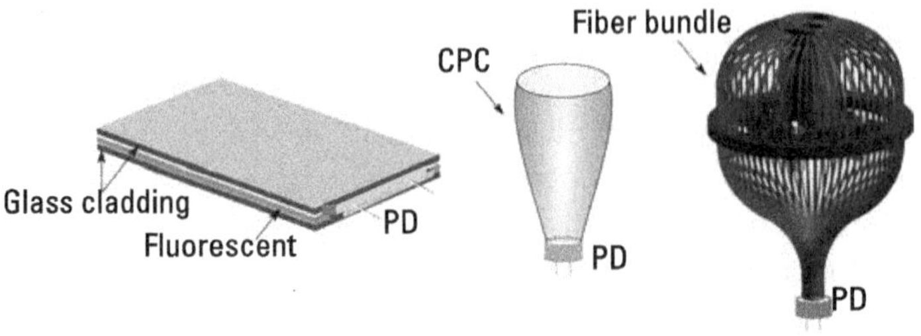

Figure 6.8 (a) A fluorescent optical antenna sandwiched between glass claddings and coupled to a PD [78]. (b) A CPC coupled to a PD. (c) A bundle of fibers arranged in a spherical shape coupled to a PD [79].

gain and a wide FOV can be obtained at the same time [78–81]. A bundle of scintillating fibers can be arranged in a spherical shape to form an omnidirectional optical antenna that can concentrate the light onto a PD [79], as shown in Figure 6.8(c). Similarly, an optical antenna based on fluorescent material that is sandwiched between glass claddings is shown in Figure 6.8(a). Part of the light is absorbed by the fluorescent material and then remitted at a longer wavelength. Total internal refraction occurs due to the fact that the refractive index of the fluorescent material is higher than the refractive index of the surrounding claddings. The light escapes at the edge of the concentrator, where a PD would ideally be placed very close to the edge in order to improve the coupling efficiency. The fluorescent concentrator is reported to achieve a FOV of 60° at an optical gain that is an order of magnitude higher than an étendue-limited concentrator [78]. In addition, it was demonstrated that the wavelength-dependent absorption of the fluorescent layer allows this type of optical antennas to function as a demultiplexer in WDM systems [81]. Different data streams transmitted in a WDM system can be demultiplexed to different edges of the device without the need for any optical filters or dichroic mirrors.

6.2.2.2 Single-Photon APDs

APDs that can operate in the Geiger mode are termed SPADs [76]. In the Geiger mode, the SPAD is biased beyond its breakdown voltage where the absorption of a single photon can create an avalanche of photocurrent [82]. As a result, SPADs are photon-counters that have the highest sensitivity when compared with other types of PDs. SPAD receivers do not require a TIA, and their output signal is not distorted by the thermal noise as in other types of PDs. However, after each photon detection event, the SPAD needs to be reset to recover from the excess charge carriers. This process introduces a finite recovery time, known as the dead time [83], during which the device cannot detect another incident photon. The dead time restricts the maximum number of photons that the SPAD can detect and, thus, the SPAD ability to recover a signal. This means that the maximum bit rate of SPAD-based systems is defined by the structure and design of the SPAD detector. Detailed characterization and analysis of the detection statistics of a single SPAD with a constant dead time were presented in [84]. This study provides a complete analytical approach to find the exact photo-count probability of a single SPAD detector, considering the impact of the dead time.

A single SPAD can only detect a single photon at the same time instant, which leads to a nonlinear behavior when detecting more than one photon. However, an array of SPADs can be used to mitigate this issue. These arrays are usually termed multipixel photon counter (MPPC), also known as silicon photomultiplier (SiPM). MPPC enables the detection of multiple photons at the same time. Each SPAD in the array generates a pulse when it detects a single photon. The outputs of all the SPADs in the array are then superimposed to form the final output, which would be proportional to the number of incident photons. The bandwidth of an MPPC is limited by its pulse width, and the sensitivity is limited by the dark count rate [85]. A comparative study of SPAD and APD-based receivers for VLC is presented in [86]. It is suggested that the sensitivity of SPAD-based receivers can be higher than the sensitivity of APD-based receivers when the photon detection efficiency is higher than 14% [86]. A follow-up study concluded that OOK is better than PAM for MPPC [85]. The study also predicted that a data rate of 1 Gbps can be achieved using MPPC with a received optical power that is lower by four orders of magnitude than the power required by a PIN-PD to achieve the same data rate [85].

The BER performance and information transfer rate of SPAD array optical receivers were investigated in [56]. SPADs are able to closely approach quantum-limited sensitivity in the detection of weak optical signals. Recent results reported 500 Mbps within 15.2-dB gap of the quantum limit using an array of 130-nm CMOS-based SPADs [57]. The quantum limit defines the minimum number of photons required to achieve a certain BER based on the particle nature of the electromagnetic radiation [87]. The high sensitivity and time resolution of SPADs have recently highlighted the potential of employing SPADs as photon-counting receivers for long-distance OWC applications. In addition, SPADs are considered for scenarios where the received optical power is severely attenuated, such as downhole monitoring [88].

6.2.2.3 Solar Cells

Solar cells, which are commonly used for energy harvesting, can also be used as Li-Fi receivers for data reception [89]. A standard commercial multicrystalline silicon solar panel could be used for this dual purpose [58]. PV-based OWC receivers offer an interesting solution for backhauling and IoT applications as they enable simultaneous wireless information and power transfer (SWIPT) [58,89–93].

Contrary to other types of PDs, PV cells usually operate in a forward-biased condition. As a result, PV cells have a lower 3-dB bandwidth than other types of PDs. However, recent results reported a data rate of 1 Gbps for a VLC system using a GaAs PV cell, with a 3-dB bandwidth of 40.3 MHz under a short-circuit condition, and a data rate of 784 Mbps while harvesting 1 mW at 41.7% efficiency [94]. In addition, organic PV cells with a 3-dB bandwidth of 1.3 MHz were reported to achieve a data rate of 34.2 Mbps while simultaneously harvesting 0.43-mW power [93]. Recently, a hybrid organic and inorganic triple-cation perovskite-based receiver was shown to enable a data rate of 56 Mbps and a power-harvesting efficiency at 20% [95]. Organic solar cells are more mechanically flexible and can be easily integrated into different devices. Therefore, they are an attractive choice for IoT applications.

Solar cells were considered as receivers for low-cost backhauling solutions within the United Kingdom 5G project Rural First [96]. A Li-Fi distribution hub was used to send Li-Fi signals over a 30-m distance to a few properties on the Orkney Islands. The properties were fitted with off-the-shelf PV panels that can harvest power up to 5W from sunlight. The 3-dB bandwidth of the panels was evaluated at 270 kHz. However, the bandwidth was improved using an analog equalization to reach 5 MHz [97]. This allowed the system to achieve a data rate of 8 Mbps. This result is significant as it paves the way to a cost-effective solution that can close the digital divide.

6.3 Channel Noise

The main noise factors at the receiver are shot noise and thermal noise [36]. Shot noise increases with the increased received intensity at the PD. The received intensity is the combination of the background illumination and the optical power of the transmitter. Shot noise can be modeled as an additive white Gaussian noise (AWGN) when the background illumination is larger than the desired signal at the receiver [98]. The power spectral density (PSD) of the shot noise is given as follows [36]:

$$N_o^{\mathrm{sh}} = 2qRM_{\mathrm{PD}}P_n \quad \mathrm{A}^2/\mathrm{Hz} \tag{6.3}$$

where q is the electron charge, R is the average responsivity of the photodetector over the used wavelength band given in A/W, P_n is the average optical power of the received signal and the background illumination, and M_{PD} is the gain of the PD, which is $M_{\mathrm{PD}} = 1$ for PIN-PD and $10^0 < M_{\mathrm{PD}} < 10^3$ for APDs [60]. Optical and electrical filters are commonly used at the receiver side to mitigate most of the background illumination noise [71]. The shot noise in APDs is gain-dependent, which increases with the increased gain. Therefore, the gain of an APD can be optimized to improve the SNR [60]. Dark current is the photocurrent generated by PDs in the absence of any incident photons. It increases with the increased bias voltage and temperature. However, it is generally considered to be low for Si-based PDs [60]. Thermal noise at the PD is generated due to the thermal generation of carriers. The thermal noise is related to the specific design of the TIA of the photodetector. However, it can also be modeled as an AWGN in most practical cases [36]. The PSD of the thermal noise is given as:

$$N_o^{\mathrm{th}} = \frac{4k_{\mathrm{B}}T_{\mathrm{K}}}{R_{\mathrm{L}}} \quad \mathrm{A}^2/\mathrm{Hz} \tag{6.4}$$

where k_{B} is the Boltzmann constant given as $k_{\mathrm{B}} = 1.381 \times 10^{-23}$J/K, T_{K} is the temperature given in K, and R_{L} is the feedback resistor in Ω. Selecting a high value for R_{L} would decrease the thermal noise. However, this would also decrease the 3-dB bandwidth of the TIA.

6.4 Modulation Techniques

It is widely believed that incoherent light sources are the preferred choices for Li-Fi transmitters due to their low cost and vast availability in the lighting infrastructure. Incoherent light sources restrict the modulation techniques in VLC to IM/DD schemes. Therefore, modulation techniques with high spectral efficiency play an essential role in designing high-speed Li-Fi systems due to the limited bandwidth of the optical front ends. The Shannon theorem for channel capacity does not apply to VLC systems due to the non-negativity constraint and illumination requirements of the different signaling schemes. However, the capacity of optical transmission systems has been extensively investigated for VLC in terms of lower and upper bounds [4,99–102].

Modulation techniques developed for Li-Fi operate in the baseband domain. This allows for a simplified system architecture when compared with coherent-based modulation schemes, which require coherent sources and external modulators at the transmitter end and local oscillators at the receiver end. Avoiding local oscillators also comes with significant energy savings since local oscillators require a significant amount of energy.

The choice of modulation techniques in Li-Fi depends on the considered application, – for example, uplink versus downlink. Modulation techniques vary in metrics such as spectral efficiency, electrical and optical power efficiencies, computational complexity, robustness to multipath propagation, and dimming capability. The dimming functionality in the context of Li-Fi refers to the flexibility of controlling the level of illumination without significantly degrading the communications performance. Two illumination modes are of interest: dimmed mode and high brightness mode. The limiting factors for these two modes are the turn-on voltage and the saturation point of the optical source, respectively. However, channel selectivity degrades the performance of modulation techniques due to the lowpass filtering effect of the front-end optical devices and the optical wireless channel, as described in Chapter 5.

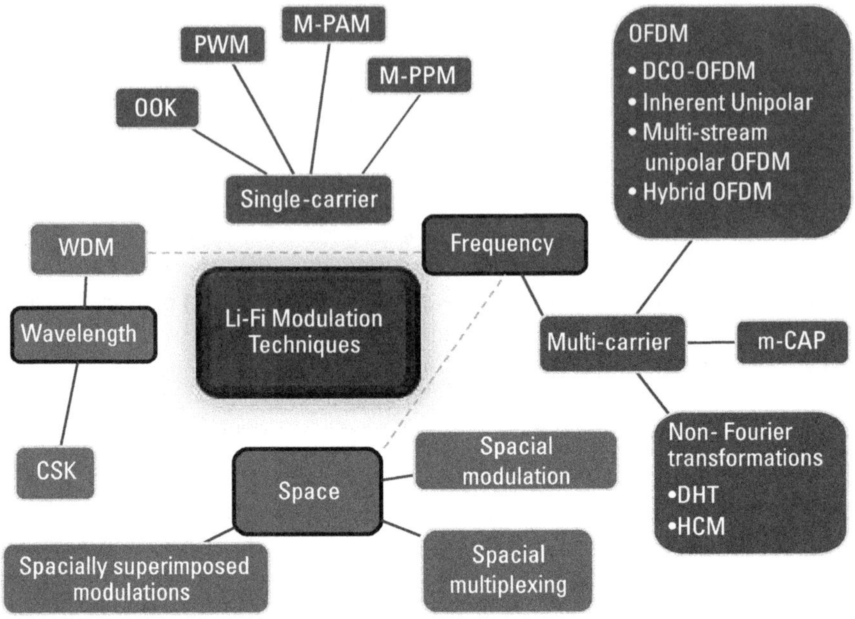

Figure 6.9 A taxonomy of the widely considered Li-Fi modulation techniques. Note that most of the space-based and WDM-based modulation techniques are also frequency-based.

Information can be encoded into different DOF including time, frequency, space, and wavelength, or a combination thereof. A taxonomy of the main Li-Fi modulation techniques categories is shown in Figure 6.9. Modulation techniques developed for OWCs can be used in Li-Fi if they satisfy the illumination requirements and if they exhibit robustness to multipath propagation. This is particularly important to high speed Li-Fi systems; the FOV of Li-Fi systems is larger than the FOV of free-space optical (FSO) and VLC point-to-point systems since the coverage in Li-Fi is an important performance metric. In this section, we survey the prominent modulation techniques. The list of the considered modulation techniques is not exhaustive. However, the list attempts to consider most of the widely used techniques.

6.4.1 Single-Carrier Modulation Techniques

Single-carrier modulation (SCM) techniques such as OOK, pulse width modulation (PWM), pulse position modulation (PPM), and PAM can be straightforwardly implemented in VLC. However, the ISI of OWC channels degrades the SCM performance at high spectral efficiency. Equalizers with a high computational complexity are usually required to mitigate the effects of ISI on the communication system performance.

Binary bits are encoded in the presence and absence of light intensity in OOK, and similarly, multilevel intensity levels can encode a stream of bits in M-ary pulse-amplitude modulation (PAM) (M-PAM) [103]. Variants of these techniques were proposed in the literature to improve the spectral and power efficiencies and adapt the modulation waveforms to the illumination requirements. The light intensity of the two binary states in OOK can be adjusted to control the illumination level without degrading the system performance. Compensation symbols that do not convey any additional information can be appended to the modulated signal to control the illumination, as proposed in the optical camera communication (OCC) standard, Institute of Electrical and Electronics Engineers (IEEE) 802.15.7. However, the illumination control is achieved here at the cost of reducing the spectral efficiency [104]. The position of the optical pulse is modulated into shorter duration pulses in PPM with a position index that varies depending on the incoming bits. PPM is more power-efficient than OOK, but it requires more bandwidth to support equivalent data rates [36]. Variable PPM (VPPM) was proposed in IEEE 802.15.7 to support dimming and to prevent any possible visible flicker. The dimming in VPPM is controlled by the width of the pulse rather than the pulse amplitude. Therefore, VPPM can be considered as a combination of PPM and PWM techniques. Multiple PPM (MPPM) was proposed as a solution to the dimming capability of PPM, where it was reported that it achieves higher spectral efficiency values than VPPM with less optical power dissipation [105].

6.4.2 OFDM-Based MCM

MCM techniques, such as OFDM, have shown great potential for high-speed VLC systems [1,10,39,106]. An optical OFDM-based modulation technique was first proposed for VLC by Tanaka et al. [107], and a first experimental proof-of-concept demonstration was presented by Afgani et al. [1]. OFDM is also considered as a promising technology by the IEEE task group focusing on light communications, namely, IEEE 802.11bb (TGbb) [108]. The data rate of an OFDM-based VLC system can be maximized by adapting the bit loading on each of the subcarriers to the frequency response of the considered channel. This can be achieved using computationally efficient single-tap equalizers in addition to adaptive bit and energy-loading algorithms. The bit and energy-loading algorithm adapts the channel usage to the estimated capacity while at the same time maintaining a correctable BER. The coincidence of multiple modulated symbols with a similar phase in the same frame results in a high peak-to-average power ratio (PAPR).[6] The peak and average values of an OFDM

[6]Note that MCM enables the use of phase modulation techniques in IM/DD systems. This includes binary phase-shift keying (BPSK), quadrature amplitude modulation (QAM), and phase shift keying (PSK).

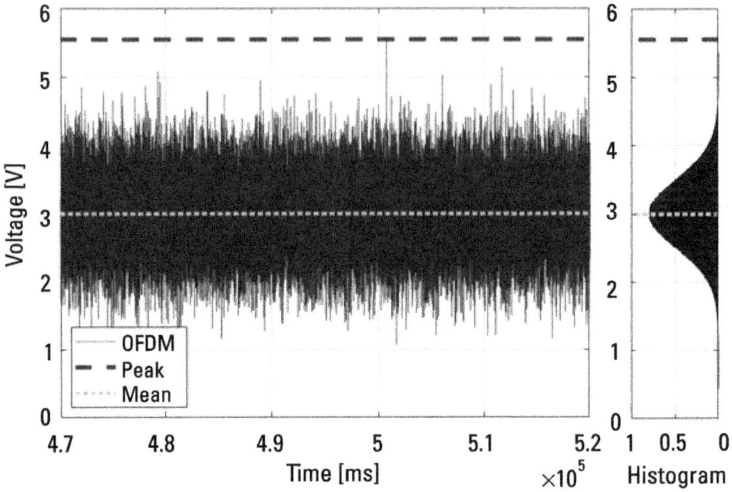

Figure 6.10 The peak and average values of an OFDM waveform shown on the time-domain signal and the corresponding PDF.

waveform are shown in Figure 6.10. High PAPR results in a degradation of the VLC performance due to the fact that the dynamic range of light sources is limited. However, several solutions were proposed to address this issue [109–111]. Most of these solutions attempt to reduce the likelihood of the existence of multiple symbols that have similar phases within the same frame.

6.4.2.1 DCO-OFDM

A block diagram of the main functions in DCO-OFDM transmitter and receiver is given in Figure 6.11. Incoming bits are modulated using techniques such as PAM or QAM. The number of bits that are encoded in each M-PAM and M-ary quadrature amplitude modulation (QAM) (M-QAM) symbols is given in b, where M is the modulation order $M = 2^b$. The modulated symbols are loaded into orthogonal subcarriers, $X[k]$, with a subcarrier spacing equal to $\Delta f = 1/(N_{\mathrm{FFT}} T_{\mathrm{s}})$, where N_{FFT} is the OFDM frame length, $T_{\mathrm{s}} = 1/f_{\mathrm{s}}$ is the sampling period, $f_{\mathrm{s}} = 2B$ is the Nyquist sampling rate and B is the single side bandwidth. Parallel to serial (P/S) transformation is then applied, usually using the inverse fast Fourier transform (IFFT) as given in (6.5). The time domain waveform output $x[n]$ can then be given as follows:

$$x[n] = \frac{1}{\sqrt{N_{\mathrm{FFT}}}} \sum_{k=0}^{N_{\mathrm{FFT}}-1} X[k] \times e^{j\frac{2\pi kn}{N_{\mathrm{FFT}}}} \tag{6.5}$$

where k is the subcarrier index. N_{FFT} is set to a value that is an even power of 2 in order to simplify the computational complexity of the FFT operation

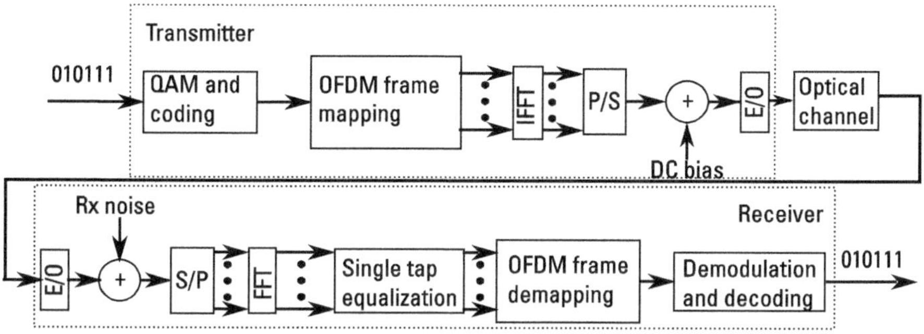

Figure 6.11 A block diagram showing the main functions in DCO-OFDM transmitter and receiver.

to $\mathcal{O}(N \log_2(N))$. The light intensity can only be represented by real and positive values. Therefore, OFDM waveforms are required to be unipolar and real-valued. Hermitian symmetry is generally imposed on the frames of the modulated symbols such as: $X[k] = X^*[N_{\text{FFT}} - k]$, for $k = 1, 2, \cdots N_{\text{FFT}} - 1$ and $X[0] = X[N_{\text{FFT}}/2] = 0$. This forces the OFDM time-domain output to be real-valued. The time-domain OFDM signal given in (6.5) is used to modulate the light source as an intensity modulation signal. The last N_{CP} symbols of each time-domain frame are repeated at the beginning of each frame. This helps in transforming the linear convolution with the optical channel into a circular convolution, which can be efficiently equalized in the frequency domain. The cyclic prefix (CP) is required to be longer than the maximum delay spread of the channel T_d so that the ISI can be mitigated using the single-tap equalizer.

Different variants of optical OFDM were proposed to achieve a unipolar OFDM output. DC bias is used in DCO-OFDM to shift the OFDM signal to the linear region of a light source dynamic range [112]. However, OFDM waveforms have a high PAPR, which makes it prone to clipping as it is practically impossible to convert all of the signal samples into unipolar ones. The PAPR is defined as:

$$\text{PAPR} \triangleq \frac{\max_{0 \leq n \leq N_{\text{FFT}}-1} |x^2[n]|}{\text{E}[x^2[n]]}, \tag{6.6}$$

where $\text{E}[\cdot]$ is the statistical expectation.

A Gaussian distribution is generally used to approximate the time-domain OFDM waveform when the length of the frame is greater than 64 [113]. The DC bias point would be ideally set within the center of the linear range between the turn-on voltage and the voltage at the saturation power. OFDM values that fall outside this linear range incur clipping distortion. The optimization of the DC bias point was studied in [4]. The DC bias can be defined as a k_{DC} multiple of the standard deviation of the time-domain OFDM signal σ_x. The electrical power penalty due to the addition of DC bias in DCO-OFDM compared with

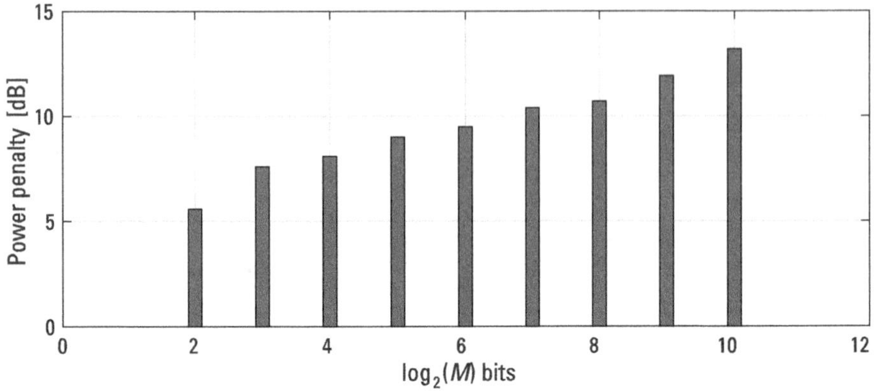

Figure 6.12 Power penalty caused by the addition of DC bias in DCO-OFDM compared with bipolar OFDM as a function of the modulation order.

bipolar OFDM is written as follows:

$$B_{\mathrm{DC}}^{\mathrm{dB}} = 10 \log_{10} \left(k_{\mathrm{DC}}^2 + 1 \right) \tag{6.7}$$

At the same clipping distortion level, the degradation in BER performance increases with the increase of the M-QAM modulation order. This is due to the fact that higher M-QAM modulated symbols have a shorter Hamming distance than lower modulation orders. Therefore, high modulation orders require additional DC bias to achieve the same BER of low modulation orders.

An estimation of the electrical power penalty caused by the addition of a DC bias is shown in Figure 6.12. The results are obtained using Monte Carlo simulations at a BER of 1×10^{-4}, based on (6.7). The inefficiency caused by the addition of large DC bias can be justified for some VLC applications where a high luminance is desired. However, an alternative modulation approach is required to reduce the electrical energy consumption. The spectral efficiency of DCO-OFDM is given by [9]:

$$\eta_{\mathrm{DCO}} = \frac{\log_2 M(N_{\mathrm{FFT}} - 2)}{2(N_{\mathrm{FFT}} + N_{\mathrm{CP}})} \quad \text{bits/s/Hz} \tag{6.8}$$

The definition of the achieved spectral efficiency is based on the definition of the bandwidth. Here it is assumed that the bandwidth is given as $2B$, where B is the single side bandwidth. In addition, the spectral efficiency given in (6.8) is based on the assumption that all the subcarriers are loaded with modulated symbols of M-QAM.

The continuous-time received OFDM waveform $y(t)$ is given as follows:

$$y(t) = h(t) * z(x(t)) + w(t) \tag{6.9}$$

where $x(t)$ is the transmitted time-domain signal, $h(t)$ is the channel impulse response of the VLC system, $z(\cdot)$ is the nonlinear transformation of the used transmitter, and $w(t)$ is the noise process at the receiver. This can be approximated as a Gaussian distribution with a variance $\sigma_n^2 = BN_o$, where B and N_o are the single side bandwidth and single side noise PSD, respectively. The light source is assumed to be the main source of nonlinearity in VLC systems due to the relatively limited dynamic range, compared to other system components. The nonlinear transformation $z(\cdot)$ can be described using the Bussgang theorem, which is valid for Gaussian inputs such as the real-valued OFDM waveform. The nonlinear transformation $z(\cdot)$ is given as follows [3,4]:

$$z(x) = \alpha x + w_{\mathrm{c}} \tag{6.10}$$

where α is an attenuation factor, w_{c} is the nonlinear distortion noise. Without loss of generality, the (t) notation is dropped for presentation simplicity. The nonlinear distortion w_{c} is uncorrelated with the transmitted signal x, $\mathrm{E}[x \cdot w_{\mathrm{c}}] = 0$. The attenuation factor α is a constant that is given as [3,4]:

$$\alpha = \frac{\mathrm{E}[z(x)x]}{\sigma_x^2} \tag{6.11}$$

The nonlinear distortion w_{c} is a non-Gaussian noise. However, its representation in the frequency domain $W_{\mathrm{c}}(f)$ follows a Gaussian distribution with a mean $\mathrm{E}[w_{\mathrm{c}}]$ and a variance $\sigma_{w_{\mathrm{c}}}^2$. The standard deviation of the clipping noise is given as follows [3,4]:

$$\sigma_{w_{\mathrm{c}}}^2 = \mathrm{E}[w_{\mathrm{c}}^2] - \mathrm{E}[w_{\mathrm{c}}]^2 \tag{6.12}$$

$$\mathrm{E}[w_{\mathrm{c}}^2] = \mathrm{E}[z(x)^2] - \alpha^2 \mathrm{E}[x^2] \tag{6.13}$$

$$\mathrm{E}[w_{\mathrm{c}}] = \mathrm{E}[z(x)] \tag{6.14}$$

Bit and energy-loading algorithms can be used to improve the data rate in OFDM-based VLC systems. An experimental comparison of different bit and energy-loading algorithms is present in [114]. In most of these algorithms, the channel is first estimated using a pilot signal that is composed of multiple OFDM frames. Estimations for the channel and the SNR can be obtained using a conventional mean estimator, as described in [115]. The estimated SNR is the ratio of the signal variance to the estimated noise, which is weighted by the estimated channel gain at each of the corresponding subcarriers. The estimated SNR can then be used to adaptively load the subcarriers with different constellation sizes based on the adopted algorithm [114–116]. Most of the algorithms allow the subcarriers that require minimal additional power to be loaded with larger constellation sizes, while at the same time preserving the BER target, P_e^{T}. The adaptive bit and energy-loading applies the following optimization problem on each single subcarrier:

$$\underset{b_k}{\text{maximize}} \quad b_k = \log_2 M_k \quad \forall k = 1, 2, \cdots, \frac{N_{\text{FFT}}}{2} - 1 \tag{6.15a}$$

$$\text{subject to} \quad \text{BER}\,(M_k, \nu_k) \leq P_e^{\text{T}} \tag{6.15b}$$

where b_k is the number of bits conveyed at subcarrier k and ν_k is the SNR per bit at subcarrier k, which is given for large $N_{\text{FFT}} > 64$, as [4]:

$$\nu_k = \frac{\alpha^2}{\dfrac{\sigma_{w_c}^2 \log_2 M_k}{\sigma_x^2} + \dfrac{(1 + k_{\text{DC}}^2)}{E_{\text{b}k}/N_o |H(k)|^2}} \tag{6.16}$$

where $E_{\text{b}k}$ is the energy per bit at subcarrier k, k_{DC} defines the DC bias in terms of σ_x as given in (6.7), and $|H(k)|^2$ is the channel gain at subcarrier k assuming that a zero forcing (ZF) equalizer is used. The theoretical BER at subcarrier k is given as [117]:

$$\text{BER}(M_k, \nu_k) \cong \frac{4}{\log_2(M_k)} \left(1 - \frac{1}{\sqrt{M_k}}\right) \times \sum_{l=1}^{2} \text{Q}\left((2l-1)\sqrt{\frac{3\log_2(M_k)\nu_k}{M_k - 1}}\right) \tag{6.17}$$

where $Q(\cdot)$ is the complementary cumulative distribution function (CCDF) for the standard normal distribution. The overall system BER can be given as:

$$\text{BER} = \frac{\sum\limits_{\substack{k=1 \\ M_k>0}}^{N_{\text{FFT}}/2-1} \text{BER}(M_k, \nu_k) \log_2(M_k)}{\sum\limits_{\substack{k=1 \\ M_k>0}}^{N_{\text{FFT}}/2-1} \log_2(M_k)} \tag{6.18}$$

The theoretical limit on the spectral efficiency at subcarrier k can be defined as follows [118]:

$$\eta_k^{\text{Sh}} = \log_2 \left(1 + \frac{\alpha^2 \sigma_x^2}{\sigma_{w_c}^2 + \dfrac{\sigma_n^2 (1 + k_{\text{DC}}^2)}{|H(k)|^2}}\right) \tag{6.19}$$

The gap between the capacity limit and exact loading achieved by the bit and energy-loading algorithm is already small [10,106]. However, it can be closed when channel coding is employed.

The adaptive bit and energy-loading on each subcarrier is shown in Figure 6.13 along with the SNR of a VLC system that uses a violet m-LED

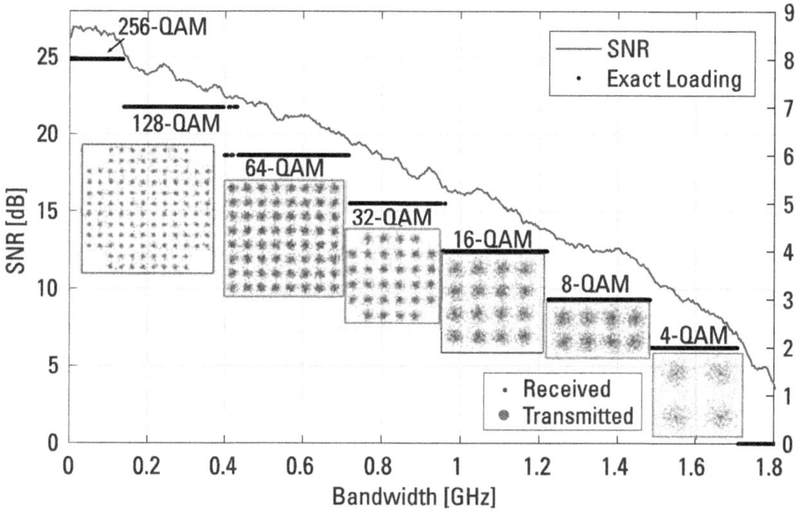

Figure 6.13 The bit-loading and the SNR of a VLC system based on a violet m-LED as presented in [10], where a data transmission rate of 7.91 Gbps is reported. The insets shows the constellations diagrams of the transmitted and received symbols.

as a transmitter [10]. The loading is shown to be dependent on the available SNR at each subcarrier, with the subcarriers achieving high SNR values being loaded with large constellation sizes. The constellation diagrams are also shown for the transmitted and received symbols at a BER below 3.8×10^{-3}.

Multiple LEDs can be used to spatially superimpose the modulation signal in the optical domain by transmitting discrete levels of the signal using multiple LEDs [119-121]. The spatial construction of the modulation signal in the optical domain overcomes the LED nonlinearity and the dynamic range limitations. In addition, the digital control of LEDs simplifies the system design by bypassing any necessity of high-resolution digital-to-analog converter (DACs). The OFDM waveform can also be transmitted by splitting the OFDM subcarriers into multiple groups of LEDs. The simplest case would be a sinusoid per each individual group of LEDs [120,122]. The OFDM subcarriers are modulated separately and transmitted through the multiple groups of LEDs. The OFDM signal can then be constructed in the optical domain through the spatial summation of streams. This approach decreases the PAPR by allowing the OFDM signal to be spatially formed in the optical domain. However, the necessity of multiple DACs and IFFT operations is a major drawback of this technique.

6.4.2.2 ACO-OFDM

A real-valued unipolar OFDM waveform can be achieved by exploiting the Fourier transformation properties. The principle of asymmetrically clipped optical

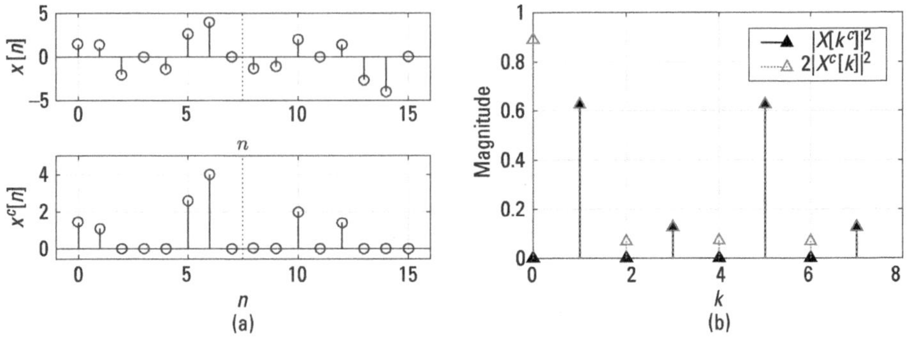

Figure 6.14 (a) Time-domain ACO-OFDM waveform before ($x[n]$) and after clipping ($x^c[n]$). (b) Frequency-domain representation of ACO-OFDM waveform before $|X[k]|^2$ and after clipping $2|X^c[k]|^2$. The factor 2 is introduced to compensate for the attenuation that occurs due to the zero-level clipping as described in (6.24).

(OFDM) (ACO-OFDM) is to skip loading the even subcarriers and only load the odd subcarriers with useful information, as shown in Figure 6.14(a) [123]. This creates symmetry in the time-domain OFDM signal, which allows the distortionless clipping of the negative samples without adding any DC bias as shown in Figure 6.14(b). The symmetry can be shown in (6.20) and (6.21), as follows:

$$x[n] = \frac{1}{\sqrt{N_{\mathrm{FFT}}}} \sum_{k=0}^{N_{\mathrm{FFT}}-1} X[k] \times e^{j\frac{2\pi kn}{N_{\mathrm{FFT}}}} \tag{6.20}$$

$$x\left[n + \frac{N_{\mathrm{FFT}}}{2}\right] = \frac{1}{\sqrt{N_{\mathrm{FFT}}}} \sum_{k=0}^{N_{\mathrm{FFT}}-1} X[k] \times e^{j\frac{2\pi kn}{N_{\mathrm{FFT}}}} \times e^{j\pi k} \tag{6.21}$$

By setting the even-indexed subcarriers to be zeros, we can write:

$$x\left[n + \frac{N_{\mathrm{FFT}}}{2}\right] = -x[n] \tag{6.22}$$

However, if only odd-indexed subcarriers are loaded with modulated symbols, the following symmetry can be achieved:

$$x\left[n + \frac{N_{\mathrm{FFT}}}{2}\right] = x[n] \tag{6.23}$$

Clipping of the negative values is distortionless since the distortion will only affect the even-indexed subcarriers. The clipping function of an OFDM waveform at zero is given as follows [9]:

$$x^{c}[n] = \mathrm{f}_{\mathrm{clip}}\{x[n]\} = \frac{1}{2}\left(x[n] + |x[n]|\right) \tag{6.24}$$

It is shown using (6.22), (6.23), and (6.24) that the clipping distortion $|x[n]|$ has the property $|x[n]| = |x[n + N_{\text{FFT}}/2]|$. This symmetry can be achieved when even-indexed subcarriers are only loaded with modulated symbols (6.23). This indicates that the clipping distorted symbols will only affect the even-indexed subcarriers [9].

Skipping half of the subcarriers reduces the spectral efficiency of ACO-OFDM to almost half of that in DCO-OFDM. The spectral efficiency of ACO-OFDM is given as follows:

$$\eta_{\text{ACO}} = \frac{\log_2 M(N_{\text{FFT}})}{4(N_{\text{FFT}} + N_{\text{CP}})} \quad \text{bits/s/Hz} \tag{6.25}$$

This is based on the assumption that all the odd-indexed subcarriers are loaded with modulated symbols of M-QAM. A penalty of 3 dB should apply to the SNR of ACO-OFDM when compared with bipolar OFDM, since half of the signal power is lost due to clipping. Hermitian symmetry is also used to guarantee a real-valued ACO-OFDM output. At the receiver, the odd subcarriers are only considered after an FFT is applied on the incoming frame.

6.4.2.3 PAM-DMT

A real unipolar optical OFDM is achieved in (PAM) discrete multitone (DMT) (PAM-DMT) by exploiting the Fourier properties of imaginary signals [124]. The real component of the subcarriers is not used in PAM-DMT, which restricts the modulation scheme used to M-PAM. An anti-symmetry in the time-domain waveform of PAM-DMT is achieved by only loading M-PAM modulated symbols on the imaginary components of the subcarriers $X[k] = jB^{\text{PAM}}[k]$, as shown in Figure 6.15. The time-domain PAM-DMT waveform is given as follows:

$$\begin{aligned} x[n] &= \frac{1}{\sqrt{N_{\text{FFT}}}} \left(\sum_{k=1}^{N_{\text{FFT}}/2-1} X[k] e^{\frac{j2\pi kn}{N_{\text{FFT}}}} + \sum_{k=1}^{N_{\text{FFT}}/2-1} X[N_{\text{FFT}} - k] e^{\frac{-j2\pi kn}{N_{\text{FFT}}}} \right) \\ &= \frac{-2}{\sqrt{N_{\text{FFT}}}} \sum_{k=1}^{N_{\text{FFT}}/2-1} B^{\text{PAM}}[k] \sin \frac{2\pi kn}{N_{\text{FFT}}} \end{aligned} \tag{6.26}$$

where $B^{\text{PAM}}[k]$ is the M-PAM modulated symbol that is loaded on the imaginary component of the subcarrier $X[k]$. The anti-symmetry property $x[n] = -x[N_{\text{FFT}} - n]$ can be proven as follows:

$$x[N_{\text{FFT}} - n] = \frac{-2}{\sqrt{N_{\text{FFT}}}} \sum_{k=1}^{N_{\text{FFT}}/2-1} B^{\text{PAM}}[k] \sin \frac{2\pi k(N_{\text{FFT}} - n)}{N_{\text{FFT}}}$$

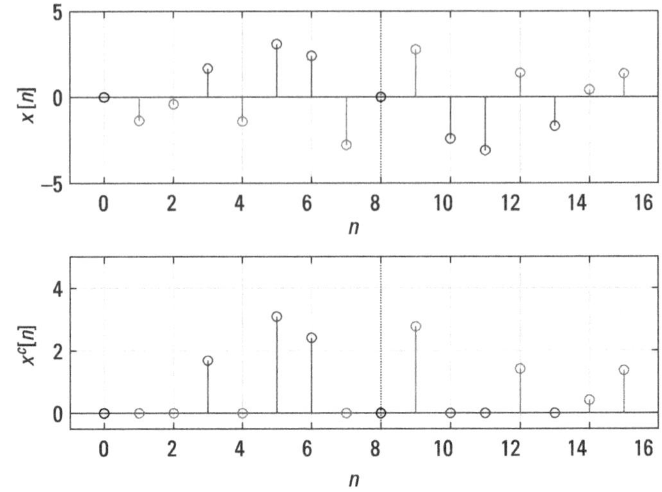

Figure 6.15 Time-domain PAM-DMT waveform before ($x[n]$) and after clipping ($x^c[n]$).

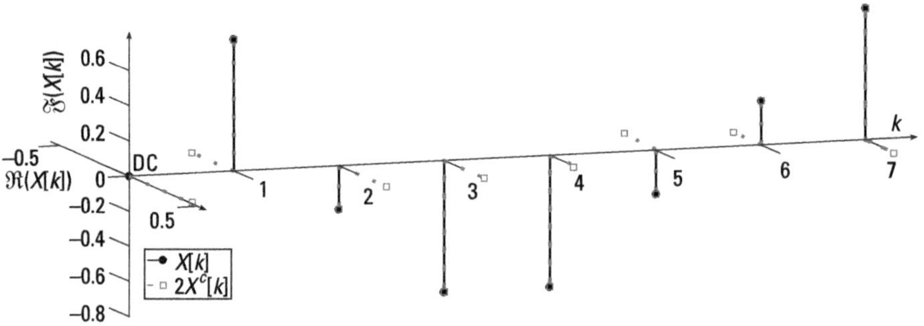

Figure 6.16 The frequency representation of the transmitted PAM-DMT waveform before ($X[k]$) and after clipping ($2X^c[k]$). The factor 2 is introduced to compensate for the attenuation that occurs due to the zero-level clipping as described in (6.24).

$$
\begin{aligned}
&= \frac{-2}{\sqrt{N_{\text{FFT}}}} \sum_{k=1}^{N_{\text{FFT}}/2-1} B^{\text{PAM}}[k] \sin\left(2\pi k - \frac{2\pi kn}{N_{\text{FFT}}}\right) \\
&= -x[n]
\end{aligned}
\tag{6.27}
$$

It can be shown using (6.24) and (6.27) that the clipping distortion $|x[n]|$ has Hermitian symmetry $|x[n]| = |x[N_{\text{FFT}} - n]|$, which means that the zero level clipping distortion will only affect the real part of the subcarriers [9], as shown in Figure 6.16. This is not to be confused with the Hermitian symmetry that is applied on the input frames of PAM-DMT to guarantee a real-valued output. PAM-DMT is more attractive than ACO-OFDM when bit-loading techniques are considered, as the PAM-DMT performance can be optimally adapted to the frequency response of the channel since all of the subcarriers are used. PAM-DMT has a 3-dB fixed penalty when compared with bipolar OFDM at an appropriate constellation size, as half of the power is also lost due to clipping. At the receiver, the imaginary part of the subcarriers is only considered, while the real part is ignored. The spectral efficiency of PAM-DMT is similar to that of DCO-OFDM and is given as follows:

$$\eta_{\text{PAM}} = \frac{\log_2 M(N_{\text{FFT}} - 2)}{2(N_{\text{FFT}} + N_{\text{CP}})} \quad \text{bits/s/Hz} \tag{6.28}$$

6.4.2.4 U-OFDM

A novel inherent optical OFDM scheme was proposed in [125,126] motivated by the fact that the polarity of a signal is only conveyed in a single bit. The concept and performance of unipolar (OFDM) (U-OFDM) [125] and flip (OFDM) (FLIP-OFDM) [126] is identical. The term U-OFDM is used in this chapter. However, the discussions and analysis apply to both schemes. Hermitian symmetry is applied to the incoming frames of M-QAM symbols as in the previously discussed schemes. Each bipolar OFDM time-domain frame is expanded into two time-domain frames in U-OFDM with similar sizes to the original OFDM frame as shown in Figure 6.17. The bipolar OFDM time-domain frame $x[n]$ is transmitted over two subsequent N_{FFT} frames. The first one, termed P, represents the positive values of $x[n]$ with zeros at the positions of the negative samples. The second, termed N, represents the absolute values of the originally negative values of $x[n]$ with zeros at the positions of the positive samples. A unipolar OFDM waveform can be achieved by clipping at the zero level. At the receiver, each second frame would be subtracted from the first frame of the same pair to reconstruct the original bipolar OFDM frame. This would double the noise at the receiver, leading to a 3-dB penalty compared with bipolar OFDM at appropriate constellation sizes. The spectral efficiency of U-OFDM is half of the spectral efficiency of DCO-OFDM since two U-OFDM frames are required to convey the same information contained in a single DCO-OFDM frame. This is given as follows:

$$\eta_{\text{U}} = \frac{\log_2 M(N_{\text{FFT}} - 2)}{4(N_{\text{FFT}} + N_{\text{CP}})} \quad \text{bits/s/Hz} \tag{6.29}$$

6.4.2.5 Hybrid OFDM Techniques

Reverse polarity optical (OFDM) (RPO-OFDM) was proposed to allow a higher degree of illumination control in the OFDM-based VLC systems [127]. RPO-OFDM combines a real-valued unipolar optical OFDM technique with a slow PWM signal to control the dimming. Therefore, the spectral efficiency of RPO-OFDM is half of that in DCO-OFDM. In addition, the PWM duty cycle is assumed to be known at the receiver, which requires perfect synchronization between the transmitting and receiving ends.

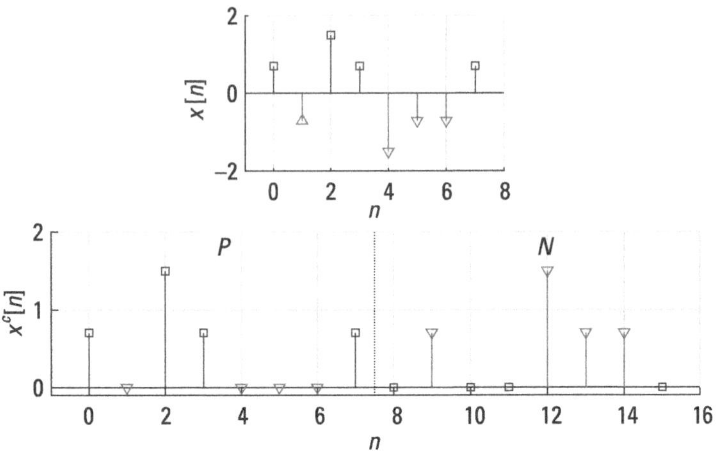

Figure 6.17 The generation of two U-OFDM frames from one bipolar OFDM frame.

The main principle of polar (OFDM) (P-OFDM) is to avoid Hermitian symmetry by converting the complex-valued output of the IFFT from Cartesian coordinates into polar coordinates [128]. The radial and angular values can be sent instead of the magnitude and phase in two successive subframes. P-OFDM allocates the QAM modulated symbols into even indexed subcarriers which results in half-wave even symmetry [129]. This means that the first and second halves of the complex-valued IFFT frame output are identical. Therefore, it is sufficient to transmit half of the IFFT output. As a result, the spectral efficiency is reduced to be identical to that of DCO-OFDM since only half of the subcarriers are used. The performance of P-OFDM was compared to that of ACO-OFDM in [129]. It was reported that P-OFDM achieves better BER performance gains than ACO-OFDM when optimal values for the power allocation of the radial and angular information are used. However, the system performance in frequency-selective channels should be investigated since any ISI between the radial and angular samples may deteriorate the BER performance.

A finite impulse response (FIR) filtering technique termed spectral factorization was used to create a unipolar optical OFDM signal [130]. The amplitude of the subcarriers in spectrally factorized optical (OFDM) (SFO-OFDM) were chosen to form an autocorrelation sequence that was shown to be sufficient to guarantee a unipolar OFDM output. The SFO-OFDM was reported to achieve 0.5-dB gain over ACO-OFDM with a 30% PAPR reduction [130].

Position modulation (OFDM) (PM-OFDM) avoids the Hermitian symmetry and splits the real and imaginary components of the OFDM output into two branches where a polarity separator is used to obtain the positive and negative samples of each branch [131]. The four frames, composed of a real positive frame, a real negative frame, an imaginary positive frame, and an imaginary negative frame, are transmitted as unipolar OFDM frames. The spectral efficiency is exactly similar to other inherent unipolar OFDM techniques discussed in Sections 6.4.2.2 through 6.4.2.4. The performance of PM-OFDM was reported to be identical to U-OFDM in flat channels. However, it was reported to have better BER performance when compared to ACO-OFDM for frequency selective channels [131].

6.4.3 Multistream Unipolar OFDM

The energy efficiency of a modulation technique refers to its capability to support a high spectral efficiency at low energy consumption. A modulation technique is more energy-efficient than another technique when it has fewer energy requirements at the same spectral efficiency. Superposition-based modulation techniques are aimed to address the high energy requirements of inherent unipolar OFDM-based modulation techniques.

Inherent unipolar OFDM modulation techniques were introduced to provide energy efficient optical OFDM alternatives to DCO-OFDM. These schemes include ACO-OFDM [123], PAM-DMT [124], FLIP-OFDM [126], and U-OFDM [125]. In order to avoid confusion with the U-OFDM modulation technique, the term inherent unipolar OFDM is used to refer to ACO-OFDM, PAM-DMT, FLIP-OFDM, and U-OFDM. Inherent unipolar OFDM modulation techniques exploit the OFDM input and output frame structures to produce a unipolar time domain waveform output. However, the energy efficiency of these techniques degrades at high modulation orders due to the restrictions imposed on their frame structures. For example, two time-domain U-OFDM frames are required to convey the information contained in a single DCO-OFDM frame. Similarly, half of the subcarriers are not loaded with modulated symbols in ACO-OFDM to guarantee a unipolar ACO-OFDM time-domain output. As a result, the performance of M-QAM DCO-OFDM has to be compared with the BER performance of M^2-QAM in ACO-OFDM and U-OFDM. Furthermore, PAM-DMT uses M-PAM on the imaginary part of the subcarriers. Therefore, the BER of PAM-DMT is similar to other inherent unipolar schemes since the BER performance of M-PAM is equivalent to the performance of M^2-QAM [117]. When compared with DCO-OFDM at the same spectral efficiency, the BER performance gain of all of the inherent unipolar OFDM techniques degrades as the constellation size of M-QAM or M-PAM increases. For example, the performance of 1,024-QAM in ACO-OFDM/U-OFDM and 32-PAM in PAM-DMT would be required to be compared with the performance of 32-QAM in DCO-OFDM.

Improved receivers for the inherent unipolar OFDM techniques were proposed in [125,132–139]. These improved receivers achieve BER performance gains at the price of additional computational complexity. Two main approaches are considered in the design of these receivers. Amplitude comparison of waveform samples based on the time-domain symmetry is considered in the first approach. The diversity between the real and imaginary components of the received symbols is exploited in the second. For example, the even-indexed subcarriers in ACO-OFDM and the real part of the subcarriers in PAM-DMT can be exploited to improve the detection performance. Inherent unipolar optical OFDM waveforms utilize the lower part of the L-V characteristic since they follow a truncated Gaussian statistical distribution. Therefore, these schemes are suitable candidates for dimming applications in VLC. However, these techniques are incapable of supporting high-speed VLC applications given that their relative BER performance gain against DCO-OFDM degrades as the spectral efficiency increases.

There have been several studies to improve the spectral and energy efficiency of OFDM-based techniques. A technique that combines ACO-OFDM on the odd-indexed subcarriers with DCO-OFDM on the even-indexed subcarriers was proposed in asymmetrically DC-biased optical OFDM (ADO-OFDM) [140]. The clipping distortion of the negative-valued samples in ADO-OFDM falls only on the even-indexed subcarriers. This distortion can be estimated and canceled at the receiver at the cost of a 3-dB penalty to the SNR. Similarly, hybrid asymmetrical clipped OFDM (HACO-OFDM) uses ACO-OFDM on the odd-indexed subcarriers and PAM-DMT on the even-indexed subcarriers to improve the spectral efficiency of inherent unipolar OFDM [141]. The symbols loaded on the odd-indexed subcarriers are demodulated first at the receiver to allow for the clipping distortion of ACO-OFDM to be estimated. The PAM-DMT symbols can then be demodulated with minimal distortion from ACO-OFDM. However, the real part of the even subcarriers remains unused in HACO-OFDM, which leads to spectral efficiency loss when compared with DCO-OFDM. Asymmetrically and symmetrically clipping optical OFDM (ASCO-OFDM) is based on combining ACO-OFDM with symmetrical clipping optical OFDM (SCO-OFDM) [142]. Modulated symbols are loaded only on the even-indexed subcarriers in SCO-OFDM, which creates a symmetry in the time-domain OFDM signal output $x[n] = x[n + N_{\mathrm{FFT}}/2]$. The clipping distortion of both ACO-OFDM and SCO-OFDM affects the even-indexed subcarriers. However, the clipping distortion of ACO-OFDM can be estimated and canceled at the receiver. The clipping noise of SCO-OFDM can be removed using U-OFDM time-domain processing techniques. The spectral efficiency of ASCO-OFDM is 75% of the spectral efficiency of DCO-OFDM. ASCO-OFDM was reported to have a better BER performance compared with ADO-OFDM since the ADO-OFDM uses the DC bias for the even-indexed subcarriers.

The principle behind superposition modulation in this section is to allow for the successful transmission of multiple OFDM-based streams by combining them into a single waveform. Additional superimposed streams are designed to avoid creating any distortion to the already existing streams. Therefore, the superposition of multiple OFDM-based streams does not hinder the recovery of any individual stream. Superposition modulation uses an inherent unipolar OFDM stream at depth $d = 1$, where d is the modulation depth index with values $d \geq 1$. Additional $D - 1$ OFDM-based streams are superimposed on top of this base stream, where D is the total number of used depths. The superimposed stream at depth d is designed to avoid distorting the $d - 1$ available streams. However, it may distort any future superimposed streams with a depth order higher than d. The receiver starts with the demodulation of the stream at depth $d = 1$. A successive clipping distortion cancellation algorithm is used to estimate the clipping distortion created by the stream at depth $d = 1$ on the rest of the streams. This permits the receiver to remove this stream from the overall received signal. The recovery of the rest of the superimposed streams continues in the same way.

Three novel superposition modulation techniques based on inherent unipolar OFDM are discussed in this section. The techniques rely on the frame structures of U-OFDM [115,143], ACO-OFDM [144–146], and PAM-DMT [147,148] to superimpose multiple unipolar streams at different modulation depths while avoiding any interdepth interference (IDI). For example, superposition modulation is achieved based on the anti-symmetry of PAM-DMT in the augmented spectral efficiency discrete multitone (ASE-DMT) [148].

6.4.3.1 Superposition Modulation Based on U-OFDM

The concept of enhanced unipolar OFDM (eU-OFDM) is to superimpose multiple U-OFDM time-domain streams in order to form a single time-domain eU-OFDM stream [143]. Four bipolar OFDM frames are used to generate eight unipolar frames at depth 1 in Figure 6.18. The subscripts of $P_{d,l}/N_{d,l}$ indicates that the current unipolar frame holds the positive/negative values of the lth bipolar frame at depth d, respectively.

Modulation Concept

The waveform at depth 1 is generated using a conventional U-OFDM modulator as described in Section 6.4.2.4. The second stream of U-OFDM frames is generated in depth 2. However, each unipolar frame in depth 2 is repeated twice. In addition, the amplitude of the U-OFDM stream at depth 2 is scaled by $1/\sqrt{2}$ in order to preserve the overall signal energy. The stream generated at depth 2 is superimposed on the previous stream at depth 1. The third stream of U-OFDM is generated in depth 3 and superimposed in a similar way to the stream at depth 2. However, each unipolar U-OFDM frame is replicated four times and scaled by $1/2$ in depth 3. Additional U-OFDM streams can be superimposed after they are replicated 2^{d-1} times and scaled by $1/\sqrt{2^{d-1}}$, where d is the depth of the

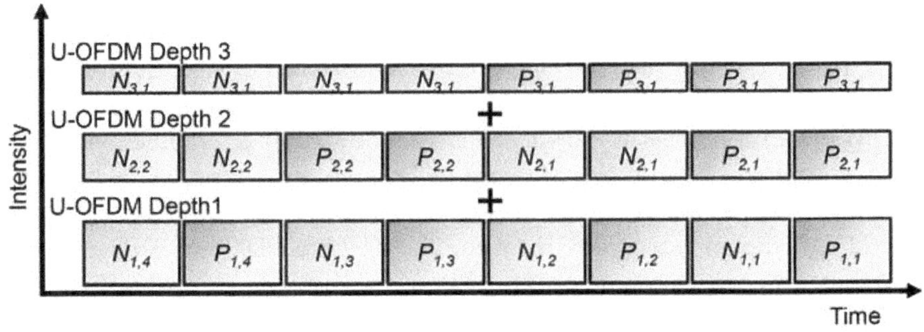

Figure 6.18 Illustration of the enhanced U-OFDM concept with three information streams. $P_{d,l}$ represents the positive U-OFDM frame, and $N_{d,l}$ represents the negative frame of U-OFDM. The subscripts denote that the frame at depth d belongs to the lth original bipolar OFDM frame.

Table 6.1
Frame Content of U-OFDM at Each Depth and Their Contribution to eU-OFDM Generated in Figure 6.18

Frame Index	Depth 1	Depth 2	Depth 3	eU-OFDM
1	$P_{1,1}$	$\frac{1}{\sqrt{2}}P_{2,1}$	$\frac{1}{2}P_{3,1}$	$P_{1,1}+\frac{1}{\sqrt{2}}P_{2,1}+\frac{1}{2}P_{3,1}$
2	$N_{1,1}$	$\frac{1}{\sqrt{2}}P_{2,1}$	$\frac{1}{2}P_{3,1}$	$N_{1,1}+\frac{1}{\sqrt{2}}P_{2,1}+\frac{1}{2}P_{3,1}$
3	$P_{1,2}$	$\frac{1}{\sqrt{2}}N_{2,1}$	$\frac{1}{2}P_{3,1}$	$P_{1,2}+\frac{1}{\sqrt{2}}N_{2,1}+\frac{1}{2}P_{3,1}$
4	$N_{1,2}$	$\frac{1}{\sqrt{2}}N_{2,1}$	$\frac{1}{2}P_{3,1}$	$N_{1,2}+\frac{1}{\sqrt{2}}N_{2,1}+\frac{1}{2}P_{3,1}$
5	$P_{1,3}$	$\frac{1}{\sqrt{2}}P_{2,2}$	$\frac{1}{2}N_{3,1}$	$P_{1,3}+\frac{1}{\sqrt{2}}P_{2,2}+\frac{1}{2}N_{3,1}$
6	$N_{1,3}$	$\frac{1}{\sqrt{2}}P_{2,2}$	$\frac{1}{2}N_{3,1}$	$N_{1,3}+\frac{1}{\sqrt{2}}P_{2,2}+\frac{1}{2}N_{3,1}$
7	$P_{1,4}$	$\frac{1}{\sqrt{2}}N_{2,2}$	$\frac{1}{2}N_{3,1}$	$P_{1,4}+\frac{1}{\sqrt{2}}N_{2,2}+\frac{1}{2}N_{3,1}$
8	$N_{1,4}$	$\frac{1}{\sqrt{2}}N_{2,2}$	$\frac{1}{2}N_{3,1}$	$N_{1,4}+\frac{1}{\sqrt{2}}N_{2,2}+\frac{1}{2}N_{3,1}$

Note: $P_{d,l}$ represents the positive U-OFDM frame, while $N_{d,l}$ represents the negative frame of U-OFDM. The subscripts indicate that the frame at depth d belongs to the lth original bipolar OFDM frame.

respective information stream. Table 6.1 illustrates the contents of U-OFDM frames at each depth, and the contents of the overall eU-OFDM frames.

At the receiver, the demodulation process starts with the demodulation of the stream at depth 1 where a conventional U-OFDM receiver is used, as described in Section 6.4.2.4. The subtraction operation removes most of the IDI caused by the superposition of multiple U-OFDM streams in the demodulation process since the interference on each positive frame at depth 1 is equivalent to the interference on the correspondent negative frame. The recovered bits are

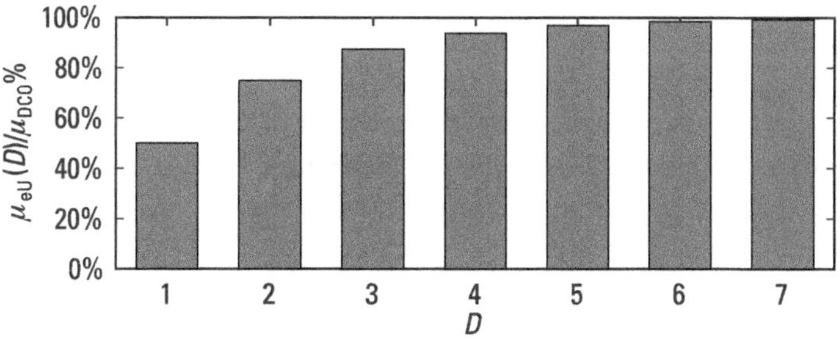

Figure 6.19 The convergence of the spectral efficiency of eU-OFDM.

remodulated at the receiver after the demodulation of the stream at depth 1. The remodulated stream of depth 1 is then subtracted from the overall received eU-OFDM signal. This results in removing most of the stream at depth 1. Afterward, every two identical frames at depth 2 are summed. The demodulation process at depth 2 continues with the conventional U-OFDM demodulator. The recovered bits at depth 2 are remodulated to allow for the information stream at depth 2 to be subtracted from the overall received information signal. The demodulation process continues in the same way for all subsequent streams until the information at all depths is recovered.

Spectral Efficiency

The spectral efficiency of eU-OFDM can be written as the sum of the spectral efficiency values of the information signals at all depths. This is expressed as follows [143]:

$$\eta_{\mathrm{eU}}(D) = \eta_{\mathrm{U}} \sum_{d=1}^{D} \frac{1}{2^{d-1}} \tag{6.30}$$

where D is the total number of the used modulation depths. Therefore, the spectral efficiency of eU-OFDM approaches the spectral efficiency of DCO-OFDM as D increases, and as illustrated in Figure 6.19. For example, more than 87.5% of the spectral efficiency of DCO-OFDM can be achieved using $D = 3$.

The spectral efficiency gap between U-OFDM and DCO-OFDM can never be completely closed with eU-OFDM, as this would require a large number of information streams to be superimposed in the modulation signal. Implementation issues, such as latency, computational complexity, and memory requirements put a practical limit on the maximum number of depths that can be superimposed. The spectral efficiency gap between U-OFDM and DCO-OFDM can be completely closed when an appropriate selection of the constellation sizes is used at the different modulation depths. This has been investigated in generalized enhanced unipolar OFDM (GREENER-OFDM) [149].

The subcarriers in eU-OFDM incorporate modulated symbols from all the modulation depths. Therefore, incorrectly decoded symbols at the receiver result in a series of errors that propagate all subcarriers in all modulation depths. This is also the case for spectrally and energy-efficient OFDM (SEE-OFDM) [145] and enhanced PAM discrete multitone (DMT) (ePAM-DMT) [147]. However, this is avoided in the techniques: layered/enhanced asymmetrically clipped optical OFDM (L/e-ACO-OFDM) and ASE-DMT, where each subcarrier is only assigned to a specific modulation depth. Incorrectly decoded symbols in L/e-ACO-OFDM and ASE-DMT would only affect a few subcarriers in all the modulation depths. The selective frequency-domain loading of subcarriers in L/e-ACO-OFDM and ASE-DMT provides a simplified and practical implementation option for OFDM-based superposition modulation.

6.4.3.2 Superposition Modulation Based on ACO-OFDM

The development of power-efficient solutions to ACO-OFDM at high spectral efficiency has been extensively considered in the last few years [140–142,150,151]. The principle of superposition modulation based on ACO-OFDM was developed independently by a number of different research groups [151]. However, it was recently revealed that the concept of superposition based on ACO-OFDM was earlier proposed by Chen et al. in 2010 [152,153]. The technique described in this section is based on layered ACO-OFDM [146] and enhanced ACO-OFDM [144,154]. Therefore, the term L/e-ACO-OFDM will be used to describe the superposition modulation technique based on ACO-OFDM.

The concept proposed by Elgala et al. [145] under the term SEE-OFDM is similar in principle to L/e-ACO-OFDM. However, SEE-OFDM is implemented by using a U-OFDM signal processing. This leads to the IDI noise falling on all subcarriers, which results in an additional SNR penalty when compared with L/e-ACO-OFDM. An interesting analysis of L/e-ACO-OFDM and other superposition techniques is presented in [151,154].

The asymmetry in the ACO-OFDM allows for multiple streams to be superimposed at multiple modulation depths without creating any IDI. The modulation concept and spectral efficiency of L/e-ACO-OFDM are presented in this section.

Modulation Concept

The main objective of the frame hierarchy design in L/e-ACO-OFDM is to convert all of the IDI of the superimposed streams into the even-indexed subcarriers at the frequency domain so that it becomes distortion-less to the ACO-OFDM stream at the first depth. A possible arrangement of the multiple ACO-OFDM signals is given in Figure 6.20. The L/e-ACO-OFDM signal generation starts at the first depth with a conventional ACO-OFDM modulator. The subframes are defined to be half of the original ACO-OFDM frames in

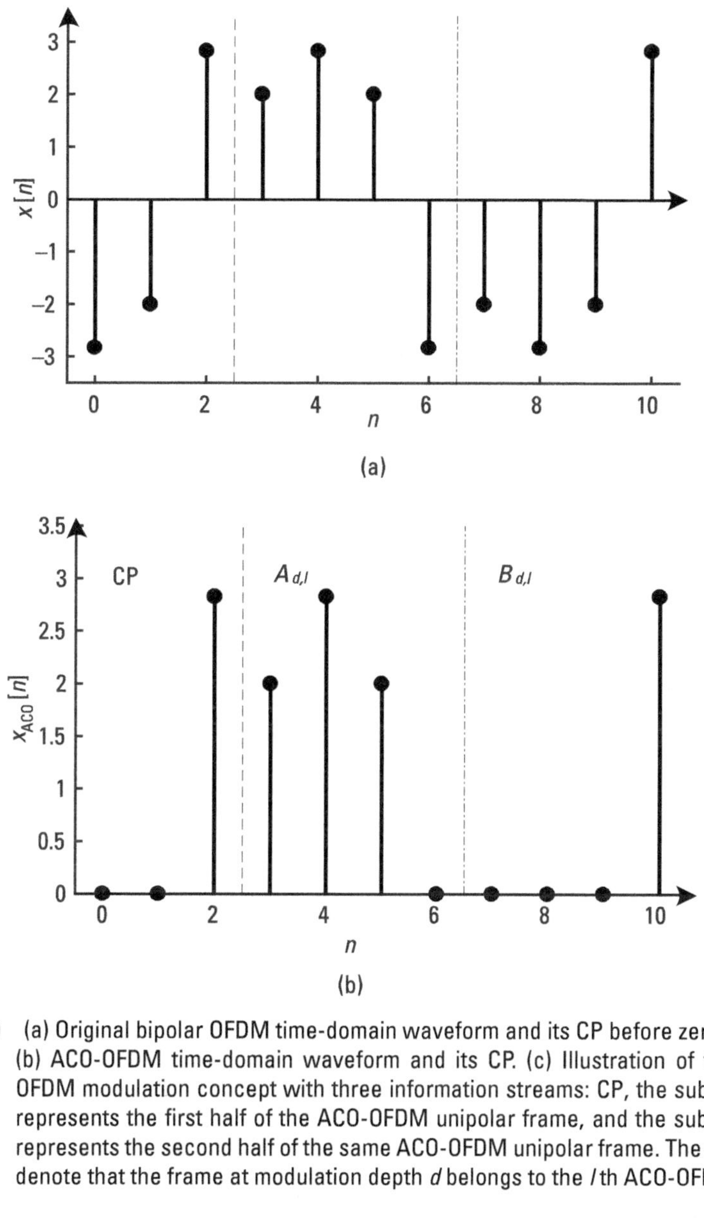

Figure 6.20 (a) Original bipolar OFDM time-domain waveform and its CP before zero clipping. (b) ACO-OFDM time-domain waveform and its CP. (c) Illustration of the eACO-OFDM modulation concept with three information streams: CP, the subframe $A_{d,l}$ represents the first half of the ACO-OFDM unipolar frame, and the subframe $B_{d,l}$ represents the second half of the same ACO-OFDM unipolar frame. The subscripts denote that the frame at modulation depth *d* belongs to the *l*th ACO-OFDM frame.

length. At depth $d = 2$, a second ACO-OFDM stream is superimposed on the first stream and generated in a similar fashion to depth $d = 1$. However, the OFDM frame length at depth $d = 2$ is $N_2^{\mathrm{eACO}} = N_1^{\mathrm{eACO}}/2$, where $N_1^{\mathrm{eACO}} = N_{\mathrm{FFT}}$ is the OFDM frame length at depth $d = 1$. The overall stream at depth $d = 2$ is scaled by $1/\sqrt{2}$ to preserve the overall signal energy at this depth and repeated two times. Furthermore, ACO-OFDM streams are

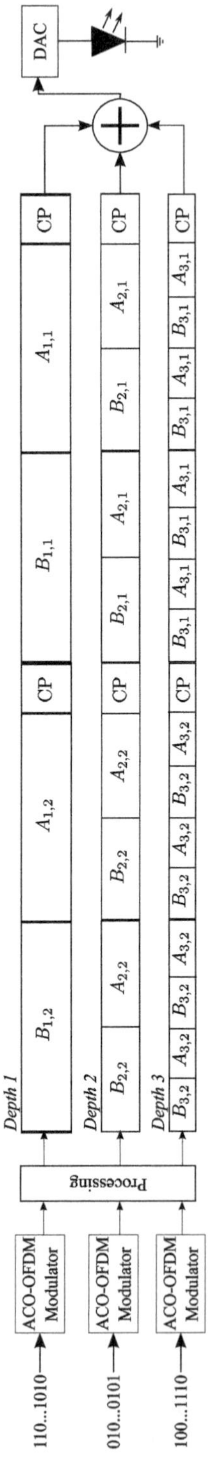

(c)

Figure 6.20 Continued.

superimposed at depth d with an OFDM frame length $N_d^{\text{eACO}} = N_{\text{FFT}}/2^{d-1}$ in a similar way as depth $d = 2$ and the overall stream at that depth is scaled by $1/\sqrt{2^{d-1}}$ and repeated 2^{d-1} times. In addition, each of the streams is scaled by a parameter $1/\gamma_d$ to facilitate the allocation of different power levels as described in Section 6.4.3.5. At the receiver, the demodulation process starts with removing the cyclic prefixes. Then the information carried at depth $d = 1$ can be extracted using a conventional ACO-OFDM demodulator. The subframes in each individual frame at depth $d = 1$ are asymmetric. Therefore, the information stream at this depth follows (6.22). At higher modulation depths, the subframes in each individual frame are identical. Therefore, the information stream at this depth follows (6.23). As a result, the odd-indexed subcarriers only convey the information of depth $d = 1$, and the IDI falls into the even-indexed subcarriers. After the information at depth $d = 1$ is demodulated, the recovered bits are remodulated at the receiver in order to reconstruct the first depth stream, which is then distorted using the channel response and subtracted from the overall L/e-ACO-OFDM signal. After the removal of the depth $d = 1$ waveform, each two identical subsequent subframes are summed. The demodulation process at depth $d = 2$ continues with the conventional ACO-OFDM demodulation process, and the recovered bits are remodulated in order to allow for the information stream at depth $d = 2$ to be subtracted from the overall received information signal. The demodulation process continues in a similar way for all subsequent streams until the information at the last depth is recovered.

The superposition of ACO-OFDM streams at multiple modulation depths is illustrated in the frequency-domain diagram at Figure 6.21 [154]. It can be shown that the subcarriers at modulation depths d' do not cause any distortion to the information-carrying subcarriers, $X_{\text{eACO},d}[k]$, at lower-order modulation depths for any $d \leq d'$. For example, the clipping distortion $X^{\text{c}}_{\text{eACO},2}[k]$ that occurs due to the zero-level clipping of the ACO-OFDM stream at modulation depth 2 falls on the subcarriers $k = \{4, 8, 12\}$, which have null information symbols at modulation depths $d \leq 2$. The information-carrying subcarriers, $X_{\text{eACO},d}[k]$, are located at subcarrier indices $k = 2^{d-1}(2k' + 1)$ for $k' = 0, 1, 2, \cdots, N_{FFT}/2 - 1$. The clipping distortion, $X^{\text{c}}_{\text{eACO},d}[k]$, falls on the subcarrier indices $k = 2^d k'$ for $k' = 0, 1, 2, \cdots, N_{FFT}/2 - 1$.

Spectral Efficiency

The spectral efficiency of L/e-ACO-OFDM is given as the summation of the spectral efficiency values of all the superimposed ACO-OFDM streams:

$$\eta_{\text{eACO}}(D) = \sum_{d=1}^{D} \eta_{\text{ACO}}(d) \quad \text{bits/s/Hz} \tag{6.31}$$

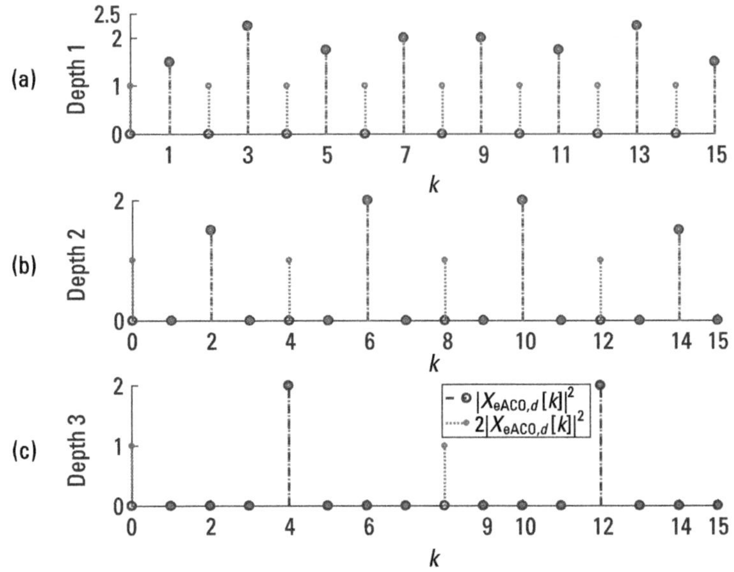

Figure 6.21 An illustration of the frequency-domain subcarrier loading at three depths of eACO-OFDM and the effects of zero clipping [154]. (a–c) The magnitude of the subcarriers before and after the zero level time-domain clipping, $|X_{eACO,d}[k]|^2$ and $2|X^{c}_{eACO,d}[k]|^2$, respectively. The factor 2 is introduced to compensate for the attenuation that occurs due to the zero-level clipping as described in (6.24).

where $\eta_{ACO}(d)$ is the spectral efficiency of ACO-OFDM at depth d:

$$\eta_{ACO}(d) = \frac{\log_2(M_{eACO,d})(N_{FFT,d})}{4(N_{FFT} + N_{CP})} \quad \text{bits/s/Hz} \tag{6.32}$$

The spectral efficiency ratio of the L/e-ACO-OFDM to an ACO-OFDM scheme with a similar constellation size is expressed as follows:

$$\alpha_{\eta}^{eACO}(D, d) = \frac{\sum_{\acute{d}=1}^{D}\left(\frac{\log_2 M_{eACO,\acute{d}}}{2^{\acute{d}-1}}\right)}{\log_2 M_{eACO,d}} \tag{6.33}$$

The spectral efficiency of L/e-ACO-OFDM is exactly equivalent to η_{DCO} for large numbers of superimposed depths and any cyclic prefix length.

6.4.3.3 Superposition Modulation Based on PAM-DMT

An earlier superposition modulation was proposed based on the anti-symmetry of PAM-DMT under the term ePAM-DMT [147]. It was shown that ePAM-DMT achieves a lower electrical energy dissipation than DCO-OFDM at the same spectral efficiency. However, the cyclic prefixes are used in preserving the anti-symmetry of the superimposed streams in ePAM-DMT, which reduces the spectral efficiency. Aligning the anti-symmetry of multiple PAM-DMT streams in the time domain is a complicated process. An alternative solution was proposed based on superimposing multiple PAM-DMT streams in the frequency domain. ASE-DMT is based on the selective loading of M-PAM symbols on multiple modulation depths. It avoids the spectral efficiency losses of ePAM-DMT and provides energy-efficiency gains over DCO-OFDM and ePAM-DMT.

The ASE-DMT technique uses most of the available subcarriers in the OFDM frame. The ASE-DMT waveform can be generated by the selective loading of imaginary and real components of the subcarriers. The waveform generation starts with a typical PAM-DMT [124] modulator at the first depth. Additional streams can only be superimposed on the first depth stream if their frequency-domain subcarriers are loaded on the real components of the subcarriers. Superposition modulation is achieved in ePAM-DMT by rearranging the PAM-DMT time-domain waveforms to have a Hermitian symmetry. A more efficient technique is achieved in ASE-DMT by using groups of subcarriers, which allows for multiple streams of PAM-DMT waveforms to be superimposed without any IDI, as will be shown in Section 6.4.3.3.

Modulation concept

The transmitter block diagram of ASE-DMT is shown in Figure 6.22. At the first depth, the imaginary components of the subcarriers are loaded with PAM symbols while the real components are kept unused $X_1[k] = jB_1^{\mathrm{PAM}}[k]$, where $B_1^{\mathrm{PAM}}[k]$ is the PAM symbol at the kth subcarrier of depth $d = 1$. Note that Hermitian symmetry is also required in the frequency domain to guarantee a real time-domain output, $B_1^{\mathrm{PAM}}[0] = B_1^{\mathrm{PAM}}[N_{\mathrm{FFT}}/2] = 0$, and $B_1^{\mathrm{PAM}}[k] = -B_1^{\mathrm{PAM}}[N_{\mathrm{FFT}} - k]$ for $k = 1, 2, \cdots, N_{\mathrm{FFT}}/2 - 1$. As a result, the time-domain PAM-DMT waveform $x_1[n]$ can be given in (6.27). The time-domain PAM-DMT waveform in (6.27) exhibits the following anti-symmetry: $x_1[n] = -x_1[N_{\mathrm{FFT}} - n]$, where $x_1[0] = x_1[N_{\mathrm{FFT}}/2] = 0$ as was shown in Section 6.4.2.3. The zero level clipping of the PAM-DMT waveform at depth d can be given in (6.24). The frequency-domain transformation of the clipped waveform, $x_d^{\mathrm{c}}[n]$, is given as follows:

$$X_d^{\mathrm{c}}[k] = \frac{X_d[k] + \mathrm{FFT}\{|x_d[n]|\}}{2} \tag{6.34}$$

where the subscripts d denotes the depth order and $X_d[k] = \mathrm{FFT}\{x_d[n]\}$. The effect of clipping is shown to be distortion-less to the information carrying subcarriers in Figure 6.23. Clipping of the negative samples at depth $d = 1$ is distortion-less to the information at the same depth because all of the distortion transforms into the real part of the subcarriers. The distortion term $|x_1[n]|$ has a Hermitian symmetry $|x_1[n]| = |x_1[N_{\mathrm{FFT}} - n]|$, which can be illustrated by the

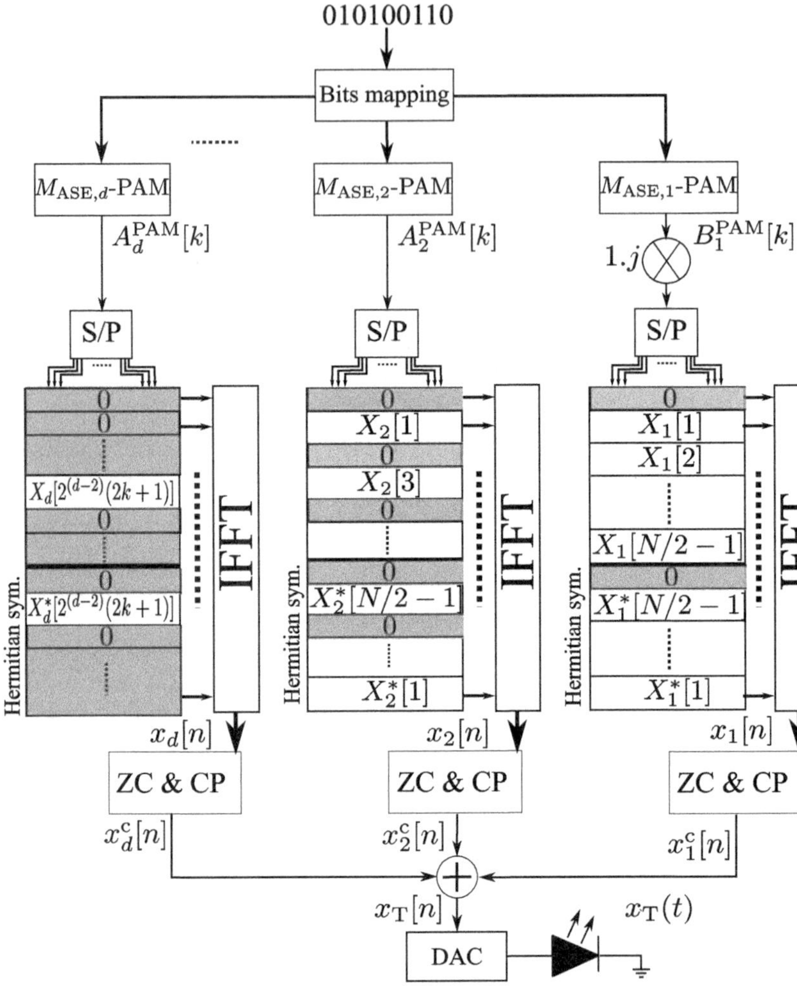

Figure 6.22 ASE-DMT transmitter block diagram. $X_d[k]$ refers to the kth subcarrier at depth d; S/P denotes for serial to parallel; DAC denotes digital-to-analog conversion; ZC denotes zero clipping; and CP refers to the cyclic prefixing.

following:

$$\text{FFT}\{|x_1[n]|\} = \frac{1}{\sqrt{N_{\text{FFT}}}} \sum_{n=0}^{N_{\text{FFT}}-1} |x_1[n]| e^{-\frac{j2\pi kn}{N_{\text{FFT}}}}$$

$$= \frac{2}{\sqrt{N_{\text{FFT}}}} \sum_{n=1}^{N_{\text{FFT}}/2-1} |x_1[n]| \cos\frac{2\pi kn}{N_{\text{FFT}}} \quad (6.35)$$

At depth $d = 2$, the odd subcarriers are loaded with real valued PAM symbols $X_2[k] = A_2^{\text{PAM}}[k]$, while all of the other subcarriers are kept unused.

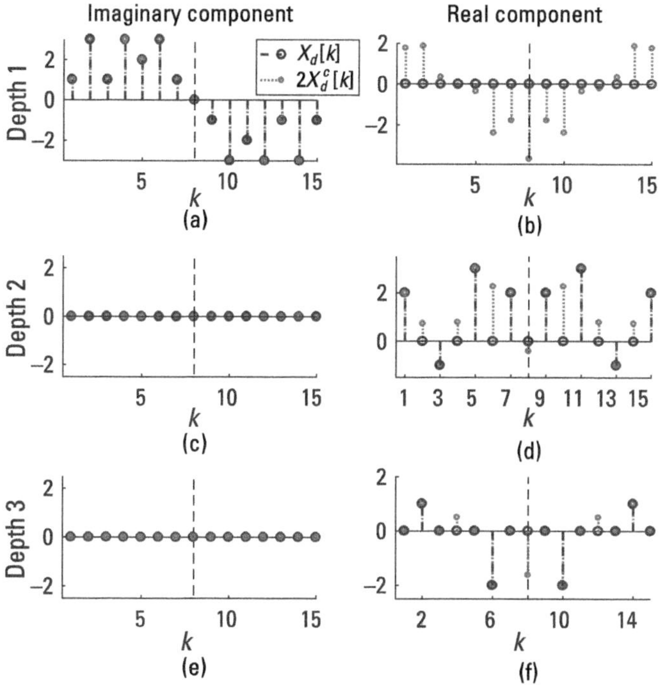

Figure 6.23 An illustration of the frequency-domain subcarrier loading at three depths ASE-DMT and the effects of zero clipping. (a,c,e) The imaginary components of the subcarriers before and after zero level time-domain clipping, $\Im[X_d[k]]$ and $\Im[2X_d^c[k]]$, respectively. (b,d,f) The real components of the subcarriers before and after zero level time-domain clipping, $\Re[X_d[k]]$ and $\Re[2X_d^c[k]]$, respectively. The factor 2 is introduced to compensate for the attenuation that occurs due to the zero-level clipping as described in (6.24).

The subcarriers at depth $d = 2$, $X_2[k]$ are given as follows:

$$X_2[k'] = \begin{cases} A_2^{\mathrm{PAM}}[k'], & \text{if } k' = 2k+1 \\ 0, & \text{Otherwise} \end{cases} \tag{6.36}$$

where $A_2^{\mathrm{PAM}}[k']$ is the PAM symbol at the k'-th subcarrier of depth $d = 2$; and $k = 0, 1, \cdots, N_{\mathrm{FFT}}/4 - 1$. Hermitian symmetry is also required to guarantee that $x_2[n]$ is real, $A_2^{\mathrm{PAM}}[k] = A_2^{\mathrm{PAM}}[N_{\mathrm{FFT}} - k]$. As a result, the time-domain waveform at depth $d = 2$, $x_2[n]$ would have the following symmetry: $x_2[n] = -x_2[n + N_{\mathrm{FFT}}/2]$. Therefore, the distortion caused by the clipping at the zero level would only affect the real domain even subcarriers. This is shown as follows:

$$\mathrm{FFT}\{|x_2[n]|\} = \frac{1}{\sqrt{N_{\mathrm{FFT}}}} \sum_{n=0}^{N_{\mathrm{FFT}}/2-1} |x_2[n]| e^{-\frac{j2\pi kn}{N_{\mathrm{FFT}}}} (1 + e^{-j\pi k}) \tag{6.37}$$

which takes values only at $X_2^c[2k]$, for $k = 0, 1, \cdots, N_{\mathrm{FFT}}/2 - 1$. Therefore, the distortion is orthogonal to the information content at both depth $d = 1$ and depth $d = 2$. Subsequent streams can be generated at depth d, where the subcarriers will be loaded with real-valued PAM symbols:

$$X_d[k'] = \begin{cases} A_d^{\mathrm{PAM}}[k'], & \text{if } k' = 2^{d-2}(2k+1) \\ 0, & \text{Otherwise} \end{cases} \tag{6.38}$$

where $A_d^{\mathrm{PAM}}[k']$ is the PAM symbol at the k'-th subcarrier of depth d; and $k = 0, 1, \cdots, N_{\mathrm{FFT}}/2^d - 1$. Hermitian symmetry is also required to guarantee that $x_d[n]$ is real, $A_d^{\mathrm{PAM}}[k] = A_d^{\mathrm{PAM}}[N_{\mathrm{FFT}} - k]$. Using (6.24) and (6.38), it can be shown that:

$$x_d[n] = -x_d[n + N_{\mathrm{FFT}}/2^{d-1}] \quad \forall d > 1 \tag{6.39}$$

Using (6.34), $X_d[k]$ is written as follows:

$$X_d[k] = \frac{1}{\sqrt{N_{\mathrm{FFT}}}} \sum_{n=0}^{N_{\mathrm{FFT}}/2^{d-1}-1} x_d[n] e^{-\frac{j2\pi kn}{N_{\mathrm{FFT}}}} \kappa (1 - e^{\frac{-j\pi k}{2^{D-2}}}) \tag{6.40}$$

and the zero-level clipping distortion effect on the subcarriers in the frequency domain is written as follows:

$$\mathrm{FFT}\{|x_d[n]|\} = \frac{1}{\sqrt{N_{\mathrm{FFT}}}} \sum_{n=0}^{N_{\mathrm{FFT}}/2^{d-1}-1} |x_d[n]| e^{-\frac{j2\pi kn}{N_{\mathrm{FFT}}}} \kappa (1 + e^{\frac{-j\pi k}{2^{D-2}}}) \tag{6.41}$$

where D is the total number of used depths, and κ is written as follows:

$$\kappa = \prod_{d=2}^{D-1} (1 + e^{\frac{-j\pi k}{2^{d-2}}}) \tag{6.42}$$

Using (6.40) and (6.41), it can be shown that the zero-level clipping is distortion-less to the information content at $X_d^c[2^{d-2}(2k+1)]$, and that all of the distortion will affect the subcarriers at $X_d^c[2^{d-1}k]$. Similar analysis can be applied to L/e-ACO-OFDM. Selective subcarrier loading at each depth will allow multiple PAM modulated waveforms to be superimposed without any IDI. The active subcarriers of each superimposed depth will not be affected by the zero-level clipping distortion of both the current depth and the higher-order depths. However, it will be affected by the distortion of the zero-level clipping of the lower-order depths. This distortion will be estimated and canceled at the receiver, as will be shown below.

After generating the time-domain waveforms of all depths, the generated waveforms are clipped and the cyclic prefixes are inserted. The overall ASE-DMT waveform can be obtained by superimposing the clipped waveforms of all depths:

$$x_{\mathrm{T}}[n] = \sum_{d=1}^{D} x_d^{\mathrm{c}}[n] \tag{6.43}$$

Using (6.24) and (6.34), the ASE-DMT subcarriers is written as follows:

$$X_{\mathrm{T}}[k] = \frac{jB_1^{\mathrm{PAM}}[k] + \sum_{d=2}^{D} A_d^{\mathrm{PAM}}[k] + \sum_{d=1}^{D} \mathrm{FFT}\{|x_d[n]|\}}{2} \tag{6.44}$$

The information content of depth $d = 1$ can be obtained by considering the imaginary-valued components and ignoring the real-valued components of the subcarriers. This can be given as $\hat{B}_1^{\mathrm{PAM}}[k] = 2\Im(X_{\mathrm{T}}[k] + W[k])$, where $W[k]$ is the frequency-domain realization of the AWGN at the receiver [155]. The information of depth $d = 1$ can then be remodulated at the receiver to obtain $\hat{x}_1[n]$, which can be subtracted from the ASE-DMT received waveform. This would result in removing the imaginary component of $X_{\mathrm{T}}[n]$ and also removing the real domain distortion caused by the zero-level clipping of the depth $d = 1$ waveform, $\mathrm{FFT}\{|x_1[n]|\}$. Subsequent depths can be demodulated by selecting the appropriate frequency subcarrier indexes at each depth. The real component of the subcarriers at $2^{d-2}(2k + 1)$ for $k = 0, 1, \cdots, N_{\mathrm{FFT}}/2^d - 1$ can then be remodulated to obtain the waveform at depth d, $\hat{x}_d[n]$, which would be subtracted from the remaining ASE-DMT waveform.

The exact process is repeated until the information at the last depth is demodulated. In this way, the distortion of the previous depths is estimated and canceled from the higher depths in the successive clipping distortion cancellation process at the receiver.

Spectral Efficiency

The spectral efficiency of the first depth of ASE-DMT is equivalent to the spectral efficiency of PAM-DMT, which is also similar to the spectral efficiency of DCO-OFDM. This is written as follows:

$$\eta_{\mathrm{ASE}}(1) = \frac{\log_2(M_{\mathrm{ASE},1})(N_{\mathrm{FFT}} - 2)}{2(N_{\mathrm{FFT}} + N_{\mathrm{CP}})} \quad \text{bits/s/Hz} \tag{6.45}$$

The spectral efficiency of higher depths $d \geq 2$ is expressed as follows:

$$\eta_{\mathrm{ASE}}(d) = \frac{\log_2(M_{\mathrm{ASE},d})N_{\mathrm{FFT}}}{2^d(N_{\mathrm{FFT}} + N_{\mathrm{CP}})} \quad \text{bits/s/Hz} \tag{6.46}$$

The spectral efficiency of ASE-DMT can then be calculated as follows:

$$\eta_{\mathrm{ASE,T}}(D) = \sum_{d=1}^{D} \eta_{\mathrm{ASE}}(d) \quad \text{bits/s/Hz} \tag{6.47}$$
$$= \frac{\log_2(M_{\mathrm{ASE},1})(N_{\mathrm{FFT}} - 2) + \sum_{d=2}^{D} \frac{\log_2(M_{\mathrm{ASE},d})(N_{\mathrm{FFT}})}{2^{d-1}}}{2(N_{\mathrm{FFT}} + N_{\mathrm{CP}})}$$

The spectral efficiency ratio of ASE-DMT to the spectral efficiency contribution of each individual depth is written as follows:

$$\alpha_{\eta}^{\mathrm{ASE}}(D,d) = \frac{\eta_{\mathrm{ASE,T}}(D)}{\eta_{\mathrm{ASE}}(d)2^{d-1}} \tag{6.48}$$

The spectral efficiency of ASE-DMT approaches twice the spectral efficiency of DCO-OFDM as the total number of depths increases. However, the energy efficiency comparison of both modulation schemes should be considered at an equivalent spectral efficiency. Therefore, the superimposed waveforms in ASE-DMT use smaller constellation sizes, $M_{\mathrm{ASE},d} = \sqrt{M_{\mathrm{DCO}}}$, which can lead to an overall spectral efficiency that is equivalent to DCO-OFDM. Since the system design of ePAM-DMT employs the cyclic prefixes in the symmetry alignment of the PAM-DMT time-domain frames, the spectral efficiency of ePAM-DMT is a function of the FFT/IFFT frame and cyclic prefix sizes.

The spectral efficiency of ePAM-DMT tends to increase as the FFT/IFFT frame length increases. Unlike ePAM-DMT, the spectral efficiency of ASE-DMT is independent of the cyclic prefix length and therefore can be employed with smaller FFT/IFFT sizes. This allows ASE-DMT to operate at a lower computational complexity while avoiding any spectral efficiency loss. The spectral efficiency of ASE-DMT is slightly higher than the spectral efficiency of L/e-ACO-OFDM at small FFT/IFFT sizes. It was shown that the spectral efficiency of ePAM-DMT, η_{ePAM}, exceeds 80% of η_{DCO} for all of the presented cyclic prefix lengths when the OFDM frame length is $N_{\mathrm{FFT}} \geq 512$ [156]. The spectral efficiency of ASE-DMT exactly matches the spectral efficiency of DCO-OFDM when $D = 6$ and $N_{\mathrm{FFT}} = 64$.

6.4.3.4 Power Efficiency

The statistics of a real-valued bipolar OFDM time-domain waveform can be approximated with a Gaussian distribution, $x(t) \sim \mathcal{N}(0, \sigma_x^2)$ when $N_{\mathrm{FFT}} \geq 64$, where σ_x is the standard deviation of $x(t)$ [155]. The waveforms of ACO-OFDM and PAM-DMT follow a truncated Gaussian distribution [9]. Therefore, the statistics of these techniques are identical. Moreover, the statistics of their superposition forms are also identical. Unless otherwise specified, the following discussion is applicable to L/e-ACO-OFDM and ASE-DMT. The superposition modulation signal $x_{\mathrm{SM}}(t)$ can be given as the summation of the individual unipolar OFDM-based waveforms, which is written as follows:

$$x_{\mathrm{SM}}(t) = \sum_{d=1}^{D} x_d^{\mathrm{SM}}(t) \tag{6.49}$$

where D is the total number of used depths. Each superimposed stream at depth d is scaled by a parameter $1/\gamma_d$ to allow for an arbitrary allocation of power, based on the associated spectral efficiency of that depth. The time-domain waveform at depth d follows a truncated Gaussian distribution with a statistical mean given as follows:

$$\mathrm{E}[x_d^{\mathrm{SM}}(t)] = \frac{\phi(0)\sigma_x}{\gamma_d\sqrt{2^{d-1}}} \tag{6.50}$$

where $\mathrm{E}[\cdot]$ is a statistical expectation and $\phi(x)$ is the PDF of the standard Gaussian distribution. As a result, the average electrical power of $x_{\mathrm{SM}}(t)$ is written as follows:

$$\begin{aligned}
P_{\mathrm{elec}}^{\mathrm{avg}}(D,\underline{\gamma}) &= \mathrm{E}\left[x_{\mathrm{SM}}^2(t)\right] = \mathrm{E}\left[\left(\sum_{d=1}^{D} x_d^{\mathrm{SM}}(t)\right)^2\right] \\
&= \sum_{d=1}^{D} \mathrm{E}\left[\left(x_d^{\mathrm{SM}}(t)\right)^2\right] + \sum_{d_1=1}^{D}\sum_{\substack{d_2=1 \\ d_1\neq d_2}}^{D} \mathrm{E}\left[\left(x_{d_1}^{\mathrm{SM}}(t)\right)^2\right]\mathrm{E}\left[\left(x_{d_2}^{\mathrm{SM}}(t)\right)^2\right] \\
&= \sigma_x^2\left(\sum_{d=1}^{D}\frac{\gamma_d^{-2}}{2^d} + 2\phi^2(0)\sum_{d_1=1}^{D}\sum_{\substack{d_2=1 \\ d_1\neq d_2}}^{D}\frac{(\gamma_{d_1}\gamma_{d_2})^{-1}}{\sqrt{2^{d_1+d_2}}}\right)
\end{aligned} \tag{6.51}$$

Similarly, the average optical power of $x_{\mathrm{T}}(t)$ is given as follows:

$$\begin{aligned}
P_{\mathrm{opt}}^{\mathrm{avg}}(D,\underline{\gamma}) &= \mathrm{E}\left[x_{\mathrm{SM}}(t)\right] = \mathrm{E}\left[\sum_{d=1}^{D} x_d^{\mathrm{SM}}(t)\right] \\
&= \sum_{d=1}^{D}\mathrm{E}\left[x_d^{\mathrm{SM}}(t)\right] \\
&= \phi(0)\sigma_x\sum_{d=1}^{D}\frac{\gamma_d^{-1}}{\sqrt{2^{d-1}}}
\end{aligned} \tag{6.52}$$

where $\underline{\gamma} = \{\gamma_d^{-1}; d = 1, 2, \ldots, D\}$ is the set of scaling factors applied to each corresponding stream. The ratio of the average electrical power of superposition modulation waveform to the average electrical power of an inherently unipolar waveform is given as follows:

$$\alpha_{\text{elec}}^{P}(D, \underline{\gamma}) = \frac{P_{\text{elec}}^{\text{avg}}(D, \underline{\gamma})}{P_{\text{elec},d}^{\text{avg}}(\gamma_d)} \tag{6.53}$$

where the average electrical power of an inherently unipolar waveform is given as follows:

$$P_{\text{elec},d}^{\text{avg}}(\gamma_d) = \frac{\sigma_x^2}{2\gamma_d^2} \tag{6.54}$$

The electrical and optical SNR of superposition modulations can be defined as [155]:

$$\frac{E_{\text{b,elec}}^{\text{SM}}}{N_o} = \frac{P_{\text{elec}}^{\text{avg}}(D, \underline{\gamma})}{2B\eta_{\text{SM}}N_o} = \frac{\text{E}[x_{\text{SM}}{}^2(t)]}{2B\eta_{\text{SM}}N_o} \tag{6.55}$$

$$\frac{E_{\text{b,opt}}^{\text{SM}}}{N_o} = \frac{P_{\text{opt}}^{\text{avg}}(D, \underline{\gamma})}{2B\eta_{\text{SM}}N_o} = \frac{\text{E}[x_{\text{SM}}(t)]}{2B\eta_{\text{SM}}N_o} \tag{6.56}$$

where $E_{\text{b,elec}}^{\text{SM}}$ is the electrical energy per bit of superposition modulation; B is the single-sided bandwidth; N_o is the single-sided PSD of the noise at the receiver; and η_{SM} is the spectral efficiency of the considered superposition modulation. For example, this can be replaced with the spectral efficiency of L/e-ACO-OFDM given in (6.32) to calculate the electrical and optical SNR for L/e-ACO-OFDM. The optical SNR for a given system configuration can be calculated from the electrical SNR using the ratio of the electrical average power given in (6.55) to the optical average power given in (6.56). This is written as follows:

$$\alpha^{\text{EO}} = \frac{P_{\text{elec}}^{\text{avg}}(D, \underline{\gamma})}{P_{\text{opt}}^{\text{avg}}(D, \underline{\gamma})} \tag{6.57}$$

6.4.3.5 Superposition Modulation with Arbitrary Constellation Sizes

Implementation issues put a practical limit on the maximum number of depths that can be used with the superposition modulation techniques. In addition, the energy per bit increases as the total number of modulation depths increases because each additional stream is added on top of an existing time-domain signal. Considering that the spectral efficiency of each additional stream decreases exponentially, it can be assumed that the practical implementation of the superposition modulation techniques is expected to be realized using only a few information streams.

Arbitrary constellation sizes were proposed to close the spectral efficiency gap between superposition modulation and DCO-OFDM with a maximum number of modulation depths $D = 3$. Moreover, the spectral efficiency achieved by non-squared constellation sizes in DCO-OFDM cannot be achieved with the constant loading of modulation depths in ASE-DMT as was shown in Section 6.4.3.3. The solution of arbitrary constellation sizes allows ASE-DMT to achieve any desired spectral efficiency value. For example, to achieve the spectral efficiency of 8-QAM DCO-OFDM, it would be possible to use 4-PAM at the first depth, 2-PAM at the second depth, and 4-PAM at the third depth of ASE-DMT. Different power allocations are required at the different modulation depths when the arbitrary constellation sizes solution is adopted. The power allocation to each modulation stream is selected to maintain an equivalent energy per bit based on the used constellation size at each modulation depth $M_{SM,d}$. Arbitrary constellation sizes offer a practical solution for superposition modulation without any spectral efficiency loss. For example, the computation complexity per bit of ASE-DMT with three modulation depths is around 1.67 times higher than the computation complexity per bit of DCO-OFDM.

6.4.4 MCM Based on Other Transformations

OFDM-based VLC systems have been mainly investigated based on FFT. Other transformations such as discrete Hartley transformation (DHT) [157], wavelet packet division multiplexing (WPDM) [158], and fast Walsh-Hadamard transformation (FWHT) [159] have also been considered for VLC. The output of a DHT transformation can be real-valued when modulation techniques such as BPSK and PAM are used. Similar to DCO-OFDM and ACO-OFDM, DC-biasing and asymmetrical clipping can also be used to achieve a unipolar output in DHT-based MCM. DHT-based MCM does not require any Hermitian symmetry. However, it fails to improve the spectral efficiency of FFT-based OFDM. High illumination in DCO-OFDM requires a high DC bias level. This results in clipping distortion due to the peak power constraint of the LED as shown in Figure 6.1. Hadamard coded modulation (HCM) was proposed based on FWHT as a solution to the limitation of DCO-OFDM at high illumination levels. HCM is reported to achieve modest performance improvement over RPO-OFDM.

6.4.5 Color-Based Transmission Techniques

Color modulations allows for a additional DOF that is unique for VLC systems. Color-tunable LEDs such as the RGB-LED can illuminate with different colors based on the intensity applied on each LED element. Color shift keying (CSK) was proposed in the IEEE 802.15.7 standard for VLC [104]. The incoming bits are mapped into a constellation of colors from the chromatic Commission Internationale de l'Eclairage (CIE 1930) color space, as shown in Figure 6.24. The color perception of the human eye is modeled in the CIE 1930 color space [160]. Any color in the model can be represented by the chromaticity dimension $[x_C, y_C]$.

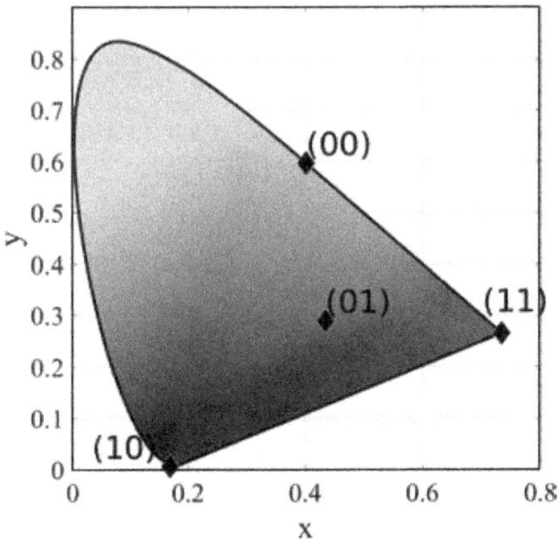

Figure 6.24 The constellation of 4-CSK on the CIE color space.

The color variations of a multicolored LEDs can be used to encode information. The average intensity of multicolored LEDs remains constant, which allows for a consistent optical power and color output throughout the transmission duration. Therefore, the instantaneous color variations cannot be observed by a human eye. At the receiver side, the two coordinates can be recovered by detecting the instantaneous power of the received multicolor intensities. The color dimension in Li-Fi can also be used to drive a multicolor LED with different streams of data. The optical summation may turn these colored parallel streams into a single color stream output that can be optically filtered at the receiver into the original transmitted colored streams.

6.5 MIMO Technologies

The basic idea of MIMO techniques is to equip a wireless communication system with multiple antennas to enable it to transmit/receive multiple streams of signals. The communication spectral efficiency and the received signal quality can be improved with appropriate signal processing. MIMO techniques have been widely used in RF wireless communications to improve the transmission data rate and reliability [161]. In Li-Fi networks, MIMO techniques can be used by deploying multiple LEDs and PDs on access points (APs) and user equipment (UE) [162]. In this section, we consider MIMO systems in Li-Fi. We will focus on the categorization of various MIMO scenarios. In addition, we highlight the unique characteristics of IM/DD-based MIMO channels in Li-Fi and introduce methods to improve the channel quality.

6.5.1 MIMO Systems in Li-Fi Networks

MIMO systems include single-user (SU)-MIMO systems and multiuser (MU)-MIMO systems. They have different system models and complexities, which will be introduced in the following sections. In addition, their potential deployment in Li-Fi networks will be discussed.

6.5.1.1 Single-User MIMO

In the SU-MIMO scenario, there are N_s LEDs at the transmitter side and N_r PDs at the receiver side, as shown in Figure 6.25. The relationship between the transmitted signals and the received signals can be described by the following equation [163]:

$$\mathbf{Y} = \mathbf{H}\mathbf{X} + \mathbf{N} \tag{6.58}$$

where $\mathbf{X} \in \mathbb{R}^{N_\mathrm{s}\times 1}$ is the transmitted signal vector, $\mathbf{H} \in \mathbb{R}^{N_\mathrm{r}\times N_\mathrm{s}}$ is the channel matrix, and $\mathbf{N} \in \mathbb{R}^{N_\mathrm{r}\times 1}$ is the noise vector. Note that the n_sth element of $\mathbf{X}$ represents the transmitted optical signal from the n_sth LED. The element corresponding to the n_sth column and the n_rth row of $\mathbf{H}$ refers to the channel coefficient between the n_sth LED and the n_rth PD. Due to the use of the IM/DD-based transmission, the channel matrix $\mathbf{H}$ and the transmitted signal vector $\mathbf{X}$ have to be real and positive, while these quantities are complex in RF wireless communications. This is a major difference between the MIMO systems in RF and OWC systems. Before and after the MIMO transmission, signal processing blocks are available to shape the transmitted signals (precoding) and recover the detected signals (post-detection). Assuming N_i different data streams are conveyed in the MIMO transmission, the transmitted signals can be defined as:

$$\mathbf{X} = \mathbf{F}\mathbf{S} \tag{6.59}$$

where $\mathbf{S} \in \mathbb{R}^{N_\mathrm{i}\times 1}$ is the data stream vector and $\mathbf{F} \in \mathbb{R}^{N_\mathrm{s}\times N_\mathrm{i}}$ is a precoding matrix. At the receiver side, post-detection can be conducted to obtain the original data streams as:

$$\mathbf{Z} = \mathbf{W}\mathbf{Y} \tag{6.60}$$

where $\mathbf{W} \in \mathbb{R}^{N_\mathrm{i}\times N_\mathrm{r}}$ is the post-detection matrix. By inserting (6.58) and (6.59) into (6.60), the SU-MIMO transmission can be described by:

$$\begin{aligned}\mathbf{Z} &= \mathbf{W}\left(\mathbf{H}\mathbf{F}\mathbf{S} + \mathbf{N}\right)\\ &= \mathbf{W}\mathbf{H}\mathbf{F}\mathbf{S} + \tilde{\mathbf{N}}\end{aligned} \tag{6.61}$$

where $\tilde{\mathbf{N}} = \mathbf{W}\mathbf{N}$ is an equivalent noise vector after the post-detection. In RF MIMO systems, the transmitted signal vector is constrained by a total signal power: $\sum_{n_\mathrm{s}=1}^{N_\mathrm{s}} \mathbb{E}\left\{\left|[\mathbf{X}]_{n_\mathrm{s}}\right|^2\right\} \le P_\mathrm{th}$. Due to the nature of IM/DD and the characteristics of LEDs, each element of the transmitted signal vector is constrained by a limited dynamic range: $0 \le P_\mathrm{th,min} \le [\mathbf{X}]_{n_\mathrm{s}} \le P_\mathrm{th,max}$ for $n_\mathrm{s} = 1, 2, \cdots, N_\mathrm{s}$ [164]. In many studies, the dynamic range is set to $P_\mathrm{th,min} = 0$ and $P_\mathrm{th,max} = \infty$ for simplicity. In addition, to fulfill the lighting

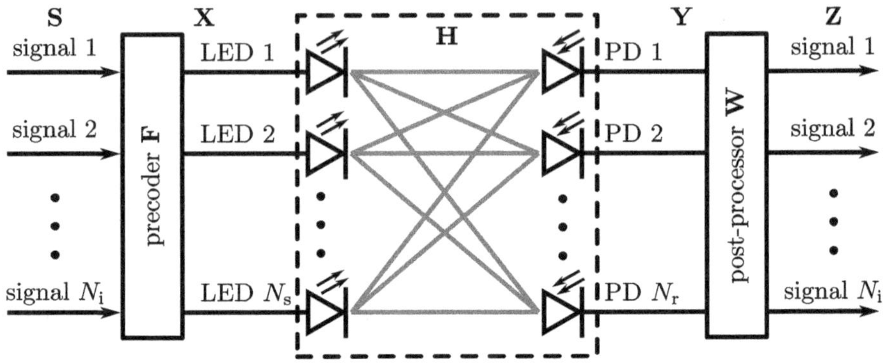

Figure 6.25 Block diagram of SU-MIMO.

requirement/photo-biological safety standard, the average optical power of each LED should be no greater than a certain threshold: $\mathbb{E}\left\{[\mathbf{X}]_{n_s}\right\} \leq P_{\text{th}}$. These differences in transmitted signal constraints make the precoder design in OWC different from those in RF MIMO systems.

A key factor in the performance of a MIMO system is the design of the precoding and post-detection matrices **F** and **W**. Various precoding/post-detection designs have been proposed in a number of studies [163–165]. In some cases, precoding or post-detection is not available due to the limited computation capability. In cases with no post-detection, the number of data streams has to be equal to the number of PDs: $N_{\text{i}} = N_{\text{r}}$ and the post-detection matrix becomes an identity matrix $\mathbf{I}^{N_{\text{r}} \times N_{\text{r}}}$. In cases with no precoding block, the number of data streams has to be equal to the number of LEDs: $N_{\text{i}} = N_{\text{s}}$ and the precoding matrix becomes an identity matrix $\mathbf{I}^{N_{\text{s}} \times N_{\text{s}}}$. In the case where $N_{\text{r}} = 1$ or $N_{\text{s}} = 1$, the considered transmission system becomes a (MISO) or (SIMO) systems, respectively [166].

6.5.1.2 Multiuser MIMO

MIMO systems can be extended to multiple user variants in Li-Fi systems serving multiple mobile users. The multiple users can be on the receiver side in the downlink (DL) scenario, where the MIMO channel is called a broadcast channel [167], as shown in Figure 6.26. The precoder is composed of multiple precoders for various users. On the receiver side, each user has its own independent post-detector. Assuming that there are N_{u} users, the detected signals of the uth user can be calculated by:

$$\mathbf{Z}_u = \mathbf{W}_u \mathbf{H}_u \left[\mathbf{F}_1, \cdots, \mathbf{F}_{N_{\text{u}}}\right] \left[\mathbf{S}_1^{\text{T}}, \cdots, \mathbf{S}_{N_{\text{u}}}^{\text{T}}\right]^{\text{T}} + \tilde{\mathbf{N}}_u \tag{6.62}$$

$$= \mathbf{W}_u \mathbf{H}_u \left(\mathbf{F}_u \mathbf{S}_u + \sum_{v \neq u} \mathbf{F}_v \mathbf{S}_v \right) + \tilde{\mathbf{N}}_u \tag{6.63}$$

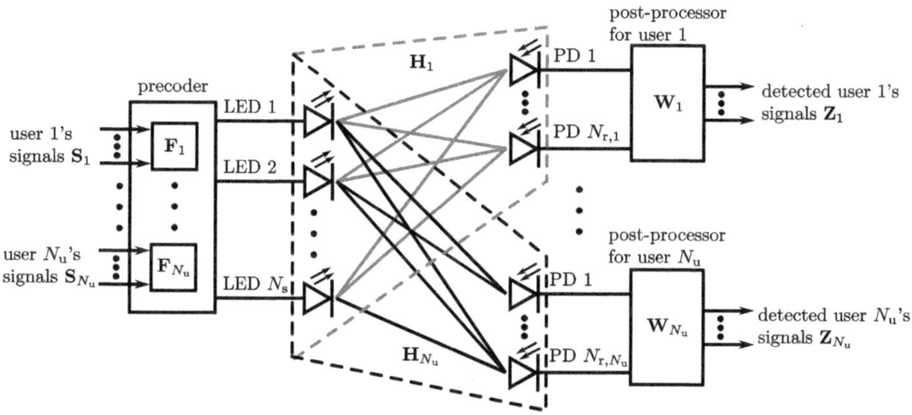

Figure 6.26 Block diagram of MU-MIMO with broadcast channel.

where $\{\cdot\}^{\mathrm{T}}$ is the transpose operator, $\mathbf{S}_u \in \mathbb{R}^{N_{\mathrm{i},u}\times 1}$ is the data stream vector for the uth user, $\mathbf{F}_u \in \mathbb{R}^{N_{\mathrm{s}}\times N_{\mathrm{i},u}}$ is the precoding matrix for the uth user, $\mathbf{H}_u \in \mathbb{R}^{N_{\mathrm{r},u}\times N_{\mathrm{s}}}$ is the channel matrix to the uth user, $\mathbf{W}_u \in \mathbb{R}^{N_{\mathrm{i},u}\times N_{\mathrm{r},u}}$ is the post-detection matrix for the uth user, and $\tilde{\mathbf{N}}_u$ denotes the equivalent noise vector: $\tilde{\mathbf{N}}_u = \mathbf{W}_u\mathbf{N}_u$ with $\mathbf{N}_u \in \mathbb{R}^{N_{\mathrm{r},u}\times 1}$ as the receiver noise vector of the uth user. Compared to the SU-MIMO, the detected signals are deteriorated by the interfering signals of the other users in addition to the noise, as shown in (6.63). Therefore, a main precoder/post-detector design target is to mitigate this interuser interference.

In the uplink (UL) scenario, multiple users are on the transmitter side, where the MIMO channel is called a multiple access channel [168]. The block diagram of a MU-MIMO system is shown in Figure 6.27. Similar to cases with broadcast channels, the detected signal for the uth user can be concluded as:

$$\mathbf{Z}_u = \mathbf{W}_u\left(\mathbf{H}_u\mathbf{F}_u\mathbf{S}_u + \sum_{v\neq u}\mathbf{H}_v\mathbf{F}_v\mathbf{S}_v\right) + \tilde{\mathbf{N}}_u \tag{6.64}$$

where $\mathbf{S}_u \in \mathbb{R}^{N_{\mathrm{i},u}\times 1}$ is the data stream vector for the uth user, $\mathbf{F}_u \in \mathbb{R}^{N_{\mathrm{s},u}\times N_{\mathrm{i},u}}$ is the precoding matrix of the uth user, $\mathbf{H}_u \in \mathbb{R}^{N_{\mathrm{r}}\times N_{\mathrm{s},u}}$ is the channel matrix with respect to the uth user, $\mathbf{W}_u \in \mathbb{R}^{N_{\mathrm{i},u}\times N_{\mathrm{r}}}$ is the post-detection matrix for the uth user, and $\tilde{\mathbf{N}}_u$ denotes the equivalent noise vector: $\tilde{\mathbf{N}}_u = \mathbf{W}_u\mathbf{N}_u$. When compared with a broadcast channel, the receiver collects a superposition of signals from all users, and uses a different post-detection matrix to detect the corresponding data streams from each user. Due to the limited studies on the Li-Fi UL systems with MIMO, this scenario is rarely considered in research.

6.5.1.3 Deployment of MIMO Systems in Li-Fi Networks

The LED-based lamp is composed of multiple LED modules to boost the total output power and to provide sufficient brightness/lighting performance.

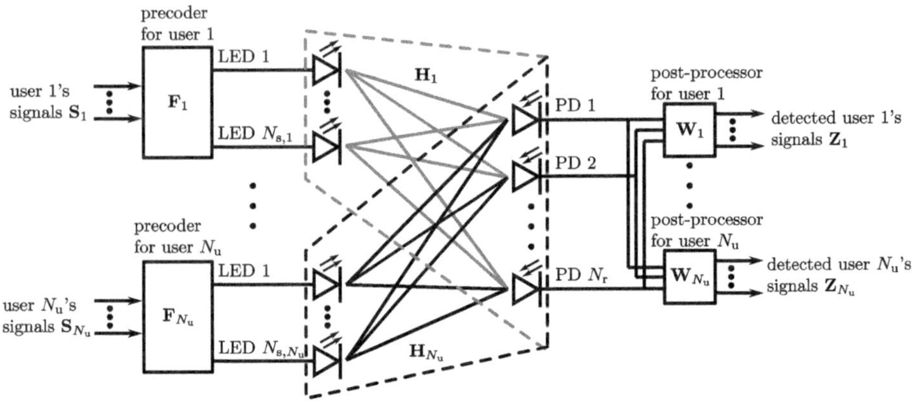

Figure 6.27 Block diagram of MU-MIMO with multiple access channel.

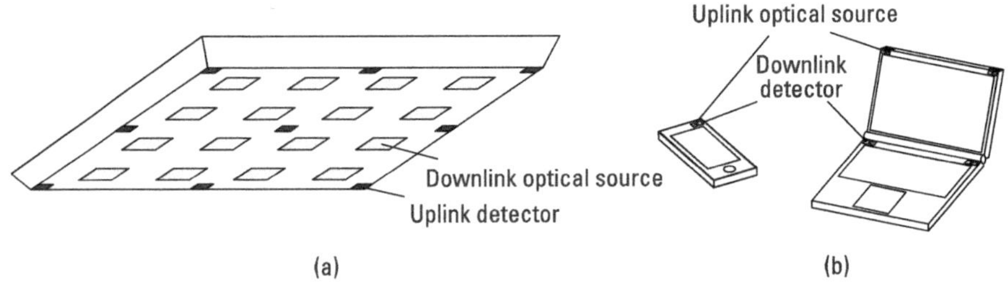

Figure 6.28 (a) MIMO elements deployments on the AP side. (b) MIMO elements deployments on the UE side.

By modulating each LED module separately, a multiple input scenario can be established for the Li-Fi DL transmission. In addition, the dimension of a Li-Fi AP ceiling unit is sufficient to accommodate multiple optical detectors to receive UL signals as shown in Figure 6.28(a). On the Li-Fi UE side, whether the device can support MIMO techniques depends on the dimension and the computational capability/battery capacity of the UE. In cases with small devices, such as smart phones or tablets, there is limited space for mounting multiple Li-Fi UL sources and DL detectors, as shown in Figure 6.28(b). In addition, the UE hardware may not be capable of supporting the necessary MIMO processing functionality. Consequently, only one pair of DL detector and UL source can be deployed on a small device. However, the MISO or SIMO scenarios can be considered to enable MIMO techniques for these devices. In cases with larger devices, such as personal laptops, there is enough space to mount multiple UL sources and DL detectors. In addition, the computational capability and battery capacity are sufficient to support real-time MIMO processing.

In cases with small office/private rooms, the number of mobile users is typically small. The Li-Fi system can be used as a personal wireless access network, where SU-MIMO techniques can be implemented. In cases with public indoor areas, such as a conference room, large office/classroom, or shopping mall, a large number of active mobile users exist in the network. Intuitively, the broadcast channel MU-MIMO techniques can be used in the Li-Fi DL transmission with MIMO capabilities, while the multiple access channel MU-MIMO techniques can be used in the Li-Fi UL transmission with MIMO capability.

6.5.2 MIMO Functions

By using various precoder and post-detector designs, various MIMO functions can be achieved. In this section, the basic MIMO functions: multiplexing, diversity, and spatial modulation are introduced for the case with a SU-MIMO system. In particular, the characteristics in IM/DD-based MIMO systems are discussed.

6.5.2.1 Multiplexing

One of the most important functions of MIMO systems is to improve the achievable transmission rate of a wireless communication system. This can be realized by establishing multiple parallel independent subchannels to transmit multiple data streams concurrently, which is known as multiplexing. Intuitively, in order to maximize the data rate, it is desirable to have as many multiplexing subchannels as possible. However, the number of supported data streams N_i is limited by the numbers of the LEDs and PDs [169]: $N_\mathrm{i} \leq \min(N_\mathrm{s}, N_\mathrm{r})$. Therefore, more than one LED and PD device is required for multiplexing transmission. In a SU-MIMO system, the detected signal vector can be written as (6.61). Thus, the ith row of the received data stream Z can be calculated as:

$$\mathfrak{z}_i = [\mathbf{WHF}]_{i,i}\, \mathfrak{s}_i + \sum_{l \neq i} [\mathbf{WHF}]_{i,l}\, \mathfrak{s}_l + \sum_{n_\mathrm{r}=1}^{N_\mathrm{r}} [\mathbf{W}]_{i,n_\mathrm{r}}\, \mathfrak{n}_{n_\mathrm{r}} \tag{6.65}$$

where $\mathfrak{s}_i$ represents the ith original data stream and $\mathfrak{n}_{n_\mathrm{r}}$ is the noise sample on the n_rth receiver. The first summation in (6.65) corresponds to the interfering signal from other subchannels and the second summation corresponds to the equivalent noise after post-detection. Based on (6.65), the corresponding signal to interference and noise ratio (SINR) of the ith subchannel can be given as follows:

$$\gamma_i = \frac{\left|[\mathbf{WHF}]_{i,i}\right|^2 \sigma_{\mathrm{s},i}^2}{\sum_{l \neq i} \left|[\mathbf{WHF}]_{i,l}\right|^2 \sigma_{\mathrm{s},l}^2 + \sum_{n_\mathrm{r}=1}^{N_\mathrm{r}} \left|[\mathbf{W}]_{i,n_\mathrm{r}}\right|^2 \sigma_{\mathrm{n},n_\mathrm{r}}^2} \tag{6.66}$$

where $\sigma_{\mathrm{s},i}^2$ refers to the signal power of the ith data stream and $\sigma_{\mathrm{n},n_\mathrm{r}}^2$ refers to the noise variance of the n_rth detector. Then the total achievable data rate can be calculated as:

$$R_{\text{total}} = \sum_{i=1}^{N_{\text{i}}} R(\gamma_i) \tag{6.67}$$

where $R(\gamma_i)$ refers to the achievable data rate of the ith data stream. The precoder $\mathbf{F}$ and post-detector $\mathbf{W}$ should be optimized to maximize the total data rate R_{total}. A straightforward solution to the appropriate $\mathbf{F}$ and $\mathbf{W}$ in a multiplexing system is based on the singular value decomposition (SVD) of the channel matrix [169]:

$$\mathbf{H} = \mathbf{U}\Lambda\mathbf{V}^{\text{H}} \tag{6.68}$$

where $\{\cdot\}^{\text{H}}$ is the conjugate transpose operator, Λ is a diagonal matrix, and $\mathbf{U}$ and $\mathbf{V}$ are two unitary matrices, which has the properties of $\mathbf{U}\mathbf{U}^{\text{H}} = \mathbf{I}$ and $\mathbf{V}\mathbf{V}^{\text{H}} = \mathbf{I}$ with $\mathbf{I}$ as the identity matrix. By letting $\mathbf{F} = \mathbf{V}$, $\mathbf{W} = \mathbf{U}^{\text{H}}$, $\sigma^2_{\text{n},n_{\text{r}}} = \sigma^2_{\text{n}}$ for $n_{\text{r}} = 1, 2, \cdots, N_{\text{r}}$ and inserting (6.68) into (6.66), the SNR can be found as:

$$\gamma_i = \left|[\Lambda]_{i,i}\right|^2 \frac{\sigma^2_{\text{s},i}}{\sigma^2_{\text{n}}} \tag{6.69}$$

which is free from interference between subchannels. However, if the original signals $\mathbf{S}$ have a low PAPR, such as OOK, the precoder $\mathbf{F} = \mathbf{V}$ will significantly increase the PAPR of the precoded signals $\mathbf{X}$. With a limited LED dynamic range, this leads to a decrease in the achievable SINR [164]. Therefore, SVD-based multiplexing is more suitable for systems with high signal PAPR, such as OFDM.

6.5.2.2 Diversity

In many applications, link reliability is more important than the transmission data rate. The MIMO system can be used to provide diversity and to improve the transmission link quality [170]. In this case, the number of link streams $N_{\text{i}} = 1$ and the precoding and post-detection matrices are reduced to two vectors: $\mathbf{f} \in \mathbb{R}^{N_{\text{s}}\times 1}$ and $\mathbf{w} \in \mathbb{R}^{1\times N_{\text{r}}}$. Thus, the detected signal (6.65) is simplified as:

$$\mathfrak{z} = \mathbf{w}\mathbf{H}\mathbf{f}\mathfrak{s} + \sum_{n_{\text{r}}=1}^{N_{\text{r}}} [\mathbf{w}]_{n_{\text{r}}} \mathfrak{n}_{n_{\text{r}}} \tag{6.70}$$

and the corresponding SNR can be found as:

$$\gamma = \frac{|\mathbf{w}\mathbf{H}\mathbf{f}|^2 \sigma^2_{\text{s}}}{\sum_{n_{\text{r}}=1}^{N_{\text{r}}} \left|[\mathbf{w}]_{n_{\text{r}}}\right|^2 \sigma^2_{\text{n},n_{\text{r}}}} \tag{6.71}$$

In a diversity system, the design objective of $\mathbf{f}$ and $\mathbf{w}$ changes to the maximization of link SNR γ.

The large size of the PD in Li-Fi relative to the wavelength of optical signals allows for a spatial averaging that makes a IM/DD-based system immune from small-scale fading effects [36]. Consequently, the intensity signals from multiple LEDs can always be combined constructively. This enables a unique diversity technique called repetition coding, which lets all LEDs transmit the same signal [171]. In cases where repetition coding is used, the channel matrix is reduced to a vector $\mathbf{h} \in \mathbb{R}^{N_r \times 1}$, where the n_rth element of $\mathbf{h}$ is equal to the summation of the entries to the n_rth row of $\mathbf{H}$:

$$[\mathbf{h}]_{n_r} = \sum_{n_s=1}^{N_s} [\mathbf{H}]_{n_r,n_s} \tag{6.72}$$

This is the equivalent to having only one transmitting element. By using (6.72) and letting $\mathbf{f} = 1$, the SNR (6.71) is simplified as:

$$\gamma = \frac{|\mathbf{wh}|^2 \sigma_s^2}{\sum_{n_r=1}^{N_r} \left|[\mathbf{w}]_{n_r}\right|^2 \sigma_{n,n_r}^2} \tag{6.73}$$

6.5.2.3 Spatial Modulation

Spatial modulation (SM) is another approach to improve the achievable data rate of the MIMO system, which uses the spatial signature to convey data symbols [172]. Without the loss of generality, the same framework introduced in Section 6.5.1.1 is also used here. Assuming that there are M_{ssk} spatial symbols, the following requirement should be fulfilled: $N_i = N_s = M_{ssk}$. When spatial symbols are used for encoding the data, the structure of the data stream vector $\mathbf{S}$ becomes time-variant. Therefore, a symbol time index n_t is used and the data stream vector at the n_tth symbol period is defined as $\mathbf{S}(n_t) \in \mathbb{R}^{M_{ssk} \times 1}$. For the convenience of representation, a set of basis vectors are defined:

$$\mathbf{e}_1 = \begin{bmatrix} 1 \\ 0 \\ \vdots \\ 0 \end{bmatrix}, \ \mathbf{e}_2 = \begin{bmatrix} 0 \\ 1 \\ \vdots \\ 0 \end{bmatrix}, \ \cdots, \ \mathbf{e}_{M_{ssk}} = \begin{bmatrix} 0 \\ 0 \\ \vdots \\ 1 \end{bmatrix}, \tag{6.74}$$

where the n_{ssk}th entry of the basis vector $\mathbf{e}_{n_{ssk}}$ is 1 and the remaining entries are all 0. In the case of space shift keying (SSK) modulation, the data stream vector at symbol period n_t is defined as:

$$\mathbf{S}(n_t) = \mathbf{e}_{n_{ssk}(n_t)} \tag{6.75}$$

where the spatial index $n_{ssk}(n_t)$ is determined by the incoming binary data combinations, as illustrated in Figure 6.29. Considering a simple zero-forcing detection, the precoding and post-detection matrices are defined as: $\mathbf{F} = \mathbf{I}$ and

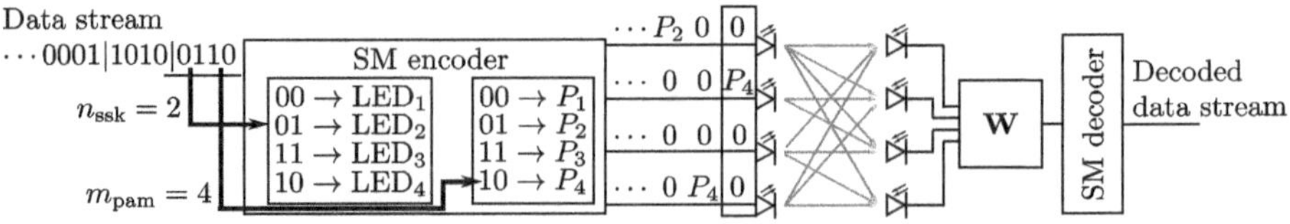

Figure 6.29 SM encoding with $N_s = N_r = 4$.

$\mathbf{W} = \left(\mathbf{H}^T\mathbf{H}\right)^{-1}\mathbf{H}^T$. In conjunction with (6.61), the detected data vector can be found as:

$$\begin{aligned}\mathbf{Z}(n_t) &= \mathbf{W}\left(\mathbf{HFS}(n_t) + \mathbf{N}\right)\\ &= \mathbf{S}(n_t) + \tilde{\mathbf{N}}\end{aligned} \tag{6.76}$$

Finally, the binary data can be decoded by:

$$\hat{n}_{ssk}(n_t) = \arg\min_{n_{ssk}} \left|\left|\mathbf{Z}(n_t) - \mathbf{e}_{n_{ssk}}\right|\right| \tag{6.77}$$

This is the basic spatial symbol encoding method. To further increase the system spectral efficiency, extra data can be encoded using the conventional IM/DD-based modulation, such as M-ary PAM, on top of the SSK modulation. Assuming a PAM modulation order of M_{PAM}, the intensity of the mth PAM symbol can be defined as:

$$P_m = \frac{2P_{opt}m}{M_{PAM} + 1} \tag{6.78}$$

where P_{opt} is the average LED optical power. Instead of using the data vector definition (6.75), the following expression is used:

$$\mathbf{S}(n_t) = P_{m(n_t)}\mathbf{e}_{n_{ssk}(n_t)} \tag{6.79}$$

At the receiver side, the PAM signal can be decoded by:

$$\hat{m}_{PAM}(n_t) = \arg\min_{m} \left|\left|\max\left(\mathbf{Z}(n_t)\right) - P_m\right|\right| \tag{6.80}$$

and the spatial symbol can be decoded by:

$$\hat{n}_{ssk}(n_t) = \arg\min_{n} \left|\left|\mathbf{Z}(n_t) - P_{\hat{m}_{PAM}(n_t)}\mathbf{e}_n\right|\right| \tag{6.81}$$

It is apparent that the number of data streams achieved by SM is smaller than the cases with multiplexing systems with the same number of LEDs and PDs. However, an SM system achieves lower energy efficiency and lower system complexity. In each symbol period, only one LED is activated in SM systems, while all LEDs are activated in multiplexing systems. In addition, the precoding matrix is an identity matrix, which lowers the complexity on the transmitter side. Improved spectral efficiency can be explored by activating multiple LEDs with various combinations during a single symbol period, in what is known as generalized spatial modulation (GSM) [173]. By doing so, a higher number of bits can be conveyed in each spatial symbol. However, the number of spatial symbols increases from $M_{\text{ssk}} = N_{\text{s}}$ to $M_{\text{gsm}} = \binom{N_{\text{s}}}{N_{\text{a}}}$.

6.5.3 MIMO Channel Decorrelation

The performance of Li-Fi networks can be improved considerably by applying various MIMO techniques. However, the performance of MIMO systems also depends on the condition of the channel matrix **H**. If the MIMO channel has a strong spatial correlation, the rank of the channel matrix **H** will be close to 1 [168]. In a multiplexing system, this leads to only a few usable subchannels, while the remaining subchannels are deemed unreliable for transmitting and receiving data. In a diversity system, this will make the characteristics of different diversity sub-channels similar to each other. Consequently, if one of them is unreliable, the remaining subchannels are likely to be unreliable as well. In an SM system, a MIMO channel with strong spatial correlation makes it hard to distinguish the difference between various spatial symbols. Therefore, it is very important to have a well-conditioned MIMO channel matrix in a Li-Fi system with MIMO capability.

In RF wireless communications, many wireless channels are based on NLOS multiple paths with small-scale fading [168]. Consequently, RF MIMO channels often have weak spatial correlation. In contrast, the IM/DD-based MIMO channels in OWC are real and non-negative, and the channel characteristics rely on the orientation/positions of LEDs/PDs. In the case of similar positions and orientations of two LEDs or PDs, the corresponding MIMO channels will have a strong spatial correlation, which is a major challenge in Li-Fi/VLC MIMO systems [162]. In this section, a number of methods to decorrelate the MIMO channel matrix are introduced.

6.5.3.1 Channel Decorrelation on the Access Point Side

On the AP side of a Li-Fi network, it is common to use LED lamps deployed on the ceiling as DL transmitters to fulfill the illumination requirement. This provides an opportunity to drive each LED module separately and form the MIMO system. However, these LED modules are often closely deployed and have the same orientation, as shown in Figure 6.30(a).

Intuitively, a good method to weaken the spatial correlation is to increase the spatial separation between different MIMO inputs. Thus, instead of deploying LED modules in the same lamp as multiple MIMO inputs, the LED lamps at different positions can function as MIMO inputs, as shown in Figure 6.30(b). This concept is similar to the distributed antenna system in RF wireless communication. However, this method raises concerns about the complexity

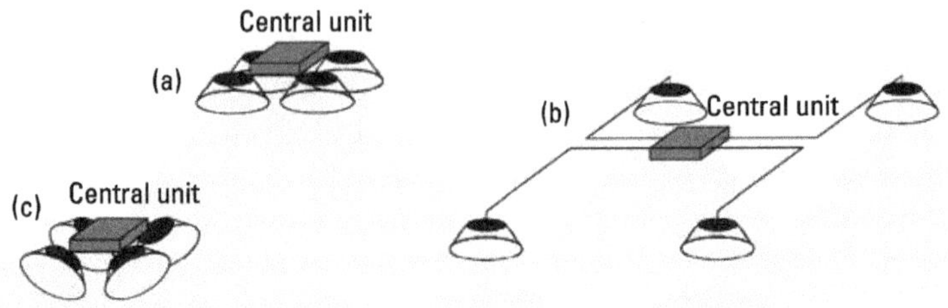

Figure 6.30 (a) Spatially correlated DL LEDs configuration on the AP side. (b) Channel decorrelation with different positions. (c) Channel decorrelation with different orientations.

and latency of backhaul connections between the distributed antennas [174]. In a case with an indoor Li-Fi system, these issues are lessened as the actual separation between lamps is much shorter than the case in RF-equivalent systems. However, the overall user-detected signal power may drop with the increase in the separations between lamps, which is a clear disadvantage of this method. In the case of closely located LED modules as MIMO inputs, a lower spatial correlation can be obtained by deploying LED modules with different orientations, as illustrated in Figure 6.30(c). This approach is also known as angular diversity transmitter [175]. To enhance the effectiveness of this approach, the beam divergence of each LED module can be decreased. In some cases, each LED has a coverage to the users in a specific direction. However, the excessive reduction of the beam divergence may raise concerns for lighting uniformity and eye-safety regulations [176]. The same channel spatial decorrelation methods are also applicable to the UL detectors on the AP side. Unlike the configurations of DL LED lamps, the configurations of UL detectors are free from any lighting requirement constraints.

6.5.3.2 Channel Decorrelation on the User Side

A MIMO channel with a significant spatial correlation will be obtained if the DL PDs/UL LEDs are closely deployed on the user side with the same orientation, as shown in Figure 6.31(a). Unlike the AP side, there is a limitation on the size of the device. Therefore, it is insufficient to decorrelate the channel by separating two DL PDs/UL LEDs on a small user equipment. However, most of the user devices have a cubic shape, such as smartphones, tablets, and laptops. These characteristics make it convenient to deploy multiple LEDs/PDs on different sides of the device with various orientations, as shown in Figure 6.31(b). This is known as angular diversity receiver, or a multidirectional receiver [175,177].

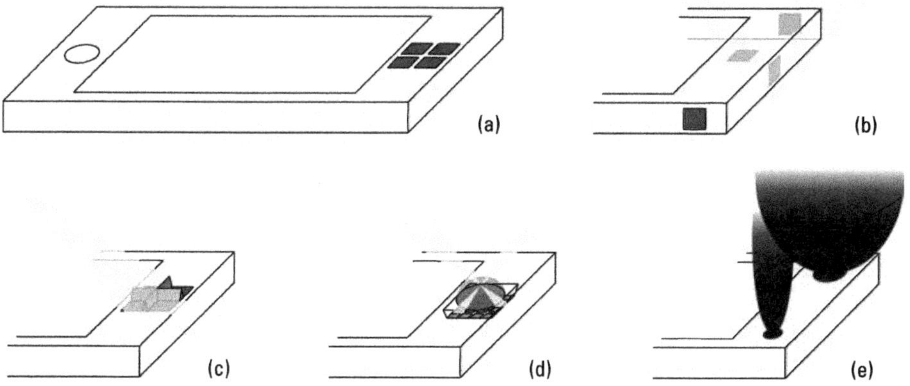

Figure 6.31 (a) Spatially correlated DL PDs configuration on the user side. (b) Channel decorrelation with different orientations. (c) Channel decorrelation with blocking units. (d) Channel decorrelation with an imaging lens. (e) Channel decorrelation with different radiation patterns.

When deploying DL PDs, another two unique methods are available if the DL PDs have to be closely located with the same orientation. The first approach is to artificially block some of the LOS paths [162]. This is achieved by placing blocking units between adjacent PDs, as shown in Figure 6.31(c). In this case, the optical power from one LED can only reach one or two of the PDs, and the MIMO channel can be effectively decorrelated. However, this approach may lower the overall received signal power, and the deployment of blocking units takes extra space on the device. The other approach is to use an imaging lens before the DL PD clusters to guide the optical power from a specific direction to a certain pixel detector [178], as shown in Figure 6.31(d). An imaging receiver is very effective in lowering the spatial correlation, but it also leads to a lower received optical power and a narrower detector FOV. A UL MIMO channel with weak spatial correlation may be obtained by using different radiation patterns for the adjacent UL LEDs, as shown in Figure 6.31(e).

6.5.3.3 Wavelength Domain Decorrelation

In cases where a MIMO channel cannot be decorrelated in the spatial domain, it is also possible to decorrelate the MIMO channel in the wavelength domain. On the transmitter side, multiple LEDs emitting light of different wavelengths are used, as shown in Figure 6.32(a). Optical filters that match the wavelength of each transmitting LED are mounted on each PD on the receiver side. Note that this approach can be used in both DL and UL transmissions. For example, visible light channels are used in the DL transmission, and IR channels are used in the UL transmission. This is similar to the conventional WDM in optical communications. However, wavelength subchannels in WDM systems are

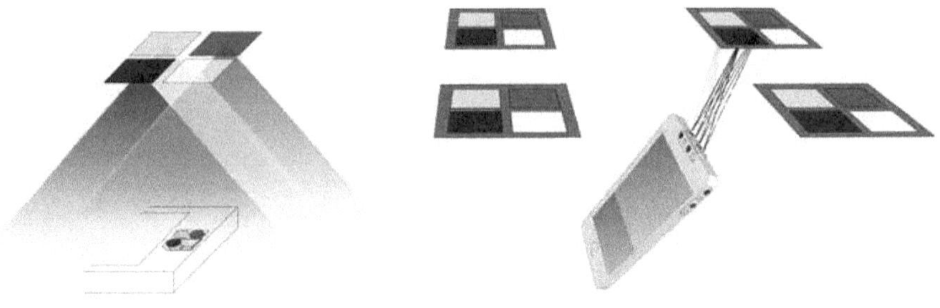

(a) (b)

Figure 6.32 (a) Channel decorrelation in wavelength domain. (b) Hybrid channel decorrelation.

considered to be independent of each other due to the negligible spectrum overlap between adjacent subchannels. However, the wavelength domain decorrelation is treated as a MIMO system with crosstalk [179]. Therefore, precoding and post-detections are required to achieve various MIMO functions as introduced in Section 6.5.2.

6.5.3.4 Hybrid Channel Decorrelation

It is also possible to use both the spatial domain and wavelength domain to decorrelate the MIMO channel. We demonstrate an example in the DL transmission with 16 LEDs and 16 PDs, as shown in Figure 6.32(b). Every four LEDs are grouped as a cluster, and the LED clusters are spatially separated to decrease the channel correlation. The LEDs in the same cluster use different wavelengths to decrease the channel correlation. On the receiver side, every four PDs form a cluster, which is mounted on one side of the mobile device. The PDs on different sides of the device have different orientations; therefore, they have a low channel correlation. The PDs on the same side of the device are decorrelated using optical filters with different passbands. From this method, a well-conditioned MIMO channel can be obtained. It was demonstrated that this hybrid channel decorrelation is very effective for VLC multiplexing systems [74].

6.6 Summary

This chapter provides a background on different types of light sources and detectors that can be used in Li-Fi to enable different applications such as downhole monitoring, high-speed indoor communications, underwater communications, and vehicular communications. The trend in optical front-end research is expected to grow in the direction of developing devices that are optimized for Li-Fi in terms of wavelength, bandwidth, and illuminance. For example, the development of compact optical front ends that can transmit and receive multiple streams of data on different wavelength bands can open great potential for future Li-Fi systems. In addition, the development of compact

LD modules that can produce a high-quality white illumination can unlock the bandwidth limitations of LED-based Li-Fi systems. There is increasing interest in developing OWC systems for mobile access beyond 1 Tbps. Therefore, the development of optical elements that can perform light collection, multiplexing, and demultiplexing is of particular importance for these systems [180].

Modulation techniques were developed to suit the Li-Fi channels with different considerations such as illumination levels, energy efficiency, and spectral efficiency. However, the choice of the modulation techniques is subject to the considered application and the desired system scenario. OFDM-based modulation techniques are promising candidates for high-speed and spectrum-efficient VLC systems. The use of a single-tap equalizer and adaptive bit and energy-loading algorithms allow OFDM-based techniques to utilize most of the available bandwidth in VLC systems. Energy efficiency is required to support dimming in VLC and uplink communications in Li-Fi systems. Inherently unipolar OFDM techniques are introduced as a solution to the significant power dissipation caused by the DC bias in DCO-OFDM. However, these techniques fail in reducing the energy requirements at high spectral efficiency values. Novel modulation techniques are discussed in this chapter based on the superposition of multiple inherently unipolar OFDM streams. These techniques offer solutions to the spectral efficiency loss in inherently unipolar OFDM without any DC bias.[7] These techniques provide significant energy-efficiency gains when compared to similar state-of-the-art IM/DD techniques at equivalent spectral efficiency values. This makes them promising candidates for future Li-Fi systems.

The trend in Li-Fi modulation techniques research shows an increasing interest in multiplexing techniques. The research into modulation techniques is expected to grow in the direction of developing schemes that are able to encode data adaptively in the multiple DOF available in Li-Fi systems. In addition, the research is expected to grow for techniques targeting specific applications, such as communications for autonomous underwater vehicles (AUVs), unmanned aerial vehicles (UAVs), and IoT devices. The limited battery life in these applications makes opportunistic communications approaches more suitable than other modulation techniques that target high-speed communications [181].

References

[1] Afgani, M., et al., "Visible Light Communication Using OFDM," *2nd International Conference on Testbeds and Research Infrastructures for the Development of Networks and Communities, TRIDENTCOM*, Barcelona, Spain, March 1–3, 2006, pp. 6–134.

[7] DC bias due to the turn-on voltage of the light source is required. However, it is neglected in our relative comparisons as it is required for all IM/DD techniques

[2] Brodrick, J., "Suggested Research Topics Supplement: Technology and Market Context," 2017, https://www.energy.gov/sites/prod/files/2017/09/f37/ssl_supplement_suggested-topics_sep2017_0.pdf.

[3] Banelli, P., and S. Cacopardi, "Theoretical Analysis and Performance of OFDM Signals in Nonlinear AWGN Channels," *IEEE Transactions on Communications*, Vol. 48, No. 3, March 2000, pp. 430–441.

[4] Dimitrov, S., and H. Haas, "Information Rate of OFDM-Based Optical Wireless Communication Systems with Nonlinear Distortion," *IEEE/OSA Journal of Lightwave Technology*, Vol. 31, No. 6, March 2013, pp. 918–929.

[5] Stepniak, G., J. Siuzdak, and P. Zwierko, "Compensation of a VLC Phosphorescent White LED Nonlinearity by Means of Volterra DFE," *IEEE Photonics Technology Letters*, Vol. 25, No. 16, August 2013, pp. 1597–1600.

[6] Zhao, W., et al., "Orthogonal Polynomial-Based Nonlinearity Modeling and Mitigation for LED Communications," *IEEE Photonics Journal*, Vol. 8, No. 4, August 2016, pp. 1–12.

[7] Inan, B., et al., "Impact of LED Nonlinearity on Discrete Multitone Modulation," *IEEE/OSA Journal of Optical Communications and Networking*, Vol. 1, No. 5, October 2009, pp. 439–451.

[8] Elgala, H., R. Mesleh, and H. Haas, "Non-Linearity Effects and Predistortion in Optical OFDM Wireless Transmission Using LEDs," *Inderscience International Journal of Ultra Wideband Communications and Systems (IJUWBCS)*, Vol. 1, No. 2, 2009, pp. 143–150.

[9] Tsonev, D., S. Sinanovic, and H. Haas, "Complete Modeling of Nonlinear Distortion in OFDM Based Optical Wireless Communication," *IEEE/OSA Journal of Lightwave Technology*, Vol. 31, No. 18, September 2013, pp. 3064–3076.

[10] Islim, M. S., et al., "Towards 10 Gb/s Orthogonal Frequency Division Multiplexing-Based Visible Light Communication Using a GaN Violet Micro-LED," *OSA Photonics Research*, Vol. 5, No. 2, April 2017, pp. A35–A43. http://www.osapublishing.org/prj/abstract.cfm?URI=prj-5-2-A35.

[11] Khalid, A. M., et al., "1-Gb/s Transmission over a Phosphorescent White LED by Using Rate-Adaptive Discrete Multitone Modulation," *IEEE Photonics Journal*, Vol. 4, No. 5, October 2012, pp. 1465–1473.

[12] Guo, Y., et al., "A Tutorial on Laser-Based Lighting and Visible Light Communications: Device and Technology," *Chinese Optics Letters*, Vol. 17, No. 4, April 2019, p. 040601. http://col.osa.org/abstract.cfm?URI=col-17-4-040601.

[13] Lee, C., et al., "Semipolar GaN-Based Laser Diodes for Gbit/s White Lighting Communication: Devices to Systems," *Proc. SPIE*, Vol. 10532, San Francisco, CA, February 23, 2018, pp. 10532–10532.

[14] Yoshida, K., et al., "245 MHz Bandwidth Organic Light-Emitting Diodes Used in a Gigabit Optical Wireless Data Link," *Nature Communications*, Vol. 11, No. 1, 2020, pp. 1–7.

[15] Wang, C., et al., "A 2.5 Gb/s Real-Time Visible-Light Communication System Based on Phosphorescent White LED," *7th International Conference on Information, Communication and Networks (ICICN)*, 2019, pp. 140–145.

[16] Xie, E., et al., "Over 10 Gbps VLC for Long-Distance Applications Using a GaN-Based Series-Biased Micro-LED Array," *IEEE Photonics Technology Letters*, Vol. 32, No. 9, 2020, pp. 499–502.

[17] Bian, R., I. Tavakkolnia, and H. Haas, "15.73 Gb/s Visible Light Communication with Off-the-Shelf LEDs," *IEEE/OSA Journal of Lightwave Technology*, Vol. 37, No. 10, May 2019, pp. 2418–2424.

[18] Lee, C., et al., "Advanced LiFi Technology: Laser Light," in *Light-Emitting Devices, Materials, and Applications XXIV*, J. K. Kim, M. R. Krames, and M. Strassburg, (eds.), Vol. 11302, International Society for Optics and Photonics/SPIE, 2020, pp. 116–123.

[19] Tsonev, D., S. Videv, and H. Haas, "Towards a 100 Gb/s Visible Light Wireless Access Network," *OSA Optics Express*, Vol. 23, No. 2, January 2015, pp. 1627–1637. http://www.opticsexpress.org/abstract.cfm?URI=oe-23-2-1627.

[20] Wei, L. -Y., et al., "Tricolor Visible-Light Laser Diodes Based Visible Light Communication Operated at 40.665 Gbit/s and 2 m Free-Space Transmission," *Optics Express*, Vol. 27, No. 18, September 2019, pp. 25072–25077. http://www.opticsexpress.org/abstract.cfm?URI=oe-27-18-25072.

[21] Huang, H., et al., "100 Tbit/s Free-Space Data Link Enabled by Three-Dimensional Multiplexing of Orbital Angular Momentum, Polarization, and Wavelength," *Optics Letters*, Vol. 39, No. 2, January 2014, pp. 197–200.

[22] Akasaki, I., H. Amano, and S. Nakamura, "Blue LEDs–Filling the World with New Light," Nobel Prize, 2014.

[23] Foucher, C., et al., "Flexible Glass Hybridized Colloidal Quantum Dots for Gb/s Visible Light Communications," *IEEE Photonics Journal*, Vol. 10, No. 1, February 2018, pp. 1–11.

[24] Leitão, M. F., et al., "Gb/s Visible Light Communications with Colloidal Quantum Dot Color Converters," *IEEE Journal of Selected Topics in Quantum Electronics*, Vol. 23, No. 5, September 2017, pp. 1–10.

[25] Ferreira, R. X. G., et al., "High Bandwidth GaN-Based Micro-LEDs for Multi-Gb/s Visible Light Communications," *IEEE Photonics Technology Letters*, Vol. 28, No. 19, October 2016, pp. 2023–2026.

[26] Rajbhandari, S., et al., "A Review of Gallium Nitride LEDs for Multi-Gigabit-per-Second Visible Light Data Communications," *IOP Semiconductor Science and Technology*, Vol. 32, No. 2, 2017, p. 023001. http://stacks.iop.org/0268-1242/32/i=2/a=023001.

[27] Samuel, I. D. W., et al., "Ultrafast OLEDs Enable Visible Light Communication at a Gigabit per Second," in *Organic and Hybrid Light Emitting Materials and Devices XXIV*, C. Adachi, J. -J. Kim, and F. So, (eds.), Vol. 11473, International Society for Optics and Photonics/SPIE, 2020.

[28] Schubert, E. F., et al., "Resonant Cavity Light-Emitting Diode," *Applied Physics Letters*, Vol. 60, No. 8, 1992, pp. 921–923.

[29] Schubert, E. F., *Communication LEDs*, 2nd ed., Cambridge, U.K.: Cambridge University Press, 2006, p. 382392.

[30] Chun, H., et al., "A Spectrally Efficient Equalization Technique for Optical Sources with Direct Modulation," *Optics Letters*, Vol. 43, No. 11, June 2018, pp. 2708–2711.

[31] Hu, F., et al., "Demonstration of a Low-Complexity Memory-Polynomial-Aided Neural Network Equalizer for CAP Visible-Light Communication with Superluminescent Diode," *Opto-Electronic Advances*, Vol. 3, No. 8, 2020, p. 200009. http://www.oejournal.org/J/ArticleQuery/A200821000003.

[32] C. Shen et al., "High-Speed 405-nm Superluminescent Diode (SLD) with 807-MHz Modulation Bandwidth," *Optical Express*, Vol. 24, No. 18, September 2016, pp. 20281–20286. http://www.opticsexpress.org/abstract.cfm?URI=oe-24-18-20281.

[33] Rashidi, A., et al., "High-Speed Nonpolar InGaN/GaN Superluminescent Diode with 2.5 GHz Modulation Bandwidth," *IEEE Photonics Technology Letters*, Vol. 32, No. 7, 2020, pp. 383–386.

[34] Cantore, M., et al., "High Luminous Flux from Single Crystal Phosphor-Converted Laser-Based White Lighting System," *Optical Express*, Vol. 24, No. 2, January 2016, pp. A215–A221. http://www.opticsexpress.org/abstract.cfm?URI=oe-24-2-A215.

[35] Lee, C., et al., "Dynamic Characteristics of 410 Nm Semipolar (2021) III-Nitride Laser Diodes with a Modulation Bandwidth of over 5 GHz," *Applied Physics Letters*, Vol. 109, No. 10, 2016, p. 101104.

[36] Kahn, J. M., and J. R. Barry, "Wireless Infrared Communications," *Proceedings of the IEEE*, Vol. 85, No. 2, February 1997, pp. 265–298.

[37] Tsai, C. -T., et al., "Toward High-Speed Visible Laser Lighting Based Optical Wireless Communications," *Progress in Quantum Electronics*, Vol. 67, 2019, p. 100225. http://www.sciencedirect.com/science/article/pii/S0079672719300254.

[38] Watson, S., et al., "Visible Light Communications Using a Directly Modulated 422nm GaN Laser Diode," *Optics Letters*, Vol. 38, No. 19, October 2013, pp. 3792–3794. http://ol.osa.org/abstract.cfm?URI=ol-38-19-3792.

[39] Viola, S., et al., "15 Gb/s OFDM-Based VLC Using Direct Modulation of 450 GaN Laser Diode," *Proceedings of SPIE*, Vol. 10437, Warsaw, Poland, 2017, pp. 10437–10437–7.

[40] Lee, C., et al., "2 Gbit/s Data Transmission from an Unfiltered Laser-Based Phosphor-Converted White Lighting Communication System," *Optics Express*, Vol. 23, No. 23, November 2015, pp. 29779–29787.

[41] BS EN 60825-1:2007, Safety of Laser Products, BSI British Standards Std., November 2007.

[42] Wang, W. -C., et al., "White-Light Color Conversion with Red/Green/Violet Laser Diodes and Yellow Light-Emitting Diode Mixing for 34.8 Gbit/s Visible Lighting Communication," *Photonics Research*, Vol. 8, No. 8, August 2020, pp. 1398–1408. http://www.osapublishing.org/prj/abstract.cfm?URI=prj-8-8-1398.

[43] Chun, H., et al., "A Wide-Area Coverage 35 Gb/s Visible Light Communications Link for Indoor Wireless Applications," *Scientific Reports*, Vol. 9, No. 1, 2019, pp. 1–8.

[44] Shen, C., et al., "Laser-Based Visible Light Communications and Underwater Wireless Optical Communications: A Device Perspective," in *Novel In-Plane Semiconductor Lasers XVIII*, A. A. Belyanin and P. M. Smowton, (eds.), Vol. 10939, International Society for Optics and Photonics/SPIE, 2019, pp. 29–37.

[45] Kuramoto, M., et al., "Watt-Class Blue Vertical-Cavity Surface-Emitting Laser Arrays," *Applied Physics Express*, Vol. 12, No. 9, August 2019, p. 091004.

[46] Ul-Masabih, S. M. M., et al., "Electrically Injected Nonpolar GaN-Based VCSELs with Lattice-matched Nanoporous Distributed Bragg Reflector Mirrors," *Applied Physics Express*, Vol. 12, No. 3, February 2019, p. 036504.

[47] Chieh Yu, H., et al., "Progress and Prospects of GaN-Based VCSEL from Near UV to Green Emission," *Progress in Quantum Electronics*, Vol. 57, 2018, pp. 1–19. http://www.sciencedirect.com/science/article/pii/S0079672718300016.

[48] Shen, C., et al., "GHz Modulation Bandwidth from Single-Longitudinal Mode Violet-Blue VCSEL Using Nonpolar InGaN/GaN QWS," *Conference on Lasers and Electro-Optics (CLEO)*, 2016, pp. 1–2.

[49] Lu, H., et al., "10 M/25 Gbps LiFi Transmission System Based on a Two-Stage Injection-Locked 680 nm VCSEL Transmitter," *Optics Letters*, Vol. 40, No. 19, October 2015, pp. 4563–4566.

[50] Tsai, W., et al., "500 Gb/s PAM4 FSO-UWOC Convergent System with a R/G/B Five-Wavelength Polarization-Multiplexing Scheme," *IEEE Access*, Vol. 8, 2020, pp. 16913–16921.

[51] Koonen, T., et al., "Optically Controlled Beam-Steering Wireless Systems," *Optical Fiber Communication Conference (OFC)*, Optical Society of America, 2020, p. M1J.1. http://www.osapublishing.org/abstract.cfm?URI=OFC-2020-M1J.1.

[52] Gomez-Agis, F., et al., "112 Gbit/s Transmission in a 2D Beam Steering AWG-Based Optical Wireless Communication System," *European Conference on Optical Communication (ECOC)*, 2017, pp. 1–3.

[53] Koonen, T., et al., "High-Capacity Optical Wireless Communication Using Two-Dimensional IR Beam Steering," *IEEE/OSA Journal of Lightwave Technology*, Vol. 36, No. 19, 2018, pp. 4486–4493.

[54] Gomez, A., et al., "Design and Demonstration of a 400 Gb/s Indoor Optical Wireless Communications Link," *IEEE/OSA Journal of Lightwave Technology*, Vol. 34, No. 22, November 2016, pp. 5332–5339. http://jlt.osa.org/abstract.cfm?URI=jlt-34-22-5332.

[55] Wei, L. -Y., et al., "2.333-Tbit/s Bi-Directional Optical Mobile Networks Using Optical Wireless Communication (OWC)," *Optics Communications*, Vol. 475, 2020, p. 126187.

[56] Sarbazi, E., M. Safari, and H. Haas, "The Bit Error Performance and Information Transfer Rate of SPAD Array Optical Receivers," *IEEE Transactions on Communications*, Vol. 68, No. 9, 2020, pp. 5689–5705.

[57] Kosman, J., et al., "Distortion Losses of High-Speed Single-Photon Avalanche Diode Optical Receivers Approaching Quantum Sensitivity," *Philosophical Transactions of the Royal Society A: Mathematical, Physical and Engineering Sciences*, Vol. 378, No. 2169, 2020, p. 20190194. https://royalsocietypublishing.org/doi/abs/10.1098/rsta.2019.0194.

[58] Wang, Z., et al., "On the Design of a Solar-Panel Receiver for Optical Wireless Communications with Simultaneous Energy Harvesting," *IEEE Journal on Selected Areas in Communications*, Vol. 33, No. 8, August 2015, pp. 1612–1623.

[59] Hamamatsu, *Photomultiplier Tubes: Basics and Applications*, 4th ed., 2017. https://www.hamamatsu.com/resources/pdf/etd/PMT_handbook_v4E.pdf.

[60] Hamamatsu Photonics, K.K, "Characteristics and Use of Si-APD (Avalanche Photodiode)," May 2004. http://www.hamamatsu.com.cn/UserFiles/DownFile/Related/20130810164103075.pdf.

[61] Armstrong, J., R. J. Green, and M. D. Higgins, "Comparison of Three Receiver Designs for Optical Wireless Communications Using White LEDs," *IEEE Communications Letters*, Vol. 16, No. 5, May 2012, pp. 748–751.

[62] Haas, H., J. Elmirghani, and I. White, "Optical Wireless Communication," *Philosophical Transactions of the Royal Society A: Mathematical, Physical and Engineering Sciences*, Vol. 378, No. 2169, 2020, p. 20200051. https://royalsocietypublishing.org/doi/abs/10.1098/rsta.2020.0051.

[63] Thorlabs, "Calibrated Photodiodes," 2020. https://www.thorlabs.com/newgrouppage-9.cfm?objectgroup id=2822.

[64] Miyazaki, E., S. Itami, and T. Araki, "Using a Light-Emitting Diode as a High-Speed, Wavelength Selective Photodetector," *Review of Scientific Instruments*, Vol. 69, No. 11, 1998, pp. 3751–3754.

[65] Liu, X., et al., "High-Bandwidth InGaN Self-Powered Detector Arrays Toward MIMO Visible Light Communication Based on Micro-LED Arrays," *ACS Photonics*, Vol. 6, No. 12, 2019, pp. 3186–3195.

[66] Ho, K. -T., et al., "3.2 Gigabit-per-Second Visible Light Communication Link with InGaN/GaN MQW Micro-Photodetector," *Optics Express*, Vol. 26, No. 3, February 2018, pp. 3037–3045.

[67] Alkhazragi, O., et al., "7.4-Gbit/s Visible-Light Communication Utilizing Wavelength-Selective Semipolar Micro-Photodetector," *IEEE Photonics Technology Letters*, Vol. 32, No. 13, 2020, pp. 767–770.

[68] Alphalas, "Ultrafast Photodetectors," UPD Series, 2020, https://www.alphalas.com/products/laser-diagnostic-tools/ultrafast-photodetectors-upd-series.html.

[69] Hamamatsu Photonics K.K., "Si Photodiodes Catalogue," 2020. https://www.hamamatsu.com/eu/en/product/optical-sensors/photodiodes/si-photodiodes/index.html.

[70] Islim, M. S., and H. Haas, "An Investigation of the Sunlight Irradiance Effect on Visible Light Communications," *IEEE 28th Annual International Symposium on Personal, Indoor, and Mobile Radio Communications (PIMRC)*, Montreal, Canada, October 2017, pp. 1–6.

[71] Islim, M. S., et al., "The Impact of Solar Irradiance on Visible Light Communications," *IEEE/OSA Journal of Lightwave Technology*, Vol. 36, No. 12, June 2018, pp. 2376–2386.

[72] Chen, Z., N. Serafimovski, and H. Haas, "Angle Diversity for an Indoor Cellular Visible Light Communication System," *Proc. IEEE Vehicular Technology Conference*, Seoul, South Korea, May 2014, pp. 1–5.

[73] Carruthers, J. B., and J. M. Kahn, "Angle Diversity for Nondirected Wireless Infrared Communication," *IEEE Transactions on Communications*, Vol. 48, No. 6, June 2000, pp. 960–969.

[74] Chen, C., et al., "Hybrid Multiplexing in OFDM-Based VLC Systems," *IEEE Wireless Communications and Networking Conference (WCNC)*, 2020, pp. 1–6.

[75] Chen, Z., D. Tsonev, and H. Haas, "Improving SINR in Indoor Cellular Visible Light Communication Networks," *IEEE International Conference on Communications (ICC)*, June 2014, pp. 3383–3388.

[76] Niclass, C., et al., "A Single Photon Avalanche Diode Implemented in 130-nm CMOS Technology," *IEEE Journal of Selected Topics in Quantum Electronics*, Vol. 13, No. 4, July 2007, pp. 863–869.

[77] Haas, H., et al., "What Is LiFi?" *IEEE/OSA Journal of Lightwave Technology*, Vol. 34, No. 6, March 2016, pp. 1533–1544.

[78] Manousiadis, P. P., et al., "Wide Field-of-View Fluorescent Antenna for Visible Light Communications Beyond the Etendue Limit," *OSA Optica*, Vol. 3, No. 7, 2016, pp. 702–706.

[79] Peyronel, T., et al., "Luminescent Detector for Free-Space Optical Communication," *OSA Optica*, Vol. 3, No. 7, July 2016, pp. 787–792.

[80] Mulyawan, R., et al., "A Comparative Study of Optical Concentrators for Visible Light Communications," Broadband Access Communication Technologies XI, SPIE, 2017, 10128, pp. 142—147.

[81] Manousiadis, P. P., et al., "Optical Antennas for Wavelength Division Multiplexing in Visible Light Communications Beyond the Etendue Limit," *Advanced Optical Materials*, Vol. 8, No. 4, 2020, p. 1901139. https://onlinelibrary.wiley.com/doi/abs/10.1002/adom.201901139.

[82] Cova, S., et al., "Avalanche Photodiode and Quenching Circuits for Single-Photon Detection," *Applied Optics*, Vol. 35, No. 12, April 1996, pp. 1956–1976.

[83] Chen, C. -C., "Effect of Detector Dead Time on the Performance of Optical Direct-Detection Communication Links," *The Telecommunications and Data Acquisition Report*, Vol. 42-93, May 1988.

[84] Sarbazi, E., M. Safari, and H. Haas, "Statistical Modeling of Single-Photon Avalanche Diode Receivers for Optical Wireless Communications," *IEEE Transactions on Communications*, 2018, pp. 1–1.

[85] Zhang, L., et al., "The Future Prospects for SiPM-Based Receivers for Visible Light Communications," *Journal of Lightwave Technology*, Vol. 37, No. 17, 2019, pp. 4367–4374.

[86] Zhang, L., et al., "A Comparison of APD- and SPAD-Based Receivers for Visible Light Communications," *IEEE/OSA Journal of Lightwave Technology*, Vol. 36, No. 12, 2018, pp. 2435–2442.

[87] Ebeling, K. J., *Integrated Optoelectronics: Waveguide Optics, Photonics, Semiconductors*, New York: Springer Science & Business Media, 2012.

[88] Li, Y., et al., "Single Photon Avalanche Diode (SPAD) VLC System and Application to Downhole Monitoring," *Proc. IEEE Global Communications Conference*, December 2014, pp. 2108–2113.

[89] Wang, Z., et al., "Towards Self-Powered Solar Panel Receiver for Optical Wireless Communication," *Proc. IEEE International Conference on Communications (ICC)*, Sydney, NSW, June 10–14, 2014, pp. 3348–3353.

[90] Zhang, R., and C. K. Ho, "MIMO Broadcasting for Simultaneous Wireless Information and Power Transfer," *IEEE Transactions on Wireless Communications*, Vol. 12, No. 5, May 2013, pp. 1989–2001.

[91] Kim, S. -M., and J. -S. Won, "Simultaneous Reception of Visible Light Communication and Optical Energy Using a Solar Cell Receiver," *Proc. IEEE International Conference on ICT Convergence (ICTC)*, Jeju Island, South Korea, October 14–16, 2013, pp. 896–897.

[92] Fakidis, J., et al., "0.5-Gb/s OFDM-Based Laser Data and Power Transfer Using a GaAs Photovoltaic Cell," *IEEE Photonics Technology Letters*, Vol. 30, No. 9, May 2018, pp. 841–844.

[93] Zhang, S., et al., "Organic Solar Cells as High-Speed Data Detectors for Visible Light Communication," *OSA Optica*, Vol. 2, No. 7, July 2015, pp. 607–610. http://www.osapublishing.org/optica/abstract.cfm?URI=optica-2-7-607.

[94] Fakidis, J., H. Helmers, and H. Haas, "Simultaneous Wireless Data and Power Transfer for a 1-Gb/s GaAs VCSEL and Photovoltaic Link," *IEEE Photonics Technology Letters*, 2020, pp. 1–1.

[95] Mica, N. A., et al., "Triple-Cation Perovskite Solar Cells for Visible Light Communications," *Photonics Research*, Vol. 8, No. 8, August 2020, pp. A16–A24. http://www.osapublishing.org/prj/abstract.cfm?URI=prj-8-8-A16.

[96] Haas, H., et al., "Solar Cell Receiver Free-Space Optical for 5G Backhaul," *Optical Fiber Communications Conference and Exhibition (OFC)*, 2019, pp. 1–3.

[97] Das, S., et al., "Towards Energy Neutral Wireless Communications: Photovoltaic Cells to Connect Remote Areas," *Energies*, Vol. 12, No. 19, October 2019, p. 3772.

[98] Davidson, F. M., and X. Sun, "Gaussian Approximation Versus Nearly Exact Performance Analysis of Optical Communication Systems with PPM Signaling and APD Receivers," *IEEE Transactions on Communications*, Vol. 36, No. 11, November 1988, pp. 1185–1192.

[99] You, R., and J. Kahn, "Upper-Bounding the Capacity of Optical IM/DD Channels with Multiple Subcarrier Modulation and Fixed Bias Using Trigonometric Moment Space Method," *IEEE Transactions on Information Theory*, Vol. 48, No. 2, February 2002, pp. 514–523.

[100] Hranilovic, S., and F. R. Kschischang, "Capacity Bounds for Power- and Band-Limited Optical Intensity Channels Corrupted by Gaussian Noise," *IEEE Transactions on Information Theory*, Vol. 50, No. 5, May 2004, pp. 784–795.

[101] Lapidoth, A., S. M. Moser, and M. A. Wigger, "On the Capacity of Free-Space Optical Intensity Channels," *IEEE Transactions on Information Theory*, Vol. 55, No. 10, October 2009, pp. 4449–4461.

[102] Farid, A. A., and S. Hranilovic, "Capacity Bounds for Wireless Optical Intensity Channels with Gaussian Noise," *IEEE Transactions on Information Theory*, Vol. 56, No. 12, December 2010, pp. 6066–6077.

[103] Randel, S., et al., "Advanced Modulation Schemes for Short-Range Optical Communications," *IEEE Journal of Selected Topics in Quantum Electronics*, Vol. 16, No. 5, March 22, 2010, pp. 1280–1289.

[104] "IEEE Standard for Local and Metropolitan Area Networks–Part 15.7: Short-Range Wireless Optical Communication Using Visible Light," IEEE Std 802.15.7-2011, 2011, pp. 1–309.

[105] Lee, S. H., S. -Y. Jung, and J. K. Kwon, "Modulation and Coding for Dimmable Visible Light Communication," *IEEE Communications Magazine*, Vol. 53, No. 2, February 2015, pp. 136–143.

[106] Tsonev, D., et al., "A 3-Gb/s Single-LED OFDM-Based Wireless VLC Link Using a Gallium Nitride LED," *IEEE Photonics Technology Letters*, Vol. 26, No. 7, April 2014, pp. 637–640.

[107] Tanaka, Y., et al., "Indoor Visible Communication Utilizing Plural White LEDs as Lighting," *Proc. of the 12th IEEE International Symposium on Personal, Indoor and Mobile Radio Communications*, Vol. 2, San Diego, CA, September 2001, pp. 81–85.

[108] Purwita, A. A., and H. Haas, "IQ-WDM for IEEE 802.11bb-Based LiFi," *IEEE Wireless Communications and Networking Conference (WCNC)*, 2020, pp. 1–6.

[109] Yang, Y., et al., "A Simple OFDM Scheme for VLC Systems Based on U-Law Mapping," *IEEE Photonics Technology Letters*, Vol. 28, No. 6, March 2016, pp. 641–644.

[110] Zhang, H., Y. Yuan, and W. Xu, "PAPR Reduction for DCO-OFDM Visible Light Communications via Semidefinite Relaxation," *IEEE Photonics Technology Letters*, Vol. 26, No. 17, September 2014, pp. 1718–1721.

[111] Popoola, W. O., Z. Ghassemlooy, and B. G. Stewart, "Pilot-Assisted PAPR Reduction Technique for Optical OFDM Communication Systems," *IEEE/OSA Journal of Lightwave Technology*, Vol. 32, No. 7, April 2014, pp. 1374–1382.

[112] Carruthers, J. B., and J. M. Kahn, "Multiple-Subcarrier Modulation for Nondirected Wireless Infrared Communication," *IEEE Journal on Selected Areas in Communications*, Vol. 14, No. 3, April 1996, pp. 538–546.

[113] Dimitrov, S., S. Sinanovic, and H. Haas, "Clipping Noise in OFDM-Based Optical Wireless Communication Systems," *IEEE Transactions on Communications*, Vol. 60, No. 4, April 2012, pp. 1072–1081.

[114] Bykhovsky, D., and S. Arnon, "An Experimental Comparison of Different Bit-and-Power-Allocation Algorithms for DCO-OFDM," *IEEE/OSA Journal of Lightwave Technology*, Vol. 32, No. 8, 2014, pp. 1559–1564.

[115] Tsonev, D., S. Videv, and H. Haas, "Unlocking Spectral Efficiency in Intensity Modulation and Direct Detection Systems," *IEEE Journal on Selected Areas in Communications*, Vol. 33, No. 9, September 2015, pp. 1758–1770.

[116] Levin, H. E., "A Complete and Optimal Data Allocation Method for Practical Discrete Multitone Systems," *IEEE Global Telecommunications Conference*, Vol. 1, San Antonio, TX, November 25–29, 2001, pp. 369–374.

[117] Xiong, F., *Digital Modulation Techniques*, 2nd ed., Norwood, MA: Artech House, 2006.

[118] Shannon, C. E., "A Mathematical Theory of Communication," *Bell System Technical Journal*, Vol. 27, July 1948, pp. 379–423 & 623–656.

[119] Fath, T., C. Heller, and H. Haas, "Optical Wireless Transmitter Employing Discrete Power Level Stepping," *IEEE/OSA Journal of Lightwave Technology*, Vol. 31, No. 11, June 2013, pp. 1734–1743.

[120] Mossaad, M., S. Hranilovic, and L. Lampe, "Visible Light Communications Using OFDM and Multiple LEDs," *IEEE Transactions on Communications*, Vol. 63, No. 11, November 2015, pp. 4304–4313.

[121] Griffiths, A. D., et al., "CMOS-Integrated GaN LED Array for Discrete Power Level Stepping in Visible Light Communications," *Optical Express*, Vol. 25, No. 8, April 2017, pp. A338–A345.

[122] Dong, H., et al., "OFDM Visible Light Communication Transmitter Based on LED Array," *Chinese Optics Letters*, Vol. 12, No. 5, May 2014, p. 052301. http://col.osa.org/abstract.cfm?URI=col-12-5-052301.

[123] Armstrong, J., and A. Lowery, "Power Efficient Optical OFDM," *IET Electronics Letters*, Vol. 42, No. 6, March 16, 2006, pp. 370–372.

[124] Lee, S. C. J., et al., "PAM-DMT for Intensity-Modulated and Direct-Detection Optical Communication Systems," *IEEE Photonics Technology Letters*, Vol. 21, No. 23, December 2009, pp. 1749–1751.

[125] Tsonev, D., S. Sinanovic, and H. Haas, "Novel Unipolar Orthogonal Frequency Division Multiplexing (U-OFDM) for Optical Wireless," *Proc. of the IEEE Vehicular Technology Conference (VTC Spring)*, Yokohama, Japan, May 6–9, 2012.

[126] Fernando, N., Y. Hong, and E. Viterbo, "Flip-OFDM for Unipolar Communication Systems," *IEEE Transactions on Communications*, Vol. 60, No. 12, December 2012, pp. 3726–3733.

[127] Elgala, H., and T. D. C. Little, "Reverse Polarity Optical-OFDM (RPO-OFDM): Dimming Compatible OFDM for Gigabit VLC links," *OSA Optics Express*, Vol. 21, No. 20, October 2013, pp. 24288–24299. http://www.opticsexpress.org/abstract.cfm?URI=oe-21-20-24288.

[128] Elgala, H., and T. Little, "P-OFDM: Spectrally Efficient Unipolar OFDM," *Optical Fiber Communications Conference and Exhibition (OFC)*, San Francisco, CA, March 2014, pp. 1–3.

[129] Elgala, H., and T. D. C. Little, "Polar-Based OFDM and SC-FDE Links Toward Energy-Efficient Gbps Transmission Under IM-DD Optical System Constraints," *IEEE/OSA Journal of Optical Communications and Networking*, Vol. 7, No. 2, February 2015, pp. A277–A284. http://jocn.osa.org/abstract.cfm?URI=jocn-7-2-A277.

[130] Asadzadeh, K., A. Farid, and S. Hranilovic, "Spectrally Factorized Optical OFDM," *12th Canadian Workshop on Information Theory (CWIT 2011)*, Kelowna, BC, Canada, May 17–20, 2011, pp. 102–105.

[131] Mao, T., et al., "PM-DCO-OFDM for PAPR Reduction in Visible Light Communications," *Opto-Electronics and Communications Conference (OECC)*, Shanghai, China, June 2015, pp. 1–3.

[132] Chen, L., B. Krongold, and J. Evans, "Diversity Combining for Asymmetrically Clipped Optical OFDM in IM/DD Channels," *IEEE Global Telecommunications Conference*, Honolulu, HI, November 2009, pp. 1–6.

[133] Dang, J., Z. Zhang, and L. Wu, "A Novel Receiver for ACO-OFDM in Visible Light Communication," *IEEE Communication Letters*, Vol. 17, No. 12, December 2013, pp. 2320–2323.

[134] Huang, N., et al., "Iterative Receiver for Flip-OFDM in Optical Wireless Communication," *IEEE Photonics Technology Letters*, Vol. 27, No. 16, August 2015, pp. 1729–1732.

[135] Zheng, Y., et al., "A Novel Receiver for Flip-OFDM in Optical Wireless Communication," *IEEE 16th International Conference on Communication Technology (ICCT)*, Hangzhou, China, October 2015, pp. 620–625.

[136] Dang, J., Z. Zhang, and L. Wu, "Frequency-Domain Diversity Combining Receiver for ACOOFDM System," *IEEE Photonics Journal*, Vol. 7, No. 6, December 2015, pp. 1–10.

[137] Xu, J., et al., "Asymmetrically Reconstructed Optical OFDM for Visible Light Communications," *IEEE Photonics Journal*, Vol. 8, No. 1, February 2016, pp. 1–18.

[138] Huang, N., et al., "Receiver Design for PAM-DMT in Indoor Optical Wireless Links," *IEEE Photonics Technology Letters*, Vol. 27, No. 2, January 2015, pp. 161–164.

[139] Xiang, N., et al., "A Novel Receiver Design for PAM-DMT in Optical Wireless Communication Systems," *IEEE Photonics Technology Letters*, Vol. 27, No. 18, September 2015, pp. 1919–1922.

[140] Dissanayake, S. D., K. Panta, and J. Armstrong, "A Novel Technique to Simultaneously Transmit ACO-OFDM and DCO-OFDM in IM/DD Systems," *IEEE GLOBECOM Workshops (GC Wkshps)*, Houston, TX, December 2011, pp. 782–786.

[141] Ranjha, B., and M. Kavehrad, "Hybrid Asymmetrically Clipped OFDM-Based IM/DD Optical Wireless System," *IEEE/OSA Journal of Optical Communications and Networking*, Vol. 6, No. 4, April 2014, pp. 387–396.

[142] Wu, N., and Y. Bar-Ness, "A Novel Power-Efficient Scheme Asymmetrically and Symmetrically Clipping Optical (ASCO)-OFDM for IM/DD Optical Systems," *EURASIP Journal on Advances in Signal Processing*, Vol. 2015, No. 1, 2015, pp. 1–10.

[143] Tsonev, D., and H. Haas, "Avoiding Spectral Efficiency Loss in Unipolar OFDM for Optical Wireless Communication," *Proc. of the IEEE International Conference on Communications (ICC)*, Sydney, Australia, June 2014.

[144] Islim, M. S., D. Tsonev, and H. Haas, "On the Superposition Modulation for OFDM-Based Optical Wireless Communication," *2015 IEEE Global Conference on Signal and Information Processing (GlobalSIP)*, December 2015, pp. 1022–1026.

[145] Elgala, H., and T. D. C. Little, "SEE-OFDM: Spectral and Energy Efficient OFDM for Optical IM/DD Systems," *2014 IEEE 25th Annual International Symposium on Personal, Indoor, and Mobile Radio Communication (PIMRC)*, September 2014, pp. 851–855.

[146] Wang, Q., et al., "Layered ACO-OFDM for Intensity-Modulated Direct-Detection Optical Wireless Transmission," *OSA Optics Express*, Vol. 23, No. 9, May 2015, pp. 12382–12393. http://www.opticsexpress.org/abstract.cfm?URI=oe-23-9-12382.

[147] Islim, M., D. Tsonev, and H. Haas, "Spectrally Enhanced PAM-DMT for IM/DD Optical Wireless Communications," *Proc. of the 25th IEEE Int. Symp. Pers. Indoor and Mobile Radio Commun. (PIMRC)*, Hong Kong, China, August 30–September 2, 2015, pp. 927–932.

[148] Islim, M. S., and H. Haas, "Augmenting the Spectral Efficiency of Enhanced PAM-DMT-Based Optical Wireless Communications," *Optical Express*, Vol. 24, No. 11, May 2016, pp. 11932–11949. http://www.opticsexpress.org/abstract.cfm?URI=oe-24-11-11932.

[149] Islim, M. S., D. Tsonev, and H. Haas, "A Generalized Solution to the Spectral Efficiency Loss in Unipolar Optical OFDM-Based Systems," *Proc. of the IEEE International Conference on Communications (ICC)*, London, U.K., June 2015.

[150] Islim, M. S., and H. Haas, "Modulation Techniques for Li-Fi," *ZTE Communications*, Vol. 14, No. 2, April 2016, pp. 29–40.

[151] Lowery, A. J., "Spectrally Efficient Optical Orthogonal Frequency Division Multiplexing," *Philosophical Transactions of the Royal Society A: Mathematical, Physical and Engineering Sciences*, Vol. 378, No. 2169, 2020, p. 20190180. https://royalsocietypublishing.org/doi/abs/10.1098/rsta.2019.0180.

[152] Chen, L., B. Krongold, and J. Evans, "Successive Decoding of Anti-Periodic OFDM Signals in IM/DD Optical Channel," *2010 IEEE International Conference on Communications (ICC)*, 2010, pp. 1–6.

[153] Chen, L., B. Krongold, and J. Evans, "Performance Analysis for Optical OFDM Transmission in Short-Range IM/DD Systems," *Journal of Lightwave Technology*, Vol. 30, No. 7, 2012, pp. 974–983.

[154] Lowery, A. J., "Comparisons of Spectrally-Enhanced Asymmetrically-Clipped Optical OFDM Systems," *Optical Express*, Vol. 24, No. 4, February 2016, pp. 3950–3966. http://www.opticsexpress.org/abstract.cfm?URI=oe-24-4-3950

[155] Dimitrov, S., and H. Haas, *Principles of LED Light Communications: Towards Networked Li-Fi*, Cambridge, U.K.: Cambridge University Press, 2015.

[156] Islim, M. S., and H. Haas, "An Experimental Demonstration of a Superposition-Based DMT Technique for IM/DD," IEEE International Conference on Communications Workshops (ICC Workshops), 2019, pp. 1–5, doi: 10.1109/ICCW.2019.8756827.

[157] Moreolo, M. S., R. M. Noz, and G. Junyent, "Novel Power Efficient Optical OFDM Based on Hartley Transform for Intensity-Modulated Direct-Detection Systems," *IEEE/OSA Journal of Lightwave Technology*, Vol. 28, No. 5, March 2010, pp. 798–805. http://jlt.osa.org/abstract.cfm?URI=jlt-28-5-798.

[158] Huang, W., C. Gong, and Z. Xu, "System and Waveform Design for Wavelet Packet Division Multiplexing-Based Visible Light Communications," *IEEE/OSA Journal of Lightwave Technology*, Vol. 33, No. 14, July 2015, pp. 3041–3051.

[159] Noshad, M., and M. Brandt-Pearce, "Hadamard-Coded Modulation for Visible Light Communications," *IEEE Transactions on Communications*, Vol. 64, No. 3, March 2016, pp. 1167–1175.

[160] International Commission on Illumination (CIE), "CIE 1931 Standard Colorimetric Observer," August 2008. http://www.cie.co.at.

[161] Stuber, G. L., et al., "Broadband MIMO-OFDM Wireless Communications," *Proceedings of the IEEE*, Vol. 92, No. 2, February 2004, pp. 271–294.

[162] Fath, T., and H. Haas, "Performance Comparison of MIMO Techniques for Optical Wireless Communications in Indoor Environments," *IEEE Transactions on Communications*, Vol. 61, No. 2, February 2013, pp. 733–742.

[163] Ying, K., et al., "Joint Optimization of Precoder and Equalizer in MIMO VLC Systems," *IEEE Journal on Selected Areas in Communications*, Vol. 33, No. 9, September 2015, pp. 1949–1958.

[164] Ying, K., et al., "MIMO Transceiver Design in Dynamic-Range-Limited VLC Systems," *IEEE Photonics Technology Letters*, Vol. 28, No. 22, 2016, pp. 2593–2596.

[165] Wang, R., et al., "Linear Transceiver Designs for MIMO Indoor Visible Light Communications Under Lighting Constraints," *IEEE Transactions on Communications*, Vol. 65, No. 6, 2017, pp. 2494–2508.

[166] Mostafa, A., and L. Lampe, "Optimal and Robust Beamforming for Secure Transmission in MISO Visible-Light Communication Links," *IEEE Transactions on Signal Processing*, Vol. 64, No. 24, 2016, pp. 6501–6516.

[167] Wang, Q., Z. Wang, and L. Dai, "Multiuser MIMO-OFDM for Visible Light Communications," *IEEE Photonics Journal*, Vol. 7, No. 6, 2015, pp. 1–11.

[168] Goldsmith, A., *Wireless Communications*, Cambridge, U.K.: Cambridge University Press, 2005.

[169] Butala, P. M., H. Elgala, and T. D. C. Little, "SVD-VLC: A Novel Capacity Maximizing VLC MIMO System Architecture Under Illumination Constraints," *2013 IEEE Globecom Workshops (GC Wkshps)*, 2013, pp. 1087–1092.

[170] Lee, E. J., and V. W. S. Chan, "Part 1: Optical Communication over the Clear Turbulent Atmospheric Channel Using Diversity," *IEEE Journal on Selected Areas in Communications*, Vol. 22, No. 9, November 2004, pp. 1896–1906.

[171] Navidpour, S. M., M. Uysal, and M. Kavehrad, "BER Performance of Free-Space Optical Transmission with Spatial Diversity," *IEEE Transactions on Wireless Communications*, Vol. 6, No. 8, August 2007, pp. 2813–2819.

[172] Mesleh, R., H. Elgala, and H. Haas, "Optical Spatial Modulation," *IEEE/OSA Journal of Optical Communications and Networking*, Vol. 3, No. 3, March 2011, pp. 234–244.

[173] Qiu, L., and M. Jiang, "A Generalized Spatial Modulation for Indoor Optical Wireless Communications," *Opto-Electronics and Communications Conference (OECC)*, 2015, pp. 1–3.

[174] Joung, J., Y. K. Chia, and S. Sun, "Energy-Efficient, Large-Scale Distributed-Antenna System (LDAS) for Multiple Users," *IEEE Journal of Selected Topics in Signal Processing*, Vol. 8, No. 5, 2014, pp. 954–965.

[175] Chen, Z., D. A. Basnayaka, and H. Haas, "Space Division Multiple Access for Optical Attocell Network Using Angle Diversity Transmitters," *Journal of Lightwave Technology*, Vol. 35, No. 11, 2017, pp. 2118–2131.

[176] BS EN 62471:2008, Photobiological Safety of Lamps and Lamp Systems, BSI British Standards Std., September 2008.

[177] Tavakkolnia, I., et al., "MIMO System with Multi-Directional Receiver in Optical Wireless Communications," *IEEE International Conference on Communications Workshops (ICC Workshops)*, 2019, pp. 1–6.

[178] Gupta, A. K., and A. Chockalingam, "Performance of MIMO Modulation Schemes with Imaging Receivers in Visible Light Communication," *Journal of Lightwave Technology*, Vol. 36, No. 10, 2018, pp. 1912–1927.

[179] Omura, N., et al., "Experimental Demonstration of OFDM Based WDM-MIMO Visible Light Communication System," *2018 Asia-Pacific Microwave Conference (APMC)*, 2018, pp. 872–874.

[180] Dadabayev, R., and D. Malka, "A Visible Light RGB Wavelength Demultiplexer Based on Polycarbonate Multicore Polymer Optical Fiber," *Optics & Laser Technology*, Vol. 116, 2019, pp. 239–245. http://www.sciencedirect.com/science/article/pii/ S0030399218318851.

[181] Basnayaka, D. A., and H. Haas, "A New Degree of Freedom for Energy Efficiency of Digital Communication Systems," *IEEE Transactions on Communications*, Vol. 65, No. 7, July 2017, pp. 3023–3036.

7

Beyond the Physical Layer

7.1 Introduction

The integration of Li-Fi systems for wireless access can be implemented by two main approaches: hybrid or heterogeneous RF/Li-Fi networks, and, homogeneous optical-only Li-Fi networks. While hybrid RF/Li-Fi networks provide the combined benefits of both wireless technologies, optical-only networks can be a viable solution in scenarios where RF connectivity is restricted or not available. To fully utilize the benefits of the ample resources of heterogeneous networks, efficient resource management at the MAC layer is required to ensure that the QoS requirements of all network users are met while maximizing the overall achievable system throughput. In this chapter, we present an overview of the MAC layer control in Li-Fi systems. This includes multiaccess techniques, resource allocation, interference management, handover, and mobility support.

7.2 Integration of Li-Fi Within Heterogeneous Networks

In scenarios with high user density where the RF network users also require high data rate connectivity, the available RF resources might fail to satisfy the capacity requirements. In this case, the possibility of exploiting Li-Fi network to offload traffic from the RF APs is critical to accommodate the high traffic demand. To this end, Li-Fi can be integrated within 5G cellular networks or, in indoor situations, with Wi-Fi networks. In fact, hybrid integration of Wi-Fi and Li-Fi in indoor situations is envisioned to boost the achievable system capacity as well as the QoS for the network users [1]. Figure 7.1 shows an illustrative diagram of a heterogeneous RF/Li-Fi network with a central RF AP and multiple neighboring Li-Fi attocells. The objective of such a hybrid configuration is to

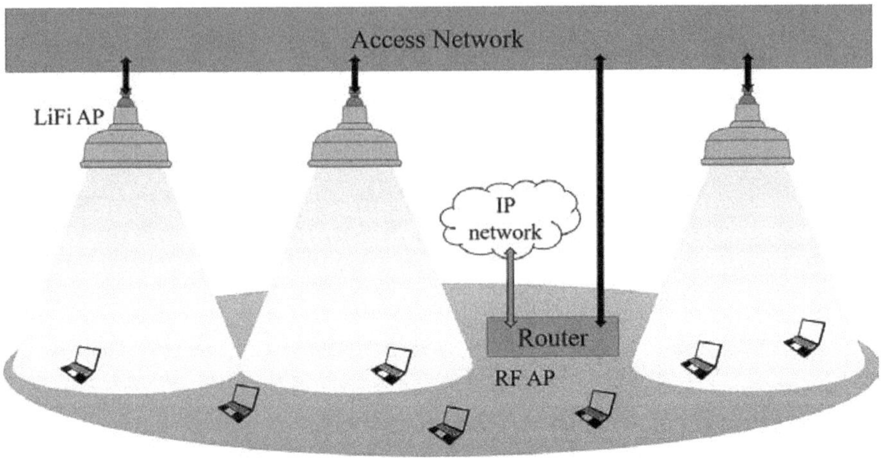

Figure 7.1 A heterogeneous network incorporating Li-Fi APs and an overshadowing RF AP.

offload some of the RF network traffic via smaller attocells in order to increase the aggregate throughput. In heterogeneous networks, each user equipment is usually associated with the AP that has the smallest coverage area, if possible, so as to reduce the resulting multiuser interference (MUI). If the user link to the Li-Fi AP becomes unreliable (i.e., due to link blockage or user mobility), the receiver device is handed over to the RF AP. The integration of Li-Fi attocells as a lower tier requires the consideration of real-time coordination and interference management among the two tiers. Since the Li-Fi APs operate in the optical spectrum (i.e., visible light for the downlink and IR for the uplink), they do not cause any intertier interference to the RF AP transmissions. It is noted that this makes the deployment of Li-Fi attocells highly advantageous compared to other small cell solutions such as femtocells, which use the licensed RF spectrum and, thus, interfere with other RF cells that use the same carrier. Intercell interference can occur between neighboring attocells with overlapping coverage areas. In fact, the severity of intercell interference depends mainly on the layout of the Li-Fi APs (i.e., the distribution of the LED light fixtures), which are mainly governed by the building infrastructure design. For this reason, interference between the Li-Fi attocells can be evaluated at the local level for each system configuration and appropriate interference management techniques can be implemented to mitigate or reduce the occurring intercell interference.

A generic heterogeneous two-tier network typically consists of an RF AP, a Li-Fi AP, and a mobile user device. There exist three possible logical topologies for the link configuration across the two tiers in such heterogeneous networks [1]. Figure 7.2 shows an illustrative diagram of the three topologies, namely: (1) symmetric links topology, (2) noninterfering asymmetric links topology, and (3) interfering asymmetric links topology. In symmetric configurations, as illustrated

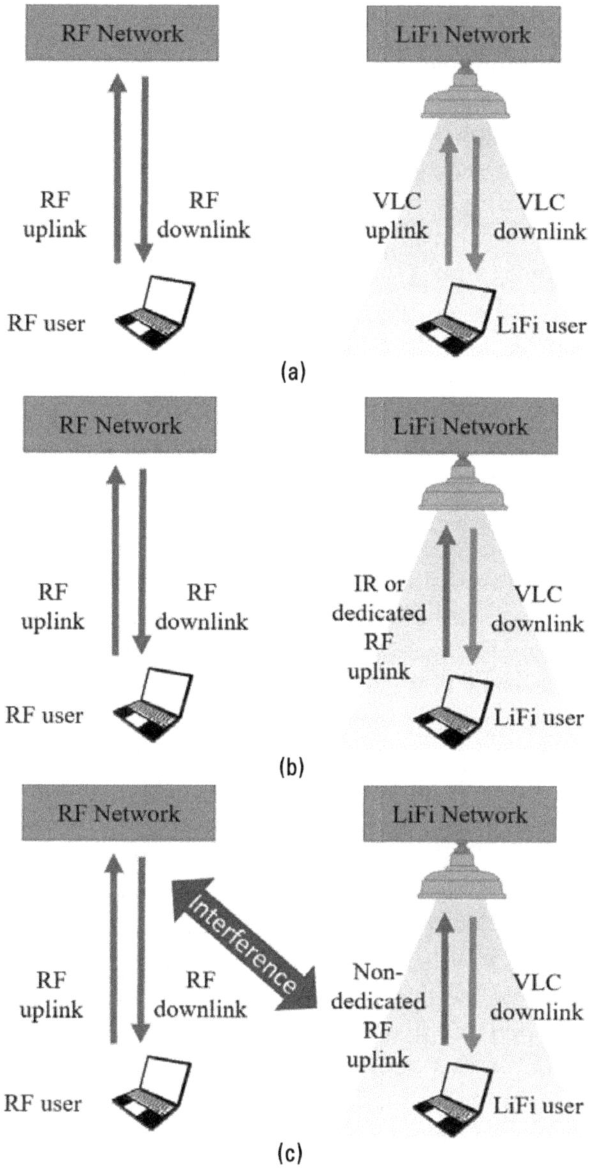

Figure 7.2 Heterogeneous Li-Fi/RF network topologies with (a) symmetric link, (b) asymmetric noninterfering link, and (c) asymmetric link with interference.

in Figure 7.2(a), visible light communication (VLC) links are used for both the downlink and uplink transmissions. As a result, there is zero interference between the RF network tier and the Li-Fi network tier. In this case, the same visible light transceiver component on the user equipment is utilized for transmission and reception. In order to avoid interference between the downlink and uplink transmissions in symmetric configurations, a duplex mechanism is needed to allow for simultaneous communications in both directions. This can be achieved by means of frequency division duplex (FDD), in which the transmitter and receiver operate at different visible light frequencies, or time division duplex (TDD), in which the transmitter and receiver are assigned alternative time slots.

In a noninterfering asymmetric configuration, illustrated in Figure 7.2(b), the downlink is implemented by a VLC transmission while the uplink is realized by a non-VLC optical link, typically IR. Since both the downlink and uplink operate in the optical spectrum, no interference exists between the Li-Fi and the RF transmissions. This also can be referred to as wavelength division duplex (WDD). Alternatively, a dedicated RF bandwidth could be allocated to the uplink transmission for the small cell users (i.e., Li-Fi users). Also, in this case, there is no interference between the two tiers. However, reserving resources for the small-cell uplink transmission may lead to a reduction in the overall achievable capacity of the RF network. The last configuration illustrated in Figure 7.2(c) is the interfering asymmetric topology, in which the Li-Fi user uplink transmission utilizes the same transmission bandwidth as the RF users. This means that there will be interference between the uplink transmission of Li-Fi users and the transmission of Wi-Fi users. To avoid interference, transmissions can be coordinated by means of TDD or FDD.

The provisioning of the available resources among the neighboring Li-Fi attocells and the RF cells in a heterogeneous two-tier network needs to be dynamically adapted according to the system variables. These variables include user density, user mobility, and user traffic and QoS demands. To this end, traffic distribution and load-balancing techniques need to be implemented at the network layer, in order to offer the flexibility to accommodate the varying network conditions and obtain an optimal distribution of the users load across the Li-Fi and the RF APs. The AP selection in a multitier heterogeneous network depends on user mobility as well as the link conditions. For example, a user moving at a high speed may experience frequent handovers between the attocells due to the small coverage area. In this case, the user is better suited for connection to the RF AP. Besides user mobility, random receiver orientation and LOS blockage are other factors that may cause sudden variations in the optical channel gain and, as a result, require the user to be connected to another Li-Fi AP or to the RF network. Consequently, these probabilistic factors need to be considered when making a decision on AP selection in heterogeneous networks. Typically, heterogeneous networks deploy a hypervisor to manage the different APs in the network. This is essential to realize efficient and adaptable load balancing, resource allocation, and mobility management. To this end, software-defined networks (SDN) can provide a platform to decouple the control plane from the data plane and hence create a practical approach to enable the coexistence of different mobile wireless technologies. An application MAC cross-layer SDN network was proposed in

[2] for the purpose of flow control in a Li-Fi/Wi-Fi heterogeneous network. The proposed network architecture employs an SDN-enabled switch that connects the different APs. Multiple SDN agents are responsible for gathering information about the current network key performance indicators (KPIs). Based on the collected information, flow control decisions are made at the SDN controller and are then executed by the SDN switches.

7.3 Multiaccess Control

As a technology enabler for 5G networks and beyond, Li-Fi is envisioned to provide ubiquitous multiuser connectivity. To enable this, the available wireless channel resources (i.e., frequency, time, power) etc need to be shared among the different network users. Conventional multiple access schemes employed in RF communications can be adapted for Li-Fi networks as long as the necessary modifications related to the restrictions of IM and DD are performed. Various optical (orthogonal and nonorthogonal) multiple access schemes have been proposed to facilitate Li-Fi multiuser access in the open literature. In orthogonal multiple access (OMA) schemes, different users are allocated orthogonal resources in either the frequency or time domain. For example, optical OFDM assigns different frequency subcarriers to different users, whereas time division multiple access (TDMA) allows users to share the same bandwidth resources by accessing the network in rapid succession during their assigned time slots. In NOMA, however, each network user can exploit the entire bandwidth for the whole time by utilizing other DOFs such as code sequences, power levels, and spatial dimensions. Moreover, Li-Fi networks offer an additional dimension for facilitating multiple access based on multiplexing the wavelength of the light signals, which can be utilized by means of the wavelength division multiple access (WDMA) technique.

In multiple access systems, it is important to maximize the system utility (in terms of efficiency, reliability, and data rates), while also maintaining fairness among the different network users. Managing the trade-off between performance and fairness depends on the specific requirements of the network users, which are basically the customers of the wireless network operators. For example, some users may require high data rates to access high-definition video streaming, while others may be idle for some time. Another factor that plays a role in the access control decision-making is the optical channel conditions for the different users. For example, users that are closer to the Li-Fi APs would usually have more favorable channel conditions compared to users existing at the edge of the attocells. Other factors such as link blockage and random receiver orientation could also affect the channel conditions of the users. Hence, these probabilistic factors need to be taken into consideration when deciding on the adopted multiple access strategy. Multiple access control also involves assessing the complexity of the used multiple access schemes, which depends on time synchronization, ease of implementation, and receiver processing. In order to reach the best possible performance, the Li-Fi AP is envisioned to dynamically monitor the users and system dynamics and requirements so as to decide on the best possible access strategy [3]. Based on the chosen multiple access scheme, logical channels are then created for each network user in the time, frequency, power, space or code domain. Resource allocation can then be optimized to reach the required KPIs of each system.

In the following sections, we discuss the main multiple access schemes that have been proposed to allow multiuser access in Li-Fi networks.

7.3.1 Time Division Multiple Access (TDMA)

TDMA is one of the simplest multiple access schemes. It is based on allowing different users to access the network in different time slots. Users transmit their data signals with rapid succession, and each user can utilize the full bandwidth and power resources in its allocated time intervals [4]. This implies that no interference exists between the signals of different users, because they access the communication link in different time intervals. However, it results in low spectral efficiency compared to other multiple access schemes. Another drawback of TDMA systems is the need for strict time synchronization between the network users and the serving APs. Ensuring such strict synchronization is considered particularly challenging for the uplink channel. This is due to the inherently different signal propagation times and transmission delays of the randomly located users' devices. Thus, there is a need to compensate for such delays by means of advanced timing synchronization [5].

7.3.2 Optical Orthogonal Frequency Division Multiple Access (O-OFDMA)

Optical OFDMA, is a multiple access scheme that is based on the popular OFDMA digital modulation technique. In optical OFDM, distinct subsets of the available subcarriers are assigned to different users, allowing multiple simultaneous transmissions by sharing the available resources in the optical bandwidth. Similar to OFDM, optical OFDMA signals need to satisfy the restrictions imposed by intensity modulation and direct detection (i.e., the modulated signals are required to be real and positive). Hence, optical OFDMA transmissions in Li-Fi systems are typically based on one of the modified OFDM techniques that are tailored to the optical channel requirements such as DC biased optical OFDM (DCO-OFDM), asymmetrically clipped optical OFDM (ACO-OFDM), asymmetrically DC-biased optical OFDM (ADO-OFDM), and enhanced unipolar OFDM (eU-OFDM). A representative block diagram of a DCO-OFDM-based optical OFDMA transceiver is shown in Figure 7.3. The parallel bit streams, denoted by $\{s_n\}_{n=1}^{N_{\mathrm{FFT}}/2-1}$, are mapped onto quadrature amplitude modulation (QAM) symbols to generate the complex valued signals $\{\tilde{X}_n\}_{n=1}^{N_{\mathrm{FFT}}/2-1}$. As discussed earlier, only a total of $(N_{\mathrm{FFT}} - 2)/2$

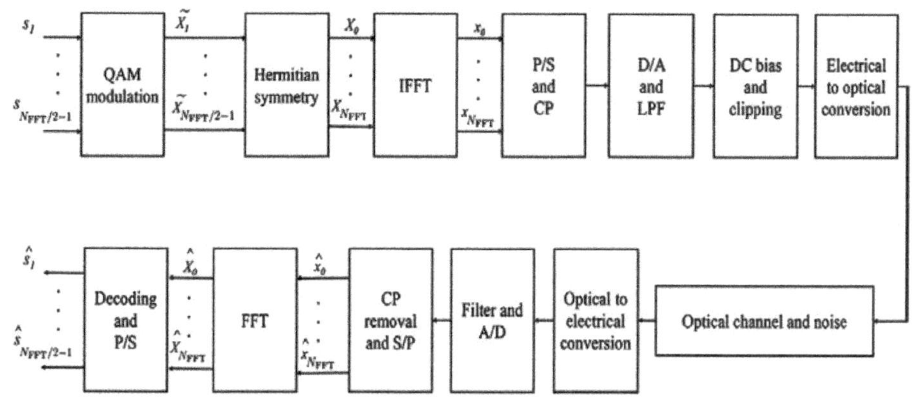

Figure 7.3 A representative block diagram of a DCO-OFDM-based optical OFDM-transceiver.

effective subcarriers out of the N_{FFT} available subcarriers can be used for data transmission so as to realize a real-valued signal. To do so, the OFDM frame is obtained by expanding $\{\tilde{X}_n\}_{n=1}^{N_{\text{FFT}}/2-1}$ to $\{X_n\}_{n=0}^{N_{\text{FFT}}}$ according to the Hermitian symmetry mapping (i.e, $X_n = \tilde{X}_n$ for $n = 1, 2, \ldots, \frac{N_{\text{FFT}}}{2} - 1$, $X_n = \tilde{X}_n^*$ for $n = \frac{N_{\text{FFT}}}{2} + 1, \frac{N_{\text{FFT}}}{2} + 2, \ldots, N_{\text{FFT}}$, and $X_0 = X_{\frac{N_{\text{FFT}}}{2}} = 0$). For $\mathcal{M} = 1, .., M$ network users, we denote the subcarriers set allocated to user m by $\mathcal{K}_m$. Since each effective subcarrier is allocated to be used by a single user only, we have $\mathcal{K}_m \cap \mathcal{K}_{m'} = \emptyset$ for $m \neq m'$. After performing IFFT, the time-domain signal $\{x_n\}_{n=1}^{N_{\text{FFT}}/2-1}$ is obtained.

It is acceptable to consider the output OFDM signal to be Gaussian according to the central limit theorem. Thus, we can say that the signal $\{x_n\}_{n=1}^{N_{\text{FFT}}/2-1}$ follows a Gaussian distribution with zero mean and variance power p_x, where p_x is the average of all effective subcarriers transmit power (i.e., $p_x = \frac{1}{N_{\text{FFT}}/2-1} \sum_{m \in \mathcal{M}} \sum_{\mathcal{K}_m} p_{k,m}$, with $p_{k,m}$ being the allocated power to subcarrier k for user m). The time-domain signal x_n is clipped at an appropriate level and DC bias is added to obtain the unipolar signal. The resulting digital signal passes through the DAC to produce an analog signal that drives the power amplifier and then the LED. At the receiver of user m, the received signal $\hat{x}_m$ observed at the PD generates a current that is proportional to the received optical power intensity. The generated signal is then passed through an ADC and then fed into discrete Fourier transform (DFT) to generate the frequency-domain signal $\hat{X}_{k,m}$, which is then demodulated to recover the QAM symbols.

The performance of optical OFDM systems depends on the way the resources are shared among the different users. This includes the joint optimization of the DC bias, bandwidth subcarriers, and power levels. Due to the nonlinear clipping distortion, the OFDM subcarriers are no longer strictly orthogonal. As a result, the subcarrier dependency and the adequate bias levels make the joint optimization problem rather challenging. Another important factor that affects the resource allocation in optical OFDMA-based Li-Fi systems is the illumination dimming requirement. This is because, in order to achieve the desired dimming level, certain additional constraints on the power allocation are put in place. Hence, it is evident that all the available system resources need to be jointly well assigned to achieve the performance requirements under the optical power constraints. An optical OFDMA resource allocation problem was considered in [6] with the aim of maximizing the weighted sum rate of the system. This optimization problem can be formulated as:

$$\text{maximize} \sum_{m \in \mathcal{M}} w_m \sum_{k \in \mathcal{K}_m} r_{k,m} \tag{7.1a}$$

$$\text{subject to } p_o \leq I_{\text{req}} \tag{7.1b}$$

$$p_s = \sum_{m \in \mathcal{M}} \sum_{k \in \mathcal{K}_m} P_{k,m}/K \tag{7.1c}$$

$$P_{k,m} \geq 0, \forall\ k, m \tag{7.1d}$$

$$\mathcal{K}_m \cap \mathcal{K}_{m'} = \emptyset, \forall\ m, m' \tag{7.1e}$$

$$\cup_{m \in \mathcal{M}} \mathcal{K}_m \subseteq \mathcal{K} \tag{7.1f}$$

In the above equation, constraint (7.1a) aims to maximize the weighted sum rate where w_m is the weight assigned to user m and $r_{k,m}$ is the data rate of user m over subcarrier k. Constraint (7.1b) implies that the average optical power p_o should be less than or equal to the required maximum luminous intensity, constraint (7.1c) implies that the time-domain power is the average of all the activated subcarriers' power, (7.1d) denotes that the subcarriers' power is unipolar, and (7.1e) and (7.1f) characterize the orthogonality between the network users. It is worth noting that this joint optimization problem was proved to be NP-hard and challenging to solve [7]. As a result, several joint resource allocation algorithms have been proposed in the related literature in order to satisfy the various complexity and accuracy requirements of optical OFDMA systems [8–11].

7.3.3 Optical Code Division Multiple Access (O-CDMA)

O-CDMA is a nonorthogonal multiple access scheme that can provide high spectral efficiency compared to orthogonal multiple access schemes such as TDMA and optical OFDMA. O-CDMA allows simultaneous transmissions to multiple network users by using signature coding sequences. The advantages of O-CDMA include supporting multiple bursty transmissions with minimal delay, easy synchronization, and inherent high levels of security. Moreover, O-CDMA

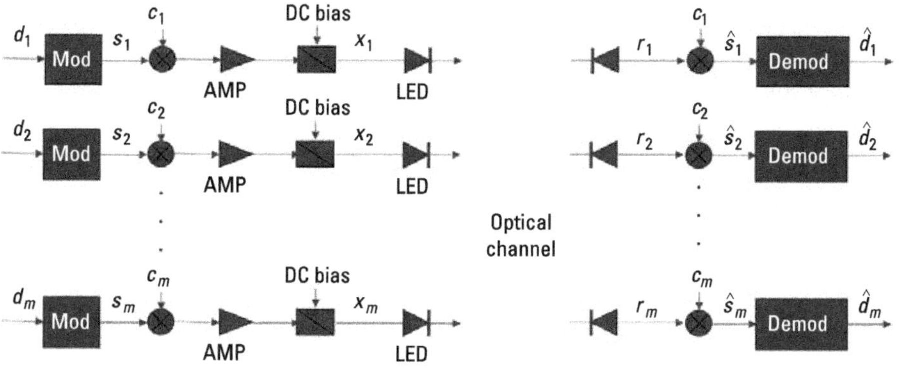

Figure 7.4 A representative block diagram of an O-CDMA transceiver.

systems require low complexity network control, due to the ease of generation of the involved optical signature codes. The performance of O-CDMA systems is highly reliant on the used optical sequences. Bipolar codes, usually used in RF-based CDMA systems, have the advantage of good correlation characteristics. This makes the separation of the signals of different users easy and effective. Such bipolar codes, however, cannot be used directly in Li-Fi systems due to the restrictions of IM and DD. One solution to tackle this problem is to modify the bipolar code sequences by replacing -1 with 0 to ensure that the signal is unipolar. This solution, however, is not always feasible because it could destroy the original orthogonality between the original codes, which makes it extremely difficult to mitigate multiuser interference and extract the desired signals at the users' terminals. Another solution to enable the use O-CDMA in optical systems is to use unipolar optical codes instead of bipolar sequences. Examples of these include random orthogonal codes (ROCs) [12], optical orthogonal codes (OOCs) [13], and prime codes (PCs) [14].

Figure 7.4 shows a conceptional block diagram for an O-CDMA system with unipolar codes [15]. At the transmitter side, the data streams $d_1, ..., d_N$ are modulated to generate $s_1, ..., s_N$ and then multiplied by the optical signature codes, $c_1, ..., c_N$. The signals are then amplified and a DC bias is added to generate the unipolar signals, $x_1, ..., x_N$, which are used to drive the LEDs' current. After propagating through the optical channel, the optical signals are detected at the receivers' PDs to generate the signals $\hat{x}_1, ..., \hat{x}_N$, which are then multiplied with the same set of signature codes to produce $\hat{s}_1, ..., \hat{s}_N$. The date sequences $\hat{d}_1, ..., \hat{d}_N$ are then recovered after demodulation. The choice of unipolar codes in O-CDMA systems depends on several factors such as the channel model, the transceiver structures, and the used modulation schemes. Typically, it is desired for optical unipolar codes to have low auto-correlation in order to facilitate frame synchronization as well as low cross-correlation in order to reduce the interference between the different code sequences.

7.3.3.1 O-CDMA with Unipolar Codes

In the following, we introduce the most commonly used types of unipolar codes in O-CDMA Li-Fi systems:

Random Optical Codes: ROCs are suited for scenarios with a large number of users sharing a common optical channel and relatively low data-rate requirements. Such requirements can be seen in wireless sensor-based networks. In an O-CDMA system with N-length ROCs, each symbol duration is divided into N slots and only $w < N$ slots out of the total N slots are actually utilized to transmit nonzero signals, where w denotes the code weight. The arrangement of the active w positions within the total available N slots is chosen and changed based on a pseudorandom sequence. This can be achieved by the use of a pseudorandom number generator in the transmitter and the receiver or, alternatively, by using a randomly built-in, predefined look-up table [16]. In order to increase the number of unique code sequences, ROCs can be modified by adding a cyclic shift, where distinct shifts are used for different transmitters. The user connected to each transmitter will then need to use a shift register in order to match-filter the received signals and recover the original data sequences.

Optical Orthogonal Codes: OOCs are basically binary sequences designed with certain auto-correlation and cross-correlation properties. Let $\mathcal{C} = (n, w, \lambda_a, \lambda_c)$ be a family of OOCs of length n, constant Hamming weight w, and auto-correlation λ_a. For any codewords $\mathbf{x} = (x_0, x_1, ..., x_{n-1}) \in \mathcal{C}$ and $\mathbf{y} = (y_0, y_1, ..., y_{n-1}) \in \mathcal{C}$, the following properties are correct: (1) $\sum_{i=0}^{n-1} x_i x_{i\oplus\tau} \leq \lambda_a$, and (2) $\sum_{i=0}^{n-1} x_i y_{i\oplus\tau} \leq \lambda_c$ hold for any integer τ, where $\oplus$ denotes the the modulo-n addition operator. For the case when $\lambda_a = \lambda_c = \lambda$, the OOC can be denoted as $\mathcal{C} = (n, w, \lambda)$ for simplicity. The binary sequences of an OOC are called its codewords. The size of an OOC family is the number of possible codewords in it. The number of generated codewords in the OOC determines the number of possible simultaneous channel reuses. It follows from the Johnson bound for constant-weight codes that the upper bound on the size of an OOC, $\mathcal{C} = (n, w, \lambda_a, \lambda)$, is given by [17]:

$$(n, w, \lambda) \leq \left\lfloor \frac{1}{w} \left\lfloor \frac{n-1}{w-1} \left\lfloor \frac{n-2}{w-2} \left\lfloor \cdots \left\lfloor \frac{n-\lambda+1}{w-\lambda+1} \left\lfloor \frac{n-\lambda}{w-\lambda} \right\rfloor \right\rfloor \cdots \right\rfloor \right\rfloor \right\rfloor \right\rfloor \tag{7.2}$$

where $\lfloor \ \rfloor$ is the integer floor function. For example, let $\mathcal{C} = (13, 3, 1)$ be an OOC; the optimal size can be obtained as $\Phi(13, 3, 1) \leq \left\lfloor \frac{1}{3} \left\lfloor \frac{13-1}{3-1} \right\rfloor \right\rfloor = 2$, which means that a maximum of two codewords can be generated for this OOC family. In general, there are two main methods for generating an OOC sequence. The first method depends on mathematical techniques such as design theory and projective geometry, while the second method utilizes computer-based search techniques such as greedy algorithms [18]. The reader may refer to [19–22] for in-depth explanation of various OOCs' construction methods.

Prime Codes: PCs offer the advantages of easy codeword generation and relatively low-complexity encoding and decoding. However, the correlation characteristics functions of the resulting code sequences are not optimal due to high auto-correlation. This reduces the feasibility of perfect synchronization between the transmitters and the receivers in O-CDMA systems based on PCs. Thus, PCs are mainly utilized in synchronous O-CDMA systems. PCs are generated based on the Galois field GF(p) of a prime number p. A prime sequence $S, = (s_{i,0}, s_{i,1}, ..., s_{i,j}, ..., s_{i,p-1})$ is then constructed where $s_{i,j} = i.j(\text{mod } p)$ and $i, j \in \text{GF}(p)$. Each generates a unique PC sequence; S_i is then mapped into an N-length binary sequence $C_i = (c_{i,0}, c_{i,1}, ..., c_{i,k}, ..., s_{i,N-1})$. Hence, a total of p unique codeword sequences are generated, each of them of weight p and length p^2, with cross-correlation values not greater than 2 [23]. It was been reported in [14] that synchronized PCs can be generated to extended beyond prime fields based on a Galois field GF(p^m) for a prime number p and a positive integer m. As a result, a total of p^{2m} unique codewords are generated with length of p^{2m} and weight of p^m.

7.3.4 Optical Space Division Multiple Access (O-SDMA)

O-SDMA is a multiple access technique that utilizes the spatial dimension to allow multiple simultaneous transmissions. To this end, angular diversity transmitters are employed at the Li-Fi AP to allow the transmission of multinarrowband optical beams to spatially separated users. In order to avoid excessive multiuser interference in such systems, the spatial transmission profile of both the Li-Fi AP and the users' PDs need to be carefully controlled. In O-SDMA, network users are grouped into small groups, each of which is allocated different time slots. Within a single time slot, the signals of the users of one group are transmitted using the same time, power, and frequency resources but in spatially distinct optical beams. As a result, O-SDMA has the potential to provide significant throughput enhancement in Li-Fi systems. In order for the AP to support multiple parallel and directional transmissions, multielement transmitters are required. Moreover, the user grouping strategy is critical as it determines the amount of multiuser interference affecting the users. In the following, we briefly discuss angle diversity transmitters as well as spatial grouping strategies in O-SDMA systems.

Angle Diversity Transmitters: In SDMA-based RF communication systems, the AP is typically equipped with special antenna arrays that could generate multiple narrow beams simultaneously. To achieve the same function in Li-Fi systems, an angle diversity transmitter is required. Figure 7.5 shows the layout of an angular diversity transmitter consisting of 18 narrow beam LEDs fixed on a semi-sphere structure [24]. The first LED element is mounted at the center of the semi-sphere and the rest are arranged symmetrically in rings around the center. This transmitter structure allows the optical channel to be divided into a total

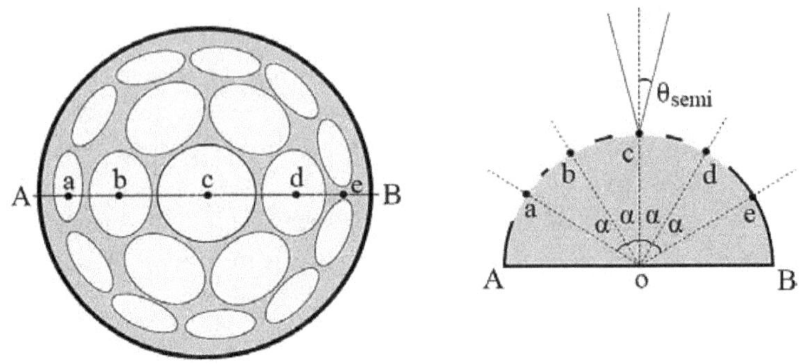

Figure 7.5 Layout of an angle diversity transmitter with 18 LED elements and angular spacing $\alpha = 18°$ between each neighboring branches of LEDs [24].

of 18 distinct narrow-beam channels, and thus allowing significant gain in the achievable spectral efficiency. It is noted that the availability of such narrow-beam LEDs make achieving spatial multiplexing in Li-Fi systems much less complicated compared to its RF equivalent since the latter requires multiple RF chains and complex beam-steering mechanisms.

Spatial Users' Grouping: The success of O-SDMA based access control is highly dependent on implementing an efficient users' grouping mechanism. Since each LED element in an angular diversity transmitter is supposed to cover a specific small area, the existence of more than one active user in the coverage area of a single LED beam necessities the use of other multiple access schemes in conjunction with O-SDMA. For example, users within the same spatial area can be multiplexed by the means of TDMA or optical OFDM. Let G_j be a spatial group of users in the ith narrow coverage area, and $\mathcal{P}_l = \{G_1, G_2, G_3, ... G_J\}$ be the set of all spatial users groups. The signal to interference and noise ratio (SINR) at user U_i in the spatial group G_j can be calculated as:

$$\text{SINR}(U_i, G_j) = \frac{(\rho P_{\text{LED}} h_{ij})^2}{\sum_{j'} (\rho P_{\text{LED}} h_{ij'})^2 + \sigma_i^2}$$

where ρ donates the optical to electric conversion efficiency, P_{LED} is the transmit optical power, h_{ij} is the channel gain between LED element serving the spatial group G_j and user U_i, σ_i^2 is the receiver noise variance, and the term $\sum_{j'} (\rho P_{\text{LED}} h_{ij'})^2$ donates the interference from other LED elements. Assuming that multiple users within a single spatial group G_j for $j = 1, 2, ..., J$ are multiplexed in the time domain, the overall system throughput can be calculated as:

$$T_{\text{SDMA}} = \frac{1}{J} \sum_{j=1}^{J} \sum_{U_i \in G_j} B \log_2(1 + \text{SINR}(U_i, G_j))$$

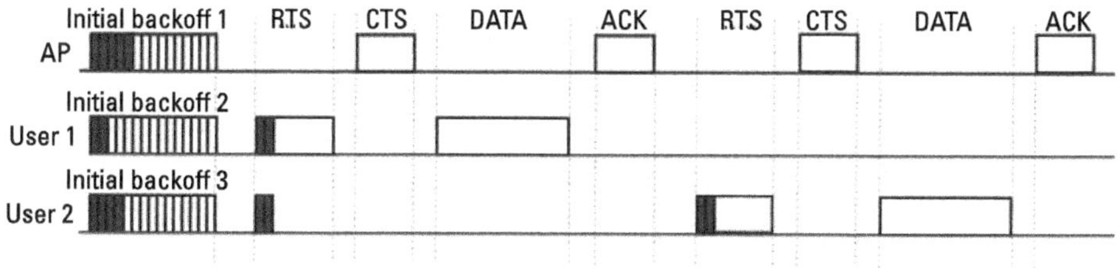

Figure 7.6 CSMA/CA with four-way handshaking RTS/CTS mechanism [26].

Hence, it is evident that the decision on the users' grouping criteria determines the achievable performance gain obtained by O-SDMA.

7.3.5 Carrier-Sense Multiple Access with Collision Avoidance (CSMA/CA)

In CSMA/CA, network users are required to sense the channel before attempting to connect to an AP and are allowed to transmit only if the channel is available. In this way, collision between the different users' signals is avoided or minimized. This process in termed as collision avoidance, and it is typically performed by means of a slotted binary exponential back-off algorithm. Once a user is allowed to access the communication link, it can utilize the entire channel bandwidth for the granted time. Typically, the user listens to the channel for a timed interval, known as distributed interframe space (DIFS). If the communication channel is found to be in an idle state, the mobile user generates random back-offs, $\mathcal{B}_i$, for $i = 1, 2, \ldots, N$, with N being the total number of network users competing to connect to the same AP. The random variables $\mathcal{B}_j$ are derived from a uniform distribution defined over the range $[0, w - 1]$, where w denotes the size of the contention window. Prior to the first attempt to transmit, the value of w is typically set to a minimum contention window size, w_{min}. Once an unsuccessful transmission attempt occurs, the contention window size, w, is doubled. This process is continued until the maximum contention window size w_{max} is reached.

The CSMA/CA protocol uses a four-way handshaking request-to-send (RTS) and clear-to-send (CTS) mechanism, as illustrated in Figure 7.6. Initially, the network user with the lowest back-off value is granted priority to access the channel, and it sends an RTS frame to the AP of interest to initiate the connection. Upon successful reception of the RTS frame, the AP replies with a CTS frame after a specific time interval, known as short interframe space (SIFS). When the user receives the CTS packet, it proceeds with its transmission after the end of the SIFS. When the AP receives the user's data packet successfully, it sends an acknowledgment (ACK) packet to notify the user of the successful delivery. This CSMA/CA protocol, which is typically used in wireless local area networks (WLANs), has been also applied for collision avoidance in Li-Fi. However, it was shown in [25] that employing conventional CSMA/CA in bidirectional Li-Fi networks could lead to a high probability of collisions. A modified CSMA/CA was proposed in [25] to address this problem. In the proposed protocol, the AP transmits a specific busy tone message to all network users in order to indicate that the communication channel is already occupied. This modified CSMA/CA was shown to reduce the probability of collision in Li-Fi systems to less than 1% [26].

7.3.6 Power-Domain Non-Orthogonal Multiple Access (NOMA)

Power-domain NOMA allows multiple users to access the network using the same bandwidth resources (i.e., subcarriers), as well as time resources (i.e., in the same time slots). Different from CDMA and O-SDMA, which allow frequency and time sharing by means of distinct optical codes and spatial beams, respectively, power-domain NOMA utilizes the power dimension. More specifically, in downlink broadcast power-domain NOMA systems, users are assigned different power levels in order to access the network simultaneously. The main idea is that users with less favorable channel conditions are allocated higher-power values and users with better channel conditions are allocated lower-power values. The signals of all users are then superimposed and transmitted as one signal. At the receiver side, each user can decode its desired signal by means of successive interference cancellation (SIC). Power-domain NOMA was originally proposed to enhance the spectral efficiency for RF systems. In RF communications, however, the application of power-domain NOMA could be challenging because of the following factors:

1. NOMA is more suited for multiplexing low numbers of users (i.e., 3 to 5 users), due to the nature of power-domain multiplexing. However, RF APs typically serve a large number of users, which results in higher error probability during the SIC process.
2. It is difficult to obtain accurate channel state information (CSI) in RF systems, which increases the error propagation in SIC.
3. Successful SIC requires high SNR values, which is not always the case in RF systems.

Power-domain NOMA in Li-Fi-based systems is less prone to these challenges due to the small cell size, the more deterministic nature of the VLC channel, and the relatively high SNR values. Furthermore, the performance of power-domain NOMA can be further enhanced by increasing the channel gain differences between the different users. This can be realized in Li-Fi systems by tuning the transmission angles of the LEDs and the FOVs of the receivers' PDs. This offers additional DOFs to optimize the performance of power-domain NOMA systems [27].

In the following sections, we introduce the main principles related to the application of power-domain NOMA in optical networks.

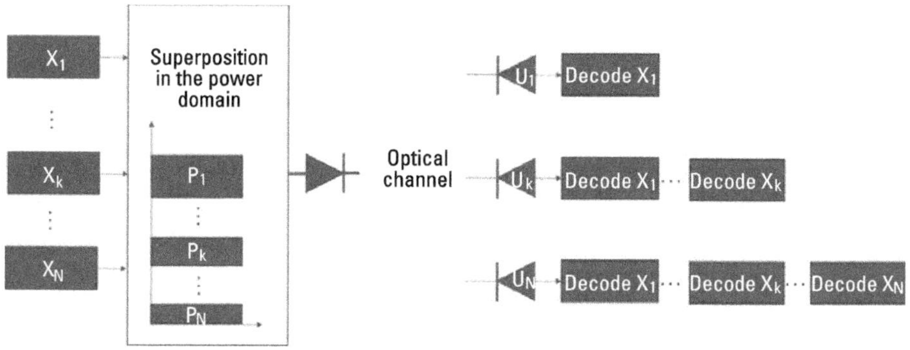

Figure 7.7 Block diagram of power-domain NOMA transceivers.

7.3.6.1 Power-Domain Superposition Coding

In power-domain superposition coding, a transmitter can simultaneously convey multiple data signals to multiple receivers by allocating distinct power values to each signal. Figure 7.7 shows an illustrative block diagram of power-domain NOMA transmission. Assuming N network users with ordered channel gain values:

$$h_1 \leq h_2 \leq .... \leq h_K ... \leq h_N$$

the power allocation will be performed such that:

$$P_1 > P_2 > ... > P_K ... > P_N$$

A total of N parallel data streams are generated, modulated, and assigned their distinct power levels. The LED then transmits the signals $s_1, s_2, ..., s_K, ..., s_N$ with associated power values $P_1, P_2, ..., P_K, ..., P_N$, where s_k denotes the signals intended for the kth sorted user. Thus, the overall transmitted signal can be expressed as:

$$x = \sum_{k=1}^{N} P_k s_k$$

such that the total power constraint is satisfied (i.e., $\sum_{k=1}^{K} P_k s_k = P_{\text{LED}}$). As a result, the signal received at each receiver terminal is the summation of all users' signals superimposed in the power domain. The received signal at the user k can be expressed as:

$$y_k = \rho h_k \sum_{k=1}^{K} P_k s_k + n_k$$

where n_k denotes the receiver noise.

7.3.6.2 Successive Interference Cancellation

In order for user k to extract its intended signal from the total received signal, SIC is performed. Based on SIC, user k decodes and subtracts the first $k-1$ data signals (i.e., $s_1, s_2, ...s_{k-1}$). Each decoding step is performed with the presence of the interference from the signals of all the users with a higher decoding order, which have lower-power values and are thus treated as noise. After successfully subtracting the first $k - 1$ signals and assuming perfect cancellation of these signals, the SINR at user k can be expressed as:

$$\text{SINR}_k = \frac{\rho P_k h_k}{\rho \sum_{j=k+1}^{K} P_j h_k + \sigma_k^2}$$

At this stage, user k can decode its own signal s_k assuming that the interference from the remaining signals with a higher decoding order (i.e., $\sum_{j=k+1}^{K} P_j s_j$), is negligible and can be treated as noise.

7.3.6.3 Power Allocation Strategies

In order to facilitate SIC at the user terminals, power-domain NOMA allocates higher power levels to users with less favorable channel conditions and vice versa. The decision of the power allocation strategy is critical to allow the desired throughput gains while maintaining fairness among users. The simplest power allocation method is fixed power allocation (FPA) [28]. In FPA, network users are sorted in ascending order according to channel gain values (i.e., $h_1 \leq h_2 \leq ... \leq h_K ... \leq h_N$), and the power assigned to the kth sorted user is calculated as:

$$P_k = \alpha P_{k-1}$$

where α is the power allocation coefficient such that $0 < \alpha < 1$. Gain ratio power allocation (GRPA) is another power allocation strategy that takes into consideration the actual channel gain values of the users and not only their order [27]. According to GRPA, the power assigned to the kth sorted user is given by:

$$P_k = \left(\frac{h_1}{h_k}\right)^k P_{k-1}$$

Figure 7.8 [29] shows the ergodic sum rate of power-domain NOMA when both FPA and GRPA strategies are employed. As shown in the figure, both power allocation strategies achieve almost similar performance when the number of multiplexed users is 2, while FPA achieves better performance when the number of multiplexed users is raised to 10. Moreover, it can be seen from the same figure that power-domain NOMA outperforms OFDMA in almost all scenarios except for GRPA with 10 users where the used power strategy does not allow successful SIC and hence the performance is hugely deteriorated.

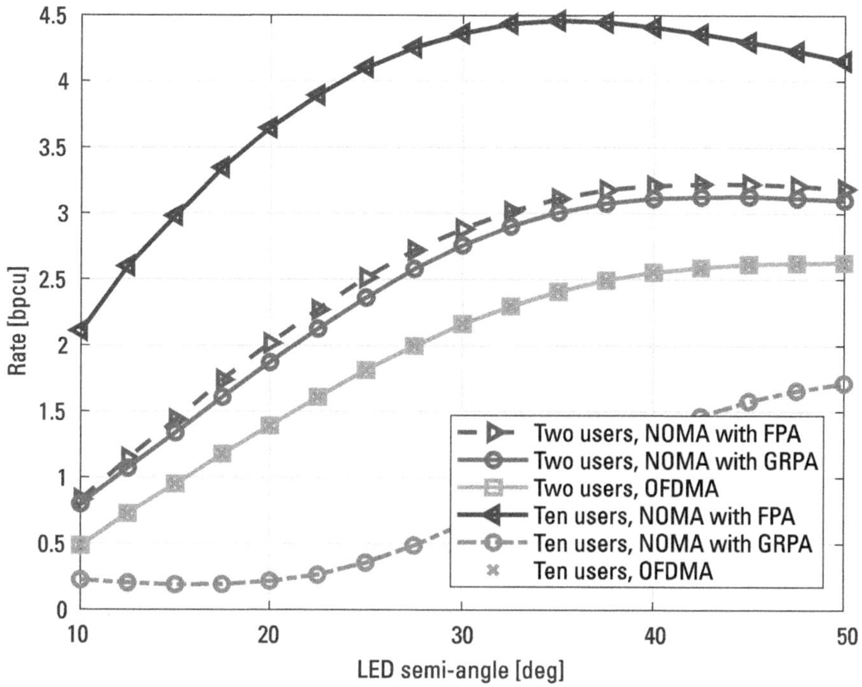

Figure 7.8 Ergodic sum rate of power-domain NOMA compared to OFDMA for different LEDs' semi-angle [29].

7.4 Resource Allocation in Li-Fi Systems

The capacity enhancement offered by Li-Fi technology is based on the spatial reuse of the optical transmission resources. To this end, Li-Fi systems are typically composed of densely deployed attocells with relatively small coverage areas. The overlapping between neighboring attocells results in intercell interference that particularly affects cell-edge users. Interference coordination in such multicell networks has the biggest impact on system performance and can be managed by employing an appropriate resource allocation strategy. Various resource allocation techniques have been widely used in conventional RF communications systems in order to distribute the available time, frequency, power, and space resources among the network users. Similar approaches have also been considered in Li-Fi networks, with the addition of the wavelength dimension to the resource partitioning capabilities.

Resource allocation can be typically performed in two different ways: static or dynamic. In static resource allocation, the decision on assigning certain resource blocks to certain users is usually predefined and maintained throughout the operation of the Li-Fi system until it is reconfigured. Dynamic resource allocation periodically reassigns the resource blocks based on the given system parameters and users' requirements. The simplest static resource allocation strategy is known as orthogonal resource partitioning. In this method, the available resources at the AP are split into multiple orthogonal resource blocks and assigned to the network users in a way that neighboring attocells always use orthogonal resource blocks. Static orthogonal resource partitioning has been applied to Li-Fi systems by splitting time resources [30], bandwidth resources [31], and wavelength resources [32]. Static orthogonal resource allocation methods offer the advantage of very low processing and configuration complexity and can effectively eliminate intercell interference. However, due to the orthogonal assignment of system resources between the adjacent attocells, each AP can only utilize a small fraction of the available system resources, which results in significant deterioration in the achievable spectral efficiency.

It is noted that, despite the simplicity and low computational complexity of static resource partitioning methods, they typically lead to underutilizing the available system resources and thus do not reach the optimal system performance. This is particularly crucial when the number of users or the QoS requirements hugely differ among the different Li-Fi APs. To this end, dynamic resource allocation [11] has been considered as a more efficient strategy. Dynamic resource allocation aims to exploit the full potential of the spatial reuse in Li-Fi networks and utilize system resources in the best possible way while maintaining fairness. For example, in [33], an optical wireless system in an aircraft cabin employed OFDMA with TDD (i.e., both time and frequency resources are split into multiple orthogonal blocks). Each network user broadcasts its signal with a fixed power level that is known network-wide, and the transmission is carried out in short time slots referred to as busy bursts. Based on these busy bursts, network users can estimate the amount of interference that they would experience and therefore can employ appropriate transmission mechanisms accordingly in a dynamic manner. It has been shown that such dynamic resource allocation can lead to significant spectral efficiency enhancements while improving fairness among network users when compared to static resource allocation. It is noted, however, that dynamic resource allocation requires the AP to acquire updated and accurate CSI for all network users, which adds to the complexity and signaling overhead.

The dynamic decision-making can extend beyond resource partitioning to also adapt the multiple access strategy based on the system requirements. Since finding an optimal solution for the multiple access selection and resource allocation problem can be difficult and computationally expensive, effective suboptimal strategies can usually be implemented. Based on this, the Li-Fi APs can dynamically assess the instantaneous demand in the network and adapt appropriate scheduling and transmission strategies. The work in [3] proposed a dynamic multiple access selection (DMAS) method to enable the Li-Fi APs to configure the adopted multiple access strategy to cater for different systems attributes such as sum data rate, average outage probability as well as fairness. In the proposed setup, the Li-Fi AP chooses between three multiple-access modes, namely: OMA, NOMA, and hybrid OMA/NOMA. Suboptimal decision-making is performed based on the technique for order of preference by similarity to ideal solution (TOPSIS) where each design attribute is given a weight, based on the current system QoS requirements. Multicell transmission coordination is another promising approach for mitigating interference affecting cell-edge users. Cell-edge users can be allowed to be associated with multiple access points so as to strengthen the desired signal strength and combat co-channel interference (CCI) interference. Multicell transmission coordination is also a viable solution for link disruptions caused by LOS blockage. If the LOS path from the user to its associated AP is blocked, it can stay connected by utilizing the link to its neighboring APs. These advancements come with the requirement of centralized coordination among the APs, which adds to the computational complexity and the associated delays [34].

While the previously discussed resource allocation and interference mitigation techniques are mainly borrowed from RF communication systems, there are unique approaches specifically tailored for Li-Fi networks. For example, user terminals can employ multiple PDs with small FOVs to serve as angular diversity receivers. In this way, the desired data signal and the interference signals from other Li-Fi APs are received by different PDs and thus the CCI can be eliminated [35]. A similar concept can be employed at the AP side by means of angular diversity transmitters. Due to the narrow beam of the multiple transmitting LEDs, the CCI is confined to a very limited area. As a result, considerable spectral efficiency gains can be realized. It is noted, however, that angular diversity-based interference mitigation in Li-Fi systems usually comes with increased hardware and computational complexity.

7.5 Interference Management in Li-Fi

In order to achieve the spectral efficiency gains offered by Li-Fi systems, it is crucial to put in place suitable interference management techniques. Since Li-Fi and RF systems operate in different spectra, no interference exists between the two different tiers in heterogeneous networks. However, there is still a need to implement interference management techniques within the Li-Fi network in order to coordinate the transmission links of multiple transmitting APs and/or multiple receiving terminals. In this section, we summarize the main interference management techniques used in Li-Fi systems.

7.5.1 Interference Cancellation Techniques

Interference cancellation can be defined as the processes or mechanisms that enable a receiving terminal to utilize its received signal together with the available

CSI in order to successfully and completely eliminate any received interference signals. After canceling the interference, the receiver terminal can then apply signal detection methods to successfully decode its desired information signal [36]. In the following, we summarize the Li-Fi interference cancellation techniques studied in the related literature.

7.5.1.1 Transmit Precoding

Various transmit precoding techniques have been proposed to mitigate the interference resulting from the simultaneous transmission of parallel data streams in MIMO Li-Fi configurations. The basic idea of transmitting precoding is to artificially create orthogonality among the MIMO subchannels. This can be achieved by multiplying the information signal with a predetermined precoding matrix. The precoding matrix is designed in a way that causes the transmitted data streams to be completely or partially hidden from the non-intended users. In general, transmit precoding techniques can be categorized into multiuser detection (MUD) and coordinated multipoint (CoMP). In MUD, the aim is to decouple the channels of different users, without canceling the interchannel interference (ICI) among the multiple PDs of a single user. As a result, the multiuser MIMO channel is decoupled into parallel single-user MIMO channels. Extra interference management techniques are then required to handle the interference between the multiple channels of a single user. Block diagonalization (BD) is an example of MUD, which was been studied and evaluated in [37].

The precoding matrix in CoMP techniques is designed to eliminate ICI between the channels of different users as well as between the multiple channels of a single user. The simplest CoMP transmit precoding technique is zero forcing (ZF), also known as channel inversion. In ZF, ICI is eliminated by multiplying the transmitted signal with a precoding matrix that is equivalent to the inverse of the downlink channel (i.e., $W_{\mathrm{ZF}} = H^{-1}$). The process of ZF results in noise amplification at the receiver's PD, leading to lower received SNR. This is because multiplying the channel matrix with its inverse significantly reduces the perceived channel gain and the received signal power at the receivers' side. Dirty paper coding (DPC) is a more computationally complex precoding technique in which the interference caused by the first $k-1$ users is eliminated during the precoding of the kth user's signal. This noncausal interference cancellation can be realized by multiplying the transmitted signal by a lower triangular precoding matrix, $W_{\mathrm{DPC}} = Q^{-1}$, where Q is an orthogonal matrix obtained by applying LQ decomposition to the downlink channel (i.e., $H = LQ$ [38]).

Transmit precoding constitutes a cost-effective solution for interference mitigation because all the required processing is performed at the transmitter side. This means that fewer computational resources are needed at the receiving terminals, which is critical for the case of low-power, battery-operated devices. Nevertheless, the implementation of transmit precoding schemes in Li-Fi systems can be particularly challenging due to the following reasons:

1. Conventional transmit precoding techniques cannot be directly applied in Li-Fi due to the restrictions of IM/DD. Thus, the design of the precoding matrix needs to be adjusted to ensure that the resulting precoded signal is real-valued and unipolar. This is usually achieved by adding a DC bias vector, which might lead to signal distortion if the DC-biased signal is out of the linear range of the transmitting LEDs.
2. The design of the precoding matrix relies on the existence of accurate CSI for all network users. This becomes more critical for mobile users because the rapid changes in the user location and/or device orientation results in fast-varying channels. Also, the possibility of LOS blockage needs to be considered in order to realize a practical and robust precoding design in Li-Fi systems.
3. Transmit precoding requires low correlation among the users' channels. This is not always possible in Li-Fi due to the symmetry of the optical channel. For example, two users that are located at different locations with the same separating distance from the AP are very likely to experience identical or very similar channel gains. Having similar channel gain values makes it extremely difficult to design a precoding matrix that nulls the undesired signals at the receiver without reducing the received strength of the desired signal. As a result, transmit precoding might lead to a significant reduction of the achievable data rates in Li-Fi MIMO configurations.
4. Transmit precoding requires coordination among different MIMO transmitters, which necessitates the deployment of high-rate backhaul solutions. This can be achieved by implementing power line communication (PLC) or dedicated optic-fiber backhaul links, adding to the complexity and cost needed for MIMO systems.

7.5.1.2 Blind Interference Alignment (BIA)

Interference alignment aims to ensure that, at each receiver, all interference is contained in a signal subspace with the smallest possible dimension. It is then possible to cancel the effect of interference by projecting the received signal onto the orthogonal subspace of the subspace containing the interference.

The idea of interference alignment is based on containing the interference signals in a subspace with the smallest possible dimension. The effect of interference can then be mitigated by projecting the desired signal onto a subspace that is orthogonal to the interference subspace. This process requires the availability of perfect CSI at the transmitter side, which is not a realistic assumption in practical implementations. BIA can be implemented in scenarios where the acquisition of perfect CSI at the transmitter is not feasible. Assuming that the receiver has perfect knowledge of the MIMO system CSI, the interference can be mitigated by reconfiguring the receivers' devices. BIA was initially proposed for RF-based MIMO systems, where the receivers' antennas can be dynamically configured. In BIA, users' devices are equipped with reconfigurable receiving antennas that can switch between a set of predefined reception modes. The receiver chooses the antenna mode that minimizes the received interference and provides a linear-independent channel response. The decision on the radiation mode is made based on exploiting a predefined pattern of channel correlations, denoted as a super symbol. Based on this, an additional degree of freedom can be obtained in order to enable parallel data transmissions for multiple network users. It is noted that the implementation of BIA requires a high channel coherence time in order to consider the channel conditions to be constant over the transmission time of the super symbol. For this reason, it proves to be best adapted to scenarios with slow-fading channel characteristics. The concept of BIA has been recently applied to Li-Fi-based MIMO systems. In [39], BIA was realized by equipping each receiver device with a single PD and multiple reconfigurable optical filters. In the proposed scheme, the AP is not required to know the CSI. Instead, the receiver device acquires the CSI of the MIMO channels and exploits this knowledge to choose among different receiver modes in order to avoid or reduce interference. It was shown in [40] that BIA provides higher spectral efficiency compared to transmit precoding. BIA was proposed for interference mitigation in a hybrid Li-Fi/RF network setup in [41]. In the proposed scheme, attocells users were divided into cell-center users and cell-edge users. Cell-center users implement BIA based on reconfigurable PDs in order to mitigate the interference that they experience from neighboring attocells. Cell-edge users, however, experience more significant interference and thus might need to be offloaded to the Wi-Fi network. Results showed that the proposed interference mitigation method leads to a sum-rate enhancement while relaxing the requirements on channel coherence time. In general, the advantages of BIA and its suitability for interference management in Li-Fi systems can be summarized in the following points:

1. The implementation of BIA techniques does not require the existence of accurate CSI at the transmitting APs. This reduces the signaling overhead as the channel gain values estimated by each user terminal are not required to be fed back to the APs.
2. Since BIA is implemented at the receiver side, it does not require extra DC bias and does not affect the illumination requirements.
3. BIA is realized by precoding matrices that are composed of zeros and ones. As a result, it inherently satisfies the IM constraints on the transmitted signal to be unipolar and real-valued.
4. The implementation of BIA schemes in Li-Fi systems is not subject to small-scale fading effects. This is due to the fact that the optical wavelength is significantly smaller than the physical PD area.

5. There is no need for high-rate backhaul links since BIA does not require cooperation between the multiple transmitters. As a result, the hardware complexity of the Li-Fi system setup is considerably reduced compared to MIMO systems utilizing transmit precoding.

However, there are some challenges associated with the implementation of BIA in Li-Fi systems, such as:

1. BIA cannot be guaranteed to completely eliminate the interference at the receiving terminal. This is particularly challenging if all the receiver's configurable PDs are exposed to interfering signals.
2. BIA inherently causes noise enhancement at the receivers' devices, leading to reduced SNR and, consequently, reduced achievable data rates.

7.5.1.3 Interference Cancellation Using Power Control

Power-control techniques can be used for interference cancellation by designing the desired signal and the interference signal to have distinctly different received power levels. For example, the receivers in power-based NOMA systems employ power-control interference cancellation when the interference power is higher than the received signal power. In such scenarios, the receiver aims to first decode the high-power interference signal, while treating its own desired low-power signal as negligible noise. Upon the successful decoding of the interference signal, it can be subtracted from the total received signal to end up with the desired signal only. In the case of multiple interfering signals, the same process is performed in a successive manner from high to low received power levels, and it is termed as SIC. It is noted that SIC can be utilized to enable multiple access, which forms the concept of the power-domain NOMA scheme discussed in Section 7.3.6. In order to facilitate the use of SIC, the transmitting AP is required to perform power allocation in a way that guarantees that the signals of different users have sufficiently distinct power levels. If the desired and interference signals have equal or very similar power levels, SIC cannot be performed unless a transmit diversity coding scheme is employed [42]. Moreover, the total transmit power needs to comply with the illumination requirements and the linear range of the LED operation.

In addition to enabling multiple access, power control can also be used to mitigate interference in MIMO systems. A power-control based inter-lighting interference cancellation technique was proposed in [43] as a solution to reduce CCI between adjacent VLC APs. In the proposed method, a user located in the overlapping area of two APs can eliminate interference by implementing the three following function blocks:

1. *Buffer memory*: The total received signal (information signal + interference) is stored upon reception.

2. *Interference component generator*: The high-power interference signal is first decoded, modulated, and then multiplied by the estimated channel gain of the interfering AP to generate replicas of the received interference signal.
3. *Subtraction module*: The generated interference component is subtracted from the total received signal in order to recover the desired information signal.

Inter-lighting interference cancellation was shown to provide significant bit error ratio (BER) enhancement when the transmit signals of the different LEDs were received with distinct power levels at the receiver PD. Nevertheless, the proposed methodology cannot be applied when the desired signal and the interference signal are received with equal power values. To address this problem, an enhanced inter-lighting interference cancellation technique was proposed in [44]. The proposed scheme employs a closed-loop power control strategy to ensure that the signals observed at the users' PDs are received with distinct power levels so as to facilitate the interference cancellation process. Moreover, a power allocation method was adopted to minimize the power consumption at the transmitting APs. Based on the system parameters and users' locations, a power weight is assigned to each transmitted signal. The power weights are recalculated when a change in the users' conditions occur, for example, due to users' mobility. When the received signal power values do not comply with the preferred thresholds, a feedback link is utilized to inform the APs in order to readjust the power weights. The presented results verified that the enhanced inter-lighting interference cancellation technique can reduce the power consumption while enhancing the average BER performance compared to the original technique in [43]. Moreover, it was shown that the best performance was obtained when the high-power signal level was set to be twice greater than the power level of the low-power signal.

Power-control-based interference cancellation has also been proposed for MIMO systems in which multiple APs are grouped and assigned to each network user. For example, the work in [45] assumed that the number of active Li-Fi APs is much higher than the number of network users. In the proposed system model, multiple highly directional LEDs were assigned to each user based on the received signal strength at the user's PD. Since each user receives a sum of signals from multiple LEDs with different locations and directions, more distinct channel gain combinations can be observed at the receivers' PDs. If the user knows exactly which combination of LEDs are transmitting its useful signal, it can apply SIC to suppress the interference from other LEDs and decode its own information signal. Simulation results showed that, even with equal power distribution among the LEDs, the combination of different spatial channel gains can still create enough distinction between the power levels of the desired signal and interference signals.

As a result, SIC is still possible without the need for optimal power control at the transmitters.

Despite its advantages, power-control-based interference cancellation has some drawbacks, which can be summarized as follows:

1. It requires dynamic optimization of the transmit power levels of each signal, which adds to the computational complexity at the transmitters.
2. It requires a closed-loop feedback between the MIMO transmitters and receivers in order to update the power allocation based on the received signal strengths. This might lead to higher latency and stability issues.
3. The computational complexity of the SIC process employed at the receivers increases with the increase in the number of users.
4. SIC is inherently susceptible to cancellation errors. For example, if an error occurs in the decoding of the first stage of SIC (i.e., decoding the signal with highest power level), all the following signals will be decoded incorrectly.

7.5.1.4 Spatial Diversity Techniques

Interference cancellation in multiuser systems requires the reduction of the degree of correlation between the channels of the different users. This can be particularly challenging in indoor Li-Fi systems due to the small coverage area and the symmetric nature of the optical channel. To address this problem, a link-blocked receiver was implemented in [46]. In the proposed technique, intentional link blockage was induced in order to stop the incoming interference signals from reaching the PD. This can be realized by installing opaque boundaries in the receiver device or by the use of narrow-FOV PDs. Even though the induced blockage can diminish the received optical power, leading to a significant reduction in the receiver's SNR, it was shown to improve the BER performance for the MIMO users. This is because the cancellation of the interference signals outweighs the loss in the received SNR, leading to an overall performance enhancement. This method, however, is challenging to implement in scenarios with high user mobility, where the locations of interfering users change frequently. Furthermore, it might be practically difficult to physically block the interference signals without blocking the desired communication link when the spatial separation between the receiving PDs is small.

A more practical solution to implement is the use of multibeam transmitting LEDs and multiple receiving PDs. Such configurations create nonoverlapping spatial links, which enables parallel interference-free transmissions in multiuser systems. For example, an array of narrow-FOV PDs can be used instead of a single wide-FOV detector to cover the same area of reception. The advantage of using an array of narrow-FOV PDs lies in the ability to use the ones that are aligned with the direction of the desired signal while rejecting the ambient light or interference signals arriving at the other PDs. Using a smaller FOV at the receivers' PDs offers the added advantage of reducing the multipath effect. This is mainly because a smaller fraction of the delayed light signal will be observed within the same FOV as the primary signal. Moreover, the use of multiple narrow-beam transmitters to emit concentrated light signals was shown to reduce the occurring channel path losses, compared to the use of a single wide-beam transmitter [47].

The orientation of the receiving PDs is another degree of freedom that can be utilized in order to reduce the correlation between the multiuser channels' Li-Fi systems. The work in [48] proposed two different receivers design in which the orientation of each PD can be dynamically configured so as to reduce channel correlation between the signals of different users. This can be realized by mechanically changing the normal vector of each receiver's PD such that the incident angles coming from the same transmitting LED are different. The proposed technique does not require spatial separation between the different PDs to achieve diversity gain. As a result, it is considered practically feasible to be implemented on handheld devices due to the small size of the receiver unit. The presented simulation results in [48] showed that the proposed technique provides better BER performance compared to the use of narrow-FOV PDs arrays. Moreover, it outperforms link-blocked receivers when the separation between the PDs is sufficiently large.

7.5.1.5 Polarization Techniques

The polarization property of the light signals can be utilized as a means for interference cancellation in multiuser Li-Fi systems. The idea is based on transmitting the information signal with specific polarization. On the receiver side, two PDs with different polarization properties are implemented such that one PD can detect the combined information plus interference signal, whereas the other PD is configured to detect only the interference. The two PDs are connected in series with opposite bias voltages. The receiver output signal is then formed by the subtraction of the two signals received by the two PDs, which results in the desired information signal. The process of subtracting the received signals is referred to as differential detection. As a result, the transmission of polarized signals combined with differential detection enables interference-free parallel transmissions. Figure 7.9 illustrates the principle of polarization-based interference cancellation [49]. Three different optical signals, namely S_1, S_2, and $S3$, are simultaneously emitted with distinct polarizations. A similar approach was proposed in [50], where a transceiver architecture was proposed to establish a short-range, interference-tolerant VLC link based on light polarization and differential detection. The transmitter consists of two LEDs for in-phase and out-of-phase data transmission. The light emitted from the first LED is polarized using a 0° linear polarizer, while the second LED employs a 90° linear polarizer. The proposed transceiver architecture was shown to be robust to interference and enabled correct decoding with BER less than 10^{-4}. It is

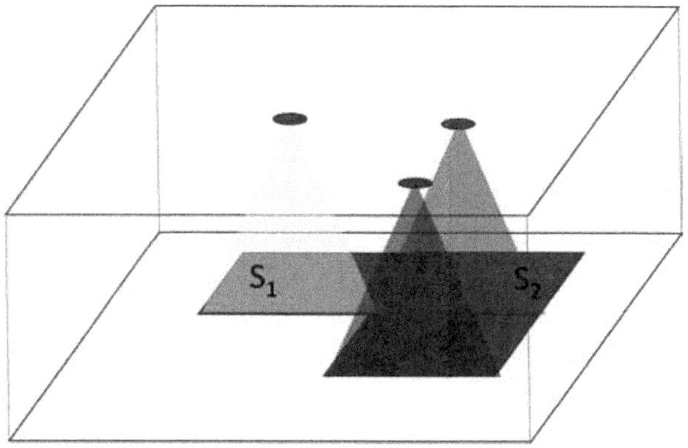

Figure 7.9 Spatial diversity of polarized light beams [49].

noted that polarization-based interference cancellation adds to the transmitter and receiver hardware complexity, due to the need for polarizers at the transmitter side and polarization filters at the receiver side. Moreover, polarization-based methods require a perfect alignment of the polarization directions between the transmitters and the receivers, which is not always feasible in practical implementations. Another downside to this method is that polarization reduces the numbers of received photons, leading to SNR loss at the receiver terminal.

7.5.2 Interference Avoidance Techniques

The interference cancellation techniques discussed in Section 7.5.1 are mainly deployed in order to mitigate the effect of the interference that already exists in the system. In this section, we discuss interference avoidance methods, which are typically applied at the transmitter side in order to prevent the transmission of interfering signals in the first place. Orthogonal multiple access schemes such as TDMA, OFDMA, and SDMA, which were discussed in Section 7.3, are considered to be interference avoidance techniques since these schemes allocate orthogonal resources to the different users, which completely eliminates the MUI. In general, interference avoidance techniques are considered more robust to the channel fluctuations that are caused by the random behavior of network users, such as mobility and device orientation. This is because interference avoidance is typically carried out at the AP (i.e., prior to signal transmission), and is not dependent on the location or channel conditions of the users. In the following, we discuss other interference avoidance techniques, namely: frequency reuse (FR) and frequency and time-hopping.

7.5.2.1 Frequency Reuse

Frequency reuse (FR) refers to the concept of utilizing the same optical frequencies within coverage areas that are separated by considerable spatial separation. Based on this, ICI between neighboring attocells can be minimized. In FR-based carrier allocation, distinct optical subcarriers are allocated for neighboring VLC cells in order to enable the receiver to distinguish its desired signal from those of adjacent cells [51]. Because the coverage area of any transmitting LED is confined within a certain coverage limit, typically a few meters, the subcarriers' allocation can be reused in a repetitive manner. While full FR allocates completely distinct subcarriers to neighboring attocells, fractional frequency reuse (FFR) divides the attocells into different zones with different frequency reuse factors. Based on this, the subcarriers allocated to the zones close to the cell edge are ensured to be orthogonal to the neighboring attocells, while the zones in the cell center can utilize the same subcarriers as in the neighboring cells because they are far away from the interference. FFR is a more cost-effective approach to reuse interference, as it provides higher spectral efficiency compared to FR [52].

In order to enable FFR, the serving Li-Fi AP needs to categorize its users into cell-center and cell-edge users. This can be realized by measuring the average power of the downlink pilot signal. If the pilot signal power is below a specified threshold, the user is considered a cell-center user. Otherwise, the user is considered a cell-edge user. Based on this, cell-center users can use the entire available transmission resources in their associated AP since they experience low to zero inter-cell interference. Cell-edge users utilize only a fraction of the transmission resource such that adjacent attocells use orthogonal resource blocks for their cell-edge users. In this way, the overall spectral efficiency and achievable system throughput can be significantly enhanced compared to orthogonal resource partitioning. FFR has the following advantages:

1. It provides high spectral efficiency compared to FR.
2. It does not require the existence of accurate CSI at the transmitters, because users can be categorized into cell-center and cell-edge users based on the average power of the downlink pilot signal.

FFR techniques can be divided into two main categories: strict fractional frequency reuse (sFFR) and soft frequency reuse (SFR). In sFFR, the available communication bandwidth is divided into multiple nonoverlapping subbands as well as a common subband. The common subband is designated to the users at the cell center since those typically experience less ICI from adjacent Li-Fi attocells. Users close to the cell edge, however, experience considerable interference and, thus, are allocated distinct protected subbands. The subbands allocation is performed in a way that guarantees minimum spatial reuse distance so as to maximize the achievable spectral efficiency within each Li-Fi attocell. SFR schemes apply even smaller reuse distance compared to sFFR and allow the users at the center to utilize the frequency subbands that are being used by edge-cell users in adjacent attocells. In order to do this, higher transmit power is required for the cell-edge users' signals so as to protect them from the interference of the cell-center transmissions. A system model was proposed in [53] based on SFR where each transmitting LED was equipped with two FOVs to produce two distinct rays' beamwidth. In order to enable SFR, one FOV is utilized to transmit to cell-center users, while the other one is dedicated to the cell-edge users.

Generally, there are two different approaches for enabling FR in the Li-Fi APs, depending on whether a single-wavelength transmission or a multiple-wavelength transmission is being used. In the first approach, a single wavelength is utilized at the transmitting AP, and thus the baseband signal bandwidth is divided between the FR zones. This baseband bandwidth is limited by the linear range of operation of the LEDs and usually ranges between 10 and 20 MHz [54]. For this reason, FR mechanisms that utilize a single wavelength lead to a further reduction in the modulation bandwidth available for the information signals and, thus, reduced achievable data rates. On the contrary, using multiple distinct wavelengths for FR enables the full utilization of the LEDs modulation bandwidth. The problem in multiwavelength FR is the need for a mechanism to allow the user to distinguish its desired wavelength and detect it with the existence of other wavelengths. Typical receivers' PDs are sensitive to a wide wavelength range from 400 nm to 1,100 nm, which covers the majority of the visible communications spectra. When a PD receives a combination of signals carried over multiple optical wavelengths, they are all converted into an electrical current. One solution to enable the receiver to extract the desired signal is to use an optical filter to filter out all the nonrelevant wavelengths as shown in [55]. Adaptive optical filters can be also employed at the receiver in order to allow the PD to dynamically select the wavelength of interest based on the FR strategy in place. A filter-based spectrum sensor array was implemented in [56] in order to allow the receiver to reject the interference signals. However, a high number of distinct wavelengths selection requires a larger receiver physical area. More research is needed to develop more compact and efficient solutions to enable adaptive wavelength filtering.

7.5.2.2 Time-Hopping and Optical Frequency Hopping

In optical frequency-hopping, the transmitted signal frequency is rapidly switched among many distinct optical frequency channels. This is done according to a pseudorandom sequence that is known to both the transmitter and the receiver. Based on this, the information signal of any system user would be modulated onto different optical carriers at different time slots. In the same manner, the time-hopping technique is used to rapidly switch the signal in the time domain. Since each user has a distinct hopping pattern, optical frequency-hopping allows multiple interference-free transmissions within the Li-Fi attocell. Furthermore, time and frequency-hopping add the advantage of enhanced system security. This is because eavesdroppers that do not acquire the exact hopping pattern cannot succeed in decoding the legitimate users' signals. The work in [57] proposed a wavelength division multiplexing (WDM)-based system with passive frequency-hopping. In the proposed system, the information signal is divided into small segments in the time domain and then modulated over distinct optical wavelengths. A field-programmable gate array (FGPA) chip was used to perform hopping, synchronization, and de-hopping functionalities. Moreover, software configuration was used to switch the hopping rate between slow frequency-hopping, fast frequency-hopping, and intermediate frequency-hopping. The maximum achievable hopping rate was, however, limited to 10 Gbps due to the chip performance. An optical frequency-hopping system was demonstrated in [58] based on frequency shift keying (FSK), which was shown to support a 10.6-Gbps error-free frequency-hopping. In the proposed system, the FSK transmitter consisted of a phase modulator driven by the 10.6-Gbps hopping sequence and a 1-bit delayed interferometer (DI). The transmitter generated two constant intensity complementary FSK signals at each output port of the DI. The FSK signal was then modulated by the hopping sequence and combined together via a 3-dB coupler. A fiber link was used to transmit the encrypted frequency-hopping sequence to the intended receiving terminal. Upon reception, the two data signals were combined together and the original user data was reconstructed. Similarly, an optical frequency-hopping system based on four-wavelength-frequency-hopping with FSK modulation was proposed in [59]. It was shown that the security of the communication links can be enhanced by fragmenting the information signals and modulating them onto different optical wavelengths. A hybrid hopping technique was demonstrated in [60], in which the intensity of the transmitted light signals was modulated along the time, space, and frequency domains, creating what is termed as frequency-hopping spatial multipulse position modulation (PPM).

7.6 Handover and Mobility Support

In wireless mobile communications, a handover occurs when a network user is transferred from its serving AP to another AP while staying in the same data session without losing connection. Handovers can be generally classified into hard handovers and soft handovers. In a hard handover, the user connection with the first AP, also called the host AP, is terminated before it starts a new session with the target serving AP. To enable this, the user and the target AP exchange signaling information via dedicated communication links. This signaling process is known as handover overhead, during which the user's data signaling is temporarily paused. In soft handover the user terminal initiates a new session with the target AP before terminating its session with the previous AP. As a result, the user can continue its data reception in parallel with performing the handover process. In Li-Fi systems, the small size of the attocells implies that mobile users could get in and out of the coverage areas more frequently than they would in an RF network.

Also, since RF signals penetrate through walls, the coverage of a Wi-Fi AP extends beyond a single confined space. The coverage area of each Li-Fi AP extends only for few square meters. This is because light signals do not penetrate through walls and do not easily bend around corners. As a result, the Li-Fi network users may experience frequent handovers as they move in the proximity of the Li-Fi APs.

Handover can also be classified into horizontal handover (HHO) and vertical handover (VHO). When the handover occurs between neighboring Li-Fi attocells, it is termed as HHO. VHO occurs when the user association is transferred from a Li-Fi AP to the Wi-Fi AP, or vice versa. It is particularly important to study the effect of the handover process in heterogeneous Li-Fi/RF networks. This is because VHO between the two tiers is a hard handover that adds to the signaling overhead and the system complexity. Moreover, efficient load balancing and user association networks are particularly critical in heterogeneous networks. This is due to the fact that the signaling needed to perform the handover procedure takes an average of 30 ms up to 3,000 ms [61]. The signaling overhead and latency performance of the handover algorithm depends on many factors involved in the user association process. These include the reauthentication mechanisms, the speed of the mobile user, the neighborhood definition, and the network link costs [62]. In conventional RF networks, handover occurs when a mobile user's received SNR drops below a certain threshold or when the packet error ratio (PER) drops below a certain value. In heterogeneous networks, however, the decision is more complicated as it does not only depend on the signal strength but also on the link stability. For example, if a mobile user is handed over from a Wi-Fi AP to a Li-Fi AP, it becomes naturally more susceptible to further handovers due to the inherently small size of Li-Fi attocell. Similarly, if a mobile user is handed over from the Li-Fi network to the Wi-Fi network, it will lead to an increase in the load on the respective Wi-Fi AP. This implies that other users that are being served by the same Wi-Fi AP might experience QoS degradation and, thus, need to be handed over to other neighboring Wi-Fi cells. Thus, efficient load-balancing algorithms are needed to minimize the handover signaling overhead while maintaining fairness and link stability in heterogeneous Li-Fi/Wi-Fi networks. Moreover, motion tracking is considered to be particularly important in the VHO assessment because it is generally more efficient to connect users with high mobility to the RF network to avoid frequent handovers, while Li-Fi links are best suited for users who are static or experience minimal mobility.

In the following, we summarize the main advances regarding the use of HHO and VHO in heterogeneous Li-Fi/RF systems.

7.6.1 Horizontal Handover (HHO)

The small coverage areas of the Li-Fi attocells, as well as the channel gain sensitivity to link blockage and random device orientation, results in frequent HHO in Li-Fi networks. In fact, the small separation distance between the adjacent Li-Fi APs can have a big influence on the achievable network throughput. A small separation distance results in enhanced Li-Fi coverage and, thus, higher area spectral efficiency. The small separation means that high overlapping exists between the adjacent APs, which results in more frequent and sudden HHO. For this reason, the layout of the Li-Fi APs needs to be carefully studied in order to find a trade-off between spectral efficiency and handover rate. There is a need for network planning tools to decide on the optimal placing of LEDs in a way that maximizes the coverage footprint and minimizes overlapping, while also taking into account the signaling overhead needed to perform HHO between adjacent attocells. It is noted, however, that it is not always feasible to configure the AP locations because they follow the structural planning of indoor lighting fixtures. In scenarios with a large number of LEDs within an open space, like museums or airports, it could be useful to decide on which LEDs are used for communications and which are kept idle so as to optimize the system performance and reduce the number of occurring handovers. Here, there is a need to evaluate the trade-off between the Li-Fi coverage and the handover overhead. The distribution of the APs in Li-Fi systems depends on the lighting design and can be generally classified into uniform lighting and spotlighting. In uniform lighting, LEDs with wide FOV are utilized and distributed uniformly in the ceiling in order to generate a continuous lighting field across the coverage area. In contrast, spotlighting employed narrow-FOV LEDs to generate intense and focused light beams. The overlapping between the Li-Fi attocells in the uniform lighting modes usually leads to frequent handovers and a ping-pong effect as the users move in and out the overlapping areas. When designing the LED placements, network planning tools need to take into account the number and moving speed of network users in order to design the system to be robust for the worst-case scenario. This problem was investigated in [63] by carrying out an optimization problem that aims to maximize the average user rate in VLC systems with overlapping attocells. In order to account for the involved handover overhead, the signaling time was added to the rate calculations.

The work in [64] proposed an HHO mechanism for seamless connectivity in a VLC network with two different lighting setups: overlapping attocells (i.e., uniform distribution of the lighting fixtures in the ceiling), and non-overlapping attocells (i.e. when spotlights with concentrated narrow beams are used). In the case of separated spotlights, it is generally difficult to ensure seamless connectivity because of the gaps between the concentrated coverage beams. In this case, the user's communication session is interrupted when it steps out the borderline of its serving LED's coverage area. The mobile user then has to walk for a certain distance without connectivity until it can connect to the nearest spotlight beam. One solution to avoid service interruption is to increase the buffer size at the receiver so as to keep the service uninterrupted until the user walks into the target attocell's coverage area. Typically, the mobile user can detect if it will find a new serving attocell before exiting its current serving attocell by performing a simple test on the received signal intensity levels. If the intensity test indicates that an overlapping attocell exists in the direction where the user is heading, there is no need to increase the receiver's buffer size.Otherwise, the buffer size is increased to enable the user to pass to the next coverage area without losing its connection. Different from the spotlighting scenario, uniform lighting distribution allows the users to perform soft handover when moving from one attocell to another. In this case, the intensity test will detect the existence of a near overlapping attocell and, hence, there is no need to increase the buffer size at the receiver terminal.

It is noted that the above-mentioned studies do not consider the effect of device orientation on the handover process. In other words, user devices are assumed to be always oriented vertically upward so that the PD's plane is parallel to the LED plane. In practice, however, the user device may rotate while moving, and the orientation changes according to the users' behavior. In fact, the change in user orientation can result in significant variations in the perceived signal strength and thus may necessitate that the user disconnects from its serving AP and connect to the AP that provides the highest signal strength for its current device orientation. Motivated by this, the work in [65] analyzed the effect of the receivers' random device orientation on the probability of handover in Li-Fi systems. The presented results demonstrated that random device orientation has a significant effect on the handover probability and that the handover rate is the highest when the receiver's device is tilted between 60° and 80°.

In order to analyze the interplay between the handover cost and the users' QoS, there is a need to consider the random LEDs' deployment as well as the random locations and orientations of the network users. Stochastic geometry methods can be utilized in order to average the system performance over all possible realizations of the LEDs' locations and user statistics. Based on this, a general idea on the optimal APs' layout can be achieved including the optimal number and locations of the active APs. Although the optimal planning of the LEDs' layout can help in reducing the rate of HHO, fast-moving users will still experience frequent and sudden handovers. The information on the APs' distribution as well as the available estimates on the locations and device orientation of network users can be exploited to introduce smart handover management. In this regard, the concept of handover skipping can be utilized to enable fast-moving users to be transferred between non-adjacent attocells so as to reduce the handover signaling overhead. This decision can be made based on the coverage size of neighboring APs as well as the speed of the mobile user. Handover skipping has been intensively applied for handover management in RF systems. The topology of Li-Fi systems, however, is dynamic and highly dependent on the user behavior (i.e., user location and orientation). Moreover, the information about the users' trajectory in Li-Fi is prone to measurement errors caused by multipath reflection, and forwarding this information to the APs required an extra feedback overhead. To overcome these challenges, it is possible to exploit the rate of change in the reference signal received power in order to indicate whether a user is traveling towards the center or the edge of its current serving attocell [66]. Since the rate of change of the reference signal received power reflects the changes in the user location and device orientation, there is no need for extra feedback between the user device and the AP. Based on this information, and by exploiting the knowledge on the LEDs deployment, a decision can be made to determine whether or not to skip a certain AP.

7.6.2 Vertical Handover

Vertical handover (VHO) occurs when a user needs to switch from the Li-Fi network to the RF network, or vice versa [61]. The handover from Li-Fi to RF typically occurs due to human activities causing one of the following: (1) a blockage of the LOS channel between the user and its serving AP, (2) a change in the receiver's device orientation such that it significantly deviates from the serving AP, and (3) a change in the user location such that it is out of the coverage area of the Li-Fi AP. When the Li-Fi service is interrupted due to one of these reasons, the user needs to be transferred to a backup RF connection. The VHO process needs more processing due to the change of air interfaces from optical to RF. As a result, VHO causes higher delays and signaling overhead, compared to HHO. When the optical link is recovered again, it is usually preferable to switch back to the Li-Fi network when it provides better performance. It is noted that link disruption due to user behavior is often a temporary phenomenon lasting for a very short period. This is particularly true for link blockage caused by the human movement. As a result, frequent handovers and ping-pong effect can be caused by switching the user back and forth between the Li-Fi AP and the RF network. In order to avoid this, it is sometimes preferable to wait until the blocker moves and the LOS is restored. Therefore, efficient VHO schemes need to assess the frequency and expected duration of the link blockages before deciding whether to switch the connection or not.

An effective VHO scheme needs to take into account the switching signaling cost as well as the QoS of the network users. This could be rather challenging in heterogeneous Li-Fi/ RF networks due to the following reasons:

1. The decision-making is governed by the allowed latency and signaling overhead. For example, even if the user can achieve a better link performance by switching back to Li-Fi when the link disruption ends, this might lead to extra signaling overhead and extra delay.
2. For temporary link disruption (i.e., temporary LOS blockage), it is better to wait until the blocker moves and the link is restored. Although this is a cost-effective solution, it is not feasible in critical applications with real-time performance requirements.

For these reasons, the decision-making mechanism is highly dependent on the specific requirements of each individual system setup (i.e., switching cost, maximum allowed latency, and QoS requirements). One possible solution to reduce the handover costs is the use of multiple APs' association, in which a single user is simultaneously served by multiple APs, and possibly different air interfaces. In this regard, AI can be utilized in order to enable a degree of autonomy. By utilizing machine learning techniques and assessing the various system parameters, autonomous and dynamic decision-making for the optimal handover execution in heterogeneous networks can be obtained.

7.7 Summary

This chapter has discussed the various techniques needed to enable a seamless user experience in Li-Fi systems. In Section 7.2, we discussed the integration of Li-Fi within heterogeneous networks. In this regard, Li-Fi systems can provide an additional small-cell layer to support and complement the existing RF wireless infrastructure, which requires the implementation of appropriate load-balancing techniques. In Section 7.3, we introduced the different multiple access schemes that can be utilized in order to provide multiuser access in Li-Fi systems. These can be divided into orthogonal multiple access schemes, such as TDMA, OFDMA, and SDMA, as well as nonorthogonal multiple access schemes, such as power-domain NOMA and CDMA. We have discussed that Li-Fi systems can have limitations when it comes to the application of some multiple access techniques such as OFDMA, due to the requirements of IM/DD, as well as CDMA, due to the nature of the unipolar optical codes. However, the nature of Li-Fi transmissions and the optical channel facilitate efficient multiuser access by means of power-domain NOMA and SDMA. In Section 7.4, we discussed the performance trade-offs related to resource allocation in Li-Fi systems. While static resource allocation requires lower computational complexity and processing, it does not guarantee optimal utilization of the available resources. Dynamic resource allocation can better exploit the available resources to accommodate the varying traffic conditions and QoS demands. Dynamic decision-making requires the development of efficient algorithms to reach optimal or suboptimal solutions in real time. We also showed that, aside from the conventional resource partitioning techniques borrowed from RF communications, there are unique approaches specifically tailored for Li-Fi networks including the exploitation of angular diversity transmitters and receivers. In Section 7.5, we introduced the different interference cancellation and interference avoidance techniques that are used to enable multiuser transmissions, while Section 7.6 discussed user mobility in Li-Fi and the associated handover mechanisms. The advancements of the various techniques discussed in this chapter enable the integration of Li-Fi as a key technology for 5G and beyond wireless connectivity.

References

[1] Rahaim, M. B., and T. D. C. Little, "Toward Practical Integration of Dual-Use VLC within 5G Networks," *IEEE Wireless Communications*, Vol. 22, No. 4, 2015, pp. 97–103.

[2] Alshaer, H., and H. Haas, "SDN-Enabled Li-Fi/Wi-Fi Wireless Medium Access Technologies Integration Framework," *IEEE Conference on Standards for Communications and Networking (CSCN)*, 2016, pp. 1–6.

[3] Abumarshoud, H., H. Alshaer, and H. Haas, "Dynamic Multiple Access Configuration in Intelligent LiFi Attocellular Access Points," *IEEE Access*, Vol. 7, 2019, pp. 62126–62141.

[4] Wang, Y., X. Wu, and H. Haas, "Load Balancing Game with Shadowing Effect for Indoor Hybrid LiFi/RF Networks," *IEEE Transactions on Wireless Communications*, Vol. 16, No. 4, 2017, pp. 2366–2378.

[5] Zhuang, Y., et al., "A Survey of Positioning Systems Using Visible LED Lights," *IEEE Communications Surveys Tutorials*, Vol. 20, No. 3, 2018, pp. 1963–1988.

[6] Ling, X., et al., "Efficient OFDMA for LiFi Downlink," *IEEE/OSA Journal of Lightwave Technology*, Vol. 36, No. 10, 2018, pp. 1928–1943.

[7] Hoo, L. M. C., J. Tellado, and J. M. Cioffi, "FDMA-Based Multiuser Transmit Optimization for Broadcast Channels," *IEEE Wireless Communications and Networking Conference, Conference Record* (Cat. No.00TH8540), Vol. 2, 2000, pp. 597–602.

[8] Bao, X., et al., "Protocol Design and Capacity Analysis in Hybrid Network of Visible Light Communication and OFDMA Systems," *IEEE Transactions on Vehicular Technology*, Vol. 63, No. 4, 2014, pp. 1770–1778.

[9] Kazemi, H., and H. Haas, "Downlink Cooperation with Fractional Frequency Reuse in DCOOFDMA Optical Attocell Networks," *IEEE International Conference on Communications (ICC)*, 2016, pp. 1–6.

[10] Chen, C., D. Tsonev, and H. Haas, "Joint Transmission in Indoor Visible Light Communication Downlink Cellular Networks," *IEEE Globecom Workshops (GC Wkshps)*, 2013, pp. 1127–1132.

[11] Bykhovsky, D., and S. Arnon, "Multiple Access Resource Allocation in Visible Light Communication Systems," *Journal of Lightwave Technology*, Vol. 32, No. 8, 2014, pp. 1594–1600.

[12] Guerra-Medina, M. F., et al., "Experimental Optical Code-Division Multiple Access System for Visible Light Communications," *Wireless Telecommunications Symposium (WTS)*, 2011, pp. 1–6.

[13] Noshad, M., and M. Brandt-Pearce, "Application of Expurgated PPM to Indoor Visible Light Communications—Part II: Access Networks," *IEEE/OSA Journal of Lightwave Technology*, Vol. 32, No. 5, 2014, pp. 883–890.

[14] Matsushima, T. K., et al., "A Visible-Light Communication System Using Optical CDMA with Inverted MPSC," *The Sixth International Workshop on Signal Design and Its Applications in Communications*, 2013, pp. 52–55.

[15] Qiu, Y., et al., "Visible Light Communications Based on CDMA Technology," *IEEE Wireless Communications*, Vol. 25, No. 2, 2018, pp. 178–185.

[16] Martin-Gonzalez, J. A., E. Poves, and F. J. Lopez-Hernandez, "Random Optical Codes Used in Optical Networks," *IET Communications*, Vol. 3, No. 8, 2009, pp. 1392–1401.

[17] Salehi, J. A., "Emerging OCDMA Communication Systems and Data Networks," *Journal of Optical Communications and Networking*, Vol. 6, No. 9, September 2007, pp. 1138–1178. http://jon.osa.org/abstract.cfm?URI=jon-6-9-1138.

[18] Chung, F. R. K., J. A. Salehi, and V. K. Wei, "Optical Orthogonal Codes: Design, Analysis and Applications," *IEEE Transactions on Information Theory*, Vol. 35, No. 3, 1989, pp. 595–604.

[19] Chu, W., and S. W. Golomb, "A New Recursive Construction for Optical Orthogonal Codes," *IEEE Transactions on Information Theory*, Vol. 49, No. 11, 2003, pp. 3072–3076.

[20] Ding, C., and C. Xing, "Cyclotomic Optical Orthogonal Codes of Composite Lengths," *IEEE Transactions on Communications*, Vol. 52, No. 2, 2004, pp. 263–268.

[21] Azizoglu, M., J. A. Salehi, and Y. Li, "Optical CDMA via Temporal Codes," *IEEE Transactions on Communications*, Vol. 40, No. 7, pp. 1162–1170, 1992.

[22] Luo, X., J. Yin, and F. Yue, "Multilength Optical Orthogonal Codes: New Upper Bounds and Optimal Constructions," *IEEE Transactions on Information Theory*, Vol. 61, No. 6, 2015, pp. 3305–3315.

[23] Yang, G., and W. C. Kwong, "Performance Analysis of Optical CDMA with Prime Codes," *Electronics Letters*, Vol. 31, No. 7, 1995, pp. 569–570.

[24] Chen, Z., and H. Haas, "Space Division Multiple Access in Visible Light Communications," *IEEE International Conference on Communications (ICC)*, 2015, pp. 5115–5119.

[25] Dehghani Soltani, M., et al., "Bidirectional User Throughput Maximization Based on Feedback Reduction in LiFi Networks," *IEEE Transactions on Communications*, Vol. 66, No. 7, 2018, pp. 3172–3186.

[26] Wu, X., M. D. Soltani, L. Zhou, M. Safari, and H. Haas, "Hybrid LiFi and WiFi Networks: A Survey," *IEEE Communications Surveys & Tutorials*, Vol. 23, No. 2, 2021, pp. 1398–1420, doi: 10.1109/COMST.2021.3058296.

[27] Marshoud, H., et al., "Non-Orthogonal Multiple Access for Visible Light Communications," *IEEE Photonics Technology Letters*, Vol. 28, No. 1, 2016, pp. 51–54.

[28] Kizilirmak, R. C., C. R. Rowell, and M. Uysal, "Non-Orthogonal Multiple Access (NOMA) for Indoor Visible Light Communications," *4th International Workshop on Optical Wireless Communications (IWOW)*, 2015, pp. 98–101.

[29] Yin, L., et al., "Performance Evaluation of Non-Orthogonal Multiple Access in Visible Light Communication," *IEEE Transactions on Communications*, Vol. 64, No. 12, 2016, pp. 5162–5175.

[30] Li, X., R. Zhang, and L. Hanzo, "Cooperative Load Balancing in Hybrid Visible Light Communications and WiFi," *IEEE Transactions on Communications*, Vol. 63, No. 4, 2015, pp. 1319–1329.

[31] Kim, H. -S., et al., "Inter-Cell Interference Mitigation and Indoor Positioning System Based on Carrier Allocation Visible Light Communication," *5th International Conference on Signal Processing and Communication Systems (ICSPCS)*, 2011, pp. 1–7.

[32] Cui, K., J. Quan, and Z. Xu, "Performance of Indoor Optical Femtocell by Visible Light Communication," *Optics Communications*, Vol. 298-299, 2013, pp. 59–66. http://www.sciencedirect.com/science/article/pii/S0030401813002046.

[33] Ghimire, B., and H. Haas, "Self-Organising Interference Coordination in Optical Wireless Networks," *EURASIP Journal on Wireless Communications and Networking*, Vol. 131, 2012.

[34] Zhang, R., et al., "Visible Light Communications in Heterogeneous Networks: Paving the Way for User-Centric Design," *IEEE Wireless Communications*, Vol. 22, No. 2, 2015, pp. 8–16.

[35] Park, K., and M. Alouini, "Optimization of an Angle-Aided Mirror Diversity Receiver for Indoor MIMO-VLC Systems," *IEEE Global Communications Conference (GLOBECOM)*, 2016, pp. 1–6.

[36] Andrews, J. G., "Interference Cancellation for Cellular Systems: A Contemporary Overview," *IEEE Wireless Communications*, Vol. 12, No. 2, 2005, pp. 19–29.

[37] Hong, Y., et al., "Performance of a Precoding MIMO System for Decentralized Multiuser Indoor Visible Light Communications," *IEEE Photonics Journal*, Vol. 5, No. 4, 2013, pp. 7800211–7800211.

[38] Marshoud, H., et al., "MU-MIMO Precoding for VLC with Imperfect CSI," *4th International Workshop on Optical Wireless Communications (IWOW)*, 2015, pp. 93–97.

[39] Wu, L., et al., "Blind Interference Alignment for Multiuser MISO Indoor Visible Light Communications," *IEEE Communications Letters*, Vol. 21, No. 5, 2017, pp. 1039–1042.

[40] Morales-Cespedes, M., et al., "Aligning the Light without Channel State Information for Visible Light Communications," *IEEE Journal on Selected Areas in Communications*, Vol. 36, No. 1, 2018, pp. 91–105.

[41] Qidan, A. A., M. Morales-Cespedes, and A. G. Armada, "The Role of WiFi in LiFi Hybrid Networks Based on Blind Interference Alignment," *IEEE 87th Vehicular Technology Conference* (VTC Spring), 2018, pp. 1–5.

[42] Chi, N., and J. Shi, "Investigation on Overlapping Interference on VLC Networks Consisting of Multiple LEDs," *ICT Express*, Vol. 1, No. 2, 2015, pp. 63–66. http://www.sciencedirect.com/science/article/pii/S2405959515000053.

[43] Kim, K., K. Lee, and K. Lee, "An Inter-Lighting Interference Cancellation Scheme for MISO-VLC Systems," *International Journal of Electronics*, Vol. 104, No. 8, 2017, pp. 1377–1387.

[44] Romero-Munoz, G. A., K. Kim, and K. Lee, "Enhanced Inter-Lighting Interference Cancellation Using Power Control to Improve the Performance of Indoor Visible Light Communication Systems," *IET Optoelectronics*, Vol. 12, No. 6, 2018, pp. 273–279.

[45] Eroglu, Y. S., et al., "Multi-Element VLC Networks: LED Assignment, Power Control, and Optimum Combining," *IEEE Journal on Selected Areas in Communications*, Vol. 36, No. 1, 2018, pp. 121–135.

[46] Fath, T., and H. Haas, "Performance Comparison of MIMO Techniques for Optical Wireless Communications in Indoor Environments," *IEEE Transactions on Communications*, Vol. 61, No. 2, February 2013, pp. 733–742.

[47] Carruther, J. B., and J. M. Kahn, "Angle Diversity for Nondirected Wireless Infrared Communication," *IEEE Transactions on Communications*, Vol. 48, No. 6, 2000, pp. 960–969.

[48] Nuwanpriya, A., S. Ho, and C. S. Chen, "Indoor MIMO Visible Light Communications: Novel Angle Diversity Receivers for Mobile Users," *IEEE Journal on Selected Areas in Communications*, Vol. 33, No. 9, 2015, pp. 1780–1792.

[49] Ryoo, H., et al., "Differential Optical Detection in VLC for Inter-Cell Interference Reduced Flexible Cell Planning," *IEEE Photonics Technology Letters*, Vol. 28, No. 23, 2016, pp. 2728–2731.

[50] Atta, M. A., and A. Bermak, "A Polarization-Based Interference-Tolerant VLC Link for Low Data Rate Applications," *IEEE Photonics Journal*, Vol. 10, No. 2, 2018, pp. 1–11.

[51] Kim, H., et al., "Mitigation of Inter-Cell Interference Utilizing Carrier Allocation in Visible Light Communication System," *IEEE Communications Letters*, Vol. 16, No. 4, 2012, pp. 526–529.

[52] Chen, C., et al., "Fractional Frequency Reuse in DCO-OFDM-Based Optical Attocell Networks," *IEEE/OSA Journal of Lightwave Technology*, Vol. 33, No. 19, 2015, pp. 3986–4000.

[53] Van Huynh, V., et al., "Inter-Cell Interference Mitigation Using Soft Frequency Reuse with Two FoVs in Visible Light Communication," *International Conference on ICT Convergence (ICTC)*, 2012, pp. 141–144.

[54] Cossu, G., et al., "3.4 Gbit/s Visible Optical Wireless Transmission Based on RGB LED," *Optics Express*, Vol. 20, No. 26, December 2012, pp. B501–B506.

[55] Bian, R., I. Tavakkolnia, and H. Haas, "10.2 Gb/s Visible Light Communication with Off-the-Shelf LEDs," *European Conference on Optical Communication (ECOC)*, 2018, pp. 1–3.

[56] Chang, C., et al., "Interference Rejection Using Filter-Based Sensor Array in VLC Systems," *IEEE Sensors Journal*, Vol. 12, No. 5, 2012, pp. 1025–1032.

[57] Wang, S. L., et al., "A Novel Optical Frequency-Hopping Scheme for Secure WDM Optical Communications," *IEEE Photonics Journal*, Vol. 7, No. 3, 2015, pp. 1–8.

[58] Huo, L., et al., "Demonstration of an Optical Frequency-Hopping Scheme for Secure Communications," *2006 Conference on Lasers and Electro-Optics and 2006 Quantum Electronics and Laser Science Conference*, 2006, pp. 1–2.

[59] Ban, D. C., et al., "A Novel Optical Frequency-Hopping Scheme Based on a Flexible Structure for Secure Optical Communications," *IEEE Photonics Journal*, Vol. 11, No. 1, 2019, pp. 1–7.

[60] Escribano, F. J., et al., "A Spatial Time-Frequency Hopping Index Modulated Scheme in Turbulence-Free Optical Wireless Communication Channels," *IEEE Transactions on Communications*, Vol. 68, No. 7, 2020, pp. 4437–4450.

[61] Wang, Y., and H. Haas, "Dynamic Load Balancing with Handover in Hybrid Li-Fi and Wi-Fi Networks," *IEEE/OSA Journal of Lightwave Technology*, Vol. 33, No. 22, 2015, pp. 4671–4682.

[62] Kassab, M., J. M. Bonnin, and A. Belghith, "Fast and Secure Handover in WLANs: An Evaluation of the Signaling Overhead," *5th IEEE Consumer Communications and Networking Conference*, 2008, pp. 770–775.

[63] Pergoloni, S., et al., "Optimized LEDs Footprinting for Indoor Visible Light Communication Networks," *IEEE Photonics Technology Letters*, Vol. 28, No. 4, 2016, pp. 532–535.

[64] Vegni, A. M., and T. D. C. Little, "Handover in VLC Systems with Cooperating Mobile Devices," *International Conference on Computing, Networking and Communications (ICNC)*, 2012, pp. 126–130.

[65] Soltani, M. D., et al., "Handover Modeling for Indoor Li-Fi Cellular Networks: The Effects of Receiver Mobility and Rotation," *IEEE Wireless Communications and Networking Conference (WCNC)*, 2017, pp. 1–6.

[66] Wu, X., and H. Haas, "Handover Skipping for LiFi," *IEEE Access*, Vol. 7, 2019, pp. 38369–38378.

8

Standardization

8.1 Introduction

A number of light communications vendors have developed various proprietary OWC products that do not conform to a specific standard nor can they work in harmony due to the lack of interoperability. Standardization is considered as a key facilitator of the mass market of OWC, as it enables optoelectronic and chipset vendors to customize their products for OWC use cases. In addition, the availability of well-defined standards encourages mobile and laptop manufacturers to integrate enabling technologies such as high-speed optical frontends (OFEs) into their devices. In this chapter, we introduce the standardization efforts related to OWC technology and summarize the main specifications of the available standards.

Section 8.2 describes OWC standards under the IEEE 802.15 category. The IEEE 802.15.7-2011 standard is the earliest one, published in 2011. With further advances in OCC, VLC, and Li-Fi technologies in the 2010s, the IEEE 802.15.7 standard has been revised to include several OCC scenarios for indoor and vehicular applications. In addition, industrial use cases with high data rate requirements have motivated another standardization activity: IEEE 802.15.13, which is expected to deliver multigigabits per second data rate transmissions.

Section 8.3 describes the ITU Telecommunication Standardization Sector (ITU-T) G.9991 standard (also known as G.vlc), which specifies the physical (PHY) and data link layer (DLL) design and system architecture for high-speed indoor VLC transceivers. The standard is based on the ITU-T G.hn home networking standards, which support home networking solutions over different types of wiring in residential environments such as power line communication

(PLC). As a result, the existing G.hn silicon solutions can be readily used to build VLC transceivers. In Section 8.3, we summarize the development framework for this standard and discuss its main features. Moreover, we list some of the commercial Li-Fi implementations built based on this ITU-T Recommendation.

Section 8.4 discusses the IEEE 802.11bb standard amendment for fully networked light communication with mobility and handover support. This standard is based on the IEEE 802.11 standard suit. It is widely believed that standardization of the Li-Fi technology within the IEEE 802.11 standard will accelerate the adoption of Li-Fi technology and enable rapid product introduction into mass markets. The IEEE 802.11 standard offers a unique ecosystem that allows Li-Fi to reuse already matured techniques and semiconductor devices. However, the standard is still, at the time of this writing, under development. Therefore, this section is mainly focused on the publicly published material on the physical layer specifications [1]. Section 8.5 summarizes this chapter.

8.2 IEEE 802.15 Optical Wireless Communication Standards

8.2.1 Standard Development

The IEEE 802.15 working group was established in 1998 to work on the standardization of potential technologies related to wireless personal area networks (WPANs), such as the widely used Bluetooth and the emerging IOT [2]. With an increase in research and interests in the field of VLC technology, the IEEE 802.15.7 task group was formed by academic and industrial researchers from multiple countries in 2009 to standardize the VLC for WPANs. After 2 years, the standard was approved and published in June 2011. The standard is titled: "IEEE 802.15.7-2011: Short-Range Wireless Optical Communication Using Visible Light", which provides a series of technical specifications in the PHY layer and the MAC layer [3], where three operation modes for different applications have been introduced. Two major focuses of the standard are dimming support and flickering mitigation. However, some research studies show that the framework for some use cases, such as vehicular communications, is too general and suggest that some parts of the specifications will potentially increase the prototype complexity [4]. With advances and development in OCC and Li-Fi technologies, an amendment to the IEEE 802.15.7-2011 standard was initiated in 2014 by the task group IEEE 802.15.7m [5]. In the initial standard proposal, the number of operation modes increases from three to eight to cover scenarios including image-sensor-based communications, low data rate PD communications and high data rate PD communications. In addition, the operation spectrum has been extended from the visible light range to a wider optical spectrum such as near-infrared and near-ultraviolet. The revised standard was completed in 2018 and its title is: "IEEE 802.15.7-2018: Short-Range Optical Wireless Communications" [6].

However, only three extra operation modes corresponding to the OCC scenarios were added to the final version of the amended standard. Due to the major technical difference between OCC and high-speed OWC, a new dedicated task group has been formed to deliver a separate standard titled "IEEE 802.15.13 Multi-Gigabit/s Optical Wireless Communications" for very high data rate OWC systems in 2017. The IEEE 802.15.13 standard will include three high data rate operation modes primarily based on OFDM modulation schemes. The first draft of the standard was completed in 2019. The Letter Ballot was conducted in 2020 and the final standard is expected to be published in 2021 [7]. In the remainder of this section, we refer to the above three standards as the IEEE 802.15 OWC standards for convenience.

8.2.2 Standard Features

In general, the three IEEE 802.15 OWC standards 802.15.7-2011, 802.15.7-2018, and 802.15.13 describe how to use the OWC technology in WPANs for various applications. In this section, the main features of the three standards are briefly introduced, which include network topologies, dimming support, flickering mitigation, transmission reliability, transmission security, enabling image sensor detection, and enabling very high data rate transmission.

8.2.2.1 Network Topology

In the IEEE 802.15 OWC standards, the OWC devices are categorized into three classes: infrastructure, mobile device, and vehicle, as illustrated in Figure 8.1(a) [6]. Among these classes, infrastructure class refers to devices fixed on other building structures. Therefore, these devices are normally connected to the mains and equipped with intense power optical sources. Their data rate, link distance, and coverage are flexible for different applications. Infrastructure-type devices are normally the default coordinator. Devices of mobile and vehicle classes have random physical positions. Mobile devices have limited battery capacity, which leads to lower power optical sources and shorter transmission ranges. Wireless broadband and multimedia data transfer are common applications of mobile devices, which normally require high data rate transmission. Vehicle-type devices have a greater battery capacity, which means they can support high-power optical sources for transmission over longer distances. Information on traffic conditions and driving states is typically transmitted in vehicle-to-vehicle (V2V) and vehicle-to-infrastructure (V2I) applications, which normally requires low data rate transmission.

The IEEE 802.15 OWC standards use three basic network topologies: (1) peer-to-peer topology, (2) broadcast topology and (3) star topology, as illustrated in Figure 8.1(b), (c) and (d) [6]. In any topology, one of the devices functions as a coordinator that controls the communication process. The peer-to-peer

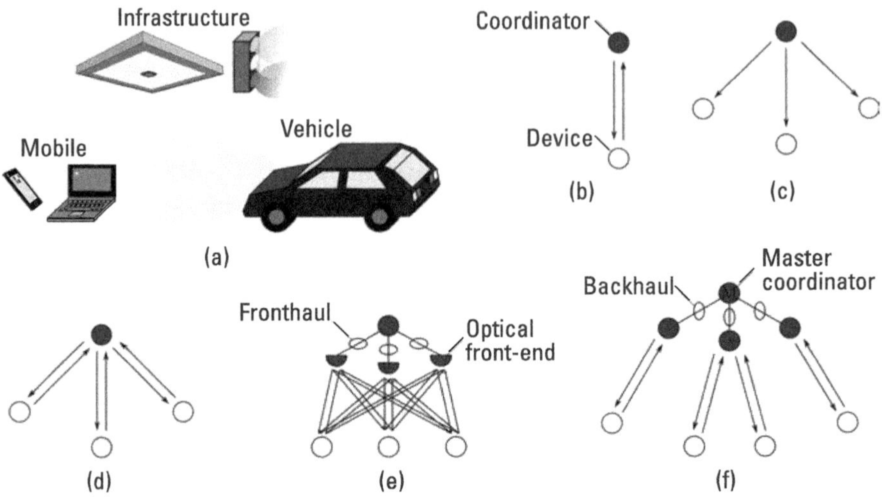

Figure 8.1 (a) Device classifications. (b) Peer-to-peer. (c) Broadcast. (d) Basic star. (e) Distributed MIMO star. (f) Coordinated star.

topology happens between two devices with bidirectional communication. In this topology, the device that initiates the communication typically functions as the coordinator. Usually, large file transfers between two mobile type devices use the peer-to-peer topology. The broadcast topology refers to a unidirectional communication from a coordinator to a number of surrounding devices. There is no need to have a network-forming procedure in this topology and the destination addresses are also not required. In cases of infrastructure-to-vehicle (I2V) applications, the infrastructure could broadcast information on traffic conditions to nearby vehicles using this topology. The star topology refers to bidirectional communications between a coordinator and multiple devices. A coordinator will organize a network, and other devices can send requests to the coordinator to join the network. It is worth noting that networks with different coordinators are independent of each other. Typically, wireless broadband services provided by an indoor infrastructure to multiple mobile devices use the star topology.

In the IEEE 802.15.13 standard, another two additional extended star topologies are introduced [8]. The first one is called the distributed MIMO topology as illustrated in Figure 8.1(e). This topology is used when multiple optical front ends are available at the coordinator device. Note that the fronthaul connection refers to the wired connections between the master coordinator and the multiple front ends that have no baseband processing capability. By using this topology, MIMO functionality can be implemented in some scenarios. The second topology is called the coordinated star topology, as shown in Figure 8.1(f). In the case of multiple coordinators located in the same area, a master coordinator is used to control all the remaining coordinators. The link between coordinators can be realized by backhaul connections, which are defined as the connection from a coordinator targeted to the wired networks. The master coordinator is responsible for the coordination between networks formed by other coordinators, which includes mobility management, handover, and interference mitigation functionalities.

8.2.2.2 Dimming Support

In many use cases, lighting may be required on top of communications, especially for infrastructure-type devices. To support changes in target brightness levels, the IEEE 802.15 OWC standards specify several dimming support approaches [6]. The first approach is called compensation symbols' insertion dimming. The basic idea is to insert a sequence of compensation symbols with one brightness level between adjacent data frames with another brightness level. By controlling the length of the compensation symbols, the average brightness level can be adjusted, as illustrated in Figure 8.2(a). The operation point of the optical source is stable when using compensation symbols, but a change in the dimming level also affects the transmission data rate. In cases that use pulse width-based modulations such as variable PPM (VPPM), pulse width control dimming can be used, as illustrated in Figure 8.2(b). The data rate of this approach is independent of the dimming level, while the pulse width-based modulation requires higher system bandwidth. Another straightforward dimming approach is controlling the amplitude of the signal directly, as shown in Figure 8.2(c). This approach is primarily used in color shift keying (CSK)-based modulations, where color coordinate shifts should be avoided when changing the signal amplitude. There is one more unique approach called out-of-band frequency control dimming, which is illustrated in Figure 8.2(d). The signal processing units and optical sources have limited operation bandwidth. By using this characteristic, a high-frequency waveform with a controllable duty cycle can be used as the input of the optical source, and only the useful low-frequency components with a lower brightness level will survive after signal processing. In the IEEE 802.15.13 standard, the dimming control function is independent of the communication functionalities [8]. This is because the dimming level can be controlled by a DC-bias current in an OFDM-based system and the signal decoding is orthogonal to the DC component of the detected waveform. However, compared to the other dimming methods, this approach will change the swing range of the communication signal. Consequently, the dimming function will affect the signal quality, if the optical source has a serious nonlinearity issue.

8.2.2.3 Flickering Mitigation and Visibility Support

The communication function of an OWC system relies on variations of light brightness. When this brightness varying frequency is lower than 200 Hz, it

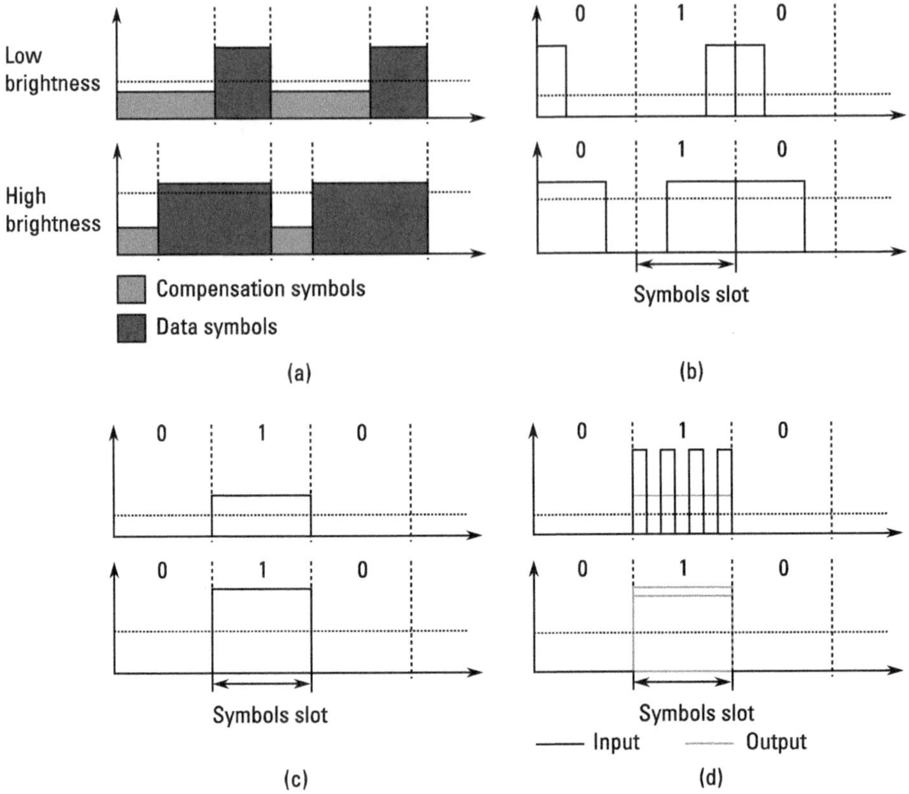

Figure 8.2 (a) Compensation symbols insertion dimming. (b) Pulse width control dimming. (c) Amplitude control dimming. (d) Out-of-band frequency control dimming.

begins to be perceivable to the human eyes and causes the flicker effect [6]. For the convenience of determination of flickering, the maximum flicker time period (MFTP) limit is used, where the brightness level should not stay the same longer than this MFTP. Various methods are specified in the IEEE 802.15 OWC standards to mitigate the flickering effect. For example, run length limited (RLL) line coding is widely used to avoid the long sequence of the same symbols. By using the RLL line code, the low-frequency component of the signal can be suppressed and the intraframe flickering effect can be mitigated. When there is no data traffic or the infrastructure-type device is in listening mode, the optical source is idle and has no data to transmit. However, the lighting function may still be required to maintain visibility. If a constant brightness level is used, it may cause interframe flickering. In order to avoid this issue, an idle pattern frame is introduced to let the optical source transmit some noninformation-carrying AC waveforms. By changing the duty cycle of the idle pattern, the dimming function can also be realized. In OFDM-based systems specified in the IEEE 802.15.13 standard, DC and very low frequency subcarriers can be disabled to avoid flickering [8].

8.2.2.4 Data Transmission Reliability

The IEEE 802.15 OWC standards also define multiple mechanisms to improve the reliability of the data transmission from various aspects. First, forward error correction (FEC) coding is used to lower the transmission bit error ratio (BER) [6]. In the case of star network topologies, multiple devices are connected to one coordinator device. To fulfill this concurrent connection, the number of multiple access schemes has been specified in the standards. In the IEEE 802.15.7 standards, the primary multiple access schemes are contention-based. In the IEEE 802.15.13 standard, a coordinated system architecture is used [8]. Therefore, the primary multiple access schemes are contention-free-based. In addition, acknowledgment frames and cyclic redundancy checks are included to verify successful data reception.

In cases where there are heavy load conditions, the coordinator may have no extra time resources for new devices entering the star network. The multiple bands (multicolor) scenarios can be used to accommodate a greater number of devices. A channel assignment look-up table is defined specifically to this unique multiple band functionality [6]. In cases where multiple cells are deployed close to each other with multiple band capability, a band-hopping scenario can be employed to avoid interchannel interference. The band-hopping pattern is prearranged in a manner that does not change the lighting performances.

In some cases, devices in the network may move from the coverage of one cell to the coverage of another cell. The IEEE 802.15 OWC standards employ mobility support mechanisms for such a situation. In IEEE 802.15.7 standards, a PHY switch can be used to control multiple infrastructure-type devices with different coverage regions to realize mobility support. In the IEEE 802.15.13 standard, a master coordinator can be used to achieve these functions. The coordinator can detect the device mobility via an uplink signal. However, an outage may still occur when there is a temporary blockage or a drop in signal strength. To mitigate transmission losses in this situation, fast link recovery algorithms have been introduced [6]. This process is triggered when the coordinator fails to receive an acknowledgment within a specified time. To guarantee the reliability of the fast link recovery signaling, the lowest transmission rate is used. The fast link recovery process can be initiated from either the coordinator or the devices in the network.

8.2.2.5 Data Transmission Security

OWC systems are different from other types of wireless networks from a security perspective. This is because the optical signal is unable to travel through opaque materials such as walls, which makes OWC transmission more secure in nature. In addition to this physical layer feature, security algorithms are provided in the IEEE 802.15 OWC standards for services including data confidentiality, data authenticity, and replay protection [6]. Data confidentiality services ensure that the transmitted information can be decrypted by the intended party. Data authenticity services ensure the received information is identical to that been transmitted. Replay protection services make sure the transmitted information is not illegally duplicated. The cryptographic mechanism uses symmetric-key cryptography and these keys are generated for processes operated in the higher layers. In addition, it also ensures the storage and security of the authentication keys. Furthermore, some compromises between the cryptographic protection level and key maintenance cost can be made to allow one key to be shared by a group of devices in the same network.

8.2.2.6 Image Sensor-Based OWC

In addition to PDs, optical cameras or image sensors can also be used to detect optical signals. An image sensor is composed of a large number of pixels, which has a high spatial resolution but a slower response speed compared to a PD. Despite the slower supported data rate achieved by OCC systems, optical cameras are available on every smart device, which makes it convenient to deploy commercial uses of OCC systems. In addition, OCC has a wide range of applications related to IOT, augmented reality (AR), and indoor positioning. Attracted by these advantages, significant advances in OCC technology have been achieved in the past decade [5]. Furthermore, with the evolution of smartphones, optical cameras with higher resolution and frames per second (FPS) have emerged every year. All these factors have motivated the standardization of OCC technology initiated in 2014 [5].

In the IEEE 802.15.7-2018 standard, many OCC applications have been considered. For the outdoor applications, the primarily considered scenarios are vehicular communications. For instance, the car taillights can be utilized to send information to other cars behind (V2V communication). Another example of this is in traffic lights, which can be used to send traffic information to vehicles (I2V communication). For indoor applications, the considered scenarios include data transmissions from lighting infrastructure or screens to cameras on mobile devices. The IEEE 802.15.7-2018 standard specifications include a wide variety of OCC modulation schemes and MIMO techniques, which include region-of-interest signaling techniques, rolling shutter techniques, and screen modulation techniques [9]. In addition, multiple operation modes have been provided to support a data rate range from tens of bits per second to a few megabits per second. The specifications in the standards have taken the characteristics and constraints of cameras into account and are compatible with a wide range of cameras to avoid extra hardware modifications.

8.2.2.7 Very High Data Rate OWC

With an increasing interest in achieving ubiquitous wireless connectivity, especially in industrial environments, very high data rate and high-security transmissions (several gigabits per second) between infrastructure and mobile devices/infrastructure without using RF spectra for wireless communication are required in many use cases [7]. It has been found that the simple MAC layer specification under IEEE 802.15.7 standard is unable to fulfill the requirements of these use cases. Consequently, a new task group has been formed to deliver a new standard for this multi-gigabit-per-second OWC scenarios (802.15.13) [7]. The main features of this standard include the capability of providing data rates of up to 10 Gbps and a link distance of up to 200m with LOS path transmission. In addition, coordinated and noncoordinated topologies are both considered. Point-to-point and point-to-multipoint link designs are included. In the case of multicoordinators' system design, mobility support and switching between coordinators have been considered. The standard also includes adaptive modulation and MIMO techniques. Mechanisms to achieve coexisting operation of multiple wireless technologies in the same network have also been included.

The considered OWC systems are mostly expected to be used to control robotics in the manufacturing process and assembly lines in Industrial 4.0 use cases [10]. Secure transmissions in RF-sensitive environments, such as nuclear power plants, petrochemical plants, and hospitals, are another promising application. In addition to the above industrial use cases, some other applications in commercial/business environments include small cell backhaul connections and high-speed wireless access in aircraft cabins or other public transportation environments.

8.2.3 Standard Specification

In order to deliver the numerous features described in Section 8.2.2 in different scenarios, a series of specifications in PHY and MAC layers have been concluded in the IEEE 802.15 OWC standards. In this section, a summary of these specifications is provided to highlight the OWC system capability and performance based on these standards.

8.2.3.1 PHY Layer Specifications

The PHY layer is responsible for achieving a series of functions such as channel selection, data transmission, and reception. Multiple scenarios have been considered in the IEEE 802.15 OWC standards. Therefore, several PHY operation modes have been defined for each of these scenarios. In order to deliver the desired communication performance, the compliant developers shall implement the OWC system based on one of these operation modes.

In the IEEE 802.15.7-2011 standard, three operation modes have been concluded (PHY I to III) [3]. In these operation modes, Reed-Solomon (RS) and convolutional coding (CC) are used to provide error correction capabilities. In addition, various RLL coding configurations have been used in the PHY I and II

Table 8.1
PHY Specifications of Operation Mode I to III in IEEE 802.15.7

Mode	Modulation	Optical Clock Rate	Data Rate
I	OOK, VPPM	200, 400 kHz	11.67–266.6 kbps
II	OOK, VPPM	3.75–120 MHz	1.25–96 Mbps
III	4, 8, 16-CSK	12, 24 MHz	12–96 Mbps

modes to remove low-frequency signal components, thereby achieving flickering mitigation. The PHY I and II operation modes are oriented to applications with low and high data rates requirements, respectively. On-off keying (OOK) and VPPM modulation schemes are used in both operation modes. The PHY I mode supports data rates in the range of tens of bits per second to hundreds of kilobits per second, while the PHY II mode supports data rates in the range of a few to tens of megabits per second. The PHY III mode also supports high data rates but with the requirement of using multicolor optical sources to implement the CSK modulation. It is worth noting that the PHY II and III modes use similar signaling frequencies with intensity modulation. Therefore, the systems implementing the PHY III mode should be able to detect any operation of the PHY II device to avoid interference between the two operation modes. For this reason, all PHY III mode devices should also implement the PHY II mode for device discovery. The main PHY layer characteristics of the PHY I to III modes are concluded in Table 8.1.

In the IEEE 802.15.7-2018 standard, another three operation modes have been included for OCC in addition to the previous three operation modes (PHY IV to VI) [6]. These OCC PHY specifications use application-based configurations. These operation modes are defined to support various OCC applications with cameras as optical detectors. The first operation mode, PHY IV, is designed to support cases with multiple optical sources that are spatially separable to the camera detector, as illustrated in Figure 8.3(a). This scenario is especially suitable for vehicular communications as the two headlights/taillights are visually separable from a distant location. For this scenario, low rate modulation schemes such as undersampled frequency shift on-off keying (UFSOOK) and spatial 2-phase shift keying (S2-PSK) can be used to identify the detection region of interest (ROI) [9]. During data transmission, high data rate modulation schemes can be used such as Twinkle VPPM and hybrid spatial-phase shift keying (HS-PSK). Twinkle-VPPM is a method combining a low data rate stream using UFSOOK modulation with a high data rate stream using VPPM. Similarly, HS-PSK is proposed as a combination of a low data rate stream and a high data rate stream with S2-PSK and dimming spatial 8-phase shift keying (DS8-PSK) waveform, respectively [11]. Regarding the type of cameras, global shutter cameras are more suitable for the PHY IV operation mode. The

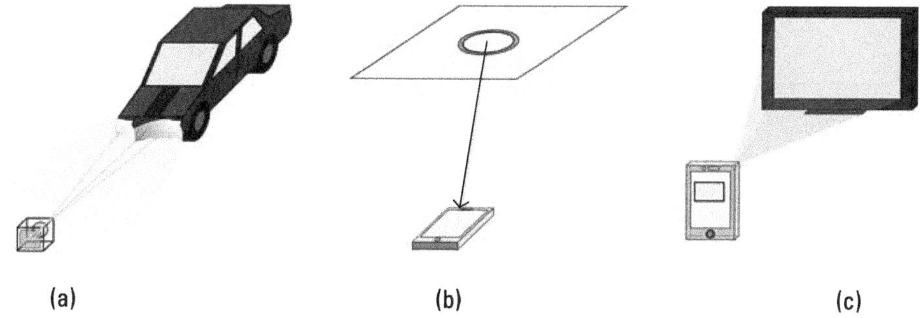

Figure 8.3 (a) Use case of PHY IV. (b) Use case of PHY V. (c) Use case of PHY VI.

achievable data rate is in the range of a few bits per second up to 22 kbps. The second operation mode PHY V is targeted at cases with diffused surface optical sources as transmitters, which can be used for various indoor applications, as shown in Figure 8.3(b). In this scenario, rolling shutter-based cameras are the desired candidates for signal detectors. This is because a rolling shutter camera is able to expose each row/column of pixels at a very high sampling rate. By using this pixel exposure feature to capture the instantaneous light intensity, it is possible to detect the fast changes of the incoming optical signal. For this operation mode, the considered time-domain modulation schemes include camera on-off keying (C-OOK) and mirror pulse modulation (MPM), while the considered frequency-domain modulation schemes include rolling shutter frequency shift keying (RS-FSK) and camera M-ary frequency shift keying (CM-FSK). Up to 5.71 kbps is achievable in the PHY V mode. The last operation mode PHY VI considers the cases using device screen/display to transmit data to cameras, as illustrated in Figure 8.3(c). This type of transmitter has a high DOF in spatial and wavelength domains. Therefore, several modulation techniques are available for this operation mode such as asynchronous quick link (A-QL), variable transparent amplitude-shape-color (VTASC), and sequential scalable 2-D color (SS2DC) [6]. Generally, these modulation schemes use various 2-D codes using DOF in space and color to embed binary information. In addition, receiver distance-adaptive data rate control has also been considered in the standard. PHY VI also includes another group of modulation schemes called invisible data embedding (IDE), which embed data symbols in the form of an invisible image or watermarking into a specific region of visual frames. Due to the limited frame rate of screen transmitters, flickering exist in PHY VI mode. However, this flickering from the screen is proved to be acceptable in practice [9]. The achievable data rate of the PHY VI mode is in the range of hundreds of bits per second to hundreds of kilobits per second. The main PHY layer characteristics of PHY IV to VI modes are concluded in Table 8.2.

The PHY specifications of IEEE 802.15.13 standard focus on delivering OWC systems with very high data rate transmissions [8]. Regarding channel

Table 8.2
PHY Specifications of Operation Mode IV to VI in IEEE 802.15.7-2018

Mode	Modulation	Optical Clock Rate	Data Rate
IV	UFSOOK, Twinkle VPPM, S2-PSK, HS-PSK, Offset-VPWM	10 Hz–10 kHz	5 bps–22 kbps
V	RS-FSK, C-OOK, CM-FSK, MPM	10 Hz–12.5 kHz	60 bps–5.71 kbps
VI	A-QL, HA-QL, VTASC, SS2DC, IDE	10–30 Hz	140 bps–512 kbps

coding, CC and low density parity check (LDPC) codes are adopted in this standard. There are three operation modes: pulsed modulation PHY (PM-PHY), low bandwidth PHY (LB-PHY), and high bandwidth PHY (HB-PHY). The PM-PHY mode uses OOK with 8B10B line coding or M-ary PAM (M-PAM) with Hadamard coding, which has improved energy efficiency and fewer nonlinearity issues. In addition, frequency-domain equalization (FDE) has been adopted in this operation mode. The PM-PHY mode is suitable for uplink transmission from mobile type devices with limited battery capacity or applications requiring energy-efficient transmissions. The achievable data rate is in the range of a few megabits per second to 150 Mbps. The LB-PHY operation mode is designed for spectral efficient OWC systems with relatively narrow modulation bandwidth due to front-end limitations (1–32 MHz). To fulfill the spectral efficiency requirement, DC biased optical OFDM (DCO-OFDM) with multilevel modulation depths is used. When energy efficiency is required, enhanced unipolar OFDM (eU-OFDM) can be adopted to achieve a balance between spectral efficiency and energy efficiency at a cost of extra complexity. The achievable data rate of the LB-PHY mode is up to 86.4 Mbps. The last operation mode HB-PHY is targeted at ultrahigh-speed OWC systems with fast optical front ends such as laser-based optical sources or micro-LEDs. By using these advanced front ends, the optical clock rate of up to 1 GHz can be used. In order to maximize the achievable data rate, DCO-OFDM with adaptive bit loading is used in the HB-PHY mode. This operation mode is suitable for applications requiring a very high data rate such as virtual reality (VR). The achievable data rate HB-PHY mode is up to 10 Gbps. The main PHY layer characteristics of PM-PHY, LB-PHY, and HB-PHY modes are concluded in Table 8.3.

High-speed OWC systems usually rely on a LOS propagation path. In order to assist the direct transmission between the coordinator and a device, relay functionality has been introduced in the IEEE 802.15.13 standard, which includes full duplexing (FD) relaying and half-duplexing (HD) relaying [8]. In addition, both amplify-and-forward (AF) and decode-and-forward (DF) modes are considered. In order to further enhance the transmission performance, the optional adaptive MIMO functionality is introduced in IEEE 802.15.13, which

Table 8.3
PHY Specifications of Operation Mode PM-PHY, LB-PHY, and HB-PHY in IEEE 802.15.13

Mode	Modulation	Optical Clock Rate	Data Rate
PM-PHY	*M*-PAM with FDE	12.5–200 MHz	9.4–150 Mbps
LB-PHY	DCO-OFDM, eU-OFDM	1–32 MHz	0.3–86.4 Mbps
HB-PHY	DCO-OFDM	25 MHz–1 GHz	11 Mbps–10 Gbps

includes two MIMO modes: repetition coding (RC) and spatial multiplexing (SMP) [12]. RC mode aims to enhance transmission reliability, while the SMP mode is designed to further improve transmission data rate on top of the spectral efficient OFDM.

In the IEEE 802.15 OWC standards, three data modes are supported by the MAC layer. For all data modes, the fast locking pattern (FLP) is used to achieve synchronization, while topology-dependent patterns (TDPs) are used to indicate the used PHY topology. The first mode is called single-mode, which has one physical protocol data unit (PPDU) in each data frame. This data mode is designed for very short data transmission such as sending acknowledgments and transmitting beacon signals. Another data mode is called the packed mode, which includes a number of PPDUs per data frame. Since more information data is packed using one copy of an overhead signal, the transmission efficiency is high, which makes the packed mode the preferred data mode for data transmission. The last data mode is called the burst mode, which uses shorter PHY preambles and reduced interframe space (RIFS) instead of short interframe space (SIFS) for even higher transmission efficiency. Intuitively, this data mode is used when there is a high traffic load to transmit a large amount of data.

8.2.3.2 MAC Layer Specifications

The IEEE 802.15 OWC standards have specified a series of MAC layer functions and frame formats. The specified basic MAC functions include optical wireless personal area network (OWPAN) management, channel access, synchronization, data transmission, data reception, and acknowledgment [6]. OWPAN management is responsible for scanning, initiating, maintaining, and terminating an OWPAN. In addition, device association and disassociation with the OWPAN are also part of the OWPAN management function. Channel access specifications provide the wireless channel accessing protocols for devices. In the IEEE 802.15.7 standards, contention-based access is the primary channel access mechanism, while a contention-free channel access scheme is also available. This is implemented by using a superframe structure that is composed of a beacon period (BP), a contention access period (CAP), and an optional contention-free period

(CFP) in a beacon-enabled OWPAN. The BP is used for network identification, user synchronization, and providing superframe structure information. The CAP is used for contention-based channel access. The channel access method includes a random access scheme and the CSMA/CA scheme. Each channel access scheme is either unslotted or slotted. CFP is used for the contention-free channel access when a guaranteed time slot (GTS) is allocated. There is also nonbeacon-enabled OWPANs, where only unslotted random access schemes can be used. In the IEEE 802.15.13 standard, both beacon-enabled and nonbeacon-enabled OWPANs are included [13]. The superframe structure in the beacon-enabled OWPANs is similar to that in IEEE 802.15.7, except that the primary channel access is contention-free-based. The contention-based access is only used during the association and transmission resource request procedures. A slotted ALOHA scheme is used as the random access mechanism. A GTS is allocated to each device during the CFP by the coordinator in an orthogonal fashion according to a dynamic TDMA scenario. To guarantee there is no interference between devices, a turnaround interframe space (TAIFS) is inserted between adjacent GTSs. In cases with nonbeacon-enabled OWPANs, a set of polling mechanisms is used to realize the contention-based and the contention-free channel access. Due to the difference between the nature of the IEEE 802.15.7 standard and that of the IEEE 802.15.13 standard, the MAC specifications in IEEE 802.15.7 standards are simpler in general, while the MAC specifications in IEEE 802.15.13 have more centralized control features [8].

In addition to the basic MAC functions discussed above, the IEEE 802.15 OWC standards feature a number of additional MAC layer functions. In the IEEE 802.15.7, a fast link recovery function has been specified to enhance the robustness against temporal link disconnections. In addition, dimming, color, and flickering mitigation control functions are specified in detail to guarantee the light visual quality of the optical signal. Furthermore, multiband resource allocation and mobility support functions are also included for the multicell scenarios. However, the IEEE 802.15.13 standard focuses on high spectral efficiency. Therefore, data unit fragmentation, reassembly, and aggregation functions are specified to support efficient data transmission [13]. In addition, adaptive transmission and multioptical front-end channel estimation functions are also specified to support adaptive bit-loading and MIMO techniques.

8.2.4 Standard Implementations

With the publication of standards IEEE 802.15.7-2011 and IEEE 802.15.7-2018, various industrial companies and research organizations have attempted implementation of OWC systems based on these standards for validation, demonstration, or commercialization purposes.

8.2.4.1 Implementations Based on the IEEE 802.15.7-2011 Standard

In 2014, transceiver units based on the standard IEEE 802.15.7-2011 PHY-I operation mode have been implemented by researchers from the Hong Kong University of Science and Technology [14]. This initial design is implemented on field-programmable gate arrays (FPGAs) with a target data rate of 11.67 to 266.6 kbps. In particular, the layout of the transmitter design covers only a small area of 1.56 mm × 1.56 mm, which demonstrates the viability for device integration. The same research group has reported a fully integrated implementation of a similar system design on a chip in 2015 [15]. The main feature of this implementation is that the entire transmitter unit is integrated on a printed circuit board (PCB) with an area of 4 mm × 2 mm. Another major achievement of this implementation is its high power efficiency. The system achieves 80% to 92% power efficiency for output power up to 8W. The power consumption of the digital signal processing (DSP) processing unit is only 1.25 mW [15].

Another OWC system implementation based on IEEE 802.15.7-2011 considers the applications of vehicular communications [16]. This implementation also complies with the standard IEEE 802.15.7-2011 PHY-I operation mode. It is featured with a relay configuration of an infrastructure-to-vehicle-to-vehicle (I2V2V) scenario and an ultralow latency, which is down to submillisecond level. The tested link distance is up to 50m. This implementation demonstrates the potential of using the IEEE 802.15.7 standard in future safety-critical ITS. Regarding commercial products based on the IEEE 802.15.7-2011 standard, LG Electronics and Kimin Electronics have attempted to develop a VLC connector for mobile users [17]. This device has a maximum link distance of 13m with an FOV of 25°, which is designed for indoor use cases only.

8.2.4.2 Implementations Based on the IEEE 802.15.7-2018 Standard

An early-stage OWC system implementation based on the IEEE 802.15.7-2018 PHY operation mode IV was reported in [11]. It considers a use case with the two headlights of a vehicle as transmitters and an image sensor as the detector. The implementation uses ROI signaling techniques with S2-PSK modulations. In the implementation, both rolling shutter and global shutter cameras are used to measure the achievable SNR. The measurement distance is in the range of 5 to 200m. The resulting SNR is in the range of 10 to 55 dB. This implementation demonstrates a potential application of OCC specifications in the IEEE 802.15.7-2018 standard in vehicular communications.

Panasonic has introduced a product named LinkRay, which is a direct application of the IEEE 802.15.7-2018 standard [18]. This product applies the PHY-V operation mode of the standard. LinkRay uses a smartphone camera to capture data from specific light ID transmitters to provide interactive signage communication services. The product is able to work in crowded spaces and in distance and support capturing multiple IDs simultaneously. To support the application of LinkRay, the operating system, modulators, and displays are available from the company. A limitation of the system is that the data reception may be affected by direct sunlight, which implies that it is more suitable for indoor applications. Another implementation of the OCC system based on neuromorphic vision sensor (NVS) has been reported by Intel Corporation, where NVS is a type of optical sensor with asynchronous pixels operated independently [19]. In this implementation, an LED tag prototype with multiple LEDs with different bands has been developed.

8.3 ITU-T G.9991 Standard

The development of the ITU-T G.9991 Recommendation (also known as G.vlc) is considered to be a major milestone, paving the way for wide adoption of VLC technology in a wide range of applications. In this section, we discuss the development efforts, main features, and the standard specifications of this recommendation. We also shed light on some of the real-life VLC implementations that are being carried out based on this standard.

8.3.1 Standard Development

The ITU is the United Nations specialized agency in information and communication technology (ICT) sector. The ITU-T is the body responsible for coordinating standards and issuing recommendations on emerging telecommunication technologies. The framework for the ITU-T is composed of the following entities:

- *World Telecommunication Standardization Assembly (WTSA):* This assembly takes place every 4 years to decide on the topics to be studied by ITU-T, to establish study groups and to define general policies.
- *Telecommunication Standardization Advisory Group (TSAG)*: The TSAG takes place in between the WTSAs and aims to review the progress of the work program. It also advises on financial matters, strategies, and working procedures.
- *Technical Study Groups:* The ITU-T study groups carry the technical work as recommended by the WTSA and the TSAG. Based on the carried work, they publish recommendations for the development of telecommunications technologies. The ITU-T G.9991 Recommendations were developed by Study Group 15: Networks, Technologies and Infrastructures for Transport, Access and Home. This study group focuses on optical communication technologies, including

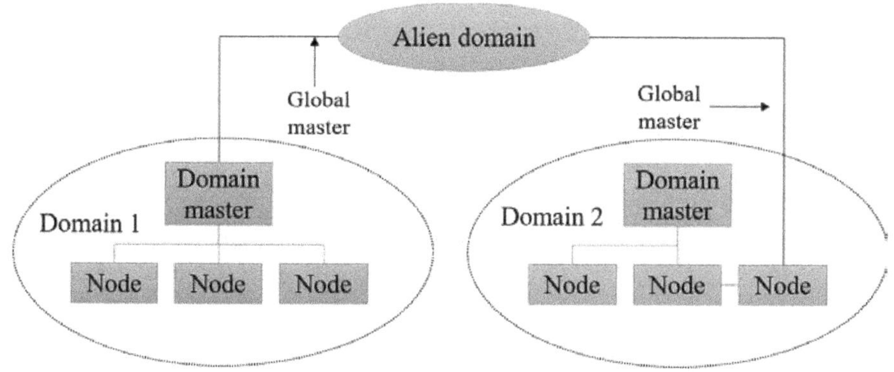

Figure 8.4 ITU-T G.9991 home network architecture reference model.

the development of standards for optical transport networks, access networks, home networks, and power utility networks.

- *Workshops and Seminars:* The ITU-T runs a number of technical events that are free and open to the public. These aim to reveal the latest standardization efforts, discuss the ongoing developments, and explore new directions.
- *Technology Watch:* The ITU-T Technology Watch publications provide up-to-date reports on the latest information and communication technologies in a language that is accessible to nonspecialists.

The development of the ITU-T G.9991 standard started in 2015 by Study Group 15 within ITU-T is considered to be one of the earliest efforts to standardize the technology on which indoor VLC systems can be built. The standardization efforts carried out by the ITU-T has resulted in the new Recommendation G.9991: "High-Speed Indoor Visible Light Communication Transceiver - System Architecture, Physical Layer And Data Link Layer Specification," which was finalized and published in 2019.

8.3.2 Standard Features and Specifications

8.3.2.1 Architecture and Network Topologies

The ITU-T G.9991 home network architecture is shown in Figure 8.4. The model consists of one or more optical wireless domains that are connected to an alien home access network. The home access network could be PLC, digital subscriber line (DSL), passive optical network (PON), or Ethernet network. Each domain comprises a domain master with one registered node or multiregistered nodes. Nodes inside the same optical wireless domain connect via VLC (typically for the downlink) or IR (typically for the uplink). A global master logical function is responsible for coordinating the interdomain operations such as bandwidth allocation and interdomain handover, as well as the intradomain operations between the optical network and the alien domain. The global master function is based on the ITU-T G.9960 Recommendation: Unified High-Speed Wireline-Based Home Networking Transceivers. The global master exchanges information with the domain masters of all coordinated domains via logical interfaces and the information exchange protocols between the global master and the domain masters are unified for all coordinated domains. Typically, the nodes are not visible to the global master except in cases where the global master is colocated with a node.

The communication medium defined in the ITU-T G.9991 Recommendation compromises the visible light band (wavelength from 380 nm to 780 nm), and the IR band (wavelength from 800 nm to 1,675 nm). The transmitter nodes support the transmission of visible light signals, IR signals, or both visible light and IR. The receiver nodes are capable of receiving either one band (visible light, IR) or both of the two bands. In the case of receiving two bands, it can be achieved by using a single PD or using different PDs. A node can join a domain only if it supports the capability to receive signals from the domain master (i.e., if the node and the domain master uses matching wavelength bands).

The ITU-T G.9991 Recommendation defines two modes of operation for the domains, namely: the centralized mode and the unified mode. In the centralized mode, the domain master can directly communicate with any node within the domain, whereas communication between the end-point nodes is not allowed. Hence, the end-point nodes can only communicate via the domain master. In the unified mode, nodes can communicate with each other directly or indirectly (via relay nodes). Besides these two communication modes, broadcast and multicast modes are supported in any domain.

8.3.2.2 Reference Model

The ITU-T G.9991 standard presents a comprehensive description of the DLL and PHY layers for high-speed indoor OWC using visible light and IR. The ITU-T G.9991 Recommendation includes one architecture for the DLL layer and two different architectures for the PHY layer: (1) a DCO-OFDM PHY layer based on G.9960/61 for which silicon solutions are readily available, and (2) an asymmetrically clipped optical OFDM (ACO-OFDM) PHY layer in which the lower layers of G.9960/61 are modified to satisfy extreme lighting constraints such as deep dimming. A device is considered to comply with this recommendation if it implements the common DLL layer and any of the two PHY layers. The reference model used for the ITU-T G.9991 is illustrated in Figure 8.5. It follows the protocol reference model of a home network transceiver as defined in the ITU-T G.9960 Recommendation. The reference model includes three main reference points: application interface (A-interface), medium-dependent interface (MDI), and physical medium-independent interface (PMI). The MDI

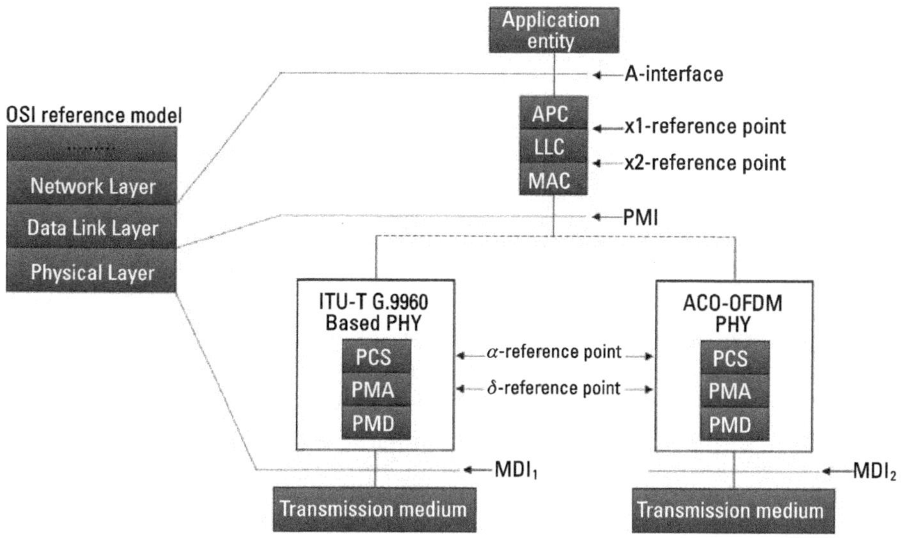

Figure 8.5 Reference model for ITU-T G.9991 Recommendation.

is a physical interface defined in terms of the physical signals transmitted over a specific medium and a mechanical connection to the medium, while the PMI interface is both medium-independent and application-independent. The A-interface is user application-protocol specific. Moreover, the reference model defines two intermediate reference points, x_1 and x_2, for the DLL as well as and two intermediate reference points, α, and δ, for the PHY layer. All intermediate reference points are independent of the type of medium and are defined as logical interfaces in terms of the functional flows and logical signals.

The DLL in the reference model comprises three sublayers: the application protocol convergence (APC) sublayer, the logical link control (LLC) sublayer, and the MAC sublayer. The APC sublayer's role is to provide an interface to the application entity as well as providing data rate adaptation between the application entity and the home network transceiver. The LLC sublayer is responsible for coordinating all connections of the nodes inside the domain according to the domain master's requests and the QoS requirements for the different connections. Moreover, the reference model defines three sublayers for the PHY layer: physical coding sublayer (PCS), physical medium attachment (PMA), and physical medium dependent (PMD). The functions of the PCS sublayer include controlling the data flow between the MAC and PHY layers, compressing transmit media access control protocol data units into the PHY layer frames and adding PHY layer control and management overhead. The PMA sublayer is responsible for encoding the PHY layer frame content before transmitting it. The PMD is responsible for modulating and demodulating the PHY frames using ACO-OFDM or DCO-OFDM.

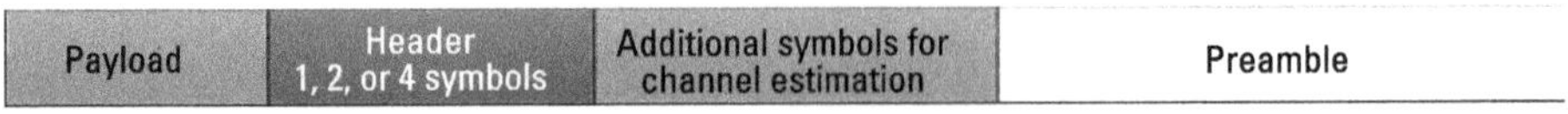

Figure 8.6 PHY frame in the ITU-T G.9991 Recommendation.

8.3.2.3 Physical Layer Specifications

Figure 8.6 shows the format of the PHY frame as defined in the ITU-T G.9991 Recommendation. The PHY frame includes a header and a payload, both carrying ACO-OFDM or DCO-OFDM data symbols. Varying bit-loading and different coding parameters can be used depending on the channel conditions, noise levels and the QoS requirements. The preamble and additional channel estimation symbols are added in the PMD sublayer. The preamble is utilized for synchronization and initial channel estimation, and it does not contain any actual data.

The ITU-T G.9991 Recommendation does not define specific values for transmitting power, since the power levels required to drive the emitting devices may differ between different implementations. For the OFDM symbols, the control parameters vary according to the transmission bands. The OFDM modulator consists of the following major parts: IDFT, clipping, cyclic extension, windowing, and frequency up-shift. The incoming signal to the modulator at the lth OFDM symbol for the ith subcarrier is the complex value $Z_{i,l}$ generated by the constellation encoder (for symbols of the header and the payload) or by preamble generator (for symbols of the preamble). In an OFDM-based VLC system, only half of the available subcarriers are used for actual data, transmission and the other half are utilized to impose Hermitian symmetry in order to satisfy the requirements of intensity modulation (i.e., to ensure that the IDFT outputs are real-valued time-domain samples). Thus, we have:

$$Z_{0,l} = Z_{N/2,l} = 0$$

and

$$Z_{i,l} = Z^{*}_{N-i,l} \quad \text{for } i = 1, 2, ..., N/2 - 1$$

where * denotes complex conjugation. After generating the time-domain samples, the real part of the resultant signal is clipped at the zero levels. Then the cyclic prefix is added. The cyclic extension provides a guard interval between adjacent OFDM symbols. This guard interval, denoted as N_{GI}, is intended to protect against ISI. The first β samples of the cyclic prefix and last β samples of the IDFT output shall be used for shaping the envelope of the transmitted signal. This process is known as windowing. The window function facilitates power spectral density (PSD) shaping and reduction of the out-of-band PSD. The number of windowed samples, β, is the same for all of the payload symbols of the same frame, as well as the PHY-header and preamble. After adding the cyclic prefix,

Table 8.4
OFDM Parameters in ITU-T G.9991

Parameter	50 MHz	100 MHz	200 MHz
Number of subcrriers N	256	512	1,024
Subcarrier frequency F_{SC}	195.3125 kHz	195.3125 kHz	195.3125 kHz
Guard interval N_{GI}	$N/32 \times k$ for $k = 1, ..., 8$	$N/32 \times k$ for $k = 1, ..., 8$	$N/32 \times k$ for $k = 1, ..., 8$
Windowed samples β	$N/32 = 8$	$N/32 = 16$	$N/32 = 32$
Up-shift frequency F_{US}	25 MHz	50 MHz	100 MHz

the time-domain samples are frequency up-shifted by F_{US}. The frequency up-shift offsets the spectrum of the transmit signal shifting it by F_{US}, where the value of F_{US} is a multiple of the subcarrier frequency F_{SC}. Table 8.4 summarizes the OFDM control parameters defined for VLC transmissions in the ITU-T G.9991 standard.

8.3.2.4 Data Link Layer Specifications

As shown in Figure 8.5, the A-interface is the demarcation point between the application entity and the DLL; the PMI interface is the demarcation point between the DLL and the PHY layer. The internal reference point x_1 provides a functional separation between the APC and the LLC, while the internal reference point x_2 provides a functional separation between the LLC and the MAC sublayer. The application data primitive (ADP) sets enter from the application entity to the DLL through the A-interface. The incoming ADP sets' format is application-specific. For example, for Ethernet application entities, the incoming ADP set follows one of the standard Ethernet formats. Upon arrival at the APC, the ADP sets are converted into APC protocol data units. The APC also identifies ADP classification primitives, which are then used by the LLC to support the specific QoS requirements. The APC is also responsible for establishing flows of the APC protocol data units between peer APCs and assigning one or more queues for these flows according to the classification information. In the LLC, the incoming data units are arranged into LLC frames and encrypted using assigned encryption keys. The LLC is also responsible for retransmission and relay operations. The MAC sublayer is responsible for concatenating the incoming data units into MAC protocol data units and then conveying these to the PHY according to the scheduling and transmission queues specifications determined by the LLC. It is noted that the functionality of the APC, LLC, and MAC remains the same across different types of transmission media. Some of the functions and control parameters, however, may be adjusted for the efficient operation of the transceiver over a particular medium.

8.3.2.5 Dimming Support and Flicker Avoidance

The ITU-T G.9991 Recommendation provides two dimming support mechanisms for the two different PHY layer models (i.e., DCO-OFDM-based PHY and ACO-OFDM-based PHY). For DCO-OFDM, the modulation does not guarantee a positive output signal and therefore clipping may occur when using dimming techniques. For this reason, the recommendation specifies adaptive mechanisms to mitigate performance losses for deep levels of dimming. When using a PHY based on ACO-OFDM, the modulation technique used guarantees a positive output signal that is added to the bias current in the LED. Thus, no additional performance loss is caused.

Regarding flicker mitigation, the ITU-T G.9991 modulates the signals at very high frequencies, leading to the absence of, or very negligible, flicker effects. However, flicker can still occur for the low-speed transitions between states (transmission state to nontransmission state) if the average current injected into the medium is different for the two states. In this case, flicker mitigation techniques will need to be put in place or borrowed from other standards.

8.3.3 Standard Implementations

The standard is based on the ITU-T G.hn home networking standards such as the G.9960 and G.9961 Recommendations. Those recommendations were developed to provide home networking solutions over different types of wiring in residential environments such as PLC, coaxial cable, and phoneline. The ITU-T G.hn series adopt a highly adaptive OFDM architecture for PLC systems that can be easily adapted to enable wireless connectivity in VLC systems. Based on this, the existing silicon solutions and hardware built for the ITU-T G.hn standards can be readily used to build VLC transceivers. This approach has led to the fast development of commercial VLC products since the approval of the recommendation by ITU-T, providing a fast and appealing route to the commercialization of VLC solutions.

One of the early adapters of the ITU-T G.9991 standard is Fraunhofer HHI. In [20], Fraunhofer HHI presented a real-time demonstrator designed to offer intrinsically high physical layer reliability for OWC in an industrial production environment. This was achieved by nondirected transmission, increased transmit signal powers, and transmit diversity. The demonstrator was built based on the ITU-T G.hn standards. The OWC link was shown to reach up to 150-Mbps data rates with an average latency of less than 1 ms for real-time mobile scenarios. Higher data rates of up to a few gigabits per second were reported in point-to-point links by also integrating the ITU-T chipsets into Fraunhofer VLC front ends [21].

Oledcomm is another company that joined the ITU-T in December 2019, aiming to contribute to the development of the standard and to propose additional features. The Oledcomm LiFiMAX products range was developed based on the ITU-TG.9991 Recommendation. The LiFiMAX system allows simultaneous transmissions for up to 16 network users, offering ultrahigh-speed connections with up to 100 Mbps for download links and up to 40 Mbps for upload links [22].

Signify, the world leader in lighting, launched its Trulifi product range in 2019, which comprises Li-Fi-enabled luminaires providing wireless connectivity serving multiple users with seamless handover. The range of Trulifi was also built based on the ITU-T G.9991 Recommendation and supports data rates of up to 150 Mbps. According to the company, that is fast enough to stream 30 1080p HDTV movies at once [23].

MaxLinear, a leading supplier of highly integrated RF analog and mixed-signal semiconductor solutions for broadband communications systems, has started providing baseband processors for VLC and OWC applications. MaxLinear chipsets are built based on the ITU-T G.9991 Recommendation and support data rates of up to 2 Gbps.

8.4 IEEE 802.11bb Standard

The IEEE 802.11bb is an amendment to the well-known IEEE 802.11 standard. More than 25 research and commercial bodies are involved in the development of the IEEE 802.11bb light communication standard [24]. The main stakeholders of this standard are lighting companies, telecom operators, internet service provider (ISPs), industrial manufacturers, and automotive and aviation companies. The IEEE 802.11bb task group, for example, is supported by pureLiFi Ltd., Signify, and Fraunhofer HHI, Oledcomm. There also exist prestandard products such as those developed by pureLiFi Ltd, which, for example, allow for the same user authentication mechanisms as available in Wi-Fi.

The IEEE 802.11bb standard amendment focuses on fully networked Li-Fi solutions with handover and mobility support. The standard targets mass-market opportunities of Li-Fi by reusing most of the mature IEEE 802.11 functions and associated services, including MAC layer. The main proposition in terms of the IEEE 802.11bb PHY layer is to shift the center frequency of the baseband waveforms output of the Wi-Fi chipsets by a few megahertz. The availability of suitable chipsets for Li-Fi-enabled devices has the potential to reduce the form factor, cost, and energy consumption of Li-Fi systems. This approach could also enable faster integration and adoption of Li-Fi technology by device manufacturers and network operators.

The IEEE 802.11bb standard amendment provides complementary features to the IEEE 802.11 standard suite, which allows it to address additional use cases such as electromagnetic interference (EMI) sensitive and highly populated environments such as petrochemical plants, conference halls, train stations, and airports [24]. Li-Fi offers a unique secure communication solution that cannot be eavesdropped on by users outside the premises where the Li-Fi AP is deployed. Moreover, Li-Fi can help in offloading some of the users' traffic by providing an alternative means to the RF-based technologies, as discussed in Chapter 4.

8.4.1 Standard Development

A topic interest group (TIG) was formed as part of the IEEE 802.11 working group to consider the technical feasibility of Li-Fi technology and to explore the economical opportunities of using light communication as part of the IEEE 802.11 standard suit. The TIG progressed into a study group and then to a task group "bb" in July 2018. The IEEE 802.11bb standard amendment aims to define a new PHY layer and to minimally modify the MAC layer of the IEEE 802.11 standard in order to allow for improved channel access for Li-Fi systems.

The main difference of the IEEE 802.11bb standard amendment to the other light communication standards is the reliance on the IEEE 802.11 standard suite and its associated services. Therefore, the standard targets mass-market opportunities for the Li-Fi technology. This can be contrasted with the specialty use cases of the OCC that are defined in the IEEE 802.15.7 standard, and the ultrahigh-speed, long-distance OWC industrial use cases that are defined in the IEEE 802.15.13 standard. However, it can be contrasted with the ITU-T G.9991 standard, which mainly focuses on indoor home-networking applications, whereas the IEEE 802.11bb focuses on mobility, networking, and integration to other RF solutions such as 5G and Wi-Fi.

Being part of the IEEE 802.11 ecosystem allows Wi-Fi chipsets to be leveraged for Li-Fi applications with minimal modifications due to the possibility of reusing most of the IEEE 802.11 PHY and MAC layers' functions and services. It is also believed that Li-Fi can gain better technology awareness, which could facilitate faster product development and market introduction. It is a well-known fact that Li-Fi technology can coexist with Wi-Fi technology since it does not create any interference with the existing RF systems. However, the adherence to the IEEE 802.11 MAC layer specifications enables the potential of seamless handover between both technologies (through the use of Fast-Session Transfer) [24]. This allows users to maintain connectivity even when one of the Wi-Fi or Li-Fi links is not available. For example, Li-Fi can offload traffic in crowded environments, when Wi-Fi technology is expected to underperform. Similarly, Wi-Fi can be used to maintain connectivity for cases where the Li-Fi link is unavailable due to a temporary link blockage. In addition, the IEEE 802.11bb standard enables the potential of channel aggregation, in which Li-Fi bandwidth resources could be aggregated with other RF bandwidth resources to improve the system throughput.

The IEEE 802.11bb standard amendment specifies one MAC layer and several PHYs specifications for fixed and mobile light-based indoor wireless connectivity. A new PHY with one mandatory mode and two optional modes is proposed, while the MAC layer will be defined based on a modified version of the 802.11 standard MAC. The main PHY specifications are given below [1]:

- Uplink and downlink communications between 800 nm and 1,000 nm.
- All modes of operations should support a minimum throughput of 10 Mbps, with at least a single operation mode supporting a minimum throughput of 5 Gbps.
- Guarantee interoperability between different lighting communication devices (dongles, access points, and modems) with varying bandwidth capabilities.

The IEEE 802.11bb MAC layer is still under development and has not yet been finalized by the standard committee. However, it is believed that the IEEE 802.11bb amendment will introduce enhancements on the IEEE 802.11 MAC layer that are specific to the Li-Fi technology. The IEEE 802.11bb MAC layer is also assumed to reuse the same authentication, security protocols, addressing, and MAC frame structure of the IEEE 802.11 standard. In addition, it is assumed that the MAC layer will enable handover between Li-Fi and Wi-Fi technologies by supporting all the considered PHY modes [25].

8.4.2 Physical Layer Specifications

The IEEE 802.11bb amendment adopts a common mandatory PHY mode based on the IEEE 802.11 standard, as well as an optional PHY mode based on the IEEE 802.11 standard, termed high efficiency (HE). All the defined PHY layer modes of the IEEE 802.11bb standard amendment are based on IM/DD with modulation bandwidths defined at 20 MHz, 40 MHz, 80 MHz, and 160 MHz. Devices compliant with the IEEE 802.11bb standard amendment should support the common PHY mode and should also support one of the two optional PHY modes [26].

The IEEE 802.11 standard specifies a frequency mask that determines the permissible power emission by an analog transmitter around each frequency band. This is an essential requirement to avoid interchannel and cochannel interference between the different terminals. The OFDM frequency mask as defined in the IEEE 802.11 standard is shown in Figure 8.7. Reference channel models were used to find the optimal central frequency by simulating the effect of the center frequency choice on the error vector magnitude (EVM) [27]. These channels were generated for multiple scenarios using ray-tracing models and validated using measurements [28]. It was found that a trade-off exists between high and low center frequency depending on the frequency profile of the reference channel.

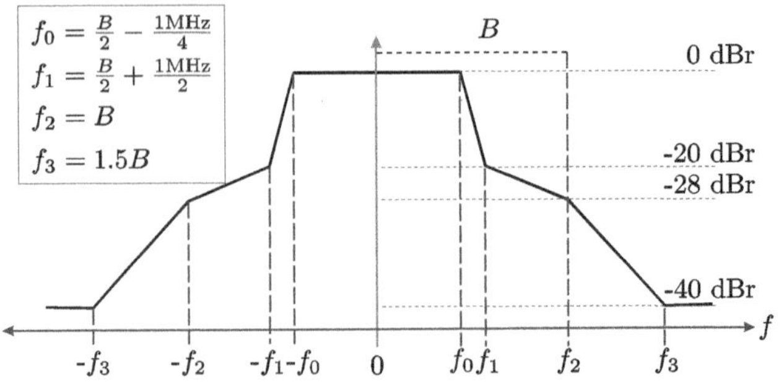

Figure 8.7 The frequency mask of the OFDM waveform based on the IEEE 802.11ax standard [30], where B denotes the single-sided signal bandwidth [29].

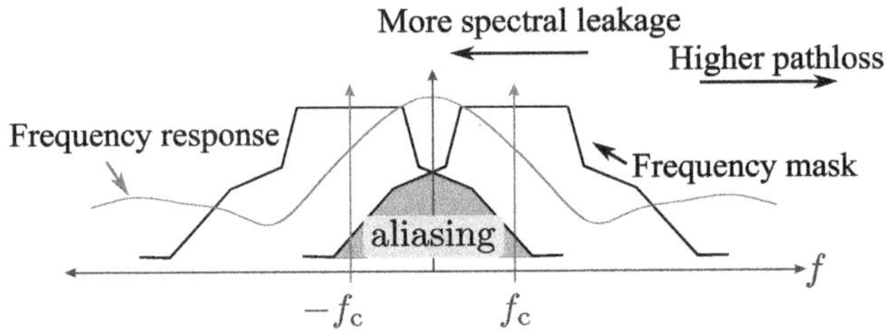

Figure 8.8 A schematic diagram of a frequency response of an off-the-shelf LED with the OFDM frequency mask.

A lower center frequency can lead to an increased spectral leakage, while a higher center frequency leads to attenuation due to the higher path loss and the lower OFE response, as shown in Figure 8.8 [29]. An exhaustive search was performed to find the optimal central frequencies for each of the reference channels and modulation and coding schemes subject to the fact that the EVM does not exceed the allowed values that are defined by the IEEE 802.11 standard [27].

There was a debate around which of the IEEE 802.11bb PHY modes should be adopted as a mandatory PHY mode [29]. The debate was mainly focused on the IEEE 802.11-compliant mode, which is based on shifting the center frequency of the baseband waveform, and the ITU-T G.9991 standard that supports DCO-OFDM with adaptive and power-loading. The IEEE 802.11-compliant mode allows Li-Fi to have a lower entry barrier to market adoption, whereas the ITU G.9991 standard allows for a better utilization of the Li-Fi frequency-selective channel. However, the advantages of using adaptive bit and power-loading depend on the flatness of the channel. These advantages quickly diminish if the channel is considered relatively flat [29].

A study of the flatness of the Li-Fi channel for the IEEE 802.11bb standard amendment was reported in [29]. The study attempted to compare the mandatory PHY IEEE 802.11bb standard and the ITU-T G.9991 standard based on the reference channel models that are defined in the IEEE 802.11bb standard amendment [28] and the reference front-end models [31]. The main finding of the study was that most of the reference channels are flat for a 20-MHz signal bandwidth [29].

The upconversion of the IEEE 802.11 baseband waveform results in wasting valuable bandwidth resources near the DC frequency component. These bandwidth resources tend to have a higher SNR due to the lowpass characteristics of LEDs. An alternative approach was proposed based on transmitting the IEEE 802.11 chipset in-phase and quadrature-phase outputs on two-wavelength channels [32]. This was shown to outperform the mandatory PHY mode of the standard with different Li-Fi channel scenarios [32].

8.4.2.1 Optical Frontend Models

The choice of the OFE imposes a significant impact on the link-level simulations, which could affect the choice of the central frequency [31]. As described in Chapter 6, the transmitter front end of a Li-Fi system is comprised of a light source, which could be based on LEDs or a light amplification by stimulated emission of radiation diodes (LDs), and its driving circuitry. The modulation bandwidth of the transmitter front end is mainly limited by the active emission area in the case of an LED. The driving circuitry is custom-built based on the used light source, since it attempts to equalize the frequency response of the light source using impedance matching techniques. Thus, the frequency profile of the transmitter front end is determined by both the light source and driving circuitry.

The transmitter front end model of the IEEE 802.11bb PHY is described in [31]. A variable gain amplifier (VGA) is used to model the variable modulation index of the input AC [A]. A lowpass filter (LPF) with a variable cutoff frequency is used to model the modulation bandwidth of the light source and a highpass filter (HPF) with a cutoff frequency specified at 100 kHz to model the attenuation at a low-frequency band [31]. A variable DC bias is added before the electrical signal is converted into an optical signal. However, it is noted that the adopted reference model does not take into account the nonlinearity of the transmitter OFE.

The receiver front-end model of the IEEE 802.11bb PHY is described in [31]. The model attempts to approximate the performance of a transimpedance amplifier (TIA) followed by a PD such as an avalanche photodiode (APD) or a P-type intrinsic N-type photodiode (PIN-PD). The PD converts the optical received signal into an electrical signal with a variable conversion ratio that depends on the wavelength of the optical signal. Shot noise is added with an RMS value, I_{Shot} [A], followed by a HPF with a low cutoff frequency at 100 kHz to reject the DC frequency component or low-frequency interference from ambient lighting. Next, thermal noise is added with a RMS of I_{thermal} [A], followed by automatic gain control (AGC) to compensate for the channel attenuation. An LPF is used to approximate the bandwidth of the PD with a variable cutoff frequency that should match the received signal bandwidth [31].

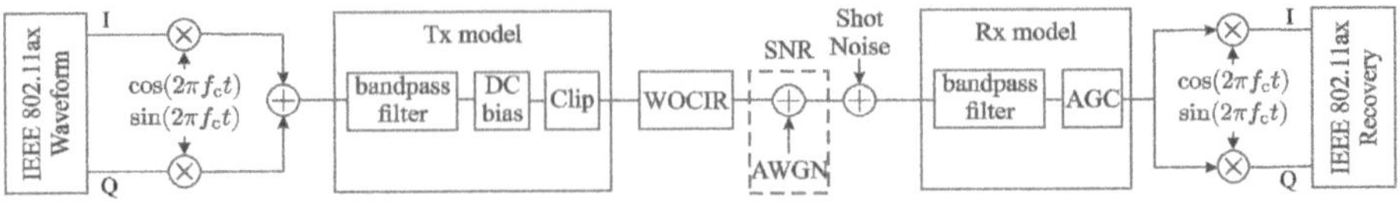

Figure 8.9 A block diagram of the IEEE 802.11bb PHY layer, showing the transmitter and receiver models based on the mandatory common mode using frequency upconversion and downconversion method [32]. Tx refers to the transmitter; Rx refers to the receiver; f_C denotes the center frequency; WOCIR refers to the wireless optical channel impulse response. AGC refers to the automatic gain control.

8.4.2.2 Common Mandatory PHY Mode

The common mode operated on a SISO channel with a 20-MHz bandwidth is based on the IEEE 802.11 standard. The block diagram of an IEEE 802.11bb-compliant transceiver is shown in Figure 8.9. At the transmitter, the in-phase and quadrature-phase baseband outputs of an IEEE 802.11ax chipset are downconverted from the passband to fit into the bandwidth of the OFE at a new center frequency, denoted f_c. A trade-off exists between choosing a low center frequency (which results in more aliasing) and choosing a high center frequency (which results in more attenuation of the frequency response of the OFE), as explained earlier in Section 8.4.2 [29].

The downconverted waveform is added to a DC bias, amplified, and clipped before it is fed to the light source. The common PHY mode operates at a center frequency of 26 MHz with a binary phase-shift keying (BPSK) modulation and an FEC rate at 1/2. The common bandwidth is specified at 20 MHz within the wavelength band between 800 nm and 1,000 nm. This center frequency shall correspond to LC channel 0 (LC0) [33].

At the receiver, the optical signal is converted to an electrical signal using a PD. The output is amplified using a TIA and filtered using a DC block filter. A downconverter is used at the same center frequency to convert the signal into baseband in-phase and quadrature-phase streams. These can finally be processed using an IEEE 802.11ax demodulator [33].

8.4.2.3 Optional Modes

The HE optional mode is also based on the IEEE 802.11 standard. However, it allows for MIMO configurations based on multiple chains of the IEEE 802.11bb common PHY mode. Multiple transmitter chains, similar to what is shown in Figure 8.9, can be driven with different information streams based on the baseband in-phase and quadrature-phase outputs of the IEEE 802.11. These transmitters are assumed to work based on spatial, wavelength, frequency division multiplexing, or a combination thereof.

Similar to the common PHY mode, the HE optional mode operates in the wavelength range between 800 nm and 1,000 nm. However, the HE mode can

Table 8.5
The Center Frequency as Defined in the IEEE 802.11bb PHY for Different Frequency Bands [33]

Channel Number	f_c [MHz]	Freq. Range [MHz]	20 MHz	40 MHz	80 MHz	160 MHz
LC0	26	16–36	20			
LC1	36	16–56		40		
LC2	46	36–56	20			
LC3	56	16–96			80	
LC4	66	56–76	20			
LC5	76	56–96		40		
LC6	86	76–96	20			
LC7	96	16–176				160
LC8	106	96–116	20			
LC9	116	96–136		40		
LC10	126	116–136	20			
LC11	136	96–176			80	
LC12	146	136–156	20			
LC13	156	136–176		40		
LC14	166	156–176	20			
LC15	176	16–336				
LC16	186	176–196	20			
LC17	196	176–216		40		
LC18	206	196–216	20			
LC19	216	176–256			80	
LC20	226	216–236	20			
LC21	236	216–256		40		
LC22	246	236–256	20			
LC23	256	176–336				160
LC24	266	256–276	20			
LC25	276	256–296		40		
LC26	286	276–296	20			
LC27	296	256–336			80	
LC28	306	296–316	20			
LC29	316	296–336		40		
LC30	326	316–336	20			

support modulation bandwidths of 20 MHz, 40 MHz, 80 MHz, and 160 MHz [26]. The list of center frequencies for each bandwidth band is shown in Table 8.5. The center frequency was determined using the following formula [26]:

$$f_c = B/2 + 16 \quad [\text{MHz}]$$

where B is the width of the frequency band.

In addition, the HE PHY mode supports BPSK modulation and M-ary quadrature amplitude modulation (QAM) (M-QAM) modulation with constellation sizes that range from $M = 4$ to $M = 256$ [26]. FEC and LDPC are also supported with coding rates of 1/2, 2/3, 3/4, and 5/6 [26].

8.5 Summary

In this chapter, we introduced several standards related to OWC technologies. The development, features, and potential of each standard have been considered. All light communication standards are also assumed to comply with the IEC 60825-1 eye-safety regulation such as those set out by the IEC 62471:2006—"Photo-Biological Safety of Lamps and Lamp Systems," the ITU-T G.664—"Optical Safety Procedures and Requirements for Optical Transmission Systems", and the IEC 60825-1 for laser light sources [34]. OWC is assumed to be an ideal solution for EMI-sensitive environments. However, any OWC standard should ensure that the associated electronic circuitry used to drive the light source or condition the received waveform are not creating any spurious RF interference beyond the transceiver package. The OWC standards discussed in this chapter give the regulators a means to standardize the access to the light spectrum for the purpose of communications.

The IEEE 802.15 OWC standards focus on applications with image sensor-based transmission and very high data rate transmission in industrial applications. In order to fulfill different requirements of different scenarios, multiple operation modes with unique specifications have been defined in these standards.

The ITU-T G.9991 Recommendation for high-speed VLC transceivers has attracted particular commercial interest in the past couple of years. This is mainly because commercial Li-Fi solutions based on G.9991 could be built using the chipsets available from the established G.hn market. As a result, some of the major Li-Fi market players have participated in the development of the standard and saw that G.9991, in combination with the available G.hn chipsets, provides a cost-effective route to the Li-Fi market growth.

The IEEE 802.11bb standard amendment targets mass markets by relying on the maturity of the IEEE 802.11 standard suit. The Li-Fi standard is being developed with numerous potential use cases in consideration. These include, but are not limited to, local area fixed and mobile wireless access in home, enterprise, industrial, medical, and EMI-sensitive environments. In addition, the standard considers use cases such as underwater and vehicular communications, cybersecurity, and defense. The IEEE 802.11bb amendment does not have the same constraints as the Infrared Data Association (IrDA) standard. There are currently improved and energy-efficient LED devices that can be simultaneously used for illumination and communication purposes. In contrast to point-to-point IR-based solutions, Li-Fi, as defined by the IEEE 802.11bb amendment, offers a light-based complete networking solution with mobility and handover support.

References

[1] "IEEE P802.11 - Light Communication (LC) Task Group (TG) - Meeting Update".

[2] "IEEE 802.15 Working Group for Wireless Specialty Networks (WSN)," https://ieee802.org/15/index.html.

[3] "IEEE Standard for Local and Metropolitan Area Networks–Part 15.7: Short-Range Wireless Optical Communication Using Visible Light," IEEE Std. 802.15.7-2011, 2011, pp. 1–309.

[4] Cailean, A., and M. Dimian, "Impact of IEEE 802.15.7 Standard on Visible Light Communications Usage in Automotive Applications," *IEEE Communications Magazine*, Vol. 55, No. 4, 2017, pp. 169–175.

[5] Cahyadi, W. A., et al., "Optical Camera Communications: Principles, Modulations, Potential and Challenges," *Electronics*, Vol. 9, No. 9, 2020, p. 1339.

[6] "IEEE Standard for Local and Metropolitan Area Networks–Part 15.7: Short-Range Optical Wireless Communications," IEEE Std. 802.15.7-2018 (Revision of IEEE Std. 802.15.7-2011), 2019, pp. 1–407.

[7] "IEEE 802.15 WPAN Task Group 13 (TG13) Multi-Gigabit/s Optical Wireless Communications," https://www.ieee802.org/15/pub/TG13.html.

[8] Bober, K. L., and V. Jungnickel, "IEEE P802.15.13 Text Proposal for Clause 4," https://mentor.ieee.org/802.15/dcn/19/15-19-0304-03-0013-text-proposal-for-clause-4.docx.

[9] Nguyen, T., et al., "Technical Issues on IEEE 802.15.7m Image Sensor Communication Standardization," *IEEE Communications Magazine*, Vol. 56, No. 2, 2018, pp. 213–218.

[10] "Criteria for Standards Development (CSD) IEEE 802.15.11 Standard for Multi-Gigabit/s Optical Wireless Communications," https://mentor.ieee.org/802.15/dcn/17/15-17-0075-01-0000-multi-gigabit-owc-csd.docx.

[11] Thieu, M. D., et al., "Optical-RoI-Signaling for Vehicular Communications," *IEEE Access*, Vol. 7, 2019, pp. 69873–69891.

[12] Baykas, T., et al., "IEEE P802.15.13 Text for Adaptive MIMO Communication." https://mentor.ieee.org/802.15/dcn/19/15-19-0414-04-0013-text-for-adaptive-mimo-communication.doc

[13] Bober, K. L., "IEEE P802.15.13 Updated Text Parts for Clause 5," https://mentor.ieee.org/802.15/dcn/19/15-19-0345-00-0013-updated-text-for-clause-5.docx

[14] Che, F., et al., "Design and Implementation of IEEE 802.15.7 VLC PHY-I Transceiver," *12th IEEE International Conference on Solid-State and Integrated Circuit Technology (ICSICT)*, 2014, pp. 1–4.

[15] Che, F., et al., "A Fully Integrated IEEE 802.15.7 Visible Light Communication Transmitter with On-Chip 8-W 85% Efficiency Boost LED Driver," *IEEE/OSA Journal of Lightwave Technology*, Vol. 34, No. 10, 2016, pp. 2419–2430.

[16] Nawaz, T., et al., "IEEE 802.15.7-Compliant Ultra-Low Latency Relaying VLC System for Safety-Critical ITS," *IEEE Transactions on Vehicular Technology*, Vol. 68, No. 12, 2019, pp. 12040–12051.

[17] LG, "VLC ZigBee Connector," https://fccid.io/BEJ9VS570V5DA/Users-Manual/User-Manual-Rev04-0512-3390900.

[18] Panasonic, "LinkRay Technology," https://bit.ly/3gn9dhm.

[19] Perez-Ramirez, J., et al., "Optical Wireless Camera Communications Using Neuromorphic Vision Sensors," *IEEE International Conference on Communications Workshops (ICC Workshops)*, 2019, pp. 1–6.

[20] Berenguer, P. W., et al., "Real-Time Optical Wireless Mobile Communication with High Physical Layer Reliability," *Journal of Lightwave Technology*, Vol. 37, No. 6, 2019, pp. 1638–1646.

[21] Jungnickel, V., "Why Fraunhofer HHI Develops and Trials 'LiFi' Systems Based on ITU Standards," *MyITU*, April 9, 2020. https://www.itu.int/en/myitu/News/2020/09/04/12/33/Why-Fraunhofer-HHI-develops-and-trials-LiFi-systems-based-on-ITU-standards.

[22] ITU News, "'LiFi' Provider Oledcomm Shares Experience of ITU Standards Process," *MyITU*, April 9, 2020. https://www.itu.int/en/myitu/News/2020/09/04/12/54/LiFi-provider-Oledcomm-shares-experience-of-ITU-standards-process.

[23] ITU News, "Why Signify Builds 'LiFi' Products to ITU Standards," September 4, 2020. https://news.itu.int/why-signify-builds-lifi-products-to-itu-standards/.

[24] Serafimovski, N., et al., "A CSD Proposal for Light Communications (LC)," IEEE Std., Rev. IEEE 802.11-17/1603r9, July 2017. https://mentor.ieee.org/802.11/dcn/17/11-17-0023-09-00lc-lc-tig-draft-report-outline.docx.

[25] Serafimovski, N., et al., "Light Communications for 802.11," Online, Std., Rev. IEEE 802.11-17/1048r4, July 2017. https://mentor.ieee.org/802.11/dcn/17/11-17-1048-04-00lc-lc-for-802-11-pdf.ppt.

[26] Jungnickel, V., et al., "Draft text for LC-Optimized PHY for TGbb D0.1," Online, IEEE Std., Rev. IEEE 802.11-19/1791r2, November 2019. https://mentor.ieee.org/802.11/dcn/19/11-19-1791-02-00bb-draft-text-for-lc-optimized-phy.docx.

[27] Purwita, A., et al., "Proposed Center Frequency for the Common Mode Mandatory PHY," Online, Std., Rev. IEEE 802.11-20/0572r1, April 2020. https://mentor.ieee.org/802.11/dcn/20/11-20-0572-01-00bb-proposed-center-frequency-mandatory-phy.pptx.

[28] Uysal, M., F. Miramirkhani, and T. Baykas, "IEEE 802.11bb Reference Channel Models for Indoor Environments," Online, IEEE Std., Rev. IEEE 802.11-18/1582r4, November 2018. https://mentor.ieee.org/802.11/dcn/18/11-18-1582-04-00bb-ieee-802-11bb-reference-channel-models-for-indoor-environments.pdf.

[29] Purwita, A. A., and H. Haas, "Studies of Flatness of LiFi Channel for IEEE 802.11bb," *IEEE Wireless Communications and Networking Conference (WCNC)*, 2020, pp. 1–6.

[30] "IEEE Standard for Information Technology – Telecommunications and Information Exchange Between Systems Local and Metropolitan Area Networks – Specific

Requirements–Part 11: Wireless LAN Medium Access Control (MAC) and Physical Layer (PHY) Specifications–Amendment 4: Enhancements for Very High Throughput for Operation in Bands Below 6 GHz." IEEE Std 802.11ac-2013 (Amendment to IEEE Std 802.11-2012, as amended by IEEE Std 802.11 ae-2012, IEEE Std 802.11aa-2012, and IEEE Std 802.11ad-2012), 2013, pp. 1–425.

[31] Hinrichs, M., et al., "Optical Frontend Model," Online, Std., Rev. IEEE 802.11-19/0087r1, January 2019. https://mentor.ieee.org/802.11/dcn/19/11-19-0087-01-00bb-optical-frontend-model-for-phy-simulation.docx.

[32] Purwita, A. A., and H. Haas, "IQ-WDM for IEEE 802.11bb-Based LiFi," *IEEE Wireless Communications and Networking Conference (WCNC)*, June 2020, Seoul, Korea, pp. 1–6.

[33] Serafimovski, N., "Proposed-Text-for-LC-Mandatory-Optional-PHY," Online, Std., Rev. IEEE 802.11-20/0571r5, January 2020. https://mentor.ieee.org/802.11/dcn/20/11-20-0571-05-00bb-proposed-text-for-lc-mandatory-optional-phy.docx.

[34] Serafimovski, N., et al., "Light Communications (LC) for 802.11: Use Cases and Functional Requirements: Guidelines for PAR and CSD Development," Online, Std., Rev. IEEE 802.11-17/1603r9, March 2018. https://mentor.ieee.org/802.11/dcn/17/11-17-1603-09-00lc-a-csd-proposal-for-light-communications.docx.

9

Commercialization

9.1 Introduction

Li-Fi is a powerful technology that has gained tremendous interest over the last decade. The development and commercialization of Li-Fi technology are disruptive to other industries such as the lighting industry and the communications industry. In this chapter, we aim to shed light on the status of the current Li-Fi market as well as the predicted trajectory for market growth in the coming years. We also discuss the key trends and drivers that are leading to the spread and adoption of this evolving technology. These drivers include the evolution of 5G and beyond wireless networks, the emergence of IoT and Big Data as well as the opportunities and requirements of the fourth industrial revolution. To successfully commercialize Li-Fi technology, collaboration among companies in different industries is required. Therefore, we also provide a detailed introduction to the Li-Fi ecosystem and describe the roles and characteristics of the various participants. In addition, we discuss some of the challenges and strategies to establish this ecosystem. This chapter also highlights some Li-Fi use cases that have been documented in research and white papers. Examples of these include OCC deployments for indoor positioning and object identification, FSO and novel solar-cell-based OWC backhauling solutions, optical wireless access solutions for smart cities, and use cases in industrial, aerospace, and automotive sectors. The main challenges that face Li-Fi technology are also discussed in this chapter. These include the need for appropriate optoelectronic devices such as LEDs, light amplification by stimulated emission of radiation diodes (LDs), and PDs, and the challenges of integrating these devices in a mobile or a laptop device with suitable integrated analog and digital processing solutions. In addition, the dual use of LEDs for communication and illumination imposes certain restrictions related to light quality and eye safety, which need to be addressed before Li-Fi systems are deployed on a mass-market scale.

This chapter is organized as follows. Section 9.2 discusses the market potentials and opportunities of Li-Fi technology. The ecosystem that can foster Li-Fi technology is discussed in Section 9.3. A number of real-world use cases are presented in Section 9.4. Section 9.5 discusses the challenges related to mass market deployment of Li-Fi technology. Finally, Section 9.6 summarizes the chapter.

9.2 Li-Fi Market Potential and Opportunities

Recent advancements in OWC paradigms, including the emerging Li-Fi technology, have led to the development of new market perspectives. Li-Fi can unlock the potential of 5G and beyond wireless networks, contribute to the emergence of IoT, drive Industry 4.0 applications, and shape the upcoming light-as-a-service (LaaS) solution. In this section, we present an overview of the Li-Fi market outlook and its main driving factors.

9.2.1 Market Overview

The Li-Fi market was valued at $143 million in U.S. dollars in 2019, with expectations for the market size to reach up to $3.52 billion in U.S. dollars by 2025 [1]. In Europe alone, the Li-Fi market valuation reached $30 million in U.S. dollars in 2019, with the United Kingdom holding around 28% of the market share [2]. This rapid market growth is attributed to many factors such as the increasing demand for indoor connectivity, the unprecedented rise in the numbers of connected smart devices, the penetration rate of smart homes technologies, and the growing trend of utilizing Li-Fi in critical applications such as defense, healthcare, and security. All these factors provide a positive outlook for the Li-Fi market and encourage governments and businesses to invest in the research and development of this emerging technology.

In general, the Li-Fi market can be segmented by the components involved in it (i.e., LEDs, PDs, optical lenses), by the possible applications (i.e., indoor networks, localization services, underwater communications), or by the use cases (i.e., defense sector, healthcare, education, automotive). Major companies contributing to the Li-Fi market include pureLiFi Ltd., KYOCERA (which recently acquired SLD Laser), Integrated System Technologies Ltd., Lightbee S.L., Lucibel, NAV Technologies, Signify, Velmenni, VLNComm, Renesas Electronics Corporation, General Electric, Oledcomm, and Fujitsu. Tactical collaborations, partnerships, and acquisitions are the key strategies adopted by the market players in order to enhance and expand their business operations. For example, Getac, a leading manufacturer of rugged tablets and PCs for military, manufacturing, and automotive industries, partnered with pureLiFi in October 2019. The goal of this partnership is to facilitate embedding Li-Fi technology into future wireless devices. Similarly, Vodafone has joined forces with Signify to combine communication capabilities of 5G and LiFi. Given the huge volume of the LED industry on a global level, product differentiation, innovation, and applicability across a wide number of use cases are critical requirements to the success and expansion of the Li-Fi industry.

9.2.2 Key Trends and Drivers

Technology is continuously evolving to accommodate the rapidly changing world around us. This change is vital to accomplish sustainable development on a massive scale. By 2050, the world's population is expected to exceed 9 billion people, whose needs must be met without causing irreversible damage to the planet. To this end, outdated methodologies and processes need to continuously be upgraded in order to serve the new market demands. In the following sections, we outline the main global strategic trends that are driving the adoption and expansion of Li-Fi technology.

9.2.2.1 5G Networks

5G wireless solutions have already been introduced worldwide since 2019, offering connection speeds that are multitudes faster than the existing 3G and 4G networks. The development requirements of the new 5G paradigms are continuously expanding to offer unprecedented capabilities such as mass connectivity, guaranteed low latency, and ultrareliability. Implementations of Li-Fi have already started in various 5G-related applications. For example, as part of their strategy to pave the way for 5G, the U.K. mobile operator O2 Telefonica has launched a Li-Fi network trial to provide high-speed wireless connectivity in partnership with pureLiFi. Signify, a world leader in lighting has started the development of applications, use cases, and solutions for secure and reliable two-way wireless data transmission and reception technology which can be built into Philips-branded luminaires. The Chinese consumer electronics and mobile communications company, Oppo, has patented a phone with Li-Fi technology. All these recent efforts indicate that Li-Fi technology will be an integrated part of 5G networks as an alternative broadband technology.

9.2.2.2 Big Data and IoT

IoT refers to an integrated system of physical devices, networks, and software capabilities that are interconnected to embed intelligence into everyday objects. In order to achieve this seamless merge between the physical and digital worlds, IoT requires high-volume communications capabilities to facilitate the interaction between the various connected objects. For the IoT devices to function effectively, the communication links must satisfy the following requirements:

- Reliable bidirectional communications to render an easy data route between IoT devices;
- Ubiquitous detection of IoT devices in order to identify the status of all connected or dropped-off devices and to track and resolve issues in real time;
- Distributed connectivity that can handle the authentication, authorization, and connection of a huge number of IoT nodes;
- Secure data transfer between the IoT devices and the cloud, which enables the storage and transmission of confidential and sensitive information.

Li-Fi has the potential to meet the demands of IoT systems as it offers extremely high data rates, high degrees of spatial reuse allowing dense deployment, and the inherently secure nature resulting from the confinement of light signals. A key advantage of leveraging Li-Fi capabilities for IoT communications is the fact that LED lighting fixtures are being extensively adopted in indoor spaces. In 2019, LED sales reached a critical milestone, achieving a record number of more than 10 billion units sold globally. Eventually, these 10 billion lightbulbs could be converted into APs for IoT applications. Current research efforts are focusing on developing Li-Fi capabilities to face the rapid growth of IoT and Big Data. This involves the development of smarter, faster, and more reliable data transmission paradigms as well as the adaptation of Li-Fi to enable specific IoT use cases and applications [3].

9.2.2.3 Industry 4.0

The first industrial revolution was driven by the discovery of steam power in the late 1700s. Then came the second industrial revolution, known as the technological revolution, which introduced advances in steel production, electricity, and petroleum. leading to rapid standardization and industrialization from the late 1800s into the early 1900s. Starting in the mid-1900s, the third industrial revolution brought the integration of electronics, telecommunications, computers, and the internet, opening the door to unprecedented digital advancements. Nowadays, we are witnessing the rise of the fourth industrial revolution, which is known as Industry 4.0. Powered by data science and machine learning, Industry 4.0 is driving entirely new approaches for embedding technology within societies, factories, and even human bodies.

One of the fundamental elements of Industry 4.0 is the concept of smart factories; factories in which each layer of the manufacturing process can be automated. This includes, but is not limited to, production lines, quality monitoring, and real-time communications with customers. It is evident that the role of wireless communications in these settings is extremely crucial to gather, process, analyze information, and provide feedback in real time. Li-Fi technology has the potential to bring industries to this level of real-time connectivity and automation to realize the vision of Industry 4.0. In fact, the use of RF-based communications in these settings is not always feasible due to the undesirable interaction with the electromagnetic radiations that can occur on the manufacturing floor. These radiations can disrupt the RF communication signals for various reasons such as reflections, shadowing, diffraction, and scattering. Moreover, the reliability of RF wireless communications is not always enough for mission-critical applications (i.e., applications), where target reliability should be $> 99.999\%$. In this regard, Li-Fi can provide Industry 4.0 with the following capabilities:

- Extremely high-speed connections and reduced latency between the units to aid real-time monitoring and full automation of the manufacturing process.
- Enhanced link reliability for mission and business-critical use cases. Reliable links also facilitate informed decision-making and accurate troubleshooting, leading to improved maintenance identification and lower operating costs.
- Greater data density since each lightbulb can be used as an information hub to gather and transmit data. Such a dense data exchange allows for a better QoS.
- Enhanced security in critical industrial settings, since light signals do not travel through walls and, thus, the signals can be confined in a physical space.

9.2.3 Opportunities in 6G

While 5G standardization efforts continue to be carried out for enhanced 5G, universities and governmental institutions around the world have already kick-started the research into 6G wireless networks.

In order to support an unprecedented level of connectivity and innovative capabilities, 6G networks are envisioned to be characterized by the following paradigm shifts [4]:

- Global coverage that is not limited to terrestrial communication networks. This could be achieved by the combination of terrestrial and nonterrestrial communication technologies in order to ensure seamless and ubiquitous service availability. Examples of potential nonterrestrial solutions include satellite communications, unmanned aerial vehicle communications, and maritime communications.

- Unprecedented spectral efficiency. This necessitates the full utilization of the available spectra, including the sub-6-GHz bands, millimeter-wave bands, terahertz bands, and optical frequency bands. Moreover, innovative transmission schemes are needed to maximize the gain of all possible diversity techniques. This includes new waveform design, air interfaces, channel coding mechanisms, and multiple access approaches.
- Innovative smart applications. To enable this, AI and machine learning (ML) will come into play to automate network configuration, management, and orchestration as well as to optimize performance and resource utilization.
- Strong security for both network and physical layers. This is a critical requirement to ensure that 6G solutions will be able to securely handle the huge data sets that need to be generated, transmitted, and stored. Moreover, massive numbers of connected nodes and autonomous systems will be required to deploy efficient mechanisms for encryption, authentication, access control, and malicious behavior detection, mostly in a distributed manner [5].

It is evident that OWC technologies and Li-Fi networks will be an integrated part of future 6G networks. In fact, the utilization of optical frequencies has already been identified in early research efforts as one of the pathways to 6G. The optical spectrum offers a huge unregulated bandwidth that can be utilized to achieve the capacity gains anticipated in 6G networks. Recent studies showed that OWC networks can increase the capacity by three orders of magnitude compared to existing RF solutions, while completely eliminating interference [6]. OWC offers versatile opportunities for the global intensive coverage anticipated in 6G. For example, VLC is considered a viable candidate for underwater communications as the attenuation of light by water is lowest in the visible region [7]. Moreover, FSO is an ideal candidate for long-distance, high-speed, backhaul links between buildings or satellites. Also, one of the major advantages that OWC can bring to future 6G networks is the physical containment of light signals. Thanks to this feature, it is possible to control the size of the coverage areas up to a few square centimeters. As a result, OWC can be utilized to build ultra-small cell terabit 6G networks, offering unprecedented dense coverage in indoor locations.

9.3 Li-Fi Ecosystem

As discussed in Section 9.2, the commercialization of Li-Fi technology creates diverse market opportunities that can be explored by different related industries. In addition, this new technology provides end-users with an improved quality of experience, such as faster download speeds. Furthermore, it provides a platform for future applications and services, such as IoT and virtual reality (VR). Li-Fi

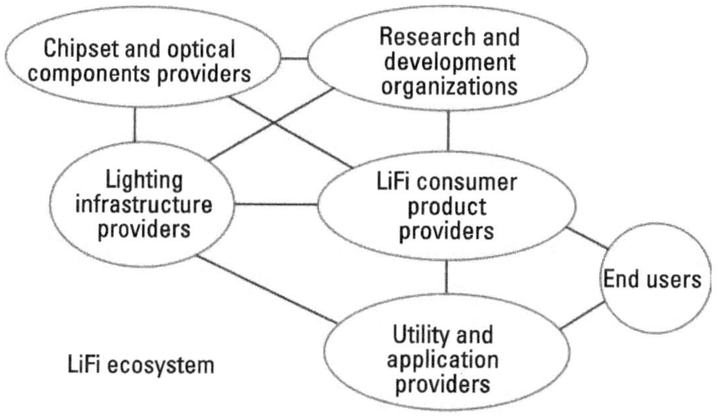

Figure 9.1 Li-Fi commercialization ecosystem. The connections between players indicate that there are interactions or collaborations between the connected players.

systems serve a dual functionality by providing both lighting and communication services. Additionally, they require an internet connection and power supply. Therefore, it is unlikely to have a single company that can provide all the required components and resources for the entire system. Alternatively, the commercialization can be achieved by the collaboration of various industrial players, which requires the conception of an innovative commercial ecosystem for Li-Fi.

The ecosystem around Li-Fi commercialization is depicted in Figure 9.1. Li-Fi Product vendors, lighting infrastructure providers, and chipset and optical component providers are three key players in the ecosystem to develop competitive Li-Fi products. In addition, research and development organizations, utility providers, and end-users also have a significant impact on this commercial ecosystem. The connections between players shown in Figure 9.1 indicate that there are interactions or collaborations between the connected players. The characteristics and roles of these participants will be introduced individually in the following sections.

9.3.1 Li-Fi Consumer Product Provider

Generally, Li-Fi consumer product providers play one of the most important roles in the Li-Fi ecosystem, since they are responsible for the design and manufacturing of the consumer product hardware. This hardware includes mobile user devices as well as the infrastructure lighting fixtures.

Regarding the mobile user devices, some products enable existing user equipment to have light communication capability. For example, different light communication companies released plug-and-play USB dongles, which can make laptops and tablets capable of accessing the internet via light communication [8].

This type of product is flexible and requires less upfront investment, which is suitable for the early stages of Li-Fi industry development.

Another way of enabling light communications at the mobile user is the integration of light communication modules on the existing user equipment. Examples include Li-Fi-enabled smart phones and laptops [9]. The dimensions of a Li-Fi module have to be compact enough to fit into the existing user equipment.

Consumer product providers have started to design and manufacture light communication infrastructures, such as Li-Fi-enabled ceiling lights and desktop lamps. Since this special lighting infrastructure is required to provide good illumination and high communication performance simultaneously, the design and development require close collaboration with lighting design companies. With further development of Li-Fi technology, there will also be options to convert existing lighting devices into a Li-Fi-enabled infrastructure.

Some light communication products require fewer connections between different players. For example, FSO products for last-mile backhaul solutions do not need to consider lighting functionality. A further example is visible light indoor positioning based on OCC, which requires no modification at the terminal as mobile phone cameras are ready to be used as optical detectors in these scenarios [10].

9.3.2 Lighting Infrastructure Provider

The primary function of Li-Fi systems is to provide a wireless connection service. At the same time, they are also required to provide sufficient lighting for their illumination functionality. Thus, lighting devices have become one of the key components in the design of Li-Fi infrastructure. Therefore, companies providing various lighting infrastructures will be another important player in the Li-Fi ecosystem, as shown in Figure 9.1.

Conventional lighting infrastructure uses incandescent/fluorescent lightbulbs as light sources, which consume a significant amount of energy. However, only a small fraction of the used power is converted to light, and the remaining energy is lost as heat. Furthermore, an incandescent/fluorescent lightbulb has a shorter lifetime, which means that its performance is reduced with age. With the emergence of high-power blue LEDs, their applications in lighting started to gain popularity. This is because of their high energy efficiency, which converts the majority of the energy into light. Therefore, to achieve the same illumination performance, much less power is required compared to incandescent/fluorescent lightbulbs. In addition, LEDs lightbulbs generally have a much longer lifetime [11]. In LED lighting, DC-electricity is used to drive an LED lamp while only AC electricity is available from the power supply. This is solved by using an LED driving circuit to convert the current from AC to DC [12]. This characteristic makes it convenient to modify LED-based infrastructures to have light communication capabilities. Therefore, this evolution in lighting technology is beneficial to both the lighting industry and Li-Fi commercialization.

Despite the high energy efficiency of LED lighting, the process of replacing inefficient lightbulbs with new ones still requires a significant upfront cost for individual and business consumers. Conventionally, consumers have to consider buying, installing, and maintaining the lighting infrastructure separately. This is not only time-consuming but also costly, especially for lighting in large public areas. In recent years, a new business model, known as LaaS, has been proposed [13]. A lighting company provides an evaluation of the lighting environment and offers an optimal lighting solution based on the evaluation. Once the plan is agreed upon by the customer, the company will install the new lighting infrastructure and will also provide regular maintenance. Instead of paying for each service separately, the customer pays a monthly subscription fee for a set period (5 to 10 years). At the end of the contract, the consumer will have ownership of the lighting infrastructure and the long-term overall energy consumption and cost will be reduced considerably.

The long lifetime of LED-based lighting is technically beneficial, but may cause issues for long-term sales as LED bulbs and lamps require less frequent replacement. This is known as socket saturation [10]. The introduction of light communication may provide customers with further motivation to replace their light sources. With further development, Li-Fi technology will transform the lighting system into a platform for future AI and various user-centric services. Following the trend of the LaaS scheme, Li-Fi capability riches the services provided by lighting companies. This new LaaS scheme will be very attractive to large retail and catering companies as the lighting company not only provides illumination-related services, but also many other services such as internet access, indoor navigation, and other IoT connectivity. Conventionally, lighting companies work on developing and designing lighting devices with a better lighting performance. With the introduction of light communication, optimization of lighting sources may also take the communication performance into account. With more efficient and fast light sources, Li-Fi systems with improved performance can be developed. In return, the more competitive Li-Fi products will create more commercial opportunities for lighting companies, which drive them to develop more advanced light sources, such as laser-based light sources [14].

9.3.3 Chipset and Optical Component Providers

Different from conventional RF wireless communication technology, unipolar and real-valued IM/DD-based signals are commonly used for OWCs. Consequently, the modules for encoding, modulation, and signal pulse-shaping/amplification are distinct from those for conventional RF communications. Therefore, chipsets that conduct these fundamental functionalities are required for Li-Fi commercialization. Various optical components, such as optical diffusers, concentrators, detectors, and sources, are also essential for Li-Fi systems. The available optical components in the market are designed for existing applications, such as optical fiber systems or other point-to-point optical systems. Consequently, the specifications of these available components may not fulfill the requirements for Li-Fi application. For example, many PDs are designed for use cases with appropriate alignment with a limited FOV. Therefore, these PDs cannot be directly used for commercial Li-Fi products as there may be significant misalignment issues.

Because there are no purposely designed chipsets and optical components for Li-Fi applications, Li-Fi consumer product manufacturers have to design and prototype their own chipsets and optical components. The cost of placing orders of customized chipsets and optical components to manufacturers is considerably high without mass-production. This is one of the main reasons why the currently available Li-Fi products in the market have a relatively high price. Therefore, it is necessary to have chipset and optical component providers in the Li-Fi ecosystem to provide standardized and low-cost Li-Fi chipsets and optical components.

Currently, various Li-Fi companies have released their Li-Fi-enabled dongles as user-side products. This type of product is valid for demonstration purposes, as this plug-and-play fashion is convenient for applications on various user devices, such as laptops, tablets, and smart phones. However, for practical applications, end-users will not be attracted to commercial Li-Fi devices with bulky dongles attached. Therefore, the final Li-Fi products have to be integrated with original equipment, such as Li-Fi-enabled laptops. This also proves the necessity of specialized Li-Fi chipsets. In addition, by optimizing the design based on a specific standardization, the physical size and power consumption of the chipsets can be minimized, which alleviates any device integration challenge.

In recent years, the photonic integrated circuit (PIC) industry has been quickly developed in the commercial market [15]. PICs are modules that integrate multiple photonic functions on a chip board made of a unique material such as silicon-on-insulator or indium phosphide (InP). Some representative products include high-speed optical transceivers, LiDAR, and sensors. The main advantages of PIC products include simplified integration, improved reliability, and lower cost per unit data rate. The development of PIC industry also brings opportunities for Li-Fi commercialization. Intuitively, by using PIC technology, all processing units and optical components can be integrated into a single Li-Fi transceiver chipset. In addition, the size, power consumption and cost of Li-Fi modules can be further reduced. Therefore, it is vital to have chipset/optical component providers (especially PIC companies) to develop and manufacture Li-Fi chipset/optical components. In return, new markets in light communication can be explored for these companies.

9.3.4 Utility and Application Providers

As a hybrid between a wireless communication system and a lighting system, the Li-Fi networks consume energy and also require internet services. Therefore, utility providers are also part of the Li-Fi ecosystem. If the internet service and the energy are provided by the same company, it will be convenient for both the utility provider and the end-user to maintain and manage the usage of energy and data. In recent years, many companies have released their smart home services to consumers, such as smart energy monitoring and control. The introduction of Li-Fi will bring a platform to provide more powerful internet applications and services. Therefore, this will broaden the market range of utility companies to provide more services and applications. This broadening of the service range may potentially introduce more competition between different utility and application providers. However, more competition will drive the service price down to eventually benefit the end-user.

9.3.5 Research and Development Organization

Research and development organizations are also essential to the Li-Fi ecosystem. As a new wireless communication technology, the related products will evolve with improved performance. This product evolution will rely on the efforts of research and development organizations. The commercial application of Li-Fi products will provide feedback and useful data to researchers as resources and motivations for improving the technology. Based on the technology development, more advanced products can be designed and released.

Another important function of research and development organizations is the standardization of technology. A number of standardization activities have been covered in Chapter 8. With a specific standard, the key technical specifications of the Li-Fi products can be unified. This means that different components from different companies are more likely to be compatible with each other. For example, a chipset design based on a specific standard can be used by many different Li-Fi consumer product providers for their own devices. This will increase the chance of massive production of the chipset and lead to reduced costs. Eventually, it will lower the price of the final Li-Fi products.

9.3.6 End-Users

End-users play one of the most important roles in the Li-Fi ecosystem, the number of end-users directly determines the market size of the entire Li-Fi industry. The scale of the Li-Fi market determines the profits of companies in the ecosystem and their investment in this industry. The key factor from the end-users' point of view is whether Li-Fi products offer significant added value. Specifically, Li-Fi products will bring a considerable QoS improvements to end-users, such as a 10 times faster data rate, a 10-fold reduction in latency, accurate indoor localization, and other convenient services.

9.3.7 Establishing the Li-Fi Ecosystem

As described in previous sections, the Li-Fi ecosystem requires collaborations between multiple parties. Without certainty on the potential market that Li-Fi networks can bring, it is challenging to persuade key players to invest in Li-Fi development. For example, lighting companies wish to install Li-Fi fixtures after they see that Li-Fi-enabled mobile devices are available [16]. Mobile device companies want their products to have Li-Fi component integration after there is Li-Fi-enabled lighting fixtures. So it becomes a chicken-and-egg problem. One of the solutions to this stalemate is to first introduce the applications requiring alterations from one side. Visible light positioning is one such example. Only minor modifications are required on the lighting equipment side, while the cameras on the user devices are ready to be used as positioning detectors with corresponding supporting software [10]. Data transfers between user devices via optical wireless technology is an example of alteration only on the user device side. For such a use case, a high-speed optical source and detector will be required on both user devices. Once this use case is adopted by the user device manufacturers, lighting companies will be more likely to follow by installing Li-Fi enabling fixtures on their lighting infrastructures. Once this type of use case receives positive feedback from consumers, it will increase the confidence of companies on Li-Fi commercialization. Thus, more companies will be attracted to the Li-Fi industry and more challenging (but more powerful) use cases will be considered for commercialization. Another key factor in Li-Fi commercialization is reducing the cost of Li-Fi products. Therefore, it is crucial to have chipset providers in the play, as discussed in Section 9.3.3. With more participants in the Li-Fi industry and therefore, an emergence of cheaper products, the Li-Fi ecosystem will be progressively established.

In this process, a number of organizations, such as the Light Communication Alliance, help to align participants in different industries. They will also develop and promote potential business models using light communications for various types of companies. In addition, these organizations deliver market education and publish white papers on the advantages and use cases of Li-Fi systems. All these activities will speed up the establishment of the Li-Fi commercial ecosystem.

9.4 Use Cases

A number of use cases from prominent Li-Fi companies and research centers are presented in this section. These include indoor positioning, backhauling solutions, and object tagging. In addition, the section presents interesting

Figure 9.2 A demonstration of a white laser-light-based 20-Gbps VLC link at the 2020 CES by the SLD Laser and Li-Fi Research and Development Centre, University of Strathclyde.

deployments in the industrial, aerospace, automotive, healthcare, IoT, and education sectors. These use cases are drawn from real-world implementations and deployments of OWC systems in different applications that have been reported in research and white papers.

9.4.1 Indoor High-Speed Networking

Indoor high-speed wireless access is one of the prominent use cases of Li-Fi technology. Most of the companies working on this technology have developed products centered around this use case with capabilities ranging from a few hundreds of megabits per second to a few gigabits per second. However, the potential of Li-Fi technology to enable communications at speeds of tens of gigabits per second was recently demonstrated using laser-based optical luminoires. A remarkable demonstration of a high-speed 20 gigabits per second VLC link was demonstrated at the 2020 Consumer Electronic Show (CES) by SLD Laser and the Li-Fi Research and Development Centre at the University of Strathclyde, as shown in Figure 9.2 [14].

Li-Fi is an appealing technology to educational establishments because it significantly limits the radiation of RF EM waves, as discussed in Chapter 4. The French parliament adopted laws to ban the use of Wi-Fi in schools in order to limit pupils exposure to excessive RF EM waves [17]. The Fraunhofer Heinrich Hertz Institute deployed a Li-Fi access network in a classroom at Hegel-Gymnasium in Stuttgart, Germany. The Li-Fi system was an interesting choice for the school due to the avoidance of RF radiation and the possibility to improve the system capacity in high-density classrooms [18].

In this section, we present the results of a real-world use case where a Li-Fi network was deployed in a classroom in addition to a Wi-Fi network. The deployment was achieved in collaboration with pureLiFi and the Li-Fi Research and Development Centre at the University of Strathclyde [6]. The network topology consists of 8 Li-Fi attocell APs as shown in Figure 9.3. The Li-Fi attocell

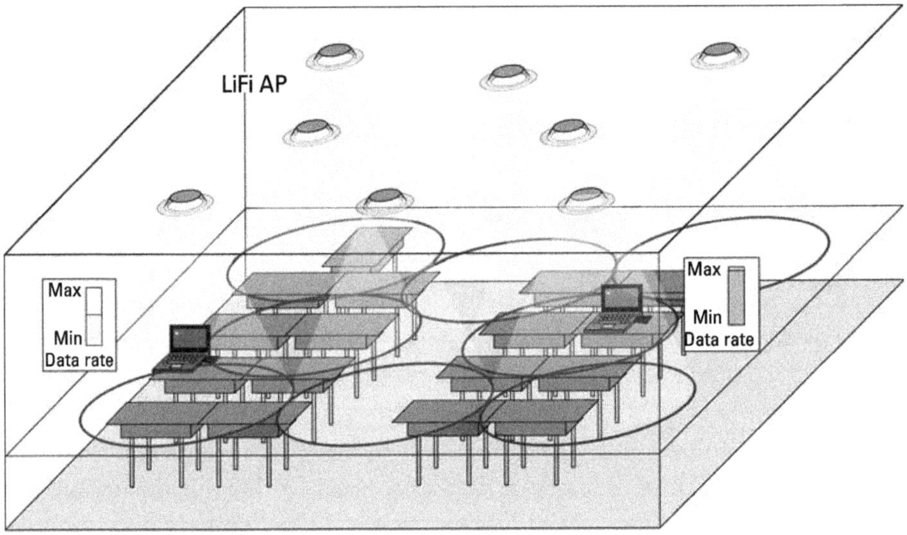

Figure 9.3 The network topology of the Li-Fi APs in classroom 2 and the Wi-Fi AP serving both classrooms [6].

APs coexist with two additional Wi-Fi APs that serve 7 classrooms. The Wi-Fi APs are commercially available from pureLiFi and based on the IEEE 802.11ac standard. Each Wi-Fi AP can support data rates between 300 Mbps and 867 Mbps depending on the mode of operation and bandwidth. The Wi-Fi APs were deployed in the corridor of the school. As a result, typically many users were served by each AP. This can lead to well-known throughput degradation because of MAC overheads as a result of contention resolution.

A commercially available Li-Fi system from pureLiFi was deployed in a single classroom. Each Li-Fi AP can support a circular coverage area with diameters ranging between 2.8m and 3.5m. The Li-Fi AP operates in full-duplex mode and supports multiuser access and an internet protocol (IP) handover between the deployed APs. Each Li-Fi AP can support a maximum of 8 users with a peak aggregate data rate of 43 Mbps. This corresponds to a total maximum aggregated data rate of 344 Mbps per classroom. We note that the data rate of an access point is significantly lower than the VLC transmission speeds reported in lab experiments. The main reason for this is that this trial uses off-the-shelf unmodified LED luminaires with an electrical bandwidth in the region of 2 MHz. The main purpose of this real-world trial was to showcase the simultaneous functions of lighting and wireless networking using the same system. Future upgrades can include the integration of bespoke optical components within the luminaires, leading to aggregate data performances of 1 Gbps and 10 Gbps. The latter will require bespoke luminaires, which are capable of WDM or laser-based lighting devices.

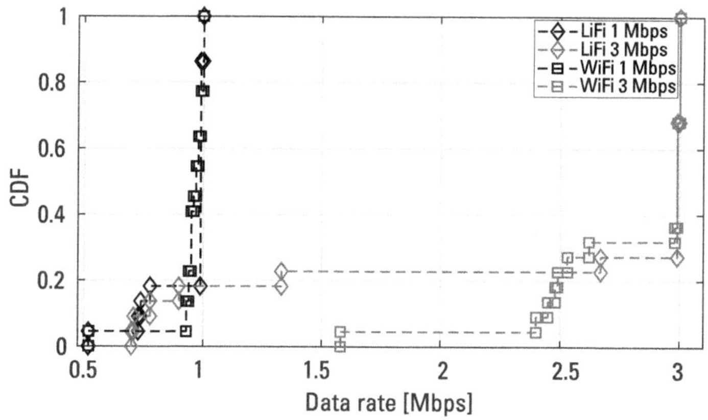

Figure 9.4 The CDF of the data rate for the Wi-Fi and Li-Fi users based on the 1-Mbps and 3-Mbps data rates targets [6].

The installation of the Li-Fi APs was constrained by the existing fluorescent lighting fixtures and the existing fire/smoke detector installation points. The Li-Fi luminaires were deployed alongside the existing infrastructure. As a result, there were some areas within the classroom that were not sufficiently covered by Li-Fi. Figure 9.3 shows the network setup. Therefore, the user performance is expected to be variable depending on the user location. Ideally, the Li-Fi AP would replace the existing lighting infrastructure and would be optimized based on the room topology to provide the best trade-off between illumination and communication [19].

A measurement campaign was carried out to compare the performance of the Li-Fi and Wi-Fi networks and to assess the total aggregate data rate. The user data rate is used as a performance metric. Twenty-two pupils simultaneously accessed the Li-Fi network and each of the two neighboring classrooms was served by Wi-Fi only. There were the same numbers of pupils in the neighboring classrooms. Two tests were conducted based on unconstrained best-effort data rates for different target data rates:

- A target data rate of 1 Mbps per user;
- A target data rate of 3 Mbps per user.

The cumulative distribution function (CDF) of the user data rate achieved by the Li-Fi and Wi-Fi network is shown in Figure 9.4 for the 1-Mbps and 3-Mbps data rate targets while in Figure 9.5 the data rate without a target is reported. The results in Figure 9.4 show that most of the users achieve the target data rate. However, some users fall short of the target data rate due to the suboptimum locations of the Li-Fi APs. Figure 9.5 demonstrates that there are some users in the Li-Fi network with considerably higher data rates, up to 20 Mbps. It also shows that the user peak data rate is higher in the Li-Fi network despite the

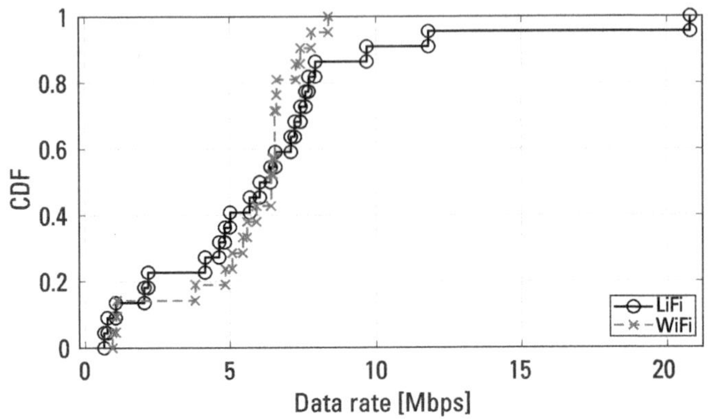

Figure 9.5 The CDF of the data rate for the Wi-Fi and Li-Fi users assuming no data rate targets [6].

Table 9.1
Average Data Rates Achieved for the Wi-Fi and Li-Fi Networks [6]

Target Data Rate	Average Data Rate [Mbps] Li-Fi	Wi-Fi
Best effort	6.24	5.57
1 Mbps	0.94	0.95
3 Mbps	2.5	2.79

fact that the maximum data rate of the given Li-Fi AP is about 10 times lower than the maximum data rate of a deployed Wi-Fi AP. The average data rates for the Li-Fi and Wi-Fi networks are shown in Table 9.1. The results show that the Li-Fi network outperforms the Wi-Fi network in terms of the 'best-effort average data rate, as also shown in Figure 9.5. However, the results also highlight that the Li-Fi network slightly underperforms compared to the Wi-Fi network at the targeted data rate of 3 Mbps. This is due to the low data rate achieved by the underperforming user equipment that is located at Li-Fi attocell border regions and in dead-spot areas of the classroom.

An indirect, but rather a significant, result of this proof-of-concept study was that there was a surge in the data rates in the neighboring Wi-Fi-only classrooms. This is because of the offload of data traffic to Li-Fi. Data rate gains in the neighboring classrooms are plotted in Figure 9.6 for different target date rates. This shows the capability of a Li-Fi network to offload traffic. This feature is particularly beneficial in dense environments like schools and airports. The results also demonstrate that frequency reuse gains are achievable within a small area, in this case, a classroom.

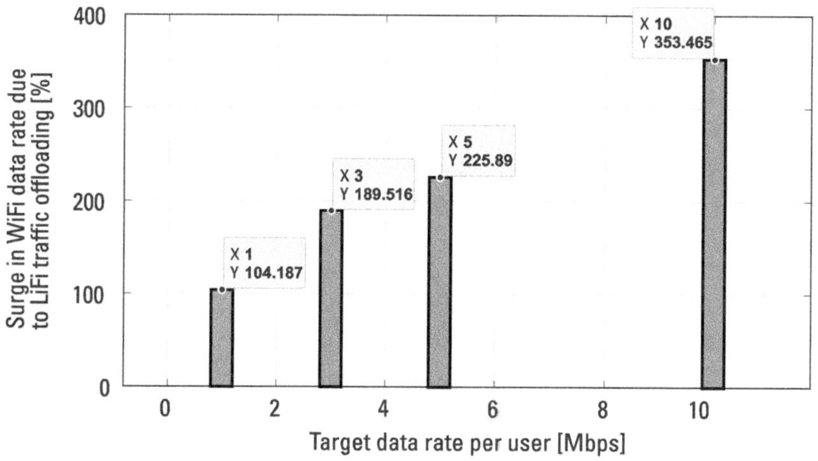

Figure 9.6 Surge in Wi-Fi aggregated data rate at neighboring classrooms [6].

Fraunhofer Heinrich Hertz Institute demonstrated a Li-Fi system in a seminar room in Mainau, Lake Constance [20]. The system was designed with two configurations for the downlink lens that can enable a single-user setting with a data rate up to 800 Mbps and a multiuser setting with a data rate up to 100 Mbps [20]. White phosphor coated (PC-LED) is used for the downlink and an IR transmitter is used for the uplink. Downlink data rate speeds are usually required to be much higher than the uplink data rate speeds in most practical use cases due to the huge differences between what users download and upload. For example, the user only needs to send a web link address to download a 4K high-definition video. However, the uplink data rate in this demonstration is shown to be lower than downlink speed due to the higher bandwidth enabled by the IR LED and the higher sensitivity of the PD at the IR wavelength band [20]. One of the main challenges in the system acceptance was the system flexibility and interface operability with the user devices. However, over a year-long evaluation of the deployed Li-Fi solution, most of the users considered the installed system as user-friendly [20].

9.4.2 Cyber Security and Defence

PureLiFi has deployed Li-Fi systems in more than 20 countries [21]. Trial deployments of Li-Fi were reported with O2 Telefonica, Babcock, and BT Defence [21]. The physical layer security of Li-Fi technology was identified as one of the main attractive features for the defense sector. The use case reported by pureLiFi and BT Defence considered assigning different security access privileges to individual light APs, which allows Li-Fi systems to exhibit sophisticated geofencing security features. In addition, pureLiFi trialled Li-Fi technology as a distaste response communication solution with Verizon, Aegex, and Nokia [22], where it was shown to provide a reliable bidirectional real-time communication in a subway disaster scenario [22]. In addition, pureLiFi collaborated with Lucibel in 2016 to manufacture Li-Fi APs with a data rate capability of 42 Mbps for both IR uplink and visible light downlink within a cell of 3m in diameter [23,24]. Data and power are fed to the AP using power over Ethernet (PoE), which enable an easy and cost-effective installation of a luminaire AP that can serve up to 8 users [24].

The US Army in Europe tested Li-Fi and FSO systems in a tactical operations centre in Germany [25]. The FSO systems connected two remote locations that are 10 miles apart. How easy it is to establish network connectivity was highlighted as one of the key features of Li-Fi. This is in addition to its more obvious physical layer security features that require users to be present in the tent where the Li-Fi AP is installed in order to gain access to the network. Unlike Wi-Fi, Li-Fi systems ensure that unlawful access from outsiders is denied due to the light confinement property within the tent. In fact, Li-Fi technology was shown to have zero digital footprints in tactical defense applications.

9.4.3 Industry, Aerospace, and Automotive

OWC technology is a promising candidate to enable communications systems for Industry 4.0 applications, as discussed in Section 9.2.2.3. Fraunhofer Heinrich Hertz Institute demonstrated an OWC system tailored for industrial applications at a BMW robot testing facility [26,27]. Multiple optical transmitters and receivers are used in a MIMO configuration to improve the diversity gain of the Li-Fi solution and to increase the reliability of the non-directed communication links [26]. Figure 9.7(a) shows the simulation of the MIMO OWC setup. The mobile end unit at the robot arm is configured in the transmitting mode, while the access points mounted on the walls are configured in the receiving mode [26]. The simulation results were used to determine the necessary number of transmitting and receiving nodes that are required to achieve a LOS and a SNR $\geq$ 5 dB, as shown in Figure 9.7(b). A total of 6 and 8 IR LEDs emitting at 850 nm were used for the uplink and downlink, respectively [26]. In addition, an analogu equal gain combining (EGC) is used to combine the signals of 8 and 6 PDs for the uplink and downlink, respectively [27]. The bandwidth of the transmitters and receivers was designed at 50 MHz. A picture of the deployment is shown in Figure 9.8(a). The mobile unit is shown in the yellow circle, while the access points are shown in red circles. The block diagram of the implemented solution is shown in Figure 9.8(b). The system was implemented on a production robot (ABB-IRB6640-235) that has a rotational speed around a single axes of up to 190°/s. OFDM-based arbitrary Ethernet traffic was generated, using Optixia XM2, based on the ITU-T G.hn standard between 2.15 MHz and 95.90 MHz. The nodes have their own full protocol stack and function as Ethernet bridges [26].

(a)

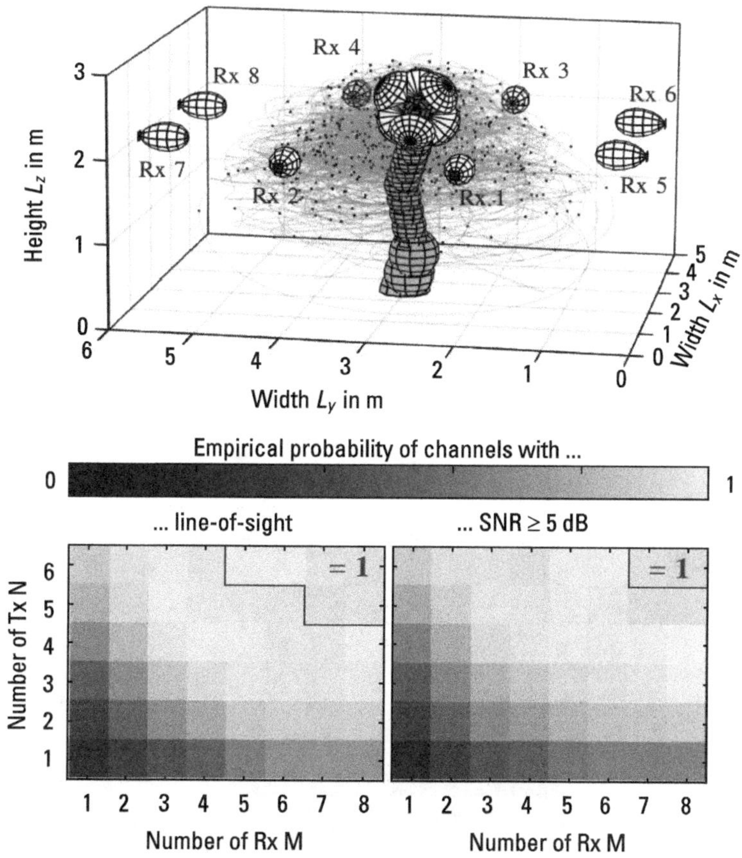

(b)

Figure 9.7 (a) The simulation setup of the uplink configuration, which shows the transmitters mounted on the robot model and the receivers mounted on the adjacent walls. The green line interpolates 500 randomly placed points that indicate the trajectory of the transmitter positions. (b) The empirical probability of achieving an LOS and a SNR $\geq$ 5 dB. (© 2019 IEEE. Reprinted, with permission, from [26].)

Figure 9.8(c) shows the SNR distribution per subcarrier frequency at the 13 testing positions indicated by the green dots in Figure 9.8(a). The 50-MHz bandwidth limit of the transceivers' bandwidth is clearly shown in the SNR results. The OWC solution was found to offer moderate data rates up to 150 Mbps, a reliable communication link, and reduced latency below 1 ms [26].

In collaboration with Air France, Oledcomm developed a Li-Fi system to enable passengers to access in-flight entertainment servers, which store movies and games [23]. The implemented solution is based on the ITU G.VLC standard that was discussed in the previous chapter. The system is based on Oledcomm's LiFiMAX product, which is reported to achieve 100 Mbps for both uplink and downlink by operating in the IR wavelength band at 940 nm [23]. A centralized

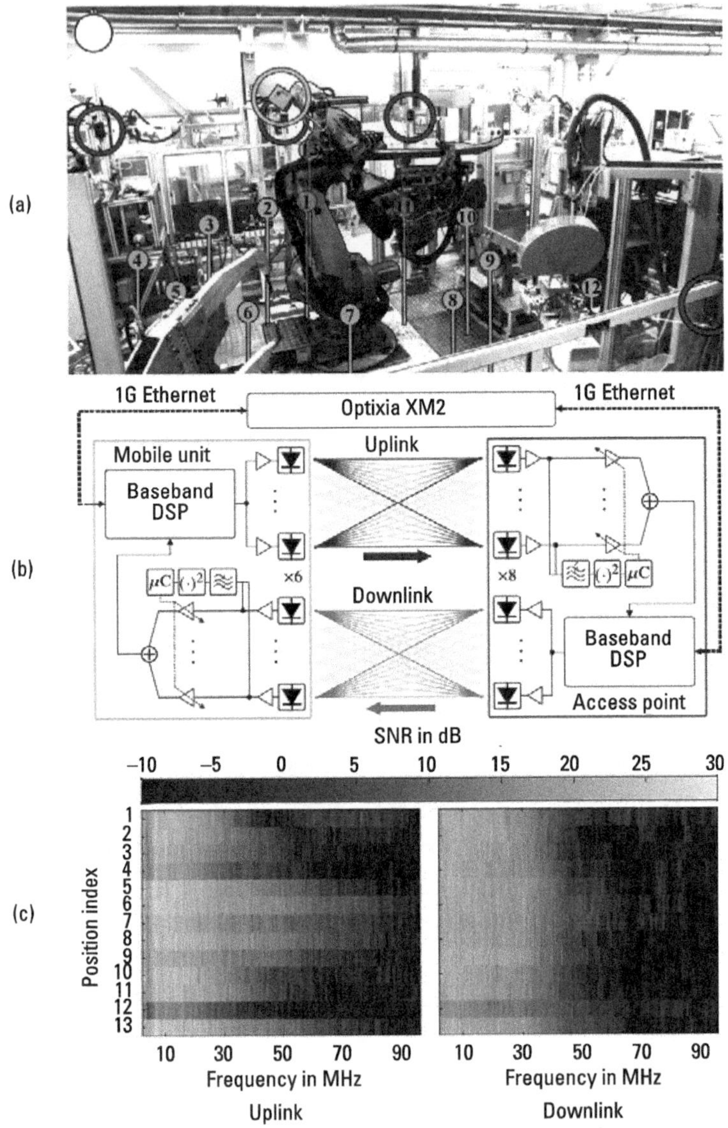

Figure 9.8 (a) A picture of the deployment of MIMO-based optical wireless mobile communication systems by the Fraunhofer Heinrich Hertz Institute. The yellow circle shows the mobile unit and the red circles show the APs. The green markers indicated 13 measurement points. (b) A block diagram of the implemented MIMO solution. (c) SNR distribution per subcarrier at the 13 measurement points marked in green in (a). (© 2019 IEEE. Reprinted, with permission, from [26].)

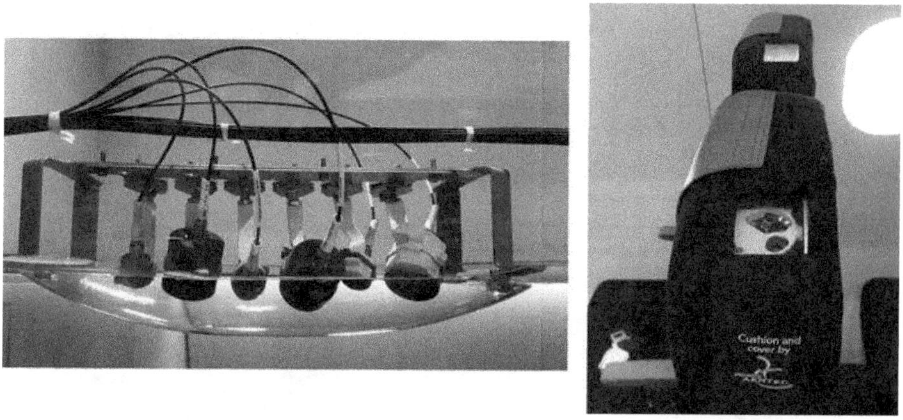

(a) (b)

Figure 9.9 (a) Fiber coupled lenses on the ceiling of the airplane acting as optical access points. (b) Transceiver integrated into the passenger seat acting as an end point. (© 2020 IEEE. Reprinted, with permission, from [23].)

Li-Fi modem is connected to multiple fiber-coupled lenses that can enable wireless multiuser access at each passenger seat. In particular, two fiber-coupled links are required to enable a full-duplex communication link at each passenger seat. The transceiver end at the user side is embedded at the top of the passenger seat, as shown in Figure 9.9(b), to enable a LOS connection with the fiber-coupled transceiver lenses at the ceiling, as shown in Figure 9.9(a). The system was demonstrated at the 2019 Paris Air Show and was installed on an Airbus A321 airplane [23].

VLC is considered as an enabling technology for ITS. Optical technologies such as high-speed cameras and light detection and ranging (LIDAR) are now considered essential elements for autonomous driving vehicles [28]. The automotive lighting firm, Magneti Marelli, collaborated with Oledcomm to incorporate an OWC into an automotive headlight that already integrates lighting functions with IR cameras and LIDAR [29]. Figure 9.10 shows the car headlight with the integrated VLC and LIDAR systems. The headlight was presented at the 2019 CES with 100 Mbps at 2m and 14 Mbps at 15m [29]. However, it is widely accepted that vehicle-to-vehicle (V2V) and vehicle-to-infrastructure (V2I) communication systems are expected to achieve transmission distances between 50m and 300m [28]. In addition, the OWC system does not reuse the visible emission of the headlight to realize a VLC link. A dedicated transmitter board is required to integrate the IR LEDs into the headlight. Figure 9.10(a–c) shows the LED, PD, and analog front-end cards that are integrated into the car headlight. The bandwidth of the used IR LEDs is extended to 30 MHz using pre-equalization and the implemented OWC is based on DC biased optical OFDM (DCO-OFDM) with adaptive bit and energy-loading [29].

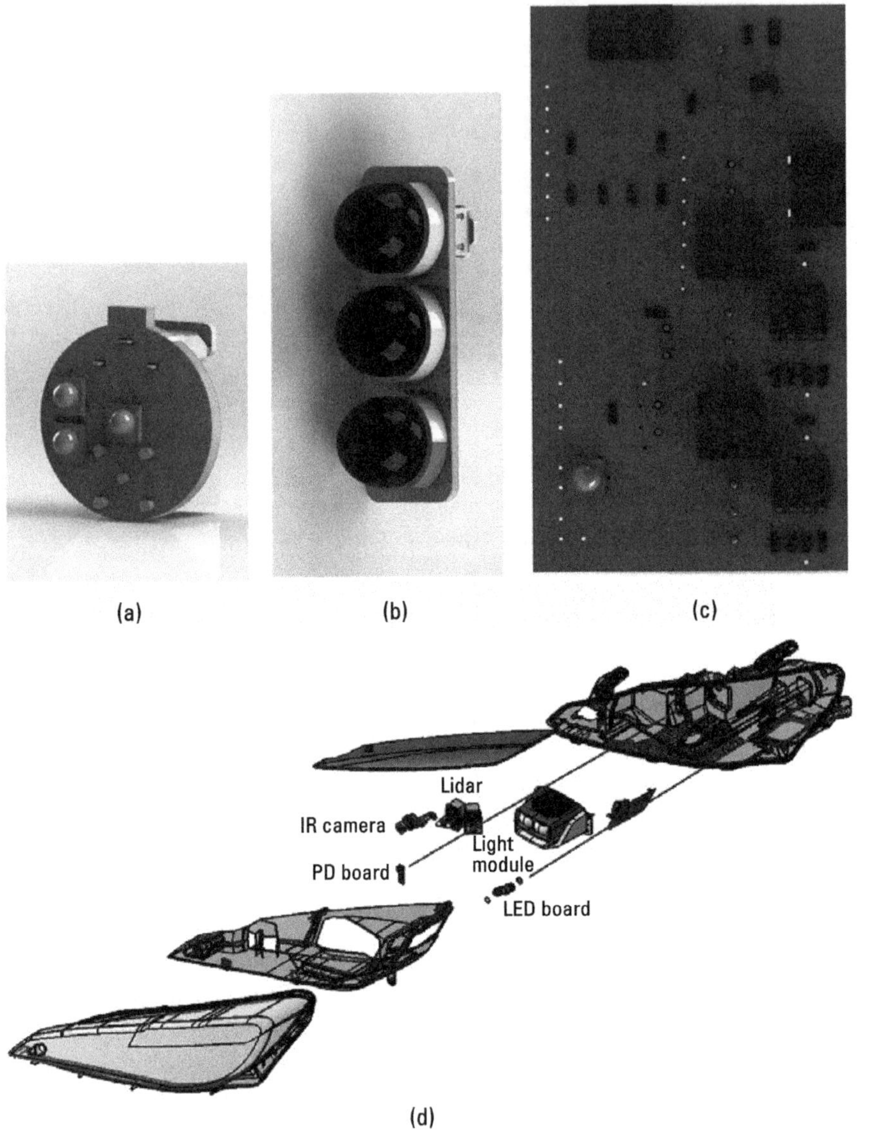

Figure 9.10 (a) LED board; (b) PD board; (c) analog front-end board; and (d) car headlight with integrated LIDAR and VLC systems. (© 2019 IEEE. Reprinted, with permission, from [29].)

9.4.4 Smart Cities

Disney Research has shown an interest in VLC to enable low-complexity IoT applications [30]. Disney Research proposed an LED-to-LED VLC system to enable low-cost transceivers without the need of PDs. However, the limited sensitivity of LEDs is a major drawback of this approach. In addition, Disney

Research also proposed a low-cost design to convert off-the-shelf LED luminaires into VLC transceivers that can support data rates below 1 kbps [31]. The main use case was targeting wirelessly connected toys that require a low bit rate of 1 kbps at transmission distances below 3m. The use case is motivated by the fact that most of the toys already have LED lights that can be used to trigger functions when communicating with other toys. The communication link is secure and free from any unnecessary RF radiation [30]. Other use cases include interactive fashion, where a piece of clothing is integrated with LEDs and PDs to enable the orchestration of lights with other users and surrounding luminaires [32].

Fraunhofer Heinrich Hertz Institute also demonstrated a Li-Fi system at the Motol University Hospital in Prague, Czech Republic [33]. Li-Fi-based solutions in the healthcare sector offer a significant advantage against RF-based communication systems due to their reduced interference with other vital health equipment, as described in Chapter 4. A multiuser MIMO Li-Fi system was implemented in a neurosurgery room at the Motol University Hospital. Multiple receivers were used to emulate a body sensor network, whereas multiple transmitters were used to improve the system performance against blockage [33]. Different multiple access techniques were compared, and it was concluded that a hybrid SDMA and TDMA approach achieves the highest system throughput [33].

9.4.5 Backhaul Solutions

OWC systems based on FSO communication are promising candidates as a solution to the 5G backhauling applications [34]. Multiple commercially available solutions in the market can support data rates up to 10 Gbps and transmission distances up to 5 km [35]. FSO systems based on coherent modulation techniques are also considered to be a prominent solution for satellite-to-ground and intersatellite communication systems [35]. However, this section discusses low-cost systems based on incoherent modulation techniques. Fraunhofer Heinrich Hertz Institute reported results for a long-term field trial of a 100-m LED-based OWC backhauling solution [34]. A low-cost, rate-adaptive, OFDM-based OWC system was developed in this trial with a commercially available LED and a PD at a reported peak data rate of 1 Gbps [34]. The trial also compared the OWC system to a 60-GHz millimeter-wave link, as shown in Figure 9.11, in terms of data rate, SNR, latency, and reliability against different weather conditions. The OWC system is claimed to achieve 99.99% service availability despite severe whether conditions such as fog, rain, and snow [34]. In addition, the OWC solution is reported to outperform the millimeter-wave solution 99.8% of the time. The OWC and RF-based millimeter-wave solutions were reported to achieve different results against the varying weather conditions. For example, the millimeter-wave link achieves a lower SNR in rainy weather conditions, whereas the OWC SNR performance is more affected due to foggy weather conditions [34]. However,

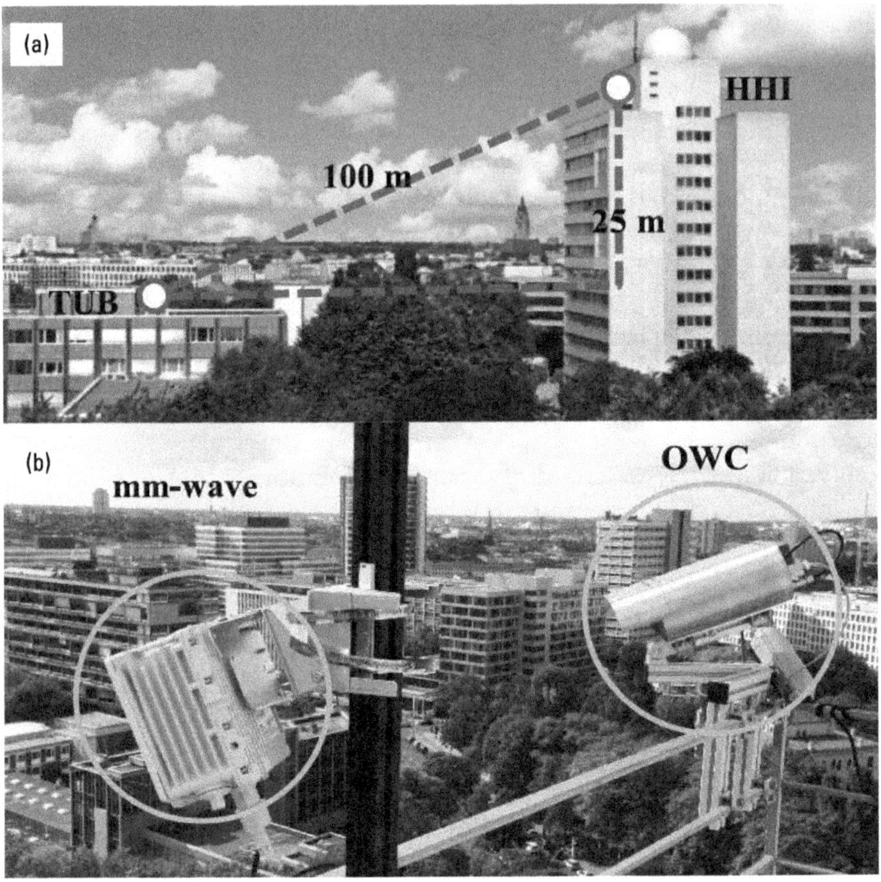

Figure 9.11 A OWC and millimeter-wave links demonstrated in an outdoor environment over a 100-m distance by the Fraunhofer Heinrich Hertz Institute. (© 2018 IEEE. Reprinted, with permission, from [34].)

the reduced visibility results in a significant reduction of the data rate of the OWC system, whereas the visibility was shown to have a negligible effect on the millimeter-wave link [34].

A novel backhauling system centered around using solar photovoltaics cells as OWC receivers were trialed on the Orkney Islands, United Kingdom, by the Li-Fi Research and Development Center at the University of Strathclyde in collaboration with pureLiFi [36,37]. The trial is part of a project termed 5G RuralFirst that is partly funded by the U.K. government's Department for Digital, Culture, Media and Sport (DCMS). The initiative is led by Cisco and involves a consortium of 28 other partners from business, government, and academia [36]. The project aims to connect remote rural areas to empower industries such as agriculture, renewable energy, and tourism. Solar cell-based OWC systems are promising candidates to close the digital divide and enable internet access in

Figure 9.12 The solar cell-based OWC system deployed at one of the residential properties on the Orkney Islands, United Kingdom Photo credit: Sovan Das from the University of Strathclyde, United Kingdom [37].

remote areas. The relatively larger detection area of solar cells, when compared with the traditionally used PDs in FSO, offers a unique advantage in terms of alignment, ease of use, and reliability. FSO systems tend to require beam-tracking and suffer from beam-wandering due to atmospheric turbulence or slight building movement effects. The large detection size of solar cells can mitigate these effects. Furthermore, the dual energy-harvesting and signal detection usage of photovoltaics devices can enable energy-neutral and self-powered receivers. This is promising for low-cost IoT applications where it is hard to power a remote IoT device. Four bidirectional OWC links were deployed at a 30-m distance to provide high-speed internet access to two residential cottages using off-the-shelf lasers and silicon photovoltaic cells. The system enables a 5-W power-harvesting and a peak data rate of 8 Mbps [37]. A picture of the deployment at one of the residential properties is shown in Figure 9.12. The figure also shows a Class 1M IR-based laser transmitter emitting at 940 nm next to the solar cell.

9.4.6 Indoor Positioning

An LED-based positioning system was deployed by Philips at the Carrefour Euralille mall, Lille, France, in 2015 [38]. Location-based services, such as offers and promotions, can be provided to shoppers using their handheld mobile devices. The traditionally used global positioning system (GPS) is inefficient for indoor scenarios as it tends to have a weak signal indoors. The system is reported to achieve a 30-cm positioning accuracy without requiring any additional hardware apart from the existing LED-based lighting fixtures, which are reported to achieve a 50% energy-saving [39]. The system uses 800 linear LED fixtures in the 7,800-m^2 shop [38]. Each LED fixture broadcasts a unique code that can be detected using a smartphone. The location of these lighting fixtures is stored alongside the corresponding codes in a cloud database. The system is based on OCC, in which the front-facing mobile device camera is used to receive the codeword information transmitted by the lighting fixtures [40]. The mobile can be connected to the internet through Wi-Fi or the mobile network. A mobile app is used to determine the indoor location of the user within a store map based on the look-up table. The system is predicted to save both customers' and staff members' time by enabling a faster interactive localization of products. In addition, it can be used to send location-based advertisements related to a nearby product. This technology can also be useful for visually impaired people or those with other types of disabilities. Additionally, the anonymized location data can be used to gain insights about users' preferences, which could influence the marketing of any future promotions.

A similar lighting positioning system was proposed by ByteLight based on the rolling shutter effect of the CMOS cameras [41]. Qualcomm Research introduced a lighting-based indoor positioning system, termed Lumicast, in collaboration with GE Lighting and Acuity Brands (which bought ByteLight) [42]. LED-based luminaires send unique binary codes that can be detected using the front camera of a mobile phone. A mobile phone application is used to identify the 3-D location of the user at accuracy in the order of 10 cm [42]. This accuracy is achieved by performing pixel-level image processing on the captured frames to determine the direction of light arriving from one or multiple light fixtures. Determining the height of a receiver can be useful for autonomous navigation applications with drones and forklifts in industrial warehouses [42]. The system also considers the use of RF technologies such as Bluetooth for cases where the mobile device is not directed at any of the lighting fixtures.

Signify (formerly Philips Lighting) has implemented similar systems through its new software brand, Interact, for connected lights. The system was deployed at Aswaaq Supermarket in Dubai, United Arab Emirates, and at MediaMarkt in Eindhoven, Netherlands [40]. Connected lights can be integrated with occupancy sensors and other IoT devices to offer energy-efficient lighting solutions that can manage and collect data on space utilization, traffic patterns, and cyber security applications. In addition to OCC, these implementations integrate Bluetooth to facilitate customers pathway tracking within the shop when the mobile is not in the LOS with the luminaires [43]. Oledcomm demonstrated a number of case studies based on an OCC indoor positioning system in retail stores to facilitate navigation and understand shoppers' habits [40]. In addition,

Oledcomm demonstrated a case study at the Grand Curtius Museum where an LED-based luminaire was used to send a location tag that could be used by a mobile app to guide visitors and to display information related to a corresponding exhibit [40,44].

A visible light positioning system termed Epsilon was developed by Microsoft Research in China [45]. The system is based on the triangulation of a few LED luminaires, and it was reported to achieve accuracy between 0.4m and 1.1m [45].

9.4.7 OCC Deployments

Fujitsu Laboratories developed an OCC system called FlowSign Light that can be used to send specific information about an exhibit [46]. A red, green, and blue LED (RGB-LED) is used to illuminate an object. The intensity of each of the RGB colors can then be modulated at a rate between 10 bps and 16 bps with a unique codeword that can be detected using a dedicated mobile application when users point their mobile front camera towards the object. The codeword corresponds to specific information stored on a cloud-based server or a web link. The service was demonstrated at a few Japanese fashion retailer stores and was also targeted at museums, galleries, and exhibition centers [47]. A similar principle was also developed by Fujitsu Laboratories using the screen to camera communications [47,48]. When interested in an advertised product on a television commercial, viewers tend to look that product up on the internet. Fujitsu Laboratories developed a system based on OCC that can provide these viewers with what they are searching for simply by pointing their mobile phones to the screen. Areas with changing colors are embedded in the video, and the codeword is encoded in the color variations of these areas. The system is reported to preserve the quality of the broadcasted videos since the size of these areas is small enough to be detected by the human eye. This system was demonstrated on a home shopping television channel that was broadcast in Japan [48].

9.5 Commercialization Challenges

9.5.1 Optical Front Ends in Transmitters and Receivers

One of the major challenges of Li-Fi commercialization is the development of optical front-end devices in the transmitter and receiver, where the achievable data rates are currently limited by the electrical bandwidth of the optical front ends. In order to fully utilize the optical spectrum resources, it is critical to continue with the development of new electrical-to-optical (E-O) (optical source) and optical-to-electrical (O-E) (optical detector) conversion devices with high bandwidth in the order of hundreds of gigahertz.

The transmitter should be comprised of at least one optical source with its driving circuitry and a compact optical system to control the light propagation.

Regarding the optical sources, the desired devices have to achieve a good balance between output power, modulation bandwidth, and cost. The existing white LED-based devices have sufficient output power but with very limited bandwidth (a few megahertz [49]). In order to significantly enhance the bandwidth of optical sources, gallium nitride (GaN) micro LEDs have been considered for VLC. Micro LEDs are able to achieve a modulation bandwidth greater than 1 GHz [50]. Note the modulation bandwidth is different from the 3-dB bandwidth because with bit-loading and power-loading techniques optical transmitters can be modulated at frequencies much higher than the 3-dB bandwidth. However, currently, micro LEDs devices exhibit limited output power and low energy efficiency. An alternative is to use laser-based devices, such as a blue laser with a phosphor reflector [14] and vertical-cavity surface-emitting laser (VCSEL) devices [51], both of which are able to achieve high output power and high bandwidth simultaneously. Commercially available LD-based lighting devices have been recently considered and demonstrated for high-speed VLC [14].

The receiver should be composed of at least one optical detector with its supporting electronics such as transimpedance amplifier (TIA) and compact optics to focus the collected light on top of the detector. On the optical detector side, the desired device should have high sensitivity, high modulation bandwidth and a fair price. Detectors' sensitivity is one of the main factors that limit the achievable capacity. A straightforward way to improve the sensitivity is to use PDs with a large active area. However, this will dramatically limit the available bandwidth and increase the cost of the optical detector. This is because PDs are capacitive elements. A PD with a larger active area has a larger capacitance of the depletion region, which leads to a greater RC time constant and narrower bandwidth. An alternative solution to avoid bandwidth limitation is to use a smaller-size avalanche photodiode (APD) with optical concentrators, such as compound parabolic concentrator (CPC). However, this will lead to bulky detector components, which will make it challenging for device integration. Currently available optical detectors are square-law devices (i.e., the produced output current is proportional to the square of the input signal), which results in low receiver sensitivity. Higher-sensitivity optical detectors are required in order to boost the performance of optical communications links in 6G networks. In recent years, SPADs have become potential candidates as optical receivers due to their sufficient sensitivity, high detection efficiency, high power efficiency, and high timing resolution. The deployment of SPAD receivers in OWC requires the design of efficient modulation schemes and operating conditions tailored to the unique characteristics of SPAD arrays [52]. However, SPAD detectors generally have a low modulation bandwidth and nonlinearity issues, which are also challenging for Li-Fi commercialization. Therefore, further efforts on finding the optimal balance between various performance and cost are still required.

9.5.2 Integration to Mobile Devices

Optical front-end and chipset integration into mobile and laptop devices is one of the main barriers stopping Li-Fi technology from becoming part of a mass market. For a mobile device or a laptop to support Li-Fi, the device needs to be equipped with an optical front end that functions as an optical antenna. In addition, placing the current driver and the TIA close to the optical emitter and detector, respectively, can be helpful to improve the impedance mismatch with these devices as they tend to have lower impedance. The optical system design of the integrated front end is required to manipulate the light propagation and spectrum such that it can increase the signal to interference and noise ratio (SINR). Therefore, the optical system can include, but is not limited to, optical filters, lenses, concentrators, and diffusers. The optical system should offer a reliable light collection despite a user's movement and orientation. In addition, the optical system design should be resilient against interference and background illumination. Multiple solutions were proposed to enable a wide FOV detector for Li-Fi systems [53–55]. The concept of using multiple detectors on different facets of the mobile device offers a spatial diversity that can enable a reliable Li-Fi system, although it might pose additional complexity for the optical front-end integration. Nowadays, mobile phone manufacturers tend to equip devices with multiple camera sensors to impress customers with higher quality and professional cameras. Therefore, equipping a mobile device with multiple high-speed detectors should be technically feasible. In fact, all of the aforementioned optical front-end components do exist sparsely in the market with high-performance metrics, such as energy efficiency, bandwidth, and sensitivity. The main challenge is integrating all of these enabling technologies into a compact, energy-efficient, high-performance solution that could fit into mobile phones.

In addition, a mobile device requires both analog and digital capabilities to process the optical signaling and modulation techniques of a Li-Fi system. The optical signals emitted by off-the-shelf LEDs are based on IM/DD and thus they are unipolar and real-valued. The chipsets that are currently available to mobile devices are unable to directly support these signaling requirements. Application-specific integrated circuit (ASIC) solutions based on the existing 802.11 standard are available in the market. PureLiFi developed integrated Li-Fi solutions for optical front ends that can work with 802.11 baseband chipset to ease the integration with mobile and laptop devices [56], as shown in Figure 9.13. However, a dedicated ASIC solution is required to enable mass-market opportunities for Li-Fi [23].

9.5.3 Lighting Considerations

Intensity modulation implies that the LED driving current is varied rapidly according to the information signal. The degree of variation in the driving current

(a) (b)

Figure 9.13 (a) Laptop with an integrated optical front end from pureLiFi. (b) Mobile device with a Li-Fi transceiver dongle integrated to its case. (Photo credit: pureLiFi, United Kingdom.)

depends on the type and order of the adopted modulation scheme. Although these subtle variations cannot be perceived by the human eye, they cause a deviation from the typical requirements of the LEDs operation, which could affect the lighting quality and impose certain eye safety concerns.

9.5.3.1 Lighting Quality

Any variations in the driving current directly affect the chromaticity and color-rendering properties of the LEDs, resulting in a change in the lighting quality. The lighting quality requirements are critically important, not only visually, but also for the associated health and psychological effects [57,58]. Moreover, undesirable changes in the LED chromaticity can also lead to a drop in the LED efficiency due to the droop effect [59].

Because of the above-mentioned considerations, the large-scale adoption of Li-Fi technology requires the realization of fast data communications through intensity modulation with minimal impact on the lighting quality. In general, the lighting quality can be measured by the following photometric properties:

1. *Correlated color temperature (CCT)*: Color temperature indicates the color appearance of the light emitted by LEDs and is measured in degrees Kelvin. For example, an extra warm light is said to have a CCT of about 2,700K, while a cool white light has higher CCT values up to 5,000K. For most residential properties, the desired LED CCT values are in the range of 2,700K to 3,000K, since these light temperatures provide a warm ambiance that creates a quiet and relaxing atmosphere. For commercial properties or retail outlets, CCT values around 4,000K are a popular option to produce clearer, more powerful white light.
2. *Color rendering index (CRI)*: The CRI score of an LED fixture indicates the ability of the produced light to reveal the real colors of the lit objects.

The highest possible CRI value is 100, which indicates that the light source renders the colors in a way that is similar to natural daylight. Most standard LED lights have a CRI value of 70 or more. It is normal to have a high CRI for indoor lighting, particularly in places such as museums and art galleries, where there needs to be enough light to enable high precision and perception of details.

3. *Chromaticity*: Chromaticity describes the stimulus of the eyes by a specific light and is represented by a three-coordinate value that is mapped onto a color space. The spectral power distribution of the emitted light is weighted by three color-matching functions and then integrated over the visible light wavelengths. Changes in light chromaticity of an LED could lead to efficacy degradation and increased energy consumption.

In order to satisfy the expected lighting quality in Li-Fi systems, it is essential to ensure that the adopted modulation schemes do not result in big variations in the three light quality measures (i.e., CCT, CRI, and chromaticity). It was shown in [60], that the light quality metrics of an LED fixture are determined by the average driving current, not by the instantaneous driving current. As a result, the modulating signals are required to be adequately DC-balanced in order to avoid degradation in the lighting quality. It was also shown in [60] that using DCO-OFDM as a modulation scheme results in no significant impact on the lighting quality of the LEDs with a maximum of 0.1-point CRI shift compared to using a constant current with no modulation. This is because DCO-OFDM signals have a constant average value over a given data symbol. It is noted that a CRI shift of less than 5 points is considered acceptable as it does not result in significant changes in the chromaticity coordinates and the CCT. As the data rates required in Li-Fi systems continue to increase, new and innovative modulation and coding schemes need to be developed. Given that high-order modulation requires larger variations in the driving current, finding the optimal trade-off between the achievable data rates and the effect on lighting quality remains an open challenge.

9.5.3.2 Eye Safety

Fluctuations in the brightness of the emitted light can cause flicker, which could have seriously damaging effects on the eye. According to the IEEE 802.15.7 standard [61], any changes in light brightness over periods that are longer than the maximum flicker time period (MFTP) must be avoided. MFTP is defined as the maximum period over which the light intensity can change, but for which the resulting flicker is not perceivable by the human eye. Generally, an MFTP of 5 ms or less is considered to be imperceptible to the human eye. Flicker is typically caused by long runs of 0s and 1s in the codewords. This can be mitigated by the use of various coding methods such as run length limited (RLL) codes and

Manchester codes, which maintain a semi-constant brightness of about 50% of the full LED brightness by ensuring a balanced repetition of 0s and 1s in the codewords. It is noted, however, that these coding methods result in reduced transmission efficiency as well as increased processing and hardware complexity. In the next-generation era of wireless communications, Li-Fi systems will be used for IoT applications with strict requirements for minimal complexity and maximum transmission efficiency. As a result, there is a need for new and innovative flicker mitigation techniques to enable the anticipated connectivity gains of Li-Fi without hindering the safety of the emitted eyes to the human eye [62].

Another factor that is critical for eye safety consideration in Li-Fi systems is the use of laser diodes for transmissions. For example, VCSELs have been recently identified as one of the strongest contenders to be used as optical transmitters due to their high modulation bandwidths, high-power conversion efficiency, reduced production cost, and their small sizes, which allow for dense arrangements as 2-D arrays. Eye safety is a critical factor when it comes to determining the maximum permissible exposure (MPE) of a VCSEL transmitter. The decision on the MPE value depends on the operating wavelength and the angular substance, which is defined by the visual angle subtended by a source at the point of measurement. When it comes to using laser diodes in indoor Li-Fi systems, there are still many constraints and open questions related to eye safety regulations that need to be considered in order to cut the laser-light power down to eye-safe levels while maintaining the advantages of lasers over LEDs.

9.6 Summary

After short-range IR data association (IrDA) was suppressed by Bluetooth technology, OWC systems have faced a setback in capturing the interest of end-users and the attention of the mass market. However, Li-Fi-based systems offer something different compared to what IrDA has promised to achieve. Li-Fi systems offer a complete networking solution that can enable optical mobile wireless links with seamless handover between optical attocell networks and other integrated RF networks such as 5G and Wi-Fi. Moreover, Li-Fi systems are envisaged to cover the visible and invisible optical spectrum regions between 400 nm and 1,550 nm to enable downlink and uplink communications.

In this chapter, we discussed Li-Fi market opportunities, key drivers, and future outlook for market growth. It has been noted that the commercialization of Li-Fi technology requires collaboration among participants in multiple industries. Therefore, the ecosystem of Li-Fi commercialization has been introduced. Furthermore, a discussion on the establishment of such an ecosystem has been provided. There have been a number of real-world demonstrations and use cases for Li-Fi technology in different sectors and applications. OCC has proved its suitability and high performance for indoor positioning systems and object identification. FSO communication systems can provide a solution to 5G backhauling applications. Li-Fi demonstrations in the residential, commercial, health, and industrial sectors highlight its potential to provide an additional means of high-speed wireless access that is free from interference and secure from eavesdropping. Finally, the challenges of realizing Li-Fi commercialization have been discussed in terms of lighting quality, device availability, and integration barriers.

References

[1] Mordor Intelligence, "Li-Fi (Light Fidelity) Market - Analysis of Growth, Trends and Forecast (2020-2025)," Tech. Rep., 2019. https://www.mordorintelligence.com/industry-reports/li-fi-market.

[2] Wadhwani, P., and P. Saha, "Li-Fi Market Forecast 2030 - Global Statistics Report," Global Market Insights, Tech. Rep., 2020. https://www.gminsights.com/industry-analysis/lifi-market.

[3] Demirkol, I., et al., "Powering the Internet of Things Through Light Communication," *IEEE Communications Magazine*, Vol. 57, No. 6, June 2019, pp. 107–113.

[4] You, X., et al., "Towards 6G Wireless Communication Networks: Vision, Enabling Technologies, and New Paradigm Shifts," *Sci. China Inf. Sci.*, Vol. 64, No. 110301, November 24, 2020.

[5] Wang, M., et al., "Security and Privacy in 6G Networks: New Areas and New Challenges," *Digital Communications and Networks*, Vol. 6, No. 3, 2020, pp. 281–291. http://www.sciencedirect.com/science/article/pii/S2352864820302431.

[6] Haas, H., et al., "Introduction to Indoor Networking Concepts and Challenges in LiFi," *IEEE/OSA Journal of Optical Communications and Networking*, Vol. 12, No. 2, 2020, pp. A190–A203.

[7] Arvanitakis, G. N., et al., "Gb/s Underwater Wireless Optical Communications Using Series-Connected GaN Micro-LED Arrays," *IEEE Photonics Journal*, Vol. 12, No. 2, 2020, pp. 1–10.

[8] Scace, S., "LiFi-XC Starter Kits Now Available to Academics," pureLiFi.com, June 20, 2018. https://purelifi.com/lifi-xc-starter-kits-now-available-to-academics/.

[9] Fenwick, S., "LiFi Integrated into Devices," *Land Mobile*, March 1, 2018. https://www.land-mobile.co.uk/news/lifi-integrated-into-devices-1/.

[10] Jovicic, A., J. Li, and T. Richardson, "Visible Light Communication: Opportunities, Challenges and the Path to Market," *IEEE Communications Magazine*, Vol. 51, No. 12, 2013, pp. 26–32.

[11] Pimputkar, S., et al., "Prospects for LED Lighting," *Nature Photonics*, Vol. 3, No. 4, 2009, pp. 180–182.

[12] Hu, Y., L. Huber, and M. M. Jovanovic, "Single-Stage, Universal-Input AC/DC LED Driver with Current-Controlled Variable PFC Boost Inductor," *IEEE Transactions on Power Electronics*, Vol. 27, No. 3, 2012, pp. 1579–1588.

[13] Halper, M., "Lighting as a Service Makes It Easier to Raise Money for Other Things, Even in 2019," *LEDs Magazine*, September 12, 2018. https://www.ledsmagazine.com/architectural-lighting/indoor-lighting/article/16695783/lighting-as-a-service-makes-it-easier-to-raise-money-for-other-things-even-in-2019-magazine.

[14] Lee, C., et al., "Advanced LiFi Technology: Laser Light," in *Light-Emitting Devices, Materials, and Applications XXIV*, J. K. Kim, M. R. Krames, and M. Strassburg (eds.), Vol. 11302, International Society for Optics and Photonics/SPIE, 2020, pp. 116–123.

[15] Absil, P. P., et al., "Silicon Photonics Integrated Circuits: A Manufacturing Platform for High Density, Low Power Optical I/O's," *Optics Express*, Vol. 23, No. 7, 2015, pp. 9369–9378.

[16] "It's the Chicken or the Egg at the 2nd Annual Global LiFi Congress," *LiFi Report*, June 2019. https://lifireport.com/its-the-chicken-or-the-egg-at-the-2nd-annual-global-lifi-congress/.

[17] Bouchet, O., et al., "Acemind New Indoor Full Duplex Optical Wireless Communication Prototype," in *Laser Communication and Propagation Through the Atmosphere and Oceans*, A. M. J. van Eijk, C. C. Davis, and S. M. Hammel, (eds.), Vol. 9979, International Society for Optics and Photonics/SPIE, September 2016, pp. 117–126.

[18] Apoorv, S., S. K. Bhowmick, and E. Annadevi, "Implementation of Li-Fi Technology in Classrooms," *IOP Conference Series: Materials Science and Engineering*, Vol. 590, October 2019, p. 012044.

[19] Stefan, I., and H. Haas, "Analysis of Optimal Placement of LED Arrays for Visible Light Communication," *IEEE 77th Vehicular Technology Conference (VTC Spring)*, 2013, pp. 1–5.

[20] Paraskevopoulos, A., et al., "Pilot Li-Fi Installation in a Conference Room in Mainau, Lake Constance: Realization and Outlook," *Global LIFI Congress (GLC)*, 2018, pp. 1–4.

[21] pureLiFi, "PureLiFi Case Studies," https://purelifi.com/case-studies/.

[22] Ventulett, T. P., et al., "Protecting Assets, Protecting People: Exploring IoT Solutions to Drive Improved Asset Integrity and HSE Performance in Oil & Gas," *Abu Dhabi International Petroleum Exhibition & Conference*, Society of Petroleum Engineers, 2019.

[23] Béchadergue, B., and B. Azoulay, "An Industrial View on LiFi Challenges and Future," *12th International Symposium on Communication Systems, Networks and Digital Signal Processing (CSNDSP)*, 2020, pp. 1–6.

[24] Jurczak, C., "LiFi: Enlightening Communications," arXiv: 1802.01471 v2, physics.app.ph, 2018.

[25] Perez, S. S., "U.S. Army Europe Soldiers Test Li-Fi," *DVIDS*, October 2019. https://www.dvidshub.net/news/347697/us-army-europe-soldiers-test-li-fi.

[26] Berenguer, P. W., et al., "Real-Time Optical Wireless Mobile Communication with High Physical Layer Reliability," *Journal of Lightwave Technology*, Vol. 37, No. 6, 2019, pp. 1638–1646.

[27] Berenguer, P. W., et al., "Optical Wireless MIMO Experiments in an Industrial Environment," *IEEE Journal on Selected Areas in Communications*, Vol. 36, No. 1, 2018, pp. 185–193.

[28] Harding, J., et al., "Vehicle-to-Vehicle Communications: Readiness of V2V Technology for Application," National Highway Traffic Safety Association, Tech. Rep., 2014.

[29] Béchadergue, B., et al., "Vehicle-to-Vehicle Optical Wireless Communication with the Smart Corner™Automotive Headlamp," *Global LIFI Congress (GLC)*, 2019, pp. 1–5.

[30] Corbellini, G., et al., "Connecting Networks of Toys and Smartphones with Visible Light Communication," *IEEE Communications Magazine*, Vol. 52, No. 7, 2014, pp. 72–78.

[31] Schmid, S., et al., "Using Consumer LED Light Bulbs for Low-Cost Visible Light Communication Systems," *Proceedings of the 1st ACM MobiCom Workshop on Visible Light Communication Systems, Ser. VLCS '14*, New York: Association for Computing Machinery, 2014, p. 9–14.

[32] Corbellini, G., et al., "LED-to-LED Visible Light Communication for Mobile Applications," *Demo at ACM SIGGRAPH Mobile*, 2012.

[33] Mana, S. M., et al., "LiFi Experiments in a Hospital," *Optical Fiber Communication Conference (OFC), Optical Society of America*, 2020, p. M3I.2. http://www.osapublishing.org/abstract.cfm?URI=OFC-2020-M3I.2.

[34] Schulz, D., et al., "Outdoor Measurements Using an Optical Wireless Link for Fixed-Access Applications," *Journal of Lightwave Technology*, Vol. 37, No. 2, January 2019, pp. 634–642. http://jlt.osa.org/abstract.cfm?URI=jlt-37-2-634.

[35] Trichili, A., et al., "Roadmap to Free Space Optics," *Journal of the Optical Society of America B*, Vol. 37, No. 11, November 2020, pp. A184–A201. http://josab.osa.org/abstract.cfm?URI=josab-37-11-A184.

[36] Haas, H., et al., "Solar Cell Receiver Free-Space Optical for 5G Backhaul," *Optical Fiber Communications Conference and Exhibition (OFC)*, 2019, pp. 1–3.

[37] Das, S., et al., "Towards Energy Neutral Wireless Communications: Photovoltaic Cells to Connect Remote Areas," *Energies*, Vol. 12, No. 19, October 2019, p. 3772.

[38] "Where Are the Discounts? Carrefour's LED Supermarket Lighting from Philips Will Guide You," Signify, News Release, May 21, 2015. https://www.signify.com/global/our-company/news/press-release-archive/2015/20150521-where-are-the-discounts-carrefours-led-supermarket-lighting-from-philips-will-guide-you.

[39] Wright, M., "Philips Lighting Deploys LED-Based Indoor Positioning in Carrefour Hypermarket," *LEDs Magazine*, Vol. 21, 2015.

[40] Hamza, A., and T. Tripp, "Optical Wireless Communication for the Internet of Things: Advances, Challenges, and Opportunities," TechRxiv, July 2020. https://www.techrxiv.org/articles/preprint/Optical_Wireless_Communication_for_the_Internet_of_Things_Advances_Challenges_and_Opportunities/12659789/1.

[41] Ganick, A., and D. Ryan, "Light Positioning System Using Digital Pulse Recognition," U.S. Patent 8,248,467, August 21 2012.

[42] Jovicic, A., "Qualcomm Lumicast: A High Accuracy Indoor Positioning System Based on Visible Light Communication," White Paper, Qualcomm, April 2016.

[43] Interact Lighting, "Unlocking The Value of Mobileapps Through Connected Lighting," White Paper, 2018. https://www.interact-lighting.com/b-dam/b2b-li/en AA/interact/what-is-possible/interact-retail/software/indoor-navigation/Rebrand-IPS-Whitepaper.pdf.

[44] Albraheem, L. I., et al., "Toward Designing a Li-Fi-Based Hierarchical IoT Architecture," *IEEE Access,* Vol. 6, 2018, pp. 40811–40825.

[45] Li, L., et al., "Epsilon: A Visible Light Based Positioning System," *11th USENIX Symposium on Networked Systems Design and Implementation (NSDI 14),* Seattle, WA: USENIX Association, April 2014, pp. 331–343. https://www.usenix.org/conference/nsdi14/technical-sessions/presentation/li.

[46] Kuraki, K., K. Kato, and R. Tanaka, "ID-Embedded LED Lighting Technology for Providing Object-Related Information," *Fujitsu Scientific & Technical Journal,* Vol. 52, No. 1, 2016, pp. 85–90.

[47] Figueiredo, M., L. N. Alves, and C. Ribeiro, "Lighting the Wireless World: The Promise and Challenges of Visible Light Communication," *IEEE Consumer Electronics Magazine,* Vol. 6, No. 4, 2017, pp. 28–37.

[48] Kuraki, K., et al., "Data Transfer Technology to Enable Communication Between Displays and Smart Devices," *Fujitsu Scientific & Technical Journal,* Vol. 50, No. 1, 2014, pp. 40–45.

[49] Le Minh, H., et al., "100-Mb/s NRZ Visible Light Communications Using a Postequalized White LED," *IEEE Photonics Technology Letters,* Vol. 21, No. 15, August 2009, pp. 1063–1065.

[50] Islim, M. S., et al., "Towards 10 Gb/s Orthogonal Frequency Division Multiplexing Based Visible Light Communication Using a GaN Violet Micro-LED," *OSA Photonics Research,* Vol. 5, No. 2, April 2017, pp. A35–A43. http://www.osapublishing.org/prj/abstract.cfm?URI=prj-5-2-A35.

[51] Lu, H., et al., "A 56 Gb/s PAM4 VCSEL-Based LiFi Transmission with Two-Stage Injection-Locked Technique," *IEEE Photonics Journal,* Vol. 9, No. 1, 2017, pp. 1–8.

[52] Sarbazi, E., M. Safari, and H. Haas, "The Bit Error Performance and Information Transfer Rate of SPAD Array Optical Receivers," *IEEE Transactions on Communications,* Vol. 68, No. 9, 2020, pp. 5689–5705.

[53] Islim, M. S., and H. Haas, "Receiver and Associated Mobile Communications Device," U.S. Patent Application 16/472,829, March 19, 2020.

[54] Chen, C., et al., "An Omnidirectional User Equipment Configuration to Support Mobility in LiFi Networks," *IEEE International Conference on Communications Workshops (ICC Workshops),* 2019, pp. 1–6.

[55] Chen, C., et al., "Hybrid Multiplexing in OFDM-Based VLC Systems," *IEEE Wireless Communications and Networking Conference (WCNC),* 2020, pp. 1–6.

[56] pureLiFi, "Why the Next Generation of Wireless Devices Need LiFi," 2019. https://purelifi.com/wp-content/uploads/2019/04/Why-Design-In-Gigabit-LiFi-pureLiFi.pdf.

[57] Riemersma-van der Lek, R. F., et al., "Effect of Bright Light and Melatonin on Cognitive and Noncognitive Function in Elderly Residents of Group Care Facilities: A Randomized Controlled Trial," *JAMA,* Vol. 299, No. 22, June 2008, pp. 2642–2655.

[58] Partonen, T., and J. Lonnqvist, "Bright Light Improves Vitality and Alleviates Distress in Healthy People," *Journal of Affective Disorders,* Vol. 57, No. 1, pp. 55–61, 2000.

[59] Tuttle, R. C., "White LED Chromaticity Control -The State of the Art," *9th Transformations Lighting DOE Solid-State Lighting R&D Workshop*, Atlanta, GA, 2011.

[60] Popoola, W. O., "Impact of VLC on Light Emission Quality of White LEDs," *IEEE/OSA Journal of Lightwave Technology*, Vol. 34, No. 10, 2016, pp. 2526–2532.

[61] "IEEE Standard for Local and Metropolitan Area Networks–Part 15.7: Short-Range Optical Wireless Communications," IEEE Std 802.15.7-2018 (Revision of IEEE Std 802.15.7-2011), 2019, pp. 1–407.

[62] Mejia, C. E., et al., "Code Design for Flicker Mitigation in Visible Light Communications Using Finite State Machines," *IEEE Transactions on Communications*, Vol. 65, No. 5, 2017, pp. 2091–2100.

10

Summary and Outlook

This book provided an up-to-date overview of the advancements in mobile OWC technologies. Particularly, the book aimed to shed light on the current issues related to wireless communications and how optical transmission solutions could be the main enabler for future wireless networks. In the following, we provide a summary of the content of the different chapters and how they contribute to providing a clear picture of the concepts, drivers, and challenges related to OWC technologies.

The physical interpretation of light has puzzled scientists for thousands of years. Chapter 2 briefly explored the different theories of optics, vision, and light from Archimedes to Albert Einstein. The current understanding of light as a dual-wave particle dates back to centuries of theoretical and experimental work. In addition, Chapter 2 discussed the early history of visual communications, which, as a concept, started more than 2,200 years ago when fire and smoke were used for signaling. Notable advancements were discussed, such as the mechanical-based visual communication system developed by Claude Chappe in 1792 and the heliograph, which was based on a shutter-aided mirror to reflect sunlight. Chapter 2 also showed how the development of the Photophone by Charles Sumner Tainter and Alexander Graham Bell marked a milestone in the history of practical OWC systems. Moving forward, the invention of the blue LED has enabled energy-efficient semiconductor-based white illumination solutions. Chapter 2 reviewed the development of the efficient blue LED and its impact on general lighting and VLC. The history of VLC is reviewed in Chapter 2 by exploring some of the early work by Komine and Nakagawa in Japan until the development and commercialization of high-speed VLC systems and subsystems. The development of VLC systems is enabled by the advances in digital modulation techniques, optics, and semiconductor technology. In addition, Chapter 2 reviewed the development and history of other OWC technologies, namely FSO and OCC. We coined the term Li-Fi in our 2011 TED talk to describe a high-speed fully networked mobile solution based on light [1]. Li-Fi offers unprecedented advantages over existing RF solutions, which has led to an increased research and commercial interest, as discussed in Chapter 2.

Chapter 3 discussed how the never-ending quest for higher data rates and better coverage has encouraged the integration of various wireless connectivity paradigms. More specifically, we discussed the fundamentals, applications, challenges, and future trends of different communications technologies within the realm of OWC. The chapter explained how each technology can be characterized based on four metrics: transmission speeds, transmission range, duplex mode, and communication mode. Moreover, we presented the fundamentals of FSO communications systems, which offer high data rate solutions for long-range, bidirectional, point-to-point connectivity. We also discussed the FSO channel characteristics and how it is prone to high distortion and attenuation due to atmospheric and misalignment factors and described different FSO communications techniques (i.e. MIMO FSO, OAM FSO, and FSO based on Kramers-Kronig (KK) receivers). The chapter moved to discuss another OWC technology, namely UV communications. UV links enable high data rate, long-distance, bidirectional NLOS wireless connectivity through atmospheric scattering. As a result, UV links are particularly robust to link blockage and shadowing effects, which makes this technology extremely desirable when precise pointing and tracking is not feasible. We shed light on the state-of-the-art advancements in UV sources and detectors and discussed the UV channel model. Moreover, we discussed some related open issues such as the need for link efficiency enhancements and more experimental validation. The chapter also discussed OCC, which, as the name indicates, employs cameras as photo-receivers. OCC offers low-range, low-speed, unidirectional point-to-point links with data rates below 10 Mbps. We defined different types of sensor images used in OCC and discussed the principle of operation. We also listed some of the common OCC applications and related challenges.

In order to justify the necessity of Li-Fi technology in future wireless communication systems, a detailed introduction to the motivation and potential of Li-Fi technology was provided in Chapter 4. First, we presented our vision on futuristic smart connected systems by comparing the structure of an AI system to the structure of a human neuron system. In order to achieve seamless interaction, future AI-enabled systems will require reliable and efficient wireless connectivity. The upcoming spectrum crunch, however, may create a connectivity bottleneck and limit the achievable capacity. On top of this challenge, we have considered a series of key performance indicators that may be required in future wireless communications systems. Then we presented the advantages that Li-Fi technology can offer compared to other RF-based wireless technologies. These advantages include abundant optical spectrum, extreme densification, and physical layer security. Also, we showed that Li-Fi can be used in places where RF signals are restricted and that the installation of Li-Fi systems can be integrated within the existing lighting infrastructure. All these advantages enable Li-Fi to aid and complement future wireless networks to fulfill the critical connectivity demands. The last part of Chapter 4 was dedicated to demonstrating many potential applications of Li-Fi technology in various wireless scenarios. In this part of the discussion, applications in defense/cybersecurity, aerospace/aeronautical communication, underwater communication, data-driven smart cities, and vehicular communication were considered. At the end of Chapter 4, we discussed several common misconceptions about Li-Fi technology. These misconceptions include the relationship between Li-Fi and Wi-Fi technologies, link blockage issues, uplink (UL) solutions, flickering issues, dimming issues, and background light interference issues.

In Chapter 5, we provided a comprehensive overview of optical wireless communication channels in indoor scenarios. At the beginning of the chapter, some major channel effects, such as path loss, multipath fading, multipath distortion, and wavelength selectivity, were discussed. In addition, several popular channel evaluation metrics were systematically introduced. The fundamental channel metrics include channel impulse response and channel frequency response. In order to reflect the effect of multipath distortion more simply, the RMS-delay spread and 3-dB bandwidth were introduced. Similarly, the channel DC gain was introduced to reflect the path loss characteristic. In the next part, we presented the decomposition of OWC channel as the front-end channel and the optical wireless channel. Regarding the front-end channel, a number of popular lowpass filter models were introduced. Considering the optical wireless channel, we focused on the characteristics of an NLOS channel and its evaluation techniques. This is because the presence of a NLOS channel is one of the main causes of multipath distortion, and the evaluation of NLOS is more complicated than the LOS counterpart. In the introduction of NLOS channel evaluation methods, we considered four deterministic approaches, three Monte Carlo approaches, and three analytical approaches. In addition, the performance of these evaluation approaches in terms of accuracy and simulation time were presented and compared. At the end of Chapter 5, the effects of varying key factors on OWC channel characteristics were demonstrated with a number of configurations. These key effects include link distance, receiver orientation, receiver FOV, transmitter beamwidth, room dimension, and surface reflectivity. Furthermore, the significance of the NLOS channel under the conditions with and without the front-end channel was evaluated.

The increasing interest in OWC was driven by the relative advantages of OWC systems against other RF solutions. However, the performance of these OWC systems was mainly enabled by a wealth of high-performance optoelectronic devices. It is fair to claim that the advances in high-performance inexpensive optoelectronic devices have fueled interest in OWC and enabled high-speed energy-efficient, light-based communication solutions. Chapter 6 surveyed, in detail, the various types of optoelectronic devices that have been previously considered for OWC transceivers. The advances in digital modulation techniques have also played an important role in enabling high-speed, energy-efficient OWC systems. Incoherent IM/DD offers low-cost and low-complexity signaling when compared to coherent-based signaling, which usually requires local oscillators and external modulators. RF-based modulation schemes, such as pulsed modulation techniques, were early adopted with minimal modifications. However, due to the restrictions imposed by the IM/DD signaling scheme, the modulation signals must be real-valued and unipolar. This led to the development of numerous modulation techniques that mainly aim to improve the energy and spectral efficiency of the OWC system. Chapter 6 featured a detailed taxonomy of these modulation techniques. The chapter also discussed the unique modulation and multiplexing techniques that are enabled by the color domain. These include CSK and WDM. Future OWC solutions are expected to benefit from the unique properties enabled by the natural beamforming of the OFEs and the accessibility of using multiple DOF and MIMO configurations.

Chapter 7 gave an idea of how Li-Fi can provide a fully bidirectional multi-user networking solution that can be seamlessly integrated with future wireless networks. We started by discussing the integration of Li-Fi within heterogeneous networks, which is envisioned to be a key characteristic of 5G and beyond wireless solutions. We then moved to discuss multiaccess control in Li-Fi. As a complete networking solution, Li-Fi is required to support simultaneous multiple access. This could be achieved by means of OMA techniques such as TDMA and O-OFDMA, or by means of NOMA techniques such as O-CDMA and power-domain NOMA. We have discussed the principal and state-of-the-art research of each of these multiple access schemes while giving an insight into the performance trade-offs. The fact that Li-Fi systems are typically composed of densely deployed attocells with relatively small coverage areas necessities efficient resource allocation techniques. Thus, it is critical to maximize the utilization of the available system resources while keeping the interference at minimal levels. The chapter included extensive discussions on the various resource allocation and interference mitigation strategies that are proposed for Li-Fi systems. These include static and dynamic resource allocation methods as well as interference cancellation and interference avoidance techniques. As a fully networked solution, Li-Fi is also required to support user mobility seamlessly. The chapter covers the different handover mechanisms in homogeneous and heterogeneous networks and discusses the related overhead.

The standardization of OWC allows for improved interoperability between devices from different manufacturers. This paves the way for laptop and mobile device manufacturers to integrate standard-compliant OFE devices into their systems. In Chapter 8, we discussed different standardization efforts for OWC. The IEEE 802.15.7-2018: Short-Range Optical Wireless Communications Standard was recently revised to target low data rate OCC systems, whereas the IEEE 802.15.13: Multi-Gigabit/s Optical Wireless Communications Standard is focused on high-speed OWC solutions. The ITU-T G.9991 (G.vlc) standard is aimed at VLC systems. The main features of these standards were discussed alongside the most notable system implementations based on these standards specifications. The IEEE 802.11bb Li-Fi standard is currently being developed based on the matured IEEE 802.11 standard suite. The standard is aimed at fully networked Li-Fi solutions with mobility and handover support, based on reusing most of the associated services and functions enabled by the available IEEE 802.11 standard. This allows for improved interoperability and integration with other RF based Wi-Fi systems. In particular, the adoption of the MAC layer of the existing IEEE 802.11 standard suite has the potential to enable handover between Li-Fi and Wi-Fi. This standard reduces the market entry barrier for future Li-Fi solutions and paves the way to mass markets due to the adoption of the matured IEEE 802.11 chipset solutions. It is worth noting that some of the standards are still under development and have not been published yet at the time of this writing (IEEE 802,15,13, IEEE 802.11bb). Therefore, the specifications in the final published standards might differ from the discussions provided in this book. The publication of these standards will motivate more OWC commercialization activities in the near future.

In Chapter 9, we focused on the commercialization aspect of Li-Fi technology. At the beginning of the chapter, the Li-Fi market potential in terms of size and value were briefly analyzed. A detailed discussion on the market trend and commercial opportunities in 5G networks, Big Data/IoT, Industry 4.0, and future 6G networks were provided. In the next section, an overview of the unique Li-Fi commercial ecosystem was presented and discussed. It is envisioned that there will be many players participating in this ecosystem, including Li-Fi consumer product providers, lighting infrastructure providers, chipsets and optical components providers, utility and application providers, research and development organization, and end-users. Although the variety of the participants implies that there will be challenges to overcome while this ecosystem forms, it allows for huge commercial opportunities in many different industries. Next, a series of commercial use cases, demonstrated by several pioneering companies and institutes, were presented in detail. Some featured practical use case demonstrations include indoor high-speed networking by pureLiFi, a MIMO system deployment in an industry application by the Fraunhofer Heinrich Hertz Institute, and a long-distance backhauling demonstration using solar photovoltaics cells by Li-Fi Research and Development Centre. Despite the promising commercial opportunities offered by Li-Fi technology, the current Li-Fi commercialization activities are still very limited. At the end of Chapter 9, many associated commercialization challenges were listed, and corresponding potential solutions were discussed.

In this book, we introduced several high data rate VLC studies with experimental demonstrations based on wideband frontend devices, such as micro LEDs and blue lasers. These studies demonstrate the potential transmission speed that can be achieved in Li-Fi networks. However, most of these studies are limited to point-to-point links with alignment. In order to validate the feasibility of using these high-performance front ends in Li-Fi networks, more technical work on demonstrations with diffused sources based on high-performance frontends is expected in future research. In addition, we introduced many studies on Li-Fi networks beyond the physical layer, such as multiple access schemes [2], interference mitigation techniques [3], and handover schemes [4]. Most of these studies are based on computer simulations. Only a few of them provide experimental demonstrations [5]. Despite the implementation complexity, an increase in Li-Fi studies beyond the physical layer with experimental demonstrations would make this technology more attractive and motivate more commercialization activities. In recent years, many researchers started to consider using ML techniques in OWC studies [6]. ML techniques are well-known to be successful in solving many prediction and classification problems with high complexity in the fields such as image processing and speech recognition. A number of studies demonstrated that algorithms based on ML methods provide promising performance improvement against conventional approaches in several research problems, such as nonlinearity mitigation, jitter compensation, modulation format identification, and phase estimation [6]. However, compared to the applications in other fields, the use of ML techniques in OWC research is still very limited. More studies on many other potential applications such as channel prediction or link outage prediction can be expected in the near future [7]. Similarly, applications of ML techniques to realize more efficient and smart Li-Fi networks are also promising future research direction.

References

[1] Haas, H., "Wireless Data from Every Light Bulb," *Ted Talks*, August 2011. https://www.ted.com/talks/harald_haas_wireless_data_from_every_light_bulb.

[2] Abumarshoud, H., H. Alshaer, and H. Haas, "Dynamic Multiple Access Configuration in Intelligent LiFi Attocellular Access Points," *IEEE Access*, Vol. 7, 2019, pp. 62126–62141.

[3] Wu, L., et al., "Blind Interference Alignment for Multiuser MISO Indoor Visible Light Communications," *IEEE Communications Letters*, Vol. 21, No. 5, 2017, pp. 1039–1042.

[4] Wu, X., and H. Haas, "Handover Skipping for LiFi," *IEEE Access*, Vol. 7, 2019, pp. 38369–38378.

[5] Chen, C., et al., "Space Division Multiple Access with Distributed User Grouping for Multi-User MIMO-VLC Systems," *IEEE Open Journal of the Communications Society*, Vol. 1, 2020, pp. 943–956.

[6] Chi, N., et al., "Challenges and Prospects of Machine Learning in Visible Light Communication," *Journal of Communications and Information Networks*, Vol. 5, No. 3, 2020, pp. 302–309.

[7] Wu, Z. Y., et al., "Efficient Prediction of Link Outage in Mobile Optical Wireless Communications," *IEEE Transactions on Wireless Communications*, Vol. 20, No. 2, 2021, pp. 882–896.

Acronyms and Abbreviations

1-D	one-dimensional
1G	first generation
2-D	two-dimensional
2G	second generation
3G	third-generation
3-D	three-dimensional
4G	fourth generation
5G	fifth generation
6G	sixth generation
AC	alternating current
ACK	acknowledgment
ACO-OFDM	asymmetrically clipped optical orthogonal frequency division multiplexing
ACO-SCFDE	asymmetrically clipped optical single-carrier frequency domain equalizer
ADC	analog-to-digital converter
ADO-OFDM	asymmetrically direct current-biased optical orthogonal frequency division multiplexing
ADP	application data primitive
AF	amplify-and-forward
AGC	automatic gain control
AI	artificial intelligence
AIN	aluminum nitride
A-interface	application interface
AM	amplitude modulation

AP	access point
APC	application protocol convergence
APD	avalanche photodiode
A-QL	asynchronous quick link
AR	augmented reality
ARQ	automatic repeat request
ASCO-OFDM	asymmetrically and symmetrically clipping optical orthogonal frequency division multiplexing
ASE-DMT	augmented spectral efficiency discrete multitone
ASIC	application-specific integrated circuit
ASTM	American Society for Testing and Materials
AUV	autonomous underwater vehicle
AR/VR	augmented and virtual reality
AWG	arbitrary waveform generator
AWGN	additive white Gaussian noise
B.C.	before Christ
BD	block diagonalization
BER	bit error ratio
BIA	blind interference alignment
BP	beacon period
bps/Hz	bits per second per hertz
BPSK	binary phase-shift keying
BRDF	bidirectional reflection distribution function
BS	base station
CAGR	compound annual growth rate
CAP	carrier-less amplitude and phase
CAP	contention access period
CC	convolutional coding
CCD	charge coupled device
CCDF	complementary cumulative distribution function
CCI	cochannel interference
CCT	correlated color temperature
CDF	cumulative distribution function
CDMA	code division multiple access
CES	consumer electronic show
CFA	color filter area
CFP	contention-free period
CFR	channel frequency response
CIE 1930	Commission Internationale de l'Eclairage
CIM	color intensity modulation
CIR	channel impulse response
CLT	central limit theorem

CM-FSK	camera M-ary frequency shift keying
CMOS	complementary metal oxide semiconductor
CoMP	coordinated multipoint
C-OOK	camera on-off keying
CP	cyclic prefix
CPC	compound parabolic concentrator
CRI	color rendering index
CSI	channel state information
CSK	color shift keying
CSMA/CA	carrier-sense multiple access with collision avoidance
CTS	clear-to-send
CW Laser	continuous wave laser
DAC	digital-to-analog converter
DC	direct current
DCO-OFDM	direct current biased optical orthogonal frequency division multiplexing
DCR-HCM	direct current reduced Hadamard coded modulation
DCSK	digital color shift keying
DD	direct detection
DF	decode-and-forward
DFB	distributed feedback
DFE	decision feedback equalizer
DFT	discrete Fourier transform
DH	double-heterostructure
DHT	discrete Hartley transformation
DIFS	distributed interframe space
DL	downlink
DLL	data link layer
DMT	discrete multitone
DoF	degrees of freedom
DPC	dirty paper coding
DPPM	differential pulse position modulation
DS8-PSK	dimming spatial 8-phase shift keying
DSB	double-sided bandwidth
DSL	digital subscriber line
DSP	digital signal processing
DWDM	dense wavelength division multiplexing
L/e-ACO-OFDM	layered/enhanced asymmetrically clipped optical orthogonal frequency division multiplexing
EE	energy efficiency
EGC	equal gain combining
EM	electromagnetic

eMBB	enhanced mobile broadband
EMI	electromagnetic interference
E-O	electrical-to-optical
ePAM-DMT	enhanced pulse-amplitude modulation discrete multitone
EQE	external quantum efficiency
eU-OFDM	enhanced unipolar orthogonal frequency division multiplexing
EVM	error vector magnitude
FCC	U.S. Federal Communications Commission
FD	full duplexing
FDD	frequency division duplex
FDE	frequency-domain equalization
FDM	frequency division multiplexing
FEC	forward error correction
FFE	feed forward equalizer
FFR	fractional frequency reuse
FFT	fast Fourier transform
FPGA	field-programmable gate array
FHWT	fast Haar wavelet transform
FIR	finite impulse response
FLIP-OFDM	flip orthogonal frequency division multiplexing
FLP	fast locking pattern
FM	frequency modulation
FOV	field of view
FP	Fabry-Perot
FPA	fixed power allocation
FPGA	field-programmable gate array
fps	frames per second
FR	frequency reuse
FD	frequency domain
FSK	frequency shift keying
FSO	free-space optical
FWHM	full width at half maximum
FWHT	fast Walsh-Hadamard transformation
GaAs	gallium arsenide
GaN	gallium nitride
GCSK	generalized color shift keying
GF	Galois field
GMST	Greenwich mean sidereal time
GPS	global positioning system

GREENER-OFDM	generalized enhanced unipolar orthogonal frequency division multiplexing
GRPA	gain ratio power allocation
GSM	generalized spatial modulation
GTS	guaranteed time slot
HACO-OFDM	hybrid asymmetrical clipped orthogonal frequency division multiplexing
HB-PHY	high bandwidth physical
HCM	Hadamard coded modulation
HD	half duplexing
HHO	horizontal handover
HD	high-definition
HO	handover
HPF	highpass filter
HS	handover skipping
HS-PSK	hybrid spatial-phase shift keying
I2V	infrastructure-to-vehicle
I2V2V	infrastructure-to-vehicle-to-vehicle
ICI	interchannel interference
ICT	information and communication technology
IDE	invisible data embedding
IDFT	inverse discrete Fourier transform
IDI	interdepth interference
IEEE	Institute of Electrical and Electronics Engineers
IFFT	inverse fast Fourier transform
IFI	interframe interference
i.i.d.	independent identically distributed
IM	intensity modulation
IM/DD	intensity modulation and direct detection
InGaN	indium gallium nitride
InP	indium phosphide
IoE	internet-of-everything
IoT	internet of things
IP	internet protocol
IPI	interpixel interference
IR	infrared
IrDA	Infrared Data Association
IrLAP	Infrared Link Access Protocol
IrLMP	Infrared Link Management Protocol
IrPHY	Infrared Physical Layer Specification
ISI	intersymbol interference
ISP	internet service provider

ITS	intelligent transportation system
ITU	International Telecommunication Union
ITU-T	International Telecommunication Union Telecommunication Standardization Sector
I-V	current-voltage
KK	Kramers-Kronig
KPI	key performance indicator
LaaS	light-as-a-service
LACO-OFDM	layered asymmetrically clipped optical orthogonal frequency division multiplexing
LAN	local-area network
LB-PHY	low bandwidth physical
LCD	liquid crystal display
LD	light amplification by stimulated emission of radiation diode
LDPC	low density parity check
LED	light emitting diode
L-I	luminance-current
LIDAR	light detection and ranging
Li-Fi	light fidelity
LLC	logical link control
LMST	local mean sidereal time
LO	local oscillator
LOS	line of sight
LPF	lowpass filter
LSB	lower side band
LTE	long-term evolution
LTI	linear time-invariant
LUT	look-up table
L-V	luminance-voltage
M2M	machine-to-machine
MAC	medium access control
MC	Monte Carlo
MCM	multicarrier modulation
MCMC	Markov chain Monte Carlo
MCMCRG	Markov chain Monte Carlo ray-gathering
MCMCRS	Markov chain Monte Carlo ray-shooting
MCRG	Monte Carlo ray-gathering
MCRS	Monte Carlo ray-shooting
MDI	medium-dependent interface
MDM	mode division multiplexing
MFTP	maximum flicker time period

MIMO	multiple-input multiple-output
MISO	multiple-input single-output
ML	machine learning
MLSD	maximum likelihood sequence detection
MM	metameric modulation
MMSE	minimum mean square error
mMTC	massive machine-type communications
MOCVD	metal organic chemical vapor deposition
MP2MP	multipoint to multipoint
M-PAM	M-ary pulse-amplitude modulation
MPE	maximum permissible exposure
MPM	mirror pulse modulation
MPPM	multiple pulse position modulation
M-QAM	M-ary quadrature amplitude modulation
MRC	maximum ratio combining
MU	multiuser
MUD	multiuser detection
MUI	multiuser interference
ND	neutral density
NLOS	non-line of sight
NOMA	non-orthogonal multiple access
NVS	neuromorphic vision sensor
OAM	orbital angular momentum
OCC	optical camera communication
O-CDMA	optical code division multiple access
O-E	optical-to-electrical
OEM	original equipment manufacturer
OFDM	orthogonal frequency division multiplexing
OFDMA	orthogonal frequency division multiple access
OFE	optical front end
OLED	organic light emitting diode
OMA	orthogonal multiple access
OOC	optical orthogonal code
O-OFDMA	optical orthogonal frequency division multiple access
OOK	on-off keying
O-SDMA	optical space division multiple access
OSM	optical spatial modulation
OSTBC	orthogonal space-time block code
OWC	optical wireless communication
OWPAN	optical wireless personal area network
P2MP	point to multipoint
P2P	point-to-point

PAM	pulse amplitude modulation
PAM-DMT	pulse amplitude modulation discrete multitone
PAPM	pulse amplitude and position modulation
PAPR	peak-to-average power ratio
PC	prime code
PCB	printed circuit board
PC-LED	phosphor coated light emitting diode
PCM	pulse code modulation
PCS	physical coding sublayer
PD	photodiode
PDF	probability density function
PDM	polarization division multiplexing
PER	packet error ratio
PHY	physical
PIC	photonic integrated circuit
PIN-PD	P-type intrinsic N-type photodiode
PLC	power line communication
PLS	physical layer security
PM	phase modulation
PMA	physical medium attachment
PMD	physical medium dependent
PMI	physical medium-independent interface
PM-OFDM	position modulation orthogonal frequency division multiplexing (OFDM)
PM-PHY	pulsed modulation PHY
PMT	photo-multiplier tube
PoE	power over Ethernet
POF	plastic optical fiber
P-OFDM	polar orthogonal frequency division multiplexing
PON	passive optical network
PPDU	physical protocol data unit
PPM	pulse position modulation
PRBS	pseudorandom binary sequence
PSD	power spectral density
M-PSK	M-ary phase shift keying
PSK	phase shift keying
PTA	photon tracing algorithm
PV	photovoltaic
PWM	pulse width modulation
QAM	quadrature amplitude modulation
QLED	quad-chromatic light emitting diodes
QoS	quality of service

QR	quick response
RC	repetition coding
RC-LED	resonant cavity light emitting diode
RF	radio frequency
RGB	red, green, and blue
RGB-LD	red, green, and blue light amplification by stimulated emission of radiation diode (LDs)
RGB-LED	red, green, and blue light emitting diodes
RIFS	reduced interframe space
RLL	run length limited
RMS	root mean square
ROC	random orthogonal code
R-OFDM	reshaped orthogonal frequency division multiplexing
RoI	region of interest
RPO-OFDM	reverse polarity optical orthogonal frequency division multiplexing
RRC	root raised cosine
RS	Reed-Solomon
RS-FSK	rolling shutter frequency shift keying
RSS	received signal strength
RSU	roadside unit
RTS	request to send
S2-PSK	spatial 2-phase shift keying
S-ACO-OFDM	spatial asymmetrically clipped optical orthogonal frequency division multiplexing
SB	select best
SCFDE	single-carrier frequency-domain equalizer
SCM	single-carrier modulation
SCO-OFDM	symmetrical clipping optical orthogonal frequency division multiplexing
S-DCO-OFDM	spatial direct current biased optical orthogonal frequency division multiplexing
SDM	space division multiplexing
SDMA	space division multiple access
SDN	software-defined networks
SDR	software-defined radio
SEE-OFDM	spectrally and energy efficient orthogonal frequency division multiplexing
sFFR	strict fractional frequency reuse
SFO-OFDM	spectrally factorized optical orthogonal frequency division multiplexing
SFR	soft frequency reuse

SIC	successive interference cancellation
SIFS	short interframe space
SIM-OFDM	subcarrier-index modulation orthogonal frequency division multiplexing
SIMO	single-input multiple-output
SINR	signal to interference and noise ratio
SISO	single-input single-output
SLD	superluminescent diode
SLM	spatial light modulator
SM	spatial modulation
SMD	surface mount device
SMP	spatial multiplexing
SNR	signal-to-noise ratio
SO-OFDM	spatial optical orthogonal frequency division multiplexing
SPAD	single-photon avalanche diode
SPM	spiral phase mirror
SPP	spiral phase plate
SS2DC	sequential scalable two-dimensional color
SSB	single-sided bandwidth
SSK	space shift keying
SSL	solid-state lighting
SU	single user
SVD	singular value decomposition
SWIPT	simultaneous wireless information and power transfer
TAIFS	turnaround interframe space
TCP	Transmission Control Protocol
TD	time domain
TDD	time division duplex
TDM	time division multiplexing
TDMA	time division multiple access
TDP	topology-dependent pattern
TIA	transimpedance amplifier
TSAG	Telecommunication Standardization Advisory Group
UAV	unmanned aerial vehicle
UE	user equipment
UFSOOK	undersampled frequency shift on-off keying
UL	uplink
m-LED	micrometer-sized gallium nitride light emitting diode; micro-sized gallium nitride light emitting diode
U-OFDM	unipolar orthogonal frequency division multiplexing
UOWC	underwater optical wireless communication

URLLC	ultrareliable and low-latency communication
USB	upper side band
UV	ultraviolet
UV-A	ultraviolet A-band
UV-B	ultraviolet B-band
UV-C	ultraviolet C-band
V2I	vehicle-to-infrastructure
V2V	vehicle-to-vehicle
VCSEL	vertical-cavity surface-emitting laser
VHO	vertical handover
VLC	visible light communication
VLCC	Visible Light Communication Consortium
VPPM	variable pulse position modulation
VR	virtual reality
VTASC	variable transparent amplitude-shape-color
WDD	wavelength division duplex
WDM	wavelength division multiplexing
WDMA	wavelength division multiple access
Wi-Fi	wireless fidelity
WLAN	wireless local area network
WPAN	wireless personal area network
WPDM	wavelet packet division multiplexing
WTSA	World Telecommunication Standardization Assembly
ZF	zero forcing
ZnSe	zinc selenide

About the Authors

Harald Haas received a Ph.D. from the University of Edinburgh in 2001. He is a distinguished professor of mobile communications at the University of Strathclyde and the director of the Li-Fi Research and Development Centre. Professor Haas also set up and cofounded pureLi-Fi Ltd., in which he currently serves as the chief scientific officer. He has authored over 600 conference and journal papers. Professor Haas' main research interests are in optical wireless communications, hybrid optical wireless and RF communications, spatial modulation, and interference coordination in wireless networks. His team invented spatial modulation. He introduced Li-Fi to the public at an invited TED Global Talk in 2011. Li-Fi was listed among the 50 best inventions in *TIME* magazine in 2011. He gave a second TED Global Lecture in 2015 on the use of solar cells as Li-Fi data detectors and energy harvesters. In 2016, he received the Outstanding Achievement Award from the International Solid State Lighting Alliance. In 2019, he was recipient of IEEE Vehicular Society James Evans Avant Garde Award. Professor Haas was elected as a Fellow of the Royal Society of Edinburgh (RSE) in 2017. In the same year, he received a Royal Society Wolfson Research Merit Award and was elevated to IEEE Fellow. In 2018, he received a 3-year EPSRC Established Career Fellowship extension and was elected as a Fellow of the IET. Professor Haas was elected as a Fellow of the Royal Academy of Engineering (FREng) in 2019.

Mohamed Sufyan Islim received an M.Sc. (Distinction) in communications engineering from Aleppo University, Syria, in 2013. Among several scholarships that he received in 2013, he was awarded the Global Edinburgh Syrian Scholarship from Edinburgh University. He received both an M.Sc. in signal processing and communications in 2014 and a Ph.D. in digital communications in 2019 from the University of Edinburgh, United Kingdom. In addition, he received the *IEEE Communications* Chapter Prize for the best M.Sc. project in 2014. He has coauthored 14 conference papers, 16 journal papers, 2 book chapters, and 2 pending patents. He is currently a postdoctoral research associate at the Li-Fi Research and Development Centre, University of Strathclyde. His main research interests include optical OFDM, Li-Fi, and optical wireless communications.

Cheng Chen received a B.Eng. in electronic and electrical engineering from the University of Strathclyde, Glasgow, United Kingdom, in 2011, an M.Sc. in communications and signal processing from the Imperial College, London, United Kingdom, in 2012, and a Ph.D. in electrical engineering from the University of Edinburgh, Edinburgh, United Kingdom, in 2017. He is currently employed as a research associate with the Li-Fi Research and Development Centre, the University of Strathclyde, and is working in the field of visible light communications and Li-Fi networks. His main research interests are in the areas of optical wireless communication, visible light communication, and wireless communication for 6G. He has authored or coauthored 30 publications in these areas.

Hanaa Abumarshoud is a research associate at the Li-Fi Research and Development Centre at the University of Strathclyde, United Kingdom. She was awarded an M.Sc. in 2013 and a Ph.D. in 2017 in electrical and computer engineering from Khalifa University, UAE. Her main expertise is the application of information theory and signal processing techniques in visible light communications and Li-Fi systems, with a particular focus on intelligent communications, resource allocation, multi-user access, and physical layer security.

Index

Artech House Mobile Communications Library

William Webb, Series Editor

3G CDMA2000 Wireless System Engineering, Samuel C. Yang

3G Multimedia Network Services, Accounting, and User Profiles, Freddy Ghys, Marcel Mampaey, Michel Smouts, and Arto Vaaraniemi

5G New Radio: Beyond Mobile Broadband, Amitav Mukherjee

5G Spectrum and Standards, Geoff Varrall

802.11 WLANs and IP Networking: Security, QoS, and Mobility, Anand R. Prasad and Neeli R. Prasad

Achieving Interoperability in Critical IT and Communications Systems, Robert I. Desourdis, Peter J. Rosamilia, Christopher P. Jacobson, James E. Sinclair, and James R. McClure

Advances in 3G Enhanced Technologies for Wireless Communications, Jiangzhou Wang and Tung-Sang Ng, editors

Advances in Mobile Information Systems, John Walker, editor

Advances in Mobile Radio Access Networks, Y. Jay Guo

Artificial Intelligence in Wireless Communications, Thomas W. Rondeau and Charles W. Bostian

Broadband Wireless Access and Local Network: Mobile WiMax and WiFi, Byeong Gi Lee and Sunghyun Choi

CDMA for Wireless Personal Communications, Ramjee Prasad

CDMA RF System Engineering, Samuel C. Yang

CDMA Systems Capacity Engineering, Kiseon Kim and Insoo Koo

Cell Planning for Wireless Communications, Manuel F. Cátedra and Jesús Pérez-Arriaga

Cellular Communications: Worldwide Market Development, Garry A. Garrard

Cellular Mobile Systems Engineering, Saleh Faruque

Cognitive Radio Interoperability through Waveform Reconfiguration, Leszek Lechowicz and Mieczyslaw M. Kokar

Cognitive Radio Techniques: Spectrum Sensing, Interference Mitigation, and Localization, Kandeepan Sithamparanathan and Andrea Giorgetti

The Complete Wireless Communications Professional: A Guide for Engineers and Managers, William Webb

EDGE for Mobile Internet, Emmanuel Seurre, Patrick Savelli, and Pierre-Jean Pietri

Emerging Public Safety Wireless Communication Systems, Robert I. Desourdis, Jr., et al.

From LTE to LTE-Advanced Pro and 5G, Moe Rahnema and Marcin Dryjanski

The Future of Wireless Communications, William Webb

Geospatial Computing in Mobile Devices, Ruizhi Chen and Robert Guinness

GPRS for Mobile Internet, Emmanuel Seurre, Patrick Savelli, and Pierre-Jean Pietri

GSM and Personal Communications Handbook, Siegmund M. Redl, Matthias K. Weber, and Malcolm W. Oliphant

GSM Networks: Protocols, Terminology, and Implementation, Gunnar Heine

GSM System Engineering, Asha Mehrotra

Handbook of Land-Mobile Radio System Coverage, Garry C. Hess

Handbook of Mobile Radio Networks, Sami Tabbane

Handbook of Next-Generation Emergency Services, Barbara Kemp and Bart Lovett

High-Speed Wireless ATM and LANs, Benny Bing

Implementing Full Duplexing for 5G, David B. Cruickshank

In-Band Full-Duplex Wireless Systems Handbook, Kenneth E. Kolodziej, Editor

Inside Bluetooth Low Energy, Second Edition, Naresh Gupta

Interference Analysis and Reduction for Wireless Systems, Peter Stavroulakis

Interference and Resource Management in Heterogeneous Wireless Networks, Jiandong Li, Min Sheng, Xijun Wang, and Hongguang Sun

Internet Technologies for Fixed and Mobile Networks, Toni Janevski

Introduction to 3G Mobile Communications, Second Edition, Juha Korhonen

Introduction to 4G Mobile Communications, Juha Korhonen

Introduction to Communication Systems Simulation, Maurice Schiff

Introduction to Digital Professional Mobile Radio, Hans-Peter A. Ketterling

An Introduction to GSM, Siegmund M. Redl, Matthias K. Weber, and Malcolm W. Oliphant

Introduction to Mobile Communications Engineering, José M. Hernando and F. Pérez-Fontán

Introduction to OFDM Receiver Design and Simulation, Y. J. Liu

An Introduction to Optical Wireless Mobile Communications, Harald Haas, Mohamed Sufyan Islim, Cheng Chen, and Hanaa Abumarshoud

Introduction to Radio Propagation for Fixed and Mobile Communications, John Doble

Introduction to Wireless Local Loop, Broadband and Narrowband, Systems, Second Edition, William Webb

IS-136 TDMA Technology, Economics, and Services, Lawrence Harte, Adrian Smith, and Charles A. Jacobs

Location Management and Routing in Mobile Wireless Networks, Amitava Mukherjee, Somprakash Bandyopadhyay, and Debashis Saha

LTE Air Interface Protocols, Mohammad T. Kawser

Metro Ethernet Services for LTE Backhaul, Roman Krzanowski

Mobile Data Communications Systems, Peter Wong and David Britland

Mobile IP Technology for M-Business, Mark Norris

Mobile Satellite Communications, Shingo Ohmori, Hiromitsu Wakana, and Seiichiro Kawase

Mobile Telecommunications Standards: GSM, UMTS, TETRA, and ERMES, Rudi Bekkers

Mobile-to-Mobile Wireless Channels, Alenka Zajić

Mobile Telecommunications: Standards, Regulation, and Applications, Rudi Bekkers and Jan Smits

Multiantenna Digital Radio Transmission, Massimiliano Martone

Multiantenna Wireless Communications Systems, Sergio Barbarossa

Multi-Gigabit Microwave and Millimeter-Wave Wireless Communications, Jonathan Wells

Multiuser Detection in CDMA Mobile Terminals, Piero Castoldi

OFDMA for Broadband Wireless Access, Slawomir Pietrzyk

Practical Wireless Data Modem Design, Jonathon Y. C. Cheah

The Practitioner's Guide to Cellular IoT, Cameron Kelly Coursey

Prime Codes with Applications to CDMA Optical and Wireless Networks, Guu-Chang Yang and Wing C. Kwong

Quantitative Analysis of Cognitive Radio and Network Performance, Preston Marshall

QoS in Integrated 3G Networks, Robert Lloyd-Evans

Radio Resource Management for Wireless Networks, Jens Zander and Seong-Lyun Kim

Radiowave Propagation and Antennas for Personal Communications, Third Edition, Kazimierz Siwiak and Yasaman Bahreini

RDS: The Radio Data System, Dietmar Kopitz and Bev Marks

Resource Allocation in Hierarchical Cellular Systems, Lauro Ortigoza-Guerrero and A. Hamid Aghvami

RF and Baseband Techniques for Software-Defined Radio, Peter B. Kenington

RF and Microwave Circuit Design for Wireless Communications, Lawrence E. Larson, editor

Sample Rate Conversion in Software Configurable Radios, Tim Hentschel

Signal Processing Applications in CDMA Communications, Hui Liu

Signal Processing for RF Circuit Impairment Mitigation, Xinping Huang, Zhiwen Zhu, and Henry Leung

Smart Antenna Engineering, Ahmed El Zooghby

Software-Defined Radio for Engineers, Travis F. Collins, Robin Getz, Di Pu, and Alexander M. Wyglinski

Software Defined Radio for 3G, Paul Burns

Spread Spectrum CDMA Systems for Wireless Communications, Savo G. Glisic and Branka Vucetic

Technical Foundations of the IoT, Boris Adryan, Dominik Obermaier, and Paul Fremantle

Technologies and Systems for Access and Transport Networks, Jan A. Audestad

Third-Generation and Wideband HF Radio Communications, Eric E. Johnson, Eric Koski, William N. Furman, Mark Jorgenson, and John Nieto

Third Generation Wireless Systems, Volume 1: Post-Shannon Signal Architectures, George M. Calhoun

Traffic Analysis and Design of Wireless IP Networks, Toni Janevski

Transmission Systems Design Handbook for Wireless Networks, Harvey Lehpamer

UMTS and Mobile Computing, Alexander Joseph Huber and Josef Franz Huber

Understanding Cellular Radio, William Webb

Understanding Digital PCS: The TDMA Standard, Cameron Kelly Coursey

Understanding WAP: Wireless Applications, Devices, and Services, Marcel van der Heijden and Marcus Taylor, editors

Universal Wireless Personal Communications, Ramjee Prasad

WCDMA: Towards IP Mobility and Mobile Internet, Tero Ojanperä and Ramjee Prasad, editors

Wi-Fi 6: Protocol and Network, Susinder R. Gulasekaran and Sundar G. Sankaran

Wireless Communications in Developing Countries: Cellular and Satellite Systems, Rachael E. Schwartz

Wireless Communications Evolution to 3G and Beyond, Saad Z. Asif

Wireless Intelligent Networking, Gerry Christensen, Paul G. Florack and Robert Duncan

Wireless LAN Standards and Applications, Asunción Santamaría and Francisco J. López-Hernández, editors

Wireless Sensor and Ad Hoc Networks Under Diversified Network Scenarios, Subir Kumar Sarkar

Wireless Technician's Handbook, Second Edition, Andrew Miceli

For further information on these and other Artech House titles,
including previously considered out-of-print books now available through our
In-Print-Forever® (IPF®) program, contact:

Artech House
685 Canton Street
Norwood, MA 02062
Phone: 781-769-9750
Fax: 781-769-6334
e-mail: artech@artechhouse.com

Artech House
16 Sussex Street
London SW1V 4RW UK
Phone: +44 (0)20 7596-8750
Fax: +44 (0)20 7630-0166
e-mail: artech-uk@artechhouse.com

Find us on the World Wide Web at: www.artechhouse.com